Springer Complexity

Springer Complexity is a publication program, cutting across all traditional disciplines of sciences as well as engineering, economics, medicine, psychology and computer sciences, which is aimed at researchers, students and practitioners working in the field of complex systems. Complex Systems are systems that comprise many interacting parts with the ability to generate a new quality of macroscopic collective behavior through self-organization, e.g., the spontaneous formation of temporal, spatial or functional structures. This recognition, that the collective behavior of the whole system cannot be simply inferred from the understanding of the behavior of the individual components, has led to various new concepts and sophisticated tools of complexity. The main concepts and tools – with sometimes overlapping contents and methodologies – are the theories of self-organization, complex systems, synergetics, dynamical systems, turbulence, catastrophes, instabilities, nonlinearity, stochastic processes, chaos, neural networks, cellular automata, adaptive systems, and genetic algorithms.

The topics treated within Springer Complexity are as diverse as lasers or fluids in physics, machine cutting phenomena of workpieces or electric circuits with feedback in engineering, growth of crystals or pattern formation in chemistry, morphogenesis in biology, brain function in neurology, behavior of stock exchange rates in economics, or the formation of public opinion in sociology. All these seemingly quite different kinds of structure formation have a number of important features and underlying structures in common. These deep structural similarities can be exploited to transfer analytical methods and understanding from one field to another. The Springer Complexity program therefore seeks to foster cross-fertilization between the disciplines and a dialogue between theoreticians and experimentalists for a deeper understanding of the general structure and behavior of complex systems.

The program consists of individual books, books series such as "Springer Series in Synergetics", "Institute of Nonlinear Science", "Physics of Neural Networks", and "Understanding Complex Systems", as well as various journals.

Springer Series in Synergetics

Series Editor

Hermann Haken
Institut für Theoretische Physik
und Synergetik
der Universität Stuttgart
70550 Stuttgart, Germany

and

Center for Complex Systems
Florida Atlantic University
Boca Raton, FL 33431, USA

Members of the Editorial Board

Åke Andersson, Stockholm, Sweden
Gerhard Ertl, Berlin, Germany
Bernold Fiedler, Berlin, Germany
Yoshiki Kuramoto, Sapporo, Japan
Jürgen Kurths, Potsdam, Germany
Luigi Lugiato, Milan, Italy
Jürgen Parisi, Oldenburg, Germany
Peter Schuster, Wien, Austria
Frank Schweitzer, Zürich, Switzerland
Didier Sornette, Los Angeles, CA, USA, and Nice, France
Manuel G. Velarde, Madrid, Spain

SSSyn – An Interdisciplinary Series on Complex Systems

The success of the Springer Series in Synergetics has been made possible by the contributions of outstanding authors who presented their quite often pioneering results to the science community well beyond the borders of a special discipline. Indeed, interdisciplinarity is one of the main features of this series. But interdisciplinarity is not enough: The main goal is the search for common features of self-organizing systems in a great variety of seemingly quite different systems, or, still more precisely speaking, the search for general principles underlying the spontaneous formation of spatial, temporal or functional structures. The topics treated may be as diverse as lasers and fluids in physics, pattern formation in chemistry, morphogenesis in biology, brain functions in neurology or self-organization in a city. As is witnessed by several volumes, great attention is being paid to the pivotal interplay between deterministic and stochastic processes, as well as to the dialogue between theoreticians and experimentalists. All this has contributed to a remarkable cross-fertilization between disciplines and to a deeper understanding of complex systems. The timeliness and potential of such an approach are also mirrored – among other indicators – by numerous interdisciplinary workshops and conferences all over the world.

Till Daniel Frank

Nonlinear Fokker–Planck Equations

Fundamentals and Applications

With 86 Figures and 18 Tables

Dr. Till Daniel Frank
Universität Münster
Institut für Theoretische Physik
Wilhelm-Klemm-Strasse 9
48149 Münster, Germany

ISSN 0172-7389

ISBN 3-540-21264-7 Springer Berlin Heidelberg New York

This work is subject to copyright. All rights are reserved, whether the whole or part of the material is concerned, specifically the rights of translation, reprinting, reuse of illustrations, recitation, broadcasting, reproduction on microfilm or in any other way, and storage in data banks. Duplication of this publication or parts thereof is permitted only under the provisions of the German Copyright Law of September 9, 1965, in its current version, and permission for use must always be obtained from Springer. Violations are liable to prosecution under the German Copyright Law.

Springer is a part of Springer Science+Business Media

springeronline.com

© Springer-Verlag Berlin Heidelberg 2005
Printed in Germany

The use of general descriptive names, registered names, trademarks, etc. in this publication does not imply, even in the absence of a specific statement, that such names are exempt from the relevant protective laws and regulations and therefore free for general use.

Typesetting: by the author
Cover design: *design & production*, Heidelberg
Printed on acid-free paper   55/3141/ts   5 4 3 2 1 0

This book is dedicated to my parents, wife and daughter.

Preface

Nonlinear Fokker–Planck equations have found applications in various fields such as plasma physics, surface physics, astrophysics, the physics of polymer fluids and particle beams, nonlinear hydrodynamics, theory of electronic circuitry and laser arrays, engineering, biophysics, population dynamics, human movement sciences, neurophysics, psychology and marketing. In spite of the diversity of these research fields, many phenomena addressed therein have a fundamental physical mechanism in common. They arise due to cooperative interactions between the subsystems of many-body systems. These cooperative interactions result in a reduction of the large number of degrees of freedom of many-body systems and, in doing so, bind the subunits of many-body systems by means of self-organization into synergetic entities. These synergetic many-body systems admit low dimensional descriptions in terms of nonlinear Fokker–Planck equations that capture and uncover the essential dynamics underlying the observed phenomena. The phenomena that will be addressed in this book range from equilibrium and nonequilibrium phase transitions and the multistability of systems to the emergence of power law and cut-off distributions and the distortion of Boltzmann distributions. We will study possible asymptotic behaviors of systems such as the approach to stationary distributions and the emergence of nonstationary traveling wave distributions. We will be concerned with normal and anomalous diffusion and we will examine how correlation functions evolve with time in these kinds of synergetic systems. We will discuss a Fokker–Planck approach to quantum statistics, linear nonequilibrium thermodynamics, and generalized extensive and nonextensive thermostatistics.

The aim of this book is to provide an introduction to the theory of nonlinear Fokker–Planck equations and to highlight what systems described by nonlinear Fokker–Planck equations have in common. Theoretical considerations and concepts will be illustrated by various examples and applications.

Due to the ramifications of the theory of nonlinear Fokker–Planck equations in various scientific fields, this book is designed for graduate students and researchers in physics and related fields such as biology, neurophysics, human movement sciences, and psychology. I hope that this book will make graduate students interested in the topic of nonlinear Fokker–Planck equations and will make researchers aware of the connections between the different areas in which nonlinear Fokker–Planck equations have been applied so far.

Acknowledgments

I am indebted to Professor H. Haken for inviting me to write a book on the topic of nonlinear Fokker–Planck equations for the Springer Series of Synergetics. I am grateful to Professor P.J. Beek and Professor R. Friedrich for supporting my studies on nonlinear Fokker-Planck equations. I would like to thank Dr. A. Daffertshofer for many helpful discussions. Finally, I wish to thank Professor W. Beiglböck and Ms. B. Reichel-Mayer of the Springer-Verlag for their help in preparing this manuscript.

Münster, *Till Daniel Frank*
September 2004

Contents

1 Introduction 1
   1.1 Fokker–Planck Equations 1
      1.1.1 Brownian Particles and Langevin Equations 1
      1.1.2 Many-Body Systems and Mean Field Theory 2
   1.2 Phase Transitions and Self-Organization 3
   1.3 Stochastic Feedback 5
   1.4 Applications 7
      1.4.1 Collective Phenomena 7
      1.4.2 Multistable Systems 7
      1.4.3 Power Law and Cut-Off Distributions 10
      1.4.4 Free Energy Systems 12
      1.4.5 Anomalous Diffusion 15
   1.5 Overview 16

2 Fundamentals 19
   2.1 Stochastic Processes 19
   2.2 Nonlinear Fokker–Planck Equation 20
      2.2.1 Notation 21
      2.2.2 Stratonovich Form 21
      2.2.3 Transient Solutions 22
      2.2.4 Continuity Equation 22
      2.2.5 Boundary Conditions 23
      2.2.6 Stationary Solutions 23
   2.3 Self-Consistency Equations 24
   2.4 Multistability and Basins of Attraction 25
   2.5 Nonlinearity Dimension 25
   2.6 Classifications 26
   2.7 Derivations 28
   2.8 Numerics 28
      2.8.1 Path Integral Solutions 28
      2.8.2 Fourier and Moment Expansions 29
      2.8.3 Finite Difference Schemes 29
      2.8.4 Distributed Approximating Functionals 30

3 Strongly Nonlinear Fokker–Planck Equations 31
- 3.1 Transformation to a Linear Problem 31
- 3.2 What Are Strongly Nonlinear Fokker–Planck Equations? 33
- 3.3 Correlation Functions 36
- 3.4 Langevin Equations 36
  - 3.4.1 Two-Layered Langevin Equations 36
  - 3.4.2 Self-Consistent Langevin Equations 38
  - 3.4.3 Hierarchies and Correlation Functions 39
  - 3.4.4 Numerics 39
- 3.5 Stationary Solutions 42
- 3.6 H-Theorem for Stochastic Processes 43
- 3.7 Nonlinear Families of Markov Processes* 46
  - 3.7.1 Linear Versus Nonlinear Families of Markov Processes 46
  - 3.7.2 Linear Families of Markov Processes 47
  - 3.7.3 Nonlinear Families of Markov Diffusion Processes 48
  - 3.7.4 Markov Embedding 50
  - 3.7.5 Hitchhiker Processes 50
- 3.8 Top-Down Versus Bottom-Up Approaches* 52
- 3.9 Transient Solutions and Transition Probability Densities 55
  - 3.9.1 Nonequivalence of Transient Solutions
    and Transition Probability Densities 55
  - 3.9.2 Gaussian Distributions* 57
  - 3.9.3 Purely Random Processes 61
  - 3.9.4 Wiener Processes 62
  - 3.9.5 Ornstein–Uhlenbeck Processes 62
  - 3.9.6 Transient Solutions: Two Examples 63
- 3.10 Shimizu–Yamada Model – Transient Solutions 66
- 3.11 Fluctuation–Dissipation Theorem 70

4 Free Energy Fokker–Planck Equations 73
- 4.1 Free Energy Principle 75
- 4.2 Maximum Entropy Principle and Relationship
  between Noise Amplitude and Temperature 76
- 4.3 H-Theorem for Free Energy Fokker–Planck Equations 77
- 4.4 Boltzmann Statistics 79
- 4.5 Linear Nonequilibrium Thermodynamics 80
  - 4.5.1 Derivation of Free Energy Fokker–Planck Equations .. 80
  - 4.5.2 Drift and Diffusion Coefficients 84
  - 4.5.3 Transition Probability Densities
    and Langevin Equations 86
  - 4.5.4 Density Functions 87
  - 4.5.5 Entropy Production and Conservative Force 87
  - 4.5.6 Stationary Solutions 88
  - 4.5.7 H-Theorem for Systems with Conservative Forces
    and Nontrivial Mobility Coefficients 89

	4.6	Canonical-Dissipative Systems 90
		4.6.1 Linear Case....................................... 90
		4.6.2 Nonlinear Case.................................... 92
	4.7	Boundedness of Free Energy Functionals* 96
		4.7.1 Distortion Functionals 97
		4.7.2 Kullback Measure and Entropy Inequality 97
		4.7.3 Generic Cases
		and Schlögl's Decomposition of Kullback Measures ... 100
	4.8	First, Second, and Third Choice Thermostatistics........... 107

5 Free Energy Fokker–Planck Equations with Boltzmann Statistics 109

	5.1	Stability Analysis....................................... 110
		5.1.1 Lyapunov's Direct Method 112
		5.1.2 Linear Stability Analysis 113
		5.1.3 Self-Consistency Equation Analysis 115
		5.1.4 Shiino's Decomposition of Perturbations............. 118
		5.1.5 Generic Cases..................................... 120
		5.1.6 Higher-Dimensional Nonlinearities 126
		5.1.7 Multiplicative Noise 130
		5.1.8 Norm for Perturbations* 132
	5.2	Natural Boundary Conditions 134
		5.2.1 Shimizu–Yamada Model – Stationary Solutions 135
		5.2.2 Dynamical Takatsuji Model – Basins of Attraction.... 140
		5.2.3 Desai–Zwanzig Model 148
		5.2.4 Bounded $B(M_1)$-Model 152
	5.3	Periodic Boundary Conditions 155
		5.3.1 Cluster Amplitude and Cluster Phase 156
		5.3.2 KSS Model – Cluster Amplitude Dynamics 157
		5.3.3 Mean Field HKB Model – Cluster Phase Dynamics ... 167
	5.4	Characteristics of Bifurcations 183
		5.4.1 Stability and Disorder 183
		5.4.2 Emergence of Collective Behavior 184
		5.4.3 Multistability and Symmetry 186
		5.4.4 Continuous and Discontinuous Phase Transitions 187
	5.5	Applications 188
		5.5.1 Ferromagnetism 188
		5.5.2 Synchronization 191
		5.5.3 Isotropic-nematic Phase Transitions
		and Maier–Saupe Model........................... 195
		5.5.4 Muscular Contraction 202
		5.5.5 Network Models for Group Behavior 206
		5.5.6 Multistable Perception-Action Systems 209

6 Entropy Fokker–Planck Equations 213
- 6.1 Existence and Uniqueness of Stationary Solutions* 215
- 6.2 Entropy Increase and Anomalous Diffusion 218
  - 6.2.1 Entropy Increase 218
  - 6.2.2 Anomalous Diffusion 219
- 6.3 Drift- and Diffusion Forms 222
- 6.4 Entropy and Information Measures 224
  - 6.4.1 Properties* 224
  - 6.4.2 Examples 231
- 6.5 Examples and Applications 246
  - 6.5.1 Porous Medium Equation 246
  - 6.5.2 $^T S_q$-Entropy Fokker–Planck Equation 250
  - 6.5.3 Sharma–Mittal Entropy Fokker–Planck Equation 261
  - 6.5.4 Fokker–Planck Equations for Fermions and Bosons 280
  - 6.5.5 Multivariate Generalizations 284
  - 6.5.6 Metal Electron Model, Black Body Radiation Model, and Planck's Radiation Formula 288
  - 6.5.7 Population Dynamics 296

7 General Nonlinear Fokker–Planck Equations 299
- 7.1 Linear Stability Analysis 299
  - 7.1.1 Stationary Solutions 299
  - 7.1.2 Stability of Stationary Solutions 300
  - 7.1.3 On an Additional Stability Coefficient* 305
- 7.2 Free Energy and Lyapunov Functional Analysis 306
  - 7.2.1 Moving Frame Transformations 306
  - 7.2.2 Derivation of Entropy and Information Measures 310
  - 7.2.3 Derivation of Local Lyapunov Functionals 312
  - 7.2.4 Derivation of Lyapunov Functionals 317
- 7.3 Examples 324
  - 7.3.1 Traveling Waves 324
  - 7.3.2 Reentrant Noise-Induced Phase Transitions 329
  - 7.3.3 Systems with Multistable Variability 332
- 7.4 Applications 341
  - 7.4.1 Landau Form and Plasma Particles 341
  - 7.4.2 Bunch-Particle Distributions of Particle Beams 342
  - 7.4.3 Noise Generator 349
  - 7.4.4 Accuracy-Flexibility Trade-Off 351
- 7.5 Bibliographic Notes 360

8 Epilogue 367

References 371

Index 401

1 Introduction

1.1 Fokker–Planck Equations

Fokker–Planck equations describe the evolution of stochastic systems. For example, they describe the erratic motions of small particles that are immersed in fluids, fluctuations of the intensity of laser light, velocity distributions of fluid particles in turbulent flows, and the stochastic behavior of exchange rates. In general, Fokker–Planck equations can be applied to equilibrium and nonequilibrium systems. That is, they can be applied to systems that operate close to stationary states of thermal equilibrium [89, 490, 576] and to systems that operate far from thermal equilibrium [263].

1.1.1 Brownian Particles and Langevin Equations

Let us consider a single particle – a so-called Brownian particle – that is in contact with its environment [282, 346, 490, 619]. In general, there are interactions between a Brownian particle and its environment. That is, the environment acts on the particle and the particle acts on the environment. A helpful notion of an environment is a heat bath composed of a large number of heat bath particles. From this point of view, a complete description of the systems at hand requires taking the state variables of the Brownian particle and the state variables of all heat bath particles into account. In many cases, however, a reduced description can be obtained that involves only the state variables of the Brownian particle. The interactions between the heat bath particles and the Brownian particle then result in two forces that act on the Brownian particle: a damping force and a fluctuating force [89, 254, 308, 317, 517]. For example, let $X(t)$ denote the state variable of the Brownian particle. Then, the evolution of the Brownian particle may be given by

$$\frac{\mathrm{d}}{\mathrm{d}t}X(t) = -\gamma X(t) + \sqrt{Q}\Gamma(t) , \qquad (1.1)$$

where $-\gamma X$ is a linear damping force and γ is a positive damping constant. The function $\Gamma(t)$ describes a fluctuating force which is only defined in a statistical sense. Q describes the strength of the fluctuating force. Since $\sqrt{Q}\Gamma(t)$ describes the impact of a noise source on the Brownian particle, we also say that Q corresponds to a noise amplitude.

If $\Gamma(t)$ in (1.1) describes a particular fluctuating force, namely, a Langevin force [221, 254, 498], then (1.1) is referred to as Langevin equation. For stochastic processes described by Langevin equations we can write evolution equations for the corresponding probability densities [221, 254, 498]. For example, for the stochastic process given by (1.1) the evolution equation of the probability density P reads

$$\frac{\partial}{\partial t} P(x,t) = \gamma \frac{\partial}{\partial x} x P(x,t) + Q \frac{\partial^2}{\partial x^2} P(x,t) \ . \qquad (1.2)$$

Equation (1.2) is an example of a linear Fokker–Planck equation. Stochastic processes described by linear Fokker–Planck equations such as (1.2) are called Markov diffusion processes [221, 298].

1.1.2 Many-Body Systems and Mean Field Theory

Many-body systems such as solids, neural networks, and social societies are composed of many interacting subsystems. A complete description of many-body systems requires taking the interactions between all subsystems into account. Often, many-body systems can approximately be described in terms of the evolution of a single subsystem or a few subsystems. Let us consider many-body systems that are composed of a single type of subsystem. Then, we may describe the interactions between a single subsystem and its fellow-subsystems by the force that on the average is exerted by all fellow-subsystems on the single subsystem. This force is called a mean field force and is described by means of a statistical measure. In order to obtain self-consistent and closed single subsystem descriptions, one can express the averaging over the subsystems in terms of an averaging with respect to the statistical properties of the single subsystem under consideration [89]. Some systems to which this mean field treatment has been applied are listed in Table 1.1.

Table 1.1. Applications of mean field theory

Phenomenon	Theory	Ref.
Ferromagnetism	Weiss mean field theory	[57, 89, 229, 490, 546]
Polarization	Debye theory	[268, 319, 410]
Nematic phases	Maier–Saupe theory	[90, 129, 458, 485, 551]
Astrophysics	Vlasov approach	[58, 92, 349, 527, 615]

In quantum mechanics, many-body systems can be treated in a similar manner. Here, using the Hartree–Fock theory, Schrödiner equations for many-body problems can be reduced to single particle Schrödinger equations. The single particle Schrödinger equations, however, involve energy terms that depend on the statistical properties of the single particles that they describe [233, 269, 279, 502]. Furthermore, a mean field approach to the Bose-Einstein condensation has been developed that is centered around the Gross–Pitaevskii equation. The Gross–Pitaevskii equation is a single particle Schrödinger equation that involves a probability-dependent coefficient and describes the behavior of a many-body system: a gas composed of Bose particles [78, 121, 137, 244, 478].

The one thing that all mean field theories have in common is that they yield nonlinear descriptions for single subsystems. The nonlinearities that occur in the single subsystem descriptions reflect the interactions between the subsystems of the many-body systems. The advantage of mean field models is that, on the one hand, they often can be solved with relatively little effort and, on the other hand, they account for subsystem-subsystem interactions of many-body systems.

In line with mean field theory, one may describe many-body systems in terms of Fokker–Planck equations for single subsystem probability densities $P(\mathbf{x}, t)$. In order to account for subsystem-subsystem interactions, one may consider Fokker–Planck equations with probability-dependent coefficients that read like

$$\frac{\partial}{\partial t}P(\mathbf{x},t) = -\sum_i \frac{\partial}{\partial x_i} D_i(\mathbf{x},t,P)P(\mathbf{x},t) + \sum_{i,k} \frac{\partial^2}{\partial x_i \partial x_k} D_{ik}(\mathbf{x},t,P)P(\mathbf{x},t) \ .$$

(1.3)

We refer to (1.3) as a nonlinear Fokker–Planck equation. The distribution function $P(\mathbf{x}, t)$ has two faces. According to the μ-space description of many-body systems [97, 291, 437] (or the Maxwell-Boltzmann picture of many-body systems [490, 619]), we interpret $P(\mathbf{x}, t)$ as a function that describes the probability to find subsystems of a many-body system at time t in states $\mathbf{x}$. In particular, if $\rho(\mathbf{x}, t)$ describes the mass density (or particle density) of a system with a total mass (or total particle number) M_0, then P is given by $P = \rho/M_0$. In line with mean field theory, $P(\mathbf{x}, t)$ is regarded as probability density of a single subsystem. In sum, $P(\mathbf{x}, t)$ describes the distribution of subsystems in the μ-space as well as the stochastic properties of single subsystems. This is because if mean field theory applies, then the stochastic properties of a single subsystem reflect the way the entire ensemble of subsystems is distributed.

1.2 Phase Transitions and Self-Organization

There are numerous examples of equilibrium and nonequilibrium systems that exhibit transitions between phases and characteristic spatio-temporal patterns and behaviors. There are transitions between solid, fluid and gas phases of matter. Ferromagnets show transitions between phases with vanishing and non-vanishing magnetization. Lasers can produce ordinary light (i.e., "chaotic" light) as well as laser light (i.e., coherent light) and show transitions between these qualitatively different regimes when their pumping energies are changed. Humans and animals change gaits at critical values of locomotion speeds. In human societies, there are transitions between fashion trends every now and then.

In the following, we will refer to the phases of matter and the characteristic spatio-temporal manifestations of systems simply as phases. Transitions between phases will be referred to as (equilibrium and nonequilibrium) phase transitions. System parameters that are involved in phase transitions will be called control parameters.

In order to quantify equilibrium and nonequilibrium phase transitions it is useful to describe systems by means of variables that capture the characteristic properties of phases: the order parameters. Phase transitions can then be described in terms of bifurcations of order parameters as illustrated in Fig. 1.1.

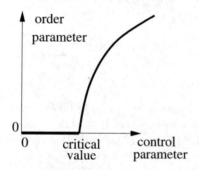

Fig. 1.1. Illustration of the bifurcation of an order parameter

The Landau theory of (second-order or continuous) phase transitions tells us that, in general, equilibrium systems at phase transition points exhibit order parameters [57, 229, 371, 372, 410, 546]. Moreover, the theory predicts that close to phase transition points the whole behavior of equilibrium systems can be described in terms of appropriately defined order parameters. That is, all quantities of interest can be expressed as functions of these order parameters. The order parameters of equilibrium systems typically correspond to mean field variables derived from mean field theory such as averaged magnetization.

In a similar spirit, many nonequilibrium systems can be described in the vicinity of phase transition points by means of a few relevant variables [104,

365] and, in particular, by means of a few order parameters [55, 207, 211, 212, 252, 254, 255, 258, 260, 419, 444, 448, 486, 637]. As pointed out by Haken [254, 267, 270, 271], it is helpful to regard order parameters of nonequilibrium systems as counterparts to the order parameters of the Landau theory.

What is the reason for the emergence of order parameters? Close to phase transition points, subsystems of many-body systems can evolve in a self-organized fashion. Due to self-organization, the degrees of freedom of many-body systems can be considerably reduced such that there are only a few degrees of freedom left. These remaining degrees of freedom then define low dimensional phase spaces in which order parameters evolve (see e.g. [254, 255]).

Self-organizing systems described by nonlinear Fokker–Planck equations of the form (1.3) typically involve coefficients D_i and D_{ik} that depend on order parameters. For example, we may deal with a random variable $\mathbf{X}(t)$ described by a nonlinear Fokker–Planck equation

$$\frac{\partial}{\partial t} P(\mathbf{x}, t) = -\sum_i \frac{\partial}{\partial x_i} D_i(\mathbf{x}, t, \mathbf{m}) P(\mathbf{x}, t) + \sum_{i,k} \frac{\partial^2}{\partial x_i \partial x_k} D_{ik}(\mathbf{x}, t, \mathbf{m}) P(\mathbf{x}, t) \tag{1.4}$$

that involves an order parameter $\mathbf{m}$ defined by $\mathbf{m} = \langle \mathbf{X} \rangle$, where $\langle \mathbf{X} \rangle$ is the average of $\mathbf{X}$.

1.3 Stochastic Feedback

In order to elucidate the notion of stochastic feedback we consider a field $v(\mathbf{x}, t)$ that describes the flow of a system. For example, $v(\mathbf{x}, t)$ may correspond to a component of a velocity vector field. We then assume that system variables $\psi_i(\mathbf{x}, t)$ change due to the flow of the system such that we obtain a linear transport equation

$$\frac{\partial}{\partial t} \psi_i(\mathbf{x}, t) = \hat{F}(\mathbf{x}, \nabla, t, v) \, \psi_i(\mathbf{x}, t) + R_i \tag{1.5}$$

involving an operator $\hat{F}(\mathbf{x}, \nabla, t, v)$ that depends on v and some source terms R_i. We say that the evolution of the field variables $\psi_i(\mathbf{x}, t)$ is determined by the flow field $v(\mathbf{x}, t)$. If $v(\mathbf{x}, t)$ satisfies (1.5) we deal with feedback. In this case (1.5) reads

$$\frac{\partial}{\partial t} v(\mathbf{x}, t) = \hat{F}(\mathbf{x}, \nabla, t, v) \, v(\mathbf{x}, t) + R_v \tag{1.6}$$

and is nonlinear with respect to v. Feedback of this kind can be found, for example, in hydrodynamics (hydrodynamic feedback). If we consider laminar flows in the direction of the x-coordinate, then v corresponds to the velocity field $v_x(\mathbf{x},t)$ of a fluid, $\hat{F}$ is given by $\hat{F} = -v_x \partial/\partial x$, and the field variables ψ_i correspond to the density field of the fluid, the temperature field, and the velocity field $v_x(\mathbf{x},t)$ [448].

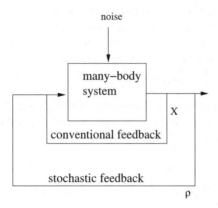

Fig. 1.2. Basic features of a stochastic feedback system

Now let us interpret a probability density P as some kind of flow field v. Let us assume that the evolution of system variables ψ_i is affected by the flow field $P(\mathbf{x},t)$ in such a way that we obtain the linear evolution equation (1.5) for $v = P$ and $R_i = 0$:

$$\frac{\partial}{\partial t}\psi_i(\mathbf{x},t) = \hat{F}(\mathbf{x},\nabla,t,P(\mathbf{x},t))\,\psi_i(\mathbf{x},t) \ . \tag{1.7}$$

If P satisfies its own evolution equation, that is, for $v = P(\mathbf{x},t)$, and if the operator $\hat{F}$ is given by

$$\hat{F}(\mathbf{x},\nabla,t,P) = -\sum_i \frac{\partial}{\partial x_i} D_i(\mathbf{x},t,P) + \sum_{i,k} \frac{\partial^2}{\partial x_i \partial x_k} D_{ik}(\mathbf{x},t,P) \ , \tag{1.8}$$

then we obtain the nonlinear Fokker–Planck equation (1.3). Consequently, we realize that (1.3) involves feedback as well. We will refer to this feedback as stochastic (or statistical) feedback [67, 188, 190, 196, 387]. In particular, as we will show in Chap. 3, for strongly nonlinear Fokker–Planck equations the transport equation (1.7) applies not only for $\psi_1 = P(\mathbf{x},t)$ but also for $\psi_2 = P(\mathbf{x},t|\mathbf{x}',t')$, that is, for a transition probability density. One may alternatively say that for strongly nonlinear Fokker–Planck equations the transport equation (1.7) with $\hat{F}$ given by (1.8) holds for an infinitely large set of field variables $\psi_1 = P(\mathbf{x},t)$, $\psi_2 = P(\mathbf{x},t;\mathbf{x}',t')$, $\psi_3 = P(\mathbf{x},t;\mathbf{x}',t';\mathbf{x}'',t'')$, and so on that describe a stochastic process.

The basic features of a stochastic feedback system are illustrated in Fig. 1.2. There is a noise source acting on the stochastic feedback system and there are two feedback loops. The conventional feedback loop involves the state variable **X** of a single representative subsystem. The conventional feedback leads to transport equations of the form

$$\frac{\partial}{\partial t}\psi_i(\mathbf{x},t) = \hat{F}'(\mathbf{x},\nabla,t)\,\psi_i(\mathbf{x},t) + R_i \ . \tag{1.9}$$

In addition, there is a feedback loop that involves the subsystem density $\rho(\mathbf{x})$. Due to this second feedback loop stochastic properties of the subsystem ensemble are fed back into the ensemble. The stochastic feedback leads to transport equations such as

$$\frac{\partial}{\partial t}\psi_i(\mathbf{x},t) = \hat{F}'(\mathbf{x},\nabla,t,\rho)\,\psi_i(\mathbf{x},t) + R_i \ . \tag{1.10}$$

If we use $P = \rho/M_0$ with M_0 defined such that P is normalized to unity, (1.10) can be cast into the form (1.7).

1.4 Applications

There are various applications of nonlinear Fokker–Planck equations, see Table 1.2. Let us briefly address some of them.

1.4.1 Collective Phenomena

Due to self-organization, many-body systems can exhibit properties that do not exist on the level of their subsystems, see Table 1.3. For example, ferromagnetic materials can show macroscopic magnetic fields. Macromolecules of liquid crystals can align themselves in a parallel fashion. In doing so, nematic phases of macroscopic orientational order can emerge. Populations of oscillatory neurons can synchronize their activity. Cross-bridges in muscle fibers can produce collective pulling forces leading to muscular contractions. Crickets can chirp in unison. Novel trends pop up in human societies every now and then and group behavior is a common feature of many societies. In these examples subsystems exhibit a collective or cooperative behavior. For this reason, we refer to the corresponding phenomena as collective or cooperative phenomena [135, 254, 267, 270, 271, 365, 546]. Collective phenomena typically involve order parameters and frequently arise due to phase transitions from phases with uncorrelated subsystem dynamics.

1.4.2 Multistable Systems

In Sect. 1.2 we have discussed the notion of phase transitions. In this context, the question naturally arises: how many different phases can we observe

Table 1.2. Some applications of nonlinear Fokker–Planck equations

		Ref./Sect.
Condensed matter and material physics		
Ferromagnetism		Sect. 5.5.1
Liquid crystals	(isotropic-nematic phase transition)	Sect. 5.5.3
Quantum statistics	(Fermi systems, Bose systems,	Sect. 6.5.4/7
	vortices of flow fields,	Sect. 6.4.2
	grains of granular matter)	[294]
Nonlinear hydrodynamics		
Flows in porous media	(porous medium equation)	Sect. 6.5.1
Flows in unsaturated media	(Richard's equation)	[395]
Polymer fluids	(consistent averaging,	[300, 458]
	Gaussian approximation)	
Plasma physics/ particle accelerator physics/ astrophysics		
Plasma particles	(Landau form of the	Sect. 7.4.1
	Fokker–Planck equation)	
Bunch-particles in particle beams	(Vlasov–Fokker–Planck equation)	Sect. 7.4.2
Self-gravitating systems		[58, 92, 94]
		[527]
Surface physics		
Surface dynamics, sedimentation	(Kadar–Parisi–Zhang equation)	Sect. 7.5
Wetting		Sect. 7.5
Statistical physics and stochastics		
Nonextensive thermostatistics	(Tsallis entropy)	Chap. 6
Power law distributions		Sect. 1.4.3, Sect. 6.5
Anomalous diffusion		Sect. 6.2.2, Sect. 6.5
Synchronization	(laser arrays, electronic circuitry)	Sect. 7.5
Biophysics		
Synchronization of brain activity	(perception, memory, task related brain activity, Parkinson's disease)	Sect. 5.5.2
Muscular contraction		Sect. 5.5.4
Multistable perception-action systems		Sect. 5.5.6
Population dynamics		Sect. 6.5.7
Flexibility-accuracy trade-off		Sect. 7.4.4
Ion channels of membranes		Sect. 7.5
Social behavior and psychology		
Behavioral synchronization	(applause, fireflies flashing, chirping crickets, group behavior)	Sect. 5.5.2 Sect. 5.5.5
Development of the self		[189]
Computational physics		
Noise generators		Sect. 7.4.3

Table 1.3. Examples of collective phenomena

Systems	Collective phenomena	Order parameter	Section
Ferromagnetic materials	ferromagnetism	magnetization	5.5.1
Brains	synchronization of oscillatory activity	non-uniformity of phase distribution	5.5.2
Human societies	mass-behavior, e.g., synchronized applause	mean normalized harmonic correlation	5.5.2
Crickets	chorusing	non-uniformity of chirp distribution	5.5.2
Fireflies	rhythmic synchronous flashing	non-uniformity of flash distribution	5.5.2
Liquid crystals	nematic phase	degree of alignment	5.5.3
Muscle fibers	muscular contraction	mean cross-bridge velocity	5.5.4

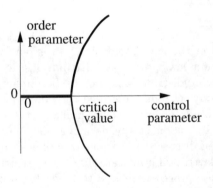

Fig. 1.3. Bifurcation of an order parameter of a system with a monostable and a bistable parameter regime

given a particular set of control parameter values? In general, the answer to this question depends on the system that we study. There are systems with unique stable states. Other systems have multiple stable states. Of particular interest are phase transitions that carry systems from monostable parameter regimes to multistable parameter regimes. In terms of order parameters, mono- and multistability means that order parameter equations have unique or multiple stationary solutions. A phase transition between a monostable and multistable parameter regime is illustrated in Fig. 1.3. Systems showing multistability are, for example, ferromagnets (see Sect. 5.5.1) and perception-action systems of humans and animals (see Sect. 5.5.6).

1.4.3 Power Law and Cut-Off Distributions

Density distributions can be classified according to their qualitative shape. In particular, we may distinguish between power law distributions and cut-off distributions.

Power Law Distributions

Power law distributions are distinguished from other distributions by means of their long range behavior, that is, their tails. The tails of power law distributions decay according to a power law. Let $\rho(\mathbf{x})$ denote a density distribution. Then, we deal with a power law distribution if for large $|\mathbf{x}|$ the distribution is described by

$$\rho(\mathbf{x}) \propto |\mathbf{x}|^{-\nu} \qquad (1.11)$$

with $\nu > 0$. Since in the asymptotic limit an exponential function decays faster than any power law, power law distributions do not decay as fast as exponentially decaying distributions such as Boltzmann distributions. Consequently, many-body systems described by power law distributions have subsystems that more frequently occupy unlikely states by comparison with systems described by exponentially decaying distributions. Power law distributions have been discussed in various disciplines, see Table 1.4. Famous examples of power law distributions are the Cauchy distribution and Lévy distributions. Stochastic processes related to Lévy distributions are also called Lévy flights and play an important role in many disciplines ranging from physics to economics [35, 533, 586] (see also Table 1.4).

Cut-Off Distributions

Cut-off distributions are finite only on subspaces of the phase spaces on which they are defined. Outside these subspaces the distributions equal zero. That is, let $\rho(\mathbf{x})$ denote a density function defined on Ω. Then, cut-off distributions satisfy

Table 1.4. Systems exhibiting power law distributions

Systems	Laws	Ref.
Earthquakes	Gutenberg–Richter law	[351]
Pulsar bursts		[95, 354]
Fragmentation		[413, 453, 543]
Laser cooling		[37]
Fully developed turbulence		[28, 41, 42, 43, 44, 45, 47]
Linguistics	Pareto's law	[587]
Urban structures		[507]
Animal behavior		[107, 612]
Protein sequences	Pareto's law	[306]
Cell aggregation		[597]
Economics		[67, 71, 231, 230]
		[402, 403, 493, 590]

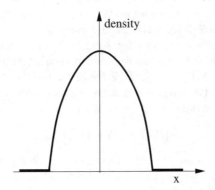

Fig. 1.4. A cut-off distribution of the state variable x defined on $\mathbb{R}$

$$\rho(\mathbf{x}) > 0 \text{ for } \mathbf{x} \in A \subset \Omega , \quad \rho(\mathbf{x}) = 0 \text{ for } \mathbf{x} \in \Omega - A , \qquad (1.12)$$

with $A \neq \Omega$. An example of a cut-off distribution is shown in Fig. 1.4. Cut-off distributions of many-body systems tell us that the subsystems of the many-body systems occupy only subspaces of phase spaces. The probability to find subsystems in states outside these subspaces is zero. Due to the decay of cut-off distributions to zero, we can say that cut-off distributions describe density and concentration fronts. Likewise, time-dependent cut-off distributions can describe distributions with moving boundaries and the propagation of concentration fronts. Therefore, they are tailored to describe, for example, density waves and the stochastic dynamics of populations. In fact, cut-off distributions have been discussed in the context of gas flow through porous media (Sect. 6.5.1), fluid flow in unsaturated media [395], and population dynamics (Sect. 6.5.7).

Classical Solutions

We will refer to infinitely differentiable solutions of Fokker–Planck equations as classical solutions. From our previous discussion it is clear that cut-off distributions are continuous functions but they are not continuously differentiable. Therefore, they do not belong to the class of classical solutions.

1.4.4 Free Energy Systems

For various equilibrium systems with constant temperatures T, the distributions of stationary states (stationary distributions) correspond to the distributions that minimize the free energy $F = U - T\,^\mathrm{G}S$, where U describes the internal energy of the systems and $^\mathrm{G}S$ denotes the Gibbs entropy [89, 509]. Nonequilibrium systems can be treated in a similar manner. To this end, one needs to employ information measures such as the Shannon information, which is (apart from a proportionality constant) equivalent to the Gibbs entropy [261, 309, 509, 518, 621]. In the context of transport phenomena, time-dependent entropy measures for single subsystem probability densities $P(\mathbf{x}, t)$ of many-body systems are frequently used. A famous example is the Boltzmann entropy $^\mathrm{B}S[P] = -\int_\Omega P(\mathbf{x}, t) \ln P(\mathbf{x}, t)\,\mathrm{d}^M x$ [291, 348, 432, 576]. In due recognition of these approaches, we will consider functionals for single subsystem probability densities $P(\mathbf{x}, t)$ of the form

$$F[P] = U[P] - QS[P]\,, \qquad (1.13)$$

where $U[P]$ measures some kind of internal energy of a subsystem, Q denotes a noise amplitude (see Sect. 1.1.1), and S is an entropy and information measure. We will see in Sect. 4.1 that the stationary solutions of appropriately defined nonlinear Fokker–Planck equations correspond to distributions that make F extremal. Moreover, we will demonstrate that stable stationary distributions of these nonlinear Fokker–Planck equations correspond to minima (or "valleys") of F (see Sects. 5.1 and 5.3.2). In line with equilibrium thermostatistics, we will therefore refer to F given by (1.13) as a free energy measure. The shape of the distributions that make the free energy measures (1.13) extremal, depends on our choice of the entropy and information measure S. There are various measures S at our disposal.

Extensive Thermostatistics and μ-Space Entropies

As mentioned above, a fundamental entropy measure for many-body systems is the Gibbs–Shannon entropy which is defined on many particle distributions of many-body systems (Γ-space description). If we describe many-body systems in terms of one-particle or one-subsystem distributions (μ-space description), then in general we need to modify the Gibbs–Shannon entropy measure in order to account for subsystem-subsystem interactions or constraints imposed on the subsystems. For the distinction between Γ and μ-space descriptions see Sect. 6.4.2. For systems with non-interacting point-like particles (e.g. the atoms of an ideal gas) for which quantum effects are irrelevant, there is no need for a modification of the Gibbs–Shannon entropy and the one-particle entropy measure is equivalent to the many-particle entropy measure. In this case, the Gibbs–Shannon entropy defined on the Γ-space gives us the famous Boltzmann entropy $^{B}S = -\int P(\mathbf{x}) \ln P(\mathbf{x}) \, \mathrm{d}^M x$ that is defined on the μ-space and related to the Maxwell–Boltzmann statistics (see Sect. 4.4). In contrast, many-body systems of quantum particles such as Bose systems and Fermi systems involve quantum mechanical constraints. These constraints can be taken into account if we consider quantum mechanical entropy measures for single particle distribution that are related to quantum statistical distributions such as the Fermi–Dirac distribution and the Bose–Einstein distribution. The quantum mechanical entropy related to Fermi–Dirac distributions has also been applied to describe the statistics of vortices of flows fields. In order to account for the fact that the free phase space of many-body systems is reduced by the space occupied by subsystems (e.g., the co-volume of real gas), in the context of so-called excluded-volume theories, once more modifications of the Boltzmann entropy measure have been considered. For example, van der Waals gases, granular matter, and self-gravitating systems have been treated in the context of μ-space entropy measures, see also Table 1.5. For self-gravitating systems generalized entropy measures have also been proposed on the basis of heuristic arguments [92, 483, 527].

Table 1.5. Quantum mechanical entropies and entropies of real gases and real matter (MB=Maxwell–Boltzmann)

Systems	Entropies	Statistics	Ref./Sect.
MB particle systems	^{B}S	Boltzmann	Sect. 4.4
Fermi systems	^{FD}S	Fermi–Dirac	Sect. 6.5.4
Bose systems	^{BE}S	Bose–Einstein	Sect. 6.5.4
Vortices of flow fields	^{FD}S	Fermi–Dirac	Sect. 6.4.2
Granular matter	^{FD}S	Fermi–Dirac	[294]
van der Waals gases			[319, 396, 408, 510, 606]
Self-gravitating systems			[615]

Temperature Fluctuations and Generalized Entropies

Under particular circumstances it is helpful to consider the temperature of systems described by the Boltzmann entropy as a fluctuating variable. The effective statistics obtained from a Boltzmann distribution involving a fluctuating temperature variable is referred to as a superstatistics. The fluctuating temperature variable can be eliminated at the cost that we need to introduce a generalized entropy measure [42, 45, 46, 98, 440, 593].

Nonextensive Thermostatistics

A popular interpretation of entropy and information measures S is that they describe the amount of chaos or the amount of disorder of systems [625]. In line with this interpretation, it is clear that the disorder of a many-body system is proportional to the size of the system provided that it is composed of non-interacting and statistically-independent subsystems (e.g. we have an ideal gas). As a result, entropy or information measures of such systems are extensive (or additive) quantities and exhibit a scaling behavior. More precisely, if we split such systems into two parts A and B, then the entropy or information $S(A + B)$ of the whole system can be computed from

$$S(A + B) = S(A) + S(B) , \qquad (1.14)$$

where $S(A)$ and $S(B)$ describe entropy or information of the parts [625]. This situation is different for many-body systems with statistically-independent but interacting subsystems. Here, we may deal with extensive measures S satisfying (1.14). Alternatively, we may deal with nonextensive (or nonadditive) measures S for which the relation

$$S(A + B) \neq S(A) + S(B) \qquad (1.15)$$

holds. Simple examples of many-body systems with statistically-independent but interacting subsystems are systems described by the Langevin equations defined in Sect. 3.4. If we interpret the realizations of random variables defined by these Langevin equations as the state variables of subsystems, we obtain subsystem ensembles with statistically-independent but interacting members. For details see Sect. 3.8.

Nonextensive entropy and information measures are of particular relevance for thermodynamics because extensive entropy and information measures in general and the Boltzmann entropy ^{B}S in particular are subjected to severe limitations. For example, extensive thermodynamics applies to systems whose total energy is the sum of the energies of their macroscopic parts. Following Terletskii [576] and Landsberg [374], systems characterized by long range forces fail to satisfy this additivity property because interaction energies between macroscopic parts of these systems can make essential contributions to the total energy. In addition, Sharma and Garg [519] pointed out that several biological and social systems should be modeled by means of nonextensive entropies and information measures. Failures of extensive thermodynamics and benefits of nonextensive thermodynamics have been nicely illustrated by means of a one-parametric entropy measure proposed by Tsallis [582], which will be addressed in Chap. 6. For reviews on this topic the reader is referred to [3, 583, 584, 587, 588, 589].

1.4.5 Anomalous Diffusion

So far, we have described systems by the number of their stationary states (e.g., unique or multiple stationary states) and the shape of their distribution functions (e.g., cut-off distributions, power law distributions, Boltzmann distributions, quantum mechanical distributions). These characteristics, however, do not provide information about the time scales on which systems evolve. Therefore, at issue is to assess how fast systems evolve with time. For this purpose, a useful measure is the variance of distributions. Let us consider the univariate case given by a random variable $X(t)$. Then, by means of the variance

$$K(t) = \langle [X(t) - \langle X(t) \rangle]^2 \rangle \;, \tag{1.16}$$

one can distinguish between normal and anomalous diffusion, see Fig. 1.5. In the former case, $K(t)$ increases linearly with time. In the latter case, there is a nonlinear increase with time. If $K(t)$ scales slower than a linear function, we refer to the diffusion process as subdiffusive (or sublinear). Likewise, if $K(t)$ scales faster than a linear function, we refer to the process as superdiffusive (or superlinear).

A system that shows normal diffusion is a Brownian particle that evolves as

$$\frac{\mathrm{d}}{\mathrm{d}t} V(t) = -\gamma V(t) + \sqrt{Q} \Gamma(t) \;, \tag{1.17}$$

where $V(t)$ denotes the particle velocity. Then $K(t)$ measures the mean squared displacement with respect to the particle position $X(t) = \int^t V(t') \, \mathrm{d}t'$ and we find $K \propto t$ [490, 498, 619]. Anomalous diffusion has been found in various systems [70, 281, 418, 533]. For example, superdiffusion has been studied in the context of turbulent diffusion [40, 213, 287, 496, 532] and Josephson junctions [223]. Similar effects have also been found in systems of polymer-like breakable micelles [457], in two-dimensional rotating flow [541, 622], and in diffusion of atom clusters on solid surfaces [397]. Using the so-called Barenblatt–Pattle solutions, polytropic gas-flow in porous media (see Sect. 6.5.1) and surface relaxations [545] have been treated as subdiffusive stochastic processes.

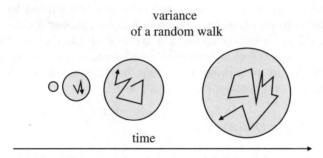

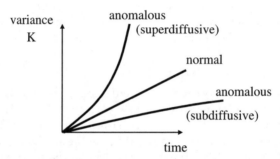

Fig. 1.5. *Upper panel*: illustration of the increase of the variance of a random walk defined on the XY-plane. *Lower panel*: anomalous diffusion versus normal diffusion

1.5 Overview

This book is organized as follows (see Table 1.6). In Chap. 2 fundamental definitions regarding Fokker–Planck equations will be given and several ways to classify Fokker–Planck equations will be addressed. Chap. 3 is concerned with Langevin equations. Langevin equations will be used in all subsequent chapters. Chapters 4, 5, and 6 deal with systems described by free energy measures. In Chap. 4 the focus is on transient and stationary solutions. We will derive an H-theorem for transient solutions and show how stationary solutions can be derived from the free energy principle mentioned in Sect. 1.4.4. Chap. 5 is about stability issues. In particular, phase transitions and multistability will be addressed. In Chap. 6 we will study the relationship between solutions of nonlinear Fokker–Planck equations and entropy and information measures. In this context, we will be concerned with normal and anomalous diffusion, on the one hand, and power law distributions, cut-off distributions, quantum statistical distributions, on the other hand. Chap. 7 is concerned with general nonlinear Fokker–Planck equations. Again, we will be concerned with stability analysis and phase transitions. Moreover, we will address several special cases for which solution methods that are similar to those developed for the free energy case can be applied. Finally, we would like to note that some of the sections have been marked with an asterisk. They contain mathematically involved proofs of results that are used in other sections of this monograph and may be skipped by readers who are not interested in the details of these proofs.

Table 1.6. Overview

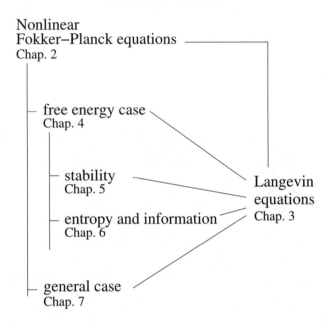

2 Fundamentals

2.1 Stochastic Processes

Let us describe a stochastic process by means of an M-dimensional time-dependent random variable $\mathbf{X}(t)$ defined on the phase space Ω, that is, we have $\mathbf{X}(t) = (X_1(t), \ldots, X_M(t)) \in \Omega$. Let us assume that $\mathbf{X}(t)$ is distributed like $u(\mathbf{x})$ at an initial time t_0. Then, the stochastic trajectory $\mathbf{X}(t)$ provides us with a complete description of the process for $t \geq t_0$. To see this, we note that from $\mathbf{X}(t)$ we can compute an infinitely large set of realizations. These realizations constitute the statistical ensemble of the stochastic process under consideration. From the statistical ensemble all quantities of interest (e.g., mean values, autocorrelations) can be obtained using ensemble averaging denoted here by $\langle \ldots \rangle$;

$$\mathbf{X}(t) \Rightarrow \langle \ldots \rangle \ . \tag{2.1}$$

In particular, the ensemble average with respect to the delta function $\delta(\cdot)$ [430] gives us the probability density of $\mathbf{X}(t)$,

$$P(\mathbf{x}, t; u) = \langle \delta(\mathbf{x} - \mathbf{X}(t)) \rangle \ , \tag{2.2}$$

which satisfies the initial condition $P(\mathbf{x}, t_0; u) = u(\mathbf{x})$. Joint probability densities can be defined in a similar manner. For example, we have $P(\mathbf{x}_2, t_2; \mathbf{x}_1, t_1; u) = \langle \delta(\mathbf{x}_1 - \mathbf{X}(t_1))\delta(\mathbf{x}_2 - \mathbf{X}(t_2)) \rangle$. By means of a hierarchy of such joint distributions, that is, by means of

$$\begin{aligned} &P(\mathbf{x}, t; u) \ , \\ &P(\mathbf{x}_2, t_2; \mathbf{x}_1, t_1; u) \ , \\ &\ldots \\ &P(\mathbf{x}_n, t_n; \ldots; \mathbf{x}_1, t_1; u) \ , \\ &\ldots \end{aligned} \tag{2.3}$$

with $t \geq t_0$ and $t_n \geq t_{n-1} \geq \ldots \geq t_1 \geq t_0$, a stochastic process is completely defined. That is, the hierarchy (2.3) provides us with a complete description of a stochastic process. Therefore, we have two alternative descriptions of a stochastic process at hand: a stochastic process can be described by means of the random variable $\mathbf{X}(t)$ or by means of the hierarchy (2.3).

2.2 Nonlinear Fokker–Planck Equation

We can now define an evolution equation for the single time-point probability density $P(\mathbf{x}, t; u)$ by

$$\frac{\partial}{\partial t} P(\mathbf{x}, t; u) = \underbrace{-\sum_{i=1}^{M} \frac{\partial}{\partial x_i} D_i(\mathbf{x}, t, P) P(\mathbf{x}, t; u)}_{\text{drift term}}$$

$$+ \underbrace{\sum_{i,k=1}^{M} \frac{\partial^2}{\partial x_i \partial x_k} D_{ik}(\mathbf{x}, t, P) P(\mathbf{x}, t; u)}_{\text{diffusion term}} \quad (2.4)$$

with $t \geq t_0$. We refer to (2.4) as a nonlinear Fokker–Planck equation because, on the one hand, the evolution equation is in general nonlinear with respect to P, and, on the other hand, for $D_i(\mathbf{x}, t, P) = D_i(\mathbf{x}, t)$ and $D_{ik}(\mathbf{x}, t, P) = D_{ik}(\mathbf{x}, t)$ it reduces to the form of the linear Fokker–Planck equation [254, 274, 498] that was originally proposed by Fokker and Planck [162, 479, 617]. In particular, we refer to D_i and D_{ik} as drift and diffusion coefficients, respectively. Accordingly, we call the two terms on the right hand side of (2.4) drift term and diffusion term. D_i and D_{ik} can depend in an arbitrary fashion on $P(\mathbf{x}, t; u)$. Some examples are

$$D_i = h_i(\mathbf{x}) + \int_{\Omega} w_i(\mathbf{x} - \mathbf{x}') P(\mathbf{x}', t; u) \, \mathrm{d}^M x , \quad (2.5)$$

$$D_i = h_i(\mathbf{x}) + \kappa \langle X_i \rangle , \quad (2.6)$$

$$D_i = D_i \left[\mathbf{x}, \partial_k^n / \partial x_k^n, \int \ldots \mathrm{d}x_l, t, P \right] , \quad (2.7)$$

$$D_{ik} = Q \delta_{ik} P , \quad (2.8)$$

$$D_{ik} = \delta_{ik} f(\langle X^2 \rangle) . \quad (2.9)$$

Here, δ_{ik} denotes the Kronecker symbol. We do not claim that (2.4) represents the most general form of a nonlinear Fokker–Planck equation. However, there are numerous nonlinear Fokker–Planck equations that have been discussed in the literature and can be cast into the form (2.4). In the univariate case, (2.4) reads

$$\frac{\partial}{\partial t} P(x, t; u) = -\frac{\partial}{\partial x} D_1(x, t, P) P(x, t; u) + \frac{\partial^2}{\partial x^2} D_2(x, t, P) P(x, t; u) , \quad (2.10)$$

with $t \geq t_0$. Just as in the case of linear Fokker–Planck equations, one usually considers nonlinear Fokker–Planck equations with symmetric and semipositive definite diffusion coefficients. Consequently, we will usually require that the inequalities $\forall \mathbf{y} \in \mathbb{R}^M : \sum_{i,k=1}^{M} D_{ik} y_i y_k \geq 0$ (multivariate case) and

$D_2 \geq 0$ (univariate case) hold and that in the multivariate case D_{ik} is symmetric (i.e., we have $D_{ik} = D_{ki}$). A more detailed discussion of this issue will be given for so-called strongly nonlinear Fokker–Planck equations, see Chap. 3.

Some systems are more conveniently described by means of density functions $\rho(\mathbf{x}, t)$ normalized to $M_0 = \int_\Omega \rho(\mathbf{x}, t) \, \mathrm{d}^M x$; M_0 may describe the total mass or the total particle number of the systems. Then, we deal with evolution equations that read as follows

$$\frac{\partial}{\partial t}\rho(\mathbf{x}, t) = -\sum_{i=1}^{M} \frac{\partial}{\partial x_i} d_i(\mathbf{x}, t, \rho)\rho(\mathbf{x}, t) + \sum_{i,k=1}^{M} \frac{\partial^2}{\partial x_i \partial x_k} d_{ik}(\mathbf{x}, t, \rho)\rho(\mathbf{x}, t) \tag{2.11}$$

in the multivariate case and

$$\frac{\partial}{\partial t}\rho(x, t) = -\frac{\partial}{\partial x} d_1(x, t, \rho)\rho(x, t) + \frac{\partial^2}{\partial x^2} d_2(x, t, \rho)\rho(x, t) \tag{2.12}$$

in the univariate case. It is clear that these evolution equations can be transformed to the nonlinear Fokker–Planck equations (2.4) and (2.10) by means of $P = \rho/M_0$, $u(\cdot) = \rho(\cdot, t_0)$,

$$D_i(\mathbf{x}, t, P) = d_i(\mathbf{x}, t, M_0 P), \quad D_{ik}(\mathbf{x}, t, P) = d_{ik}(\mathbf{x}, t, M_0 P) \tag{2.13}$$

and

$$D_1(x, t, P) = d_1(x, t, M_0 P), \quad D_2(x, t, P) = d_2(x, t, M_0 P). \tag{2.14}$$

2.2.1 Notation

In order to ease the presentation of evolution equations for $P(\mathbf{x}, t; u)$, the arguments of P will be listed only in the expression with the derivative with respect to t. For example, we will write (2.4) as

$$\frac{\partial}{\partial t}P(\mathbf{x}, t; u) = -\sum_{i=1}^{M} \frac{\partial}{\partial x_i} D_i(\mathbf{x}, t, P) P + \sum_{i,k=1}^{M} \frac{\partial^2}{\partial x_i \partial x_k} D_{ik}(\mathbf{x}, t, P) P. \tag{2.15}$$

However, if we deal with evolution equations for transition probability densities $P(\mathbf{x}, t | \mathbf{x}', t'; u)$, we will write out the arguments in full in order to distinguish between $P(\mathbf{x}, t | \mathbf{x}', t'; u)$ and $P(\mathbf{x}, t; u)$ and to avoid confusion.

2.2.2 Stratonovich Form

Alternatively to (2.4) one may study nonlinear Fokker–Planck equations of the form

$$\frac{\partial}{\partial t}P(\mathbf{x},t;u) = -\sum_{i=1}^{M}\frac{\partial}{\partial x_i}\tilde{D}_i(\mathbf{x},t,P)P + \sum_{i,k=1}^{M}\frac{\partial}{\partial x_i}\left[D_{ik}(\mathbf{x},t,P)\frac{\partial}{\partial x_k}P\right].$$
(2.16)

For $\tilde{D}_i(\mathbf{x},t,P) = \tilde{D}_i(\mathbf{x},t)$ and $D_{ik}(\mathbf{x},t,P) = D_{ik}(\mathbf{x},t)$ the evolution equation (2.16) becomes linear and corresponds to the Stratonovich form of the linear Fokker–Planck equation [221, 498]. For a state-independent diffusion matrix D_{ik}, (2.4) and (2.16) are equivalent if we put $D_i = \tilde{D}_i$. For diffusion coefficients that depend explicitly or implicitly (i.e. by means of P) on $\mathbf{x}$ one can transform (2.16) into (2.4) by means of

$$\tilde{D}_i(\mathbf{x},t,P) = D_i(\mathbf{x},t,P) - \sum_{k=1}^{M}\frac{\partial D_{ik}(\mathbf{x},t,P)}{\partial x_k}.$$
(2.17)

2.2.3 Transient Solutions

Transient solutions are time-dependent solutions of the nonlinear Fokker–Planck equation (2.4). That is, we have $\partial P/\partial t \neq 0$. Of particular interest is the behavior of transient solutions in the limit $t \to \infty$. We may distinguish between the following cases:

- transient solutions converge to stationary ones:
  $\lim_{t\to\infty} \partial P/\partial t = 0$
- transient solutions do not converge to stationary ones:
  $\lim_{t\to\infty} \partial P/\partial t \neq 0$
- transient solutions converge to solutions of stationary free energies:
  $\lim_{t\to\infty} dF[P]/dt = 0$

In the latter case, $F[P]$ denotes the free energy of a state described by P. One caveat: while the implication $\partial P/\partial t = 0 \Rightarrow \lim_{t\to\infty} dF[P]/dt = 0$ holds for free energy measures $F[P]$ that do not explicitly depend on t, the opposite implication $dF[P]/dt = 0 \Rightarrow \partial P/\partial t = 0$ is not necessarily satisfied. We will return to this issue in Chapters 4 and 6.

2.2.4 Continuity Equation

Equation (2.4) can be written as

$$\frac{\partial}{\partial t}P(\mathbf{x},t;u) = -\nabla \cdot \mathbf{J}(\mathbf{x},t,P),$$
(2.18)

where $\mathbf{J} = (J_1, \ldots, J_M)$ is defined by

$$J_i(\mathbf{x},t,P) = D_i(\mathbf{x},t,P)P - \sum_{k=1}^{M}\frac{\partial}{\partial x_k}D_{ik}(\mathbf{x},t,P)P.$$
(2.19)

In particular, (2.10) can be written as

$$\frac{\partial}{\partial t} P(x,t;u) = -\frac{\partial}{\partial x} J(x,t,P) \qquad (2.20)$$

with J given by

$$J(x,t,P) = D_1(x,t,P)P - \frac{\partial}{\partial x} D_2(x,t,P)P \; . \qquad (2.21)$$

Equations (2.18) and (2.20) are referred to as continuity equations. In analogy to the theory of linear Fokker–Planck equations [254, 274, 498], we will refer to $\mathbf{J}$ and J as probability currents.

2.2.5 Boundary Conditions

Solutions of the nonlinear Fokker–Planck equation (2.4) depend on boundary conditions. One can generalize boundary conditions defined for linear Fokker–Planck equations [221, 498] to the nonlinear case. Accordingly, we distinguish between

- natural boundary conditions:
  $\mathbf{X} \in \Omega = \mathbb{R}^M$ and $\lim_{|\mathbf{x}| \to \infty} P(\mathbf{x},t;u) = 0$
- periodic boundary conditions:
  $\mathbf{X} \in \Omega = \prod_{i=1}^{M}[a_i,b_i]$, $b_i - a_i = T_i > 0$, and $P(\ldots, x_i + T_i, \ldots) = P(\ldots, x_i, \ldots)$
- reflective boundary conditions:
  $\mathbf{X} \in \Omega = \prod_{i=1}^{M}[a_i,b_i]$, $b_i > a_i$, and $\mathbf{J}|_{\partial\Omega} = 0$, where $\mathbf{J}$ is the probability current
- mixed boundary conditions

In the case of mixed boundary conditions, we have, for example, a natural boundary in combination with a reflective boundary. Furthermore, if a multivariate random vector $\mathbf{X}$ contains components that are subjected to boundary conditions of different kinds, we deal with mixed boundary conditions.

2.2.6 Stationary Solutions

Stationary solutions of the nonlinear Fokker–Planck equation (2.4) will be denoted by P_{st} and satisfy $\partial P_{\text{st}}/\partial t = 0 \Rightarrow P_{\text{st}}(\mathbf{x},t) = P_{\text{st}}(\mathbf{x})$. Note that we will also refer to P_{st} as stationary distributions or stationary probability densities. From (2.18) we read off that stationary solutions satisfy

$$\nabla \cdot \mathbf{J}(\mathbf{x}, P_{\text{st}}) = 0 \; . \qquad (2.22)$$

That is, in the stationary case the divergence of the probability current vanishes. In the univariate case, from (2.20) it follows that $\partial J/\partial x = 0$, which implies that J corresponds to a constant. Consequently, P_{st} is implicitly given by

$$D_1(x, P_{\mathrm{st}}) P_{\mathrm{st}}(x) - \frac{\partial}{\partial x} D_2(x, P_{\mathrm{st}}) P_{\mathrm{st}}(x) = J = \mathrm{const.} \qquad (2.23)$$

The constant J is determined by the boundary conditions imposed on P. Note that in order to write (2.22) and (2.23) we have assumed that $\mathbf{J}$ and J given by (2.18) and (2.20) do not explicitly depend on t.

2.3 Self-Consistency Equations

In many cases, from (2.22) and (2.23) implicit equations for P_{st} can be derived (see, e.g., (4.20) and (7.4) below). Solutions of these implicit equations can then be found by solving appropriately defined self-consistency equations. In order to elucidate this point, we assume that the stationary solutions of a multivariate nonlinear Fokker–Planck equation satisfy the implicit equation

$$P_{\mathrm{st}}(\mathbf{x}) = f(\mathbf{x}, \langle \mathbf{A} \rangle_{\mathrm{st}}, \alpha) , \qquad (2.24)$$

with $\langle \mathbf{A} \rangle_{\mathrm{st}} = \int_\Omega \mathbf{A}(\mathbf{x}) P_{\mathrm{st}}(\mathbf{x}) \, \mathrm{d}^M x$. Here, $\mathbf{A}$ is an r-dimensional vector field $\mathbf{A}(\mathbf{x}) = (A_1(\mathbf{x}), \ldots, A_r(\mathbf{x}))$ and α is chosen such that P is normalized to unity. We will use $\alpha = \mu$ (chemical potential) or $\alpha = Z$ (partition function). Then, we define a probability density

$$P(\mathbf{x}, \mathbf{m}) = f(\mathbf{x}, \mathbf{m}, \alpha(\mathbf{m})) \qquad (2.25)$$

that depends on a vector $\mathbf{m} = (m_1, \ldots, m_r)$. Here, $\mathbf{m}$ denotes an independent variable and $\alpha(\mathbf{m})$ is chosen such that P is normalized to unity. Next, we define

$$\mathbf{R}(\mathbf{m}) = \int_\Omega \mathbf{A}(\mathbf{x}) P(\mathbf{x}, \mathbf{m}) \, \mathrm{d}^M x \qquad (2.26)$$

which is a function of $\mathbf{m}$ only. We can now define the integral equation

$$\mathbf{m} = \mathbf{R}(\mathbf{m}) . \qquad (2.27)$$

If we substitute $\langle \mathbf{A} \rangle_{\mathrm{st}} = \mathbf{m}$ into (2.24), where $\mathbf{m}$ satisfies (2.27), then we realize that (2.24) is satisfied. That is, the solutions of (2.27) determine the unknown expectation values $\langle A_i \rangle_{\mathrm{st}}$ that occur in (2.24). Therefore, the basic idea is to solve (2.27) and, subsequently, substitute the result into (2.24).

Equation (2.27) is called a self-consistency equation. In deriving (2.27), we have transformed an implicit equation for a function P_{st} into an implicit equation for a vector $\mathbf{m}$. The advantage of this procedure is that the implicit equation for $\mathbf{m}$ can usually be solved numerically with relatively little computation effort. Moreover, as we will see in Chapters 5 and 7 the stability of stationary solutions can also be determined by evaluating self-consistency equations.

Finally, note that the application of self-consistency equations is not restricted to Fokker–Planck equations with a finite nonlinearity dimension d (see Sect. 2.5 below). It can be applied to all kinds of nonlinear Fokker–Planck equations whose stationary distributions are described by implicit equations of the form (2.24).

2.4 Multistability and Basins of Attraction

Self-consistency equations can have multiple solutions. If so, we obtain multiple stationary solutions of the form (2.24) with different order parameters $\langle \mathbf{A} \rangle_{\mathrm{st}}$. In many cases, we can show (e.g., by means of H-theorems) that transient solutions of a nonlinear Fokker–Planck equation converge to stationary ones in the long time limit. If a Fokker–Planck equation is multistable the question arises as to which one of the stationary solutions does a transient solution converge. In other words: what are the basins of attraction of stationary distributions? We will not answer this question in general. However, there is a simple answer to this question for Fokker–Planck equations that have stationary solutions of the form (2.24) and give rise to closed evolution equations for order parameters $\langle \mathbf{A} \rangle(t) = \int_\Omega \mathbf{A}(\mathbf{x}) P(\mathbf{x}, t; u) \, \mathrm{d}^M x$ that read

$$\frac{\mathrm{d}}{\mathrm{d}t} \langle \mathbf{A} \rangle(t) = \mathbf{N}(\langle \mathbf{A} \rangle) \; . \tag{2.28}$$

For every time-dependent distribution P with initial distribution $u(\mathbf{x})$ and $\langle \mathbf{A} \rangle(t_0) = \int_\Omega \mathbf{A}(\mathbf{x}) u(\mathbf{x}) \, \mathrm{d}^M x$ we obtain the corresponding stationary value $\langle \mathbf{A} \rangle_{\mathrm{st}}$ by solving (2.28), which, in turn, gives us the corresponding stationary solution (2.24). In short, from $P(\mathbf{x}, t_0; u) = u$ we get $\langle \mathbf{A} \rangle(t_0)$. From $\langle \mathbf{A} \rangle(t_0)$ we get $\langle \mathbf{A} \rangle_{\mathrm{st}}$ via (2.28) and from $\langle \mathbf{A} \rangle_{\mathrm{st}}$ we get $P_{\mathrm{st}}(\mathbf{x}, u)$ via (2.24). Thus, we can determine the basins of attraction of stationary solutions and write down mappings $u \to P_{\mathrm{st}}(\mathbf{x}, u)$. For examples, see Sects. 5.2.2 and 7.3.3.

2.5 Nonlinearity Dimension

The probability density P of a stochastic process can be described by means of an infinitely large set of moments [221, 498]. In the univariate case, the moments read

$$M_n = \langle X^n \rangle \; , \tag{2.29}$$

with $n \geq 1$. Here, the functions X^n constitute a set of linearly independent functions. Since the drift and diffusion coefficients of nonlinear Fokker–Planck equations depend on the probability density, the drift and diffusion coefficients depend in general on an infinitely large set of expectation values involving linearly independent functions. In particular, this is the case for the nonlinearity shown in (2.8). In contrast, the diffusion coefficient shown in (2.9) depends only on a single expectation value, namely, $\langle X^2 \rangle$. That is, we may also deal with drift and diffusion coefficients that depend on a finite number of expectation values $\langle A_1 \rangle, \ldots, \langle A_d \rangle$, where $A_1, \ldots, A_d$ denote linearly independent functions. We refer to the number of expectation values that occur in the coefficients of a nonlinear Fokker–Planck equation and are related to linearly independent functions as the nonlinearity dimension d of the Fokker–Planck equation. Accordingly, for $0 < d < \infty$ the univariate nonlinear Fokker–Planck equation (2.10) can be written as

$$\frac{\partial}{\partial t}P(x,t;u) = -\frac{\partial}{\partial x}D_1(x,t,\langle \mathbf{A} \rangle)P + \frac{\partial^2}{\partial x^2}D_2(x,t,\langle \mathbf{A} \rangle)P , \qquad (2.30)$$

where the vector $\mathbf{A}(x)$ is given by $\mathbf{A}(x) = (A_1(x), \ldots, A_d(x))$ and the components $A_1(x), \ldots, A_d(x)$ denote linearly independent functions. In addition, there are two extreme cases. On the one hand, we may have $d = \infty$ as discussed earlier. On the other hand, $d = 0$ means that we deal with a linear Fokker–Planck equation. That is, we regard a linear Fokker–Planck equation as a nonlinear Fokker–Planck equation that has a nonlinearity of dimension zero.

2.6 Classifications

There are different ways of classifying nonlinear Fokker–Planck equations, see Fig. 2.1. First, we may distinguish between nonlinear Fokker–Planck equations to which Markov processes can be assigned and those to which Markov processes cannot be assigned or for which this issue cannot be decided. We will refer to the former kind of nonlinear Fokker–Planck equations as strongly nonlinear Fokker–Planck equations, see Chap. 3. Second, we may use free energy measures to distinguish between nonlinear Fokker–Planck equations of different kinds. If (2.4) can be written by means of a free energy functional F as

$$\frac{\partial}{\partial t}P(\mathbf{x},t;u) = -\nabla \cdot [P\mathbf{I}] + \nabla \cdot \left[\mathcal{M}P \cdot \nabla \frac{\delta F}{\delta P}\right] \qquad (2.31)$$

we say that we deal with a free energy Fokker–Planck equation. Here, $\mathcal{M}$ denotes an $M \times M$ matrix and $\mathbf{I}$ is a drift force. If the free energy functional can be decomposed into a linear internal energy functional $\langle U_0 \rangle = \int_\Omega U_0(\mathbf{x})P\,d^M x$ and an entropy or information measure S like $F = \langle U_0 \rangle - QS$, then we get

$$\frac{\partial}{\partial t}P(\mathbf{x},t;u) = -\nabla \cdot [\mathbf{I}P - P\mathcal{M} \cdot \nabla U_0(\mathbf{x})] - Q\nabla \cdot \left[\mathcal{M}P \cdot \nabla \frac{\delta S}{\delta P}\right] . \qquad (2.32)$$

We will refer to nonlinear Fokker–Planck equations (2.32) as entropy Fokker–Planck equations. If S corresponds to the Boltzmann entropy and $\mathcal{M}$ and $\mathbf{I}$ do not depend on P, we obtain a linear Fokker–Planck equation. We will

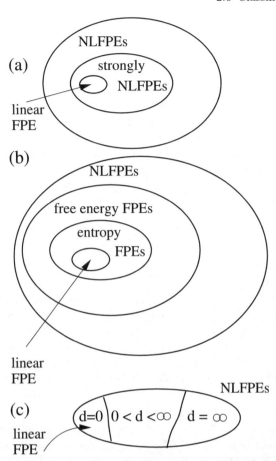

Fig. 2.1. Classifications of nonlinear Fokker–Planck equations based on (a) the Markov property, (b) free energy measures, (c) the nonlinearity dimension (FPE=Fokker–Planck equation, NLFPE=nonlinear Fokker–Planck equation)

return to free energy and entropy Fokker–Planck equations in Chapters 4 to 6. Third, we may count the expectation values $\langle A_1 \rangle, \langle A_2 \rangle, \ldots$ that involve linearly independent functions A_i and occur in the drift and diffusion coefficients of (2.4). That is, we may use the nonlinearity dimension in order to classify nonlinear Fokker–Planck equations. In Chap. 5 we will discuss nonlinear Fokker–Planck equations with $d = 1$ and $d = 2$. Furthermore, it is often helpful to distinguish between univariate and multivariate nonlinear Fokker–Planck equations. Finally, one may distinguish between nonlinear Fokker–Planck equations subjected to different boundary conditions. For example, in Chap. 5 we will distinguish between Fokker–Planck equations subjected to natural and periodic boundary conditions.

2.7 Derivations

Nonlinear Fokker-Planck equations can be derived in several ways. They are often regarded as N-dimensional linear Fokker-Planck equations describing many-body systems involving N subsystems in the limit $N \to \infty$ [60, 135, 365, 568] (see also Sect. 3.8). They may be derived from the principles of linear nonequilibrium thermodynamics [99, 167, 168] (see Sect. 4.5). They are often obtained as diffusion approximations of the Boltzmann equations [322, 323, 326, 327, 328] using, for example, the kinetic interaction principle. They can be derived from nonlinear master equations [105, 278, 428, 449, 451]. They often correspond to mean field approximations of spatially extended systems with diffusive couplings between different parts of the systems [220, 225, 404, 601, 646].

2.8 Numerics

Several techniques have been developed to solve nonlinear Fokker–Planck equations numerically. We will briefly address some of these approaches: path integral solutions, Fourier and moment expansions, finite difference schemes, and distributed approximating functionals. The solutions of nonlinear Fokker–Planck equations can also be determined using computer simulations of Langevin equations. We will return to this issue in Chap. 3.

2.8.1 Path Integral Solutions

Solutions of nonlinear Fokker–Planck equations can be written in terms of path integrals [624]. Let us illustrate the path integral method for the univariate nonlinear Fokker–Planck equation (2.10). For $t \geq t_0$ and small time steps Δt the evolution of P can approximately be determined by means of the integral equation

$$P(x, t + \Delta t; u) \approx \int_\Omega G(x, x'; t, \Delta t, P) P(x', t; u) \, dx' , \qquad (2.33)$$

where $G(x, x'; t, \Delta t; P)$ is called a short time propagator and is defined by [624]

$$G(x, x'; t, \Delta t; P) = \sqrt{\frac{1}{4\pi \Delta t D_2[x', t; P(x', t; u)]}} \exp\left\{-\frac{[x - x' - \Delta t D_1[x', t; P(x', t; u)]]^2}{4 \Delta t D_2[x', t; P(x', t; u)]}\right\} . \qquad (2.34)$$

In the limit of infinitesimal small time steps Δt we obtain

$$P(x,t;u) = \lim_{N\to\infty, \Delta t\to 0} \int_{\Omega^N} P(x_1, t_1; u) \prod_{i=1}^{N} \left\{ G(x_{i+1}, x_i; t_i, \Delta t, P) \, \mathrm{d}x_i \right\}, \tag{2.35}$$

with $\Delta t N = t - t_1$, where x corresponds to the final chain element x_{N+1} in the limit $N \to \infty$ (i.e., we have $x = \lim_{N\to\infty} x_{N+1}$). Equation (2.35) is the path integral solution of the nonlinear Fokker–Planck equation (2.10). Equation (2.35) generalizes the path integral approach for the linear Fokker–Planck equation [235, 253, 254, 297]. From (2.35) we read off that, on the one hand, the short time propagator determines the evolution of P and, on the other hand, it depends on P. Therefore, we are dealing with a feedback loop – in line with our considerations in Sect. 1.3. By means of the path integral solution (2.35) the nonlinear Fokker–Planck equation (2.10) can be solved numerically [540, 623, 624].

2.8.2 Fourier and Moment Expansions

Nonlinear Fokker–Planck equations subjected to periodic boundary conditions can often be solved numerically by means of Fourier transformations. To this end, the probability density P is expanded into Fourier modes with time-dependent amplitudes. The expansion is truncated after a particular Fourier mode. Thus, one obtains a coupled set of nonlinear ordinary differential equations for the Fourier amplitudes [365, 531]. For an example see Sect. 5.3.3.

In a similar vein, one can treat nonlinear Fokker–Planck equations subjected to natural boundary conditions. Here, the probability density is expressed in terms of time-dependent first and higher order moments. For these moments coupled nonlinear differential equations are derived and truncated such that they can be solved numerically [77].

2.8.3 Finite Difference Schemes

It has been suggested to investigate numerically nonlinear evolution equations defined by

$$\frac{\partial}{\partial t} P(x, t; u) = -\frac{1}{A(x)} \frac{\partial}{\partial x} \left[D_1'(x, t, P) P - D_2(x, t, P) \frac{\partial}{\partial x} P \right] \tag{2.36}$$

by means of a particular finite-difference scheme [141]. Using this finite-difference scheme, solutions can be obtain on an interval $x \in [x_l, x_r]$ with $x_l < x_r$ and it can be shown that the condition

$$\int_{x_l}^{x_r} A(x) P(x, t; u) \, \mathrm{d}x = \int_{x_l}^{x_r} A(x) u(x) \, \mathrm{d}x \tag{2.37}$$

is satisfied. Let us put $A(x) = 1$ and

$$D_1(x, t, P) = D'_1(x, t, P) - \frac{\partial}{\partial x} D_2(x, t, P) \tag{2.38}$$

(see also Sect. 2.2.2). Then (2.36) can be transformed into the nonlinear Fokker–Planck equation (2.10) and the conservation condition (2.37) guarantees the normalization of the time-dependent probability density: $\int_{x_l}^{x_r} P(x, t; u) \, dx = \int_{x_l}^{x_r} u(x) \, dx = 1$. We would like to mention that there are also alternative finite-difference schemes that have been applied to study the time-dependent behavior of solutions of nonlinear Fokker–Planck equations [377].

2.8.4 Distributed Approximating Functionals

The basic idea of this approach is to exploit the integral relations

$$P(x) = \int_{-\infty}^{\infty} \delta(x - x') P(x') \, dx' ,$$

$$\frac{d}{dx} P(x) = \int_{-\infty}^{\infty} \frac{d}{dx} \delta(x - x') P(x') \, dx' ,$$

$$\frac{d^2}{dx^2} P(x) = \int_{-\infty}^{\infty} \frac{d^2}{dx^2} \delta(x - x') P(x') \, dx' \tag{2.39}$$

of the delta function [430]. The delta function and its derivatives can be approximately expressed in terms of Hermite polynomials and the integrals can be discretized on a grid. Thus, the integrals become discretized functionals that involve Hermite polynomials and evaluate P at several points that are distributed along the real line. We say, we are dealing with Hermite distributed approximating functionals (Hermite DAFs). If we write the nonlinear Fokker–Planck equation (2.10) as

$$\frac{\partial}{\partial t} P(x, t; u) = \left\{ \left(\frac{\partial^2 D_2}{\partial x^2} - \frac{\partial D_1}{\partial x} \right) + \left(2 \frac{\partial D_2}{\partial x} - D_1 \right) \frac{\partial}{\partial x} + D_2 \frac{\partial^2}{\partial x^2} \right\} P$$

$$= \underbrace{\left\{ A(x, t, P) + B(x, t, P) \frac{\partial}{\partial x} + D_2(x, t, P) \frac{\partial^2}{\partial x^2} \right\}}_{\hat{N}} P , \tag{2.40}$$

then the operator $\hat{N}$ can be expressed in terms of Hermite DAFs and (2.10) can be solved by means of (2.40) on a one-dimensional grid. For details the reader is referred to [648].

3 Strongly Nonlinear Fokker–Planck Equations

Although the nonlinear Fokker–Planck equation (2.4) defines the evolution of a probability density $P(\mathbf{x}, t; u)$ it cannot be used to define a hierarchy of probability densities as shown in (2.3). That is, from (2.4) we cannot compute a stochastic process [165]. Equation (2.4) provides us only with partial information about a stochastic process. Our objective now is to define stochastic processes that are consistent with the solutions of the nonlinear Fokker–Planck equation (2.4), on the one hand, and are completely defined, on the other.

3.1 Transformation to a Linear Problem

In what follows, we will adopt an idea of McKean Jr. [165, 411]. We first illustrate this idea by means of the solution of an ordinary nonlinear first order differential equation. Let $p(t)$ be a function of time described by

$$\frac{\mathrm{d}}{\mathrm{d}t} p(t) = f(p(t)) , \qquad (3.1)$$

where $f(p)$ is a force acting on the system described by p. Then, consider another function $g(z_1, z_2)$ with $g(z, z) = f(z)$. Next, we define a function $w(t)$ by means of the differential equation

$$\frac{\mathrm{d}}{\mathrm{d}t} w(t) = g(p(t), w(t)) . \qquad (3.2)$$

We claim that a solution of this differential equation is given by $w(t) = p(t)$. In order to prove this claim, we substitute $w(t) = p(t)$ into (3.2) and exploit the relationship $g(z, z) = f(z)$ to obtain

$$\frac{\mathrm{d}}{\mathrm{d}t} p(t) = g(p(t), p(t)) = f(p(t)) . \qquad (3.3)$$

The transformation between (3.1) and (3.2) can easily be extended to one-parametric functions $p_x(t)$ and $w_x(t)$ depending on a parameter x. Furthermore, we may interpret $p_x(t)$ and $w_x(t)$ as one-variable probability densities $P(x, t; u)$ and $W(x, t; u)$, respectively. Then, our previous considerations lead to the following conclusion. Let us assume that the evolution of a probability density P is given by $P(x,t;u)$ for $t \geq t_0$. Let us further assume that $P(x,t;u)$ solves the nonlinear partial integro-differential equation

$$\frac{\mathrm{d}}{\mathrm{d}t}P(x,t;u) = \hat{N}[P], \tag{3.4}$$

where the operator $\hat{N}$ involves both differential and integral forms of P and in general is nonlinear with respect to P:

$$\hat{N}[P] = \hat{N}\left[x, \frac{\partial^n}{\partial x^n}, \int \ldots \mathrm{d}x, t, P\right]. \tag{3.5}$$

Suppose that there is another operator $\hat{L}[W_1, W_2]$ that is linear with respect to W_1, acts on the probability densities W_1 and W_2, and satisfies

$$\hat{L}[P,P] = \hat{N}[P] \tag{3.6}$$

for solutions of (3.4). Then, the partial integro-differential equation

$$\frac{\partial}{\partial t}W(x,t;u) = \hat{L}[W,P] \tag{3.7}$$

is solved by $W(x,t;u) = P(x,t;u)$ for $t \geq t_0$. In order to prove this proposition, we substitute $W(x,t;u) = P(x,t;u)$ into (3.7) and exploit the relationship (3.6) to get

$$\frac{\partial}{\partial t}P(x,t;u) = \hat{L}[P,P] = \hat{N}[P]. \tag{3.8}$$

Note that we have $W(x,t_0;u) = P(x,t_0;u) = u(x)$. It is important to realize that (3.4) describes a nonlinear evolution equation, whereas (3.7) corresponds to a linear evolution equation involving a time-dependent external force field P. That is, we have transformed a nonlinear problem into a linear one. This transformation can in particular be applied to the nonlinear Fokker–Planck equation (2.10). Then, we see that the linear evolution equation

$$\frac{\partial}{\partial t}W(x,t;u) = -\frac{\partial}{\partial x}D_1(x,t,P)\,W + \frac{\partial^2}{\partial x^2}D_2(x,t,P)\,W \tag{3.9}$$

has the same solution as (2.10). That is, (3.9) is solved by $W(x,t;u) = P(x,t;u)$. In the multivariate case, we can assign to (2.4) the linear evolution equation

$$\frac{\partial}{\partial t}W(\mathbf{x},t;u) = -\sum_{i=1}^{M}\frac{\partial}{\partial x_i}D_i(\mathbf{x},t,P)\,W + \sum_{i,k=1}^{M}\frac{\partial^2}{\partial x_i \partial x_k}D_{ik}(\mathbf{x},t,P)\,W, \tag{3.10}$$

where P is defined by (2.4) and a solution is given by $W = P$. Equations (3.9) and (3.10) are linear evolution equations for the probability density W.

3.2 What Are Strongly Nonlinear Fokker–Planck Equations?

In the preceding section, we have distinguished between the probability densities W and P in order to elucidate how to assign linear Fokker–Planck equations to nonlinear ones. In what follows, we will dispense with this distinction because it will simplify the introduction of strongly nonlinear Fokker–Planck equations. Accordingly, we assign to the multivariate nonlinear Fokker–Planck equation (2.4) the linear evolution equations

$$\frac{\partial}{\partial t} P(\mathbf{x},t;u) = -\sum_{i=1}^{M} \frac{\partial}{\partial x_i} D'_i(\mathbf{x},t,u) P + \sum_{i,k=1}^{M} \frac{\partial^2}{\partial x_i \partial x_k} D'_{ik}(\mathbf{x},t,u) P \quad (3.11)$$

and

$$\frac{\partial}{\partial t} P(\mathbf{x},t|\mathbf{x}',t';u) = -\sum_{i=1}^{M} \frac{\partial}{\partial x_i} D'_i(\mathbf{x},t,u) P(\mathbf{x},t|\mathbf{x}',t';u)$$
$$+ \sum_{i,k=1}^{M} \frac{\partial^2}{\partial x_i \partial x_k} D'_{ik}(\mathbf{x},t,u) P(\mathbf{x},t|\mathbf{x}',t';u) , \quad (3.12)$$

with

$$D'_i(\mathbf{x},t,u) = D_i(\mathbf{x},t,P(\mathbf{x},t,u)) ,$$
$$D'_{ik}(\mathbf{x},t,u) = D_{ik}(\mathbf{x},t,P(\mathbf{x},t,u)) , \quad (3.13)$$

where P is a solution of (2.4). If the linear evolution equations (3.11) and (3.12) describe a Markov diffusion process, then we refer to the corresponding nonlinear Fokker–Planck equation (2.4) as a strongly nonlinear Fokker–Planck equation [188]:

$$\begin{array}{c} \text{strongly} \\ \text{nonlinear} \\ \text{Fokker–Planck equation} \end{array} \Leftrightarrow \begin{array}{c} \text{linear} \\ \text{Fokker–Planck equation} \\ \text{with driving forces} \end{array} \quad (3.14)$$

A sufficient condition that (3.11) defines a linear Fokker–Planck equation is that (3.12) has a fundamental solution, that is, it has a solution satisfying $P(\mathbf{x},t|\mathbf{x}',t';u) \geq 0$, $\int_\Omega P(\mathbf{x},t|\mathbf{x}',t';u) \, \mathrm{d}^M x = 1$, and the limiting case $\lim_{t \to t'} P(\mathbf{x},t|\mathbf{x}',t';u) = \delta(\mathbf{x} - \mathbf{x}')$, see, for example, [305, Sect. 2.3.5] or [145, Sect. 5]. In other words, strongly nonlinear Fokker–Planck equations are defined by a pair of evolution equations given by

$$\frac{\partial}{\partial t}P(\mathbf{x},t;u) = -\sum_{i=1}^{M}\frac{\partial}{\partial x_i}D_i(\mathbf{x},t,P)P + \sum_{i,k=1}^{M}\frac{\partial^2}{\partial x_i \partial x_k}D_{ik}(\mathbf{x},t,P)P\;,$$

(3.15)

$$\frac{\partial}{\partial t}P(\mathbf{x},t|\mathbf{x}',t';u) = -\sum_{i=1}^{M}\frac{\partial}{\partial x_i}D_i(\mathbf{x},t,P(\mathbf{x},t;u))P(\mathbf{x},t|\mathbf{x}',t';u)$$

$$+ \sum_{i,k=1}^{M}\frac{\partial^2}{\partial x_i \partial x_k}D_{ik}(\mathbf{x},t,P(\mathbf{x},t;u))P(\mathbf{x},t|\mathbf{x}',t';u)\;.\quad (3.16)$$

and are

- nonlinear with respect to $P(\mathbf{x},t;u)$
- linear with respect to $P(\mathbf{x},t|\mathbf{x}',t';u)$

and

- have solutions $P(\mathbf{x},t|\mathbf{x}',t';u)$ that describe transition probability densities of Markov diffusion processes.

A necessary condition for strongly nonlinear Fokker–Planck equations is that the diffusion coefficient D_{ik} corresponds to a symmetric and semi-positive definite matrix (which is the condition satisfied by the second Kramers–Moyal coefficient [498]). For our understanding of strongly nonlinear Fokker–Planck equations it is crucial to realize that the evolution equation of the transition probability density $P(\mathbf{x},t|\mathbf{x}',t';u)$ does not read

$$\frac{\partial}{\partial t}P(\mathbf{x},t|\mathbf{x}',t';u) = -\sum_{i=1}^{M}\frac{\partial}{\partial x_i}D_i(\mathbf{x},t,P(\mathbf{x},t|\mathbf{x}',t';u))P(\mathbf{x},t|\mathbf{x}',t';u)$$

$$+ \sum_{i,k=1}^{M}\frac{\partial^2}{\partial x_i \partial x_k}D_{ik}(\mathbf{x},t,P(\mathbf{x},t|\mathbf{x}',t';u))P(\mathbf{x},t|\mathbf{x}',t';u)\;.\quad (3.17)$$

From (3.15) and (3.16) one can obtain a hierarchy of probability densities (2.3) that is consistent with the solution of the nonlinear Fokker–Planck equation (2.4) and completely describes a Markov diffusion process. The n-point joint probability density of this process reads

$$P(\mathbf{x}_n,t_n;\ldots;\mathbf{x}_1,t_1;u) =$$
$$P(\mathbf{x}_n,t_n|\mathbf{x}_{n-1},t_{n-1};u)\cdots P(\mathbf{x}_2,t_2|\mathbf{x}_1,t_1;u)P(\mathbf{x}_1,t_1;u)\;.\quad (3.18)$$

As we will see in Sect. 3.4, a Markov diffusion process defined by (3.18) can alternatively be described by means of a Langevin equation.

Stochastic Feedback

We are now in the position to prove two claims made in Sect. 1.3. First, we realize that (3.15) and (3.16) can be written as transport equations

3.2 What Are Strongly Nonlinear Fokker–Planck Equations?

$$\frac{\partial}{\partial t}\psi_i = \hat{F}(\mathbf{x}, \nabla, t, P(\mathbf{x},t;u))\,\psi_i \;, \tag{3.19}$$

where $\hat{F}$ is defined by the operator (1.8) and the field variables ψ_i are given by $\psi_1 = P(\mathbf{x},t;u)$ and $\psi_2 = P(\mathbf{x},t|\mathbf{x}',t';u)$. Second, if we multiply (3.16) with $P(\mathbf{x}',t';u)$ we obtain

$$\frac{\partial}{\partial t}P(\mathbf{x},t;\mathbf{x}',t';u) = -\sum_{i=1}^{M}\frac{\partial}{\partial x_i}D_i(\mathbf{x},t,P(\mathbf{x},t;u))P(\mathbf{x},t;\mathbf{x}',t';u)$$
$$+ \sum_{i,k=1}^{M}\frac{\partial^2}{\partial x_i \partial x_k}D_{ik}(\mathbf{x},t,P(\mathbf{x},t;u))P(\mathbf{x},t;\mathbf{x}',t';u) \;. \tag{3.20}$$

This evolution equation, in turn, can be written as (3.19) with $\psi_i = P(\mathbf{x},t;\mathbf{x}',t';u)$. If we multiply (3.16) with $P(\mathbf{x}',t'|\mathbf{x}'',t'';u)P(\mathbf{x}'',t'';u)$, we obtain (3.19) for $\psi_i = P(\mathbf{x},t;\mathbf{x}',t';\mathbf{x}'',t'';u)$. Repeating this argument, we see that for strongly nonlinear Fokker–Planck equations all members of the hierarchy (2.3) satisfy the transport equation (3.19). That is, we have

$$\frac{\partial}{\partial t}P(\mathbf{x},t;u) = \hat{F}(\mathbf{x},\nabla,t,P(\mathbf{x},t;u))\,P(\mathbf{x},t;u) \;,$$
$$\frac{\partial}{\partial t}P(\mathbf{x},t;\mathbf{x}',t';u) = \hat{F}(\mathbf{x},\nabla,t,P(\mathbf{x},t;u))\,P(\mathbf{x},t;\mathbf{x}',t';u) \;,$$
$$\frac{\partial}{\partial t}P(\mathbf{x},t;\mathbf{x}',t';\mathbf{x}'',t'';u) = \hat{F}(\mathbf{x},\nabla,t,P(\mathbf{x},t;u))\,P(\mathbf{x},t;\mathbf{x}',t';\mathbf{x}'',t'';u) \;,$$
$$\ldots \tag{3.21}$$

Density Functions

If the drift and diffusion coefficients (2.13) of the nonlinear Fokker–Planck equation (2.11) correspond to Kramers–Moyal coefficients of Markov diffusion processes, then these Markov diffusion processes are defined by the evolution equation (2.11) for $\rho(\mathbf{x},t)$ and the transition probability density

$$\frac{\partial}{\partial t}P(\mathbf{x},t|\mathbf{x}',t';u) = -\sum_{i=1}^{M}\frac{\partial}{\partial x_i}d_i(\mathbf{x},t,\rho(\mathbf{x},t))P(\mathbf{x},t|\mathbf{x}',t';u)$$
$$+ \sum_{i,k=1}^{M}\frac{\partial^2}{\partial x_i \partial x_k}d_{ik}(\mathbf{x},t,\rho(\mathbf{x},t))P(\mathbf{x},t|\mathbf{x},t';u) \;, \tag{3.22}$$

with $u(\mathbf{x}) = \rho(\mathbf{x},t_0)$. Higher order density functions such as $\rho(\mathbf{x},t;\mathbf{x}',t';u)$ and $\rho(\mathbf{x},t;\mathbf{x}',t';\mathbf{x}'',t'';u)$ may be defined in analogy to the joint probability densities discussed above:

$$\rho(\mathbf{x},t;\mathbf{x}',t') = M_0 P(\mathbf{x},t;\mathbf{x}',t';u)$$
$$= P(\mathbf{x},t|\mathbf{x}',t';u)\rho(\mathbf{x}',t') ,$$
$$\rho(\mathbf{x},t;\mathbf{x}',t';\mathbf{x}'',t'') = M_0 P(\mathbf{x},t;\mathbf{x}',t';\mathbf{x}'',t'';u)$$
$$= P(\mathbf{x},t|\mathbf{x}',t';u)P(\mathbf{x}',t'|\mathbf{x}'',t'';u)\rho(\mathbf{x}'',t'') ,$$
$$\ldots . \tag{3.23}$$

3.3 Correlation Functions

Since strongly nonlinear Fokker–Planck equations define completely stochastic processes, we can compute correlation functions such as $C_{ik}^{mn}(t,t') = \langle X_i^m(t) X_k^n(t') \rangle$ for $t \geq t'$ and $\mathbf{X} = (X_1, \ldots, X_M)$. In particular, evolution equations of the form

$$\frac{\partial}{\partial t} C_{ik}^{mn}(t,t') = f_{ik}^{mn} \tag{3.24}$$

can be obtained by multiplying (3.20) with x_i^m and $x_k'^n$ and integrating with respect to $\mathbf{x}$ and $\mathbf{x}'$. In some cases, (3.24) yields close descriptions and can be solved analytically or numerically. We would like to point out that the very same correlation functions can also be obtained by solving numerically appropriately defined Langevin equations (see next section) and computing the ensemble averages $\langle X_i^m(t) X_k^n(t') \rangle$. For examples, see Sects. 3.10, 5.2.2, and 6.5.2.

3.4 Langevin Equations

We distinguish between two-layered and self-consistent Langevin equations [165].

3.4.1 Two-Layered Langevin Equations

Due to the correspondence between (2.10) and (3.11-3.13), the Ito–Langevin equation for the random variable $X(t)$ with

$$P(x,t;u) = \langle \delta(x - X(t)) \rangle \tag{3.25}$$

is given by

$$\frac{\mathrm{d}}{\mathrm{d}t} X(t) = D_1'(X,t,u) + \sqrt{D_2'(X,t,u)}\, \Gamma(t) ,$$
$$D_1'(x,t,u) = D_1(x,t,P(x,t;u)) ,$$
$$D_2'(x,t,u) = D_2(x,t,P(x,t;u)) \tag{3.26}$$

for $t \geq t_0$, where P is a solution of (2.10). Here, $\Gamma(t)$ is a particular kind of fluctuating force: a Langevin force. The Langevin force $\Gamma(t)$ has a vanishing mean and is delta-correlated with respect to time:

$$\langle \Gamma(t) \rangle = 0 \ , \quad \langle \Gamma(t)\Gamma(t') \rangle = 2\delta(t-t') \ . \qquad (3.27)$$

For further details the reader is referred to [33, 221, 254, 274, 498, 550]. Note that (3.26) can be written in a more concise way:

$$\frac{\mathrm{d}}{\mathrm{d}t}X(t) = D_1(x,t,P)|_{x=X(t)} + \sqrt{D_2(x,t,P)}\Big|_{x=X(t)} \Gamma(t) \ . \qquad (3.28)$$

In the multivariate case, we first need to define a matrix G with elements $G_{ik}(\mathbf{x},t,P)$ that satisfy

$$\sum_{l=1}^{M} G_{il}(\mathbf{x},t,P) G_{lk}(\mathbf{x},t,P) = D_{ik}(\mathbf{x},t,P) \qquad (3.29)$$

(for an explicit derivation of G_{il} see, for example, [498]). Then, stochastic trajectories $\mathbf{X}(t)$ that correspond to the solutions $P(\mathbf{x},t;u) = \langle \delta(\mathbf{x}-\mathbf{X}(t)) \rangle$ of (2.4) and (3.11-3.13), are described by the Ito–Langevin equation

$$\frac{\mathrm{d}}{\mathrm{d}t}X_i(t) = D'_i(\mathbf{X},t,u) + \sum_{k=1}^{M} G'_{ik}(\mathbf{X},t;u)\Gamma_k(t) \ ,$$

$$D'_i(\mathbf{x},t,u) = D_i(\mathbf{x},t,P(\mathbf{x},t;u)) \ ,$$

$$G'_{ik}(\mathbf{x},t,u) = G_{ik}(\mathbf{x},t,P(\mathbf{x},t;u)) \ , \qquad (3.30)$$

for $t \geq t_0$, where P denotes a solution of (2.4). The random variables $\Gamma_k(t)$ describe an M-dimensional Langevin force with statistically independent components:

$$\langle \Gamma_k(t) \rangle = 0 \ , \quad \langle \Gamma_i(t)\Gamma_k(t') \rangle = 2\delta_{ik}\delta(t-t') \ . \qquad (3.31)$$

Just as in the univariate case, (3.30) may be written as

$$\frac{\mathrm{d}}{\mathrm{d}t}X_i(t) = D_i(\mathbf{x},t,P)|_{\mathbf{x}=\mathbf{X}(t)} + \sum_{k=1}^{M} G_{ik}(\mathbf{x},t,P)|_{\mathbf{x}=\mathbf{X}(t)} \Gamma_k(t) \ . \qquad (3.32)$$

The Ito–Langevin equations (3.28) and (3.32) involve the solution P of a nonlinear Fokker–Planck equation. As a result, we deal with two-layered systems composed of Langevin equations that involve time-dependent driving forces P and nonlinear Fokker–Planck equations that determine how these driving forces evolve.

If (2.12) corresponds to a strongly nonlinear Fokker–Planck equation for the density function $\rho(x,t) = M_0 \langle \delta(x-X(t)) \rangle$ then the corresponding two-layered Ito–Langevin equation reads

$$\frac{\mathrm{d}}{\mathrm{d}t}X(t) = d_1(x,t,\rho(x,t))|_{x=X(t)} + \sqrt{d_2(x,t,\rho(x,t))}\Big|_{x=X(t)}\Gamma(t), \quad (3.33)$$

for $t \geq t_0$, where ρ is a solution of (2.12) (see also (3.26)). The multivariate case can be treated in a similar way.

3.4.2 Self-Consistent Langevin Equations

By means of (3.25) we can eliminate P in (3.26) and write (3.26) in a self-consistent fashion [13, 88, 165, 488]. Accordingly, (3.26) becomes

$$\frac{\mathrm{d}}{\mathrm{d}t}X(t) = D'_1(X,t,u) + \sqrt{D'_2(X,t,u)}\Gamma(t),$$
$$D'_1(x,t,u) = D_1(x,t,\langle\delta(x-X(t))\rangle),$$
$$D'_2(x,t,u) = D_2(x,t,\langle\delta(x-X(t))\rangle). \quad (3.34)$$

Likewise, if (2.4) describes a strongly nonlinear Fokker–Planck equation, its self-consistent Ito–Langevin equation reads

$$\frac{\mathrm{d}}{\mathrm{d}t}X_i(t) = D'_i(\mathbf{X},t,u) + \sum_{k=1}^{M} G'_{ik}(\mathbf{X},t,u)\Gamma_k(t),$$
$$D'_i(\mathbf{x},t,u) = D_i(\mathbf{x},t,\langle\delta(\mathbf{x}-\mathbf{X}(t))\rangle),$$
$$G'_{ik}(\mathbf{x},t,u) = G_{ik}(\mathbf{x},t,\langle\delta(\mathbf{x}-\mathbf{X}(t))\rangle). \quad (3.35)$$

Equations (3.34) and (3.35) may be alternatively written as

$$\frac{\mathrm{d}}{\mathrm{d}t}X(t) = D_1(x,t,\langle\delta(x-x(t))\rangle)|_{x=X(t)}$$
$$+ \sqrt{D_2(x,t,\langle\delta(x-x(t))\rangle)}\Big|_{x=X(t)}\Gamma(t) \quad (3.36)$$

and

$$\frac{\mathrm{d}}{\mathrm{d}t}X_i(t) = D_i(\mathbf{x},t,\langle\delta(\mathbf{x}-\mathbf{X}(t))\rangle)|_{\mathbf{x}=\mathbf{X}(t)}$$
$$+ \sum_{k=1}^{M} G_{ik}(\mathbf{x},t,\langle\delta(\mathbf{x}-\mathbf{X}(t))\rangle)|_{\mathbf{x}=\mathbf{X}(t)}\Gamma_k(t). \quad (3.37)$$

The self-consistent version of (3.33) reads

$$\frac{\mathrm{d}}{\mathrm{d}t}X(t) = d_1(x,t,M_0\langle\delta(x-x(t))\rangle)|_{x=X(t)}$$
$$+ \sqrt{d_2(x,t,M_0\langle\delta(x-X(t))\rangle)}\Big|_{x=X(t)}\Gamma(t). \quad (3.38)$$

3.4.3 Hierarchies and Correlation Functions

We would like to point out that (3.32) and (3.37) are the Ito–Langevin equations of the pair of evolution equations given by (3.15) and (3.16). In particular, the transition probability density (3.16) can be computed from

$$P(\mathbf{x}, t|\mathbf{x}', t'; u) = \langle \delta(\mathbf{x} - \mathbf{X}(t)) \rangle|_{\mathbf{X}(t')=\mathbf{x}'} \ . \tag{3.39}$$

Moreover, all hierarchy members (2.3) can be computed from the Ito–Langevin equations (3.32) and (3.37). For example, the joint probability density $P(\mathbf{x}_n, t_n; \ldots; \mathbf{x}_1, t_1; u)$ is given by

$$P(\mathbf{x}_n, t_n; \ldots; \mathbf{x}_1, t_1; u) = \langle \delta(\mathbf{x}_n - \mathbf{X}(t_n)) \cdots \delta(\mathbf{x}_1 - \mathbf{X}(t_1)) \rangle \ . \tag{3.40}$$

That is, in the case of strongly nonlinear Fokker–Planck equations joint probability densities and stochastic processes can be described in two equivalent ways: by means of the Fokker–Planck equation approach (3.18) and the Langevin equation approach (3.40), see also Sect. 2.1. Furthermore, we can compute all possible correlation functions from the Langevin equations of strongly nonlinear Fokker–Planck equations. For example, in the univariate case, the autocorrelation function $\langle X(t)X(t') \rangle$ can be computed from (3.28) and (3.36), see Sects. 3.10, 5.2.2, and 6.5.2.

3.4.4 Numerics

Two-Layered Langevin Equations

Recall that the two-layered Langevin equations correspond to ordinary Langevin equations that depend on a driving force P. Consequently, the two-layered Langevin equations (3.26) and (3.30) can be solved numerically using the techniques developed for ordinary Langevin equations as described in [72, 242, 350, 498]. For example, the Ito–Langevin equation (3.26) can be solved numerically using the Euler forward scheme

$$X_{n+1}^l = X_n^l + \Delta t\, D_1'(X_n^l, t_n, u) + \sqrt{D_2'(X_n^l, t_n, u)} \sqrt{\Delta t}\, w_n^l \ , \tag{3.41}$$

where the variables X_n^l denote realizations of $X(t)$ at discrete time steps t_n and time is given by $t_n = n\Delta t + t_0$, $n = 0, 1, 2, \ldots$. Here, Δt denotes the time step of a single iteration and is a small quantity. The variables w_n^l denote realizations of the random variables w_n. The variables w_n are statistically-independent Gaussian distributed random variables with zero mean and variance 2:

$$\langle w_n \rangle = 0 \ , \ \langle w_n w_{n'} \rangle = 2\,\delta_{nn'} \ . \tag{3.42}$$

The realizations w_n^l can be obtained by means of the Box–Muller algorithm or alternative methods [242, 350, 498]. D_1' and D_2' are defined by

$$D'_1(x,t,u) = D_1(x,t,P(x,t;u)) ,$$
$$D'_2(x,t,u) = D_2(x,t,P(x,t;u)) , \qquad (3.43)$$

where P corresponds to a solution of (2.10). In some cases P can be determined analytically and the exact solution for P can be substituted in (3.43) (see e.g. [66]). In particular, for some systems in the stationary case such analytical expressions can be derived. In general, P can be determined by solving the nonlinear Fokker–Planck equation (2.10) numerically by means of the techniques discussed in Sect. 2.8. The approximation (3.41) becomes exact in the limit $\Delta t \to 0$ [498].

Self-Consistent Langevin Equations

In order to write down a simulation scheme for self-consistent Langevin equations, we exploit the fact that the probability densities P obtained from the simulations of the two-layered Langevin equations are by definition equivalent to the driving forces P involved in the drift and diffusion coefficients of the Langevin equations. Therefore, at each iteration step, the simulation output can be used to determine the simulation input for the subsequent iteration step. Accordingly, the Ito–Langevin equation (3.34) can be solved numerically using the Euler forward scheme

$$X^l_{n+1} = X^l_n + \Delta t\, D'_1(X^l_n, t_n, u) + \sqrt{D'_2(X^l_n, t_n, u)} \sqrt{\Delta t}\, w^l_n , \qquad (3.44)$$

for $t_n = n\Delta t + t_0$, $n = 0, 1, 2, \ldots$ and w^l_n as defined above. The coefficients $D'_1(X^l_n, t_n, u)$ and $D'_2(X^l_n, t_n, u)$ are now computed from the realizations X^l_n with $l = 1, \ldots, L$ by means of

$$D'_1(x, t_n, u) = D_1\left(x, t_n, \frac{1}{L}\sum_{l=1}^{L}\delta(x - X^l_n)\right) ,$$
$$D'_2(x, t_n, u) = D_2\left(x, t_n, \frac{1}{L}\sum_{l=1}^{L}\delta(x - X^l_n)\right) , \qquad (3.45)$$

where L has to be a large number. The approximation becomes exact if we carry out the limit $L \to \infty$ and subsequently the limit $\Delta t \to 0$. In the multivariate case an analogous discretization scheme can be used. In view of (3.45) the question arises as to how to evaluate numerically the expression $L^{-1}\sum_{l=1}^{L} \delta(x - X^l_n)$.

$d < \infty$ (Averaging Method)

If D_1 and D_2 depend on a finite set of expectation values, the nonlinear Fokker–Planck equation (2.10) can be cast into the form (2.30) and (3.45) becomes [13, 88, 194, 378]

3.4 Langevin Equations

$$D'_1(x, t_n, u) = D_1\left(x, t_n, \frac{1}{L}\sum_{l=1}^{L} A_1(X_n^l), \ldots, \frac{1}{L}\sum_{l=1}^{L} A_d(X_n^l)\right),$$

$$D'_2(x, t_n, u) = D_2\left(x, t_n, \frac{1}{L}\sum_{l=1}^{L} A_1(X_n^l), \ldots, \frac{1}{L}\sum_{l=1}^{L} A_d(X_n^l)\right). \quad (3.46)$$

For examples see Sects. 3.10, 5.2.2, 7.3.2, and 7.3.3.

$d = \infty$ (Histogram and Averaging Method)

In this case, we have at least two options.

Histogram Method [196]

We may approximate the time-dependent probability density $P(x, t_n; u) = \lim_{L\to\infty} L^{-1} \sum_{l=1}^{L} \delta(x - X_n^l)$ by the histogram H_n of the random variable X_n given in intervals $[x_j, x_j + \Delta x)$ with $x_j = x_0 + j\Delta x$ and $j = 1, 2, \ldots, N_j$, where Δx denotes a small quantity. That is, H_n is defined on $\Omega = [x_0, x_0 + \Delta x(N_j + 1))$. If we define the next-left-neighbor-point $\bar{x}(x)$ of a point x by $\bar{x}(x) = \{x_j : x_j \leq x \wedge x - x_j = \min\}$, we can write H_n as

$$H_n(x) = \frac{1}{L} \sum_{I_n(x)} 1, \quad I_n(x) = \{l : X_n^l \in [\bar{x}(x), \bar{x}(x) + \Delta x)\}. \quad (3.47)$$

Roughly speaking, the set $I_n(x)$ denotes the set of indices l that belong to realizations $X_n^l \approx x$. For large values of L and small values of Δx the histogram H_n can be regarded as an approximation of the density function P, that is, we have

$$\frac{1}{L} \sum_{l=1}^{L} \delta(x - X_n^l) \approx P \approx H_n(x)/\Delta x, \quad (3.48)$$

whereas in the limit $\Delta x \to 0$ and $L \to \infty$ we get $P(x, t_n; u) = H_n(x)/\Delta x$. For an application of the histogram method see Sect. 7.4.4.

Averaging Method [177]

Alternatively to the histogram method, we may use a representation of the Dirac delta function in order to compute $L^{-1} \sum_{l=1}^{L} \delta(x - X_n^l)$. For example, we may exploit the relation [430]

$$\delta(x - x') = \lim_{\Delta x \to 0} \frac{1}{\sqrt{2\pi}\Delta x} \exp\left\{-\frac{1}{2}\left[\frac{x - x'}{\Delta x}\right]^2\right\}. \quad (3.49)$$

Then, (3.45) reads

$$D_1'(x,t_n,u) = D_1\left(x,t_n,\frac{1}{\sqrt{2\pi\Delta xL}}\sum_{l=1}^{L}\exp\left\{-\frac{1}{2}\left[\frac{x-X_n^l}{\Delta x}\right]^2\right\}\cdot\right),$$

$$D_2'(x,t_n,u) = D_2\left(x,t_n,\frac{1}{\sqrt{2\pi\Delta xL}}\sum_{l=1}^{L}\exp\left\{-\frac{1}{2}\left[\frac{x-X_n^l}{\Delta x}\right]^2\right\}\right), \quad (3.50)$$

where Δx should be chosen sufficiently small. For examples see Sects. 6.5.2, 6.5.3, 6.5.6, and 7.4.2.

3.5 Stationary Solutions

Let us consider drift and diffusion coefficients that do not explicitly depend on t. Then, in the stationary case, (3.15) and (3.16) become $\partial P_{st}(\mathbf{x})/\partial t = 0$ and

$$\frac{\partial}{\partial t}P_{st}(\mathbf{x},t|\mathbf{x}',t';P_{st}(\mathbf{x})) = -\sum_{i=1}^{M}\frac{\partial}{\partial x_i}D_i(\mathbf{x},P_{st}(\mathbf{x}))P_{st}(\mathbf{x},t|\mathbf{x}',t';P_{st}(\mathbf{x}))$$

$$+\sum_{i,k=1}^{M}\frac{\partial^2}{\partial x_i \partial x_k}D_{ik}(\mathbf{x},P_{st}(\mathbf{x}))P_{st}(\mathbf{x},t|\mathbf{x}',t';P_{st}(\mathbf{x})), \quad (3.51)$$

where $P_{st}(\mathbf{x})$ can be derived from (2.22). In the univariate case, we have $\partial P_{st}(x)/\partial t = 0$ and

$$\frac{\partial}{\partial t}P_{st}(x,t|x',t';P_{st}(x)) = -\frac{\partial}{\partial x}D_1(x,P_{st}(x))P_{st}(x,t|x',t';P_{st}(x))$$

$$+\frac{\partial^2}{\partial x^2}D_{ik}(x,P_{st}(x))P_{st}(x,t|x',t';P_{st}(x)) \quad (3.52)$$

with $P_{st}(x)$ defined by (2.23). Note that the stationary transition probability densities $P_{st}(\mathbf{x},t|\mathbf{x}',t';P_{st}(\mathbf{x}))$ and $P_{st}(x,t|x',t';P_{st}(x))$ depend only on the time difference $t-t'$ (see [498] and Sect. 3.6 below).

In the stationary case, the multivariate Ito–Langevin equation (3.32) of the nonlinear Fokker–Planck equation (2.4) reads

$$\frac{d}{dt}X_i(t) = D_i(\mathbf{x},P_{st})|_{\mathbf{x}=\mathbf{X}(t)} + \sum_{k=1}^{M}G_{ik}(\mathbf{x},P_{st})|_{\mathbf{x}=\mathbf{X}(t)}\Gamma_k(t), \quad (3.53)$$

for $t \geq t_0$ and $\mathbf{X}(t_0)$ distributed like P_{st}. The univariate Ito–Langevin equation (3.28) of the nonlinear Fokker–Planck equation (2.10) becomes

$$\frac{d}{dt}X(t) = D_1(x,P_{st})|_{x=X(t)} + \sqrt{D_2(x,P_{st})}\Big|_{x=X(t)}\Gamma(t), \quad (3.54)$$

for $t \geq t_0$ and $X(t_0)$ is distributed like $P_{st}(x)$.

Multiple stationary solutions

In general, nonlinear Fokker–Planck equations can have multiple stationary solutions. Let P_{st}^i with $i = 1, 2, \ldots$ denote these solutions. Then, for every stationary solution, we obtain a stationary transition probability density that defines a stationary Markov diffusion process. In the univariate case, these stationary transition probability densities $P_{\text{st}}^i(x, t|x', t')$ can formally be written as

$$P_{\text{st}}^i(x, t|x', t') = P_{\text{st}}(x, t|x', t'; P_{\text{st}}^i) \ . \tag{3.55}$$

Here, $P_{\text{st}}^i(x, t|x', t')$ are the members of a family of stationary transition probability densities, whereas the probability density $P_{\text{st}}(x, t|x', t'; P_{\text{st}})$ describes the whole family (see Sect. 3.7). Using $P_{\text{st}}^i(x, t|x', t')$, we can define stationary Markov processes in terms of the joint probability densities

$$\begin{aligned} P_{\text{st}}^i(x_n, t_n; \ldots; x_0, t_0) = \\ P_{\text{st}}^i(x_n, t_n|x_{n-1}, t_{n-1}) \cdots P_{\text{st}}^i(x_1, t_1|x_0, t_0) P_{\text{st}}^i(x_0) \ , \end{aligned} \tag{3.56}$$

see (3.18). Alternatively, these stationary processes can be expressed by means of the Ito–Langevin equation

$$\frac{\text{d}}{\text{d}t} X(t) = D_1(x, P_{\text{st}}^i)\big|_{x=X(t)} + \sqrt{D_2(x, P_{\text{st}}^i)}\bigg|_{x=X(t)} \Gamma(t) \ , \tag{3.57}$$

for $t \geq t_0$ and $X(t_0)$ distributed like P_{st}^i.

3.6 H-Theorem for Stochastic Processes

In many cases, one can show that transient solutions $P(\mathbf{x}, t; u)$ of nonlinear Fokker–Planck equations of the form (2.4) converge to stationary ones in the long time limit (we will return to this issue in Chapters 4 and 7). Since by means of strongly nonlinear Fokker–Planck equations we can completely define stochastic processes, the question arises as to whether or not these stochastic processes converge to stationary ones when their single time-point distributions $P(\mathbf{x}, t; u)$ converge to stationary ones. That is, can we conclude from the asymptotic behavior $P(\mathbf{x}, t; u) \to P_{\text{st}}(\mathbf{x})$ that joint probability densities $P(\mathbf{x}, t; \mathbf{x}', t'), P(\mathbf{x}, t; \mathbf{x}', t'; \mathbf{x}'', t''), \ldots$ and in particular correlation functions such as $\langle X_i^n(t) X_k^m(t') \rangle$ become stationary? In what follows, we will show that this conclusion can indeed be drawn. As a result, H-theorems that state that transition solutions $P(\mathbf{x}, t; u)$ of nonlinear Fokker–Planck equations converge to stationary ones can in fact be applied to show that transient stochastic processes converge to stationary ones.

For the sake of convenience, we consider the univariate case given by (2.10) and

3 Strongly Nonlinear Fokker–Planck Equations

$$\frac{\partial}{\partial t}P(x,t|x',t';u) = -\frac{\partial}{\partial x}D_1(x,t,P(x,t;u))P(x,t|x',t';u)$$
$$+\frac{\partial^2}{\partial x^2}D_2(x,t,P(x,t;u))P(x,t|x',t';u) \quad (3.58)$$

(which is (3.16) in the univariate case). Let us assume that for a particular choice of the drift and diffusion coefficients there is a convergence theorem that states that solutions of (2.10) satisfy the limiting case

$$\lim_{t\to\infty} P(x,t;u) = P_{\mathrm{st}}(x) \ . \quad (3.59)$$

That is, they become stationary in the long time limit. Our first objective is to show that the limiting case

$$\lim_{t\to\infty} P(x,t|x',t';u) = P_{\mathrm{st}}(x,z|x';P_{\mathrm{st}}) \quad (3.60)$$

holds with $z = t - t'$. Let us consider systems with $D_1(x,t,P) = D_1(x,P)$ and $D_2(x,t,P) = D_2(x,P)$. Then, we first note that on account of (3.59) in the limit $t \to \infty$ (3.58) can be written as

$$\frac{\partial}{\partial t}P(x,t|x',t';u) = -\frac{\partial}{\partial x}D_1(x,P_{\mathrm{st}})P(x,t|x',t';u)$$
$$+\frac{\partial^2}{\partial x^2}D_2(x,P_{\mathrm{st}})P(x,t|x',t';u) \ . \quad (3.61)$$

Using the Fokker–Planck operator

$$\hat{F}(x,\partial/\partial x, P_{\mathrm{st}}) = -\frac{\partial}{\partial x}D_1(x,P_{\mathrm{st}}) + \frac{\partial^2}{\partial x^2}D_2(x,P_{\mathrm{st}}) \quad (3.62)$$

and $t = t' + z$, the formal solution of (3.61) under the initial condition $\lim_{z\to 0} P(x,t'+z|x',t';u) = \delta(x-x')$ reads

$$P(x,t'+z|x',t';u) = \exp\{\hat{F}(x,\partial/\partial x, P_{\mathrm{st}})z\}\delta(x-x')$$
$$= P_{\mathrm{st}}(x,z|x';P_{\mathrm{st}}) \ . \quad (3.63)$$

As indicated, we can read off from this formal solution that $P(x,t'+z|x',t';u)$ depends only on the time interval $z = t-t'$ and, consequently, can be cast into the form $P_{\mathrm{st}}(x,z|x';P_{\mathrm{st}})$. Equation (3.60) implies that all joint distributions $P(x_n,t_n;\ldots;x_1,t_1;u) = P(x_n,t_n|x_{n-1},t_{n-1};u)\cdots P(x_2,t_2|x_1,t_1;u)P(x_1,t_1;u)$ only depend on time differences (such as $z_1 = t_2-t_1$, $z_2 = t_3-t_2$, and so on) in the long time limit and therefore stochastic processes described by these joint distributions become stationary processes.

Our second objective is to show that stationary transition probability densities of strongly nonlinear Fokker–Planck equations converge to stationary distribution functions in the limit $z \to \infty$. That is, we will prove that the limiting case

3.6 H-Theorem for Stochastic Processes

$$\lim_{z \to \infty} P_{\text{st}}(x, z | x'; P_{\text{st}}) = P_{\text{st}}(x) \tag{3.64}$$

holds. In order to derive (3.64), we confine ourselves to systems with stationary distributions P_{st} that are defined by vanishing probability currents J. For $J = 0$, $D_1(x, t, P) = D_1(x, P)$, $D_2(x, t, P) = D_2(x, P)$, and $P = P_{\text{st}}$ from (2.23) it follows that

$$D_1(x, P_{\text{st}}) = \frac{\partial}{\partial x} D_2(x, P_{\text{st}}) + D_2(x, P_{\text{st}}) \frac{\partial}{\partial x} \ln P_{\text{st}} \ . \tag{3.65}$$

In this case, (3.61) for $P(x, t | x', t'; u) = P_{\text{st}}(x, z | x'; P_{\text{st}})$ can be written as [176]

$$\frac{\partial}{\partial z} P_{\text{st}}(x, z | x'; P_{\text{st}}) = \frac{\partial}{\partial x} D_2(x, P_{\text{st}}) P_{\text{st}}(x) \frac{\partial}{\partial x} \frac{P_{\text{st}}(x, z | x'; P_{\text{st}})}{P_{\text{st}}(x)} \ . \tag{3.66}$$

Note that (3.66) can be transformed into

$$\frac{\partial}{\partial z} P_{\text{st}}(x, z | x'; P_{\text{st}}) = \frac{\partial^2}{\partial x^2} D_2(x, P_{\text{st}}) P_{\text{st}}(x, z | x'; P_{\text{st}})$$
$$+ \frac{\partial}{\partial x} P_{\text{st}}(x, z | x'; P_{\text{st}}) \left[\frac{\partial}{\partial x} D_2(x, P_{\text{st}}) + D_2(x, P_{\text{st}}) \frac{\partial}{\partial x} \ln P_{\text{st}} \right] \ . \tag{3.67}$$

Using (3.65), we obtain (3.61). Next, we write (3.66) as

$$\frac{\partial}{\partial z} P_{\text{st}}(x, z | x'; P_{\text{st}}) =$$
$$\frac{\partial}{\partial x} D_2(x, P_{\text{st}}) P_{\text{st}}(x, z | x'; P_{\text{st}}) \frac{\partial}{\partial x} \ln \frac{P_{\text{st}}(x, z | x'; P_{\text{st}})}{P_{\text{st}}(x)} \tag{3.68}$$

and introduce the Kullback measure

$$^{\text{B}}K(z, x') = \int_\Omega P_{\text{st}}(x, z | x'; P_{\text{st}}) \ln \frac{P_{\text{st}}(x, z | x'; P_{\text{st}})}{P_{\text{st}}(x)} \, dx \geq 0 \ . \tag{3.69}$$

We will show in Sect. 4.7.2 that $^{\text{B}}K$ is semi-positive definite – as indicated in (3.69). Differentiating $K(z, x')$ with respect to z, using (3.68), and integrating by parts, we obtain

$$\frac{\partial}{\partial z} K(z, x') =$$
$$- \int_\Omega D_2(x, P_{\text{st}}) P_{\text{st}}(x, z | x'; P_{\text{st}}) \left[\frac{\partial}{\partial x} \ln \frac{P_{\text{st}}(x, z | x'; P_{\text{st}})}{P_{\text{st}}(x)} \right]^2 dx \leq 0 \ . \tag{3.70}$$

It is clear from (3.70) that $\partial K / \partial z = 0$ implies $P_{\text{st}}(x, z | x'; P_{\text{st}}) / P_{\text{st}}(x) = C(x', z)$, where C is independent of x. From (3.66) it then follows that $\partial P_{\text{st}}(x, z | x'; P_{\text{st}}) / \partial z = 0$ holds, which means that $P_{\text{st}}(x, z | x'; P_{\text{st}})$ is independent of z. This in turn implies that $C(x', z)$ is independent of z. Thus, we obtain the intermediate result: $\partial K/\partial z = 0 \Rightarrow P_{\text{st}}(x,z|x';P_{\text{st}}) = C(x')P_{\text{st}}(x)$. Integrating this result with respect to x and taking the normalization condition into account (i.e., $\int_\Omega P_{\text{st}}(x,z|x';P_{\text{st}})\,\mathrm{d}x = \int_\Omega P_{\text{st}}(x)\,\mathrm{d}x = 1$), we see that $C(x') = 1$ holds. In sum, $K(z,x')$ satisfies the relations:

$$K(z,x') \geq 0\,,$$
$$\frac{\partial}{\partial z}K(z,x') \leq 0\,,$$
$$\frac{\partial}{\partial z}K(z,x') = 0 \Leftrightarrow P_{\text{st}}(x,z|x';P_{\text{st}}) = P_{\text{st}}(x)\,. \qquad (3.71)$$

Consequently, $K(z,x')$ is a Lyapunov functional for $P_{\text{st}}(x,z|x';P_{\text{st}})$ and we conclude that (3.64) is satisfied (see also Sect. 4.3). Explicit examples that illustrate the convergence of transient stochastic processes to stationary ones will be given in Sects. 3.10 and 5.2.2 and in the context of the Gaussian entropy in Sect. 6.5.3.

3.7 Nonlinear Families of Markov Processes*

In this section, we consider (2.3) as a family of hierarchies that describes a family of stochastic processes. The family members are distinguished by means of their initial distributions. That is, $u(\mathbf{x})$ is regarded as a label that can be used to distinguish between different family members. For example, for the set of initial probability densities $A_0 = \{u_1(\mathbf{x}), u_2(\mathbf{x}),\ldots\}$, we obtain from (2.2) a set of probability densities given by $P^i(\mathbf{x},t) = P(\mathbf{x},t;u_i(\mathbf{x}))$. In general, we define the conditional probability density of a family of stochastic processes by

$$P(\mathbf{x}_n,t_n|\mathbf{x}_{n-1},t_{n-1};\ldots;\mathbf{x}_1,t_1;u) = \frac{P(\mathbf{x}_n,t_n;\ldots;\mathbf{x}_1,t_1;u)}{P(\mathbf{x}_{n-1},t_{n-1};\ldots;\mathbf{x}_1,t_1;u)}\,. \qquad (3.72)$$

3.7.1 Linear Versus Nonlinear Families of Markov Processes

If the conditional probability density (3.72) satisfies

$$P(\mathbf{x}_n,t_n|\mathbf{x}_{n-1},t_{n-1};\ldots;\mathbf{x}_1,t_1;u) = P(\mathbf{x}_n,t_n|\mathbf{x}_{n-1},t_{n-1};u) \qquad (3.73)$$

for all n, then we deal with a family of Markov processes [188]. It is important to realize that the members of the family described by the conditional probability density (3.73) depend only on two time points. For example, let us consider a family of stochastic processes characterized by the initial distributions $u_1, u_2, u_3, \ldots$. Then, we deal with a set of Markov transition probability densities given by $P^i(\mathbf{x}_n,t_n|\mathbf{x}_{n-1},t_{n-1};\ldots;\mathbf{x}_1,t_1) = P^i(\mathbf{x}_n,t_n|\mathbf{x}_{n-1},t_{n-1})$. If there is a unique transition probability density,

$$P(\mathbf{x}, t | \mathbf{x}', t'; u) = P(\mathbf{x}, t | \mathbf{x}', t') ,\qquad(3.74)$$

that describes the evolution of all members of the family, we refer to the family as a linear family of Markov processes. In other words, if the transition probability density of a family of Markov processes does not depend on the initial distribution u, we have a linear family. If (3.74) is not satisfied, we refer to the family as nonlinear, see also Table 3.1. The transition probability densities $P(\mathbf{x}_n, t_n | \mathbf{x}_{n-1}, t_{n-1}; u)$ of families of Markov processes satisfy the Chapman–Kolmogorov equation

$$P(\mathbf{x}_3, t_3 | \mathbf{x}_1, t_1; u) = \int_\Omega P(\mathbf{x}_3, t_3 | \mathbf{x}_2, t_2; u) P(\mathbf{x}_2, t_2 | \mathbf{x}_1, t_1; u) \, \mathrm{d}^M x_2 ,\qquad(3.75)$$

for $t_3 \geq t_2 \geq t_1$ because every member $u(x)$ of the family satisfies this integral relation [254, 498].

Table 3.1. Linear and nonlinear families of Markov processes

$P(\mathbf{x}_n, t_n | \mathbf{x}_{n-1}, t_{n-1}; u)$	Type of family
$\forall u : P(\mathbf{x}_n, t_n | \mathbf{x}_{n-1}, t_{n-1})$	linear
otherwise	nonlinear

3.7.2 Linear Families of Markov Processes

For linear families of Markov processes we obtain from (3.74) the propagator relation

$$P(\mathbf{x}, t; u) = \int P(\mathbf{x}, t | \mathbf{x}', t_0) \, u(\mathbf{x}') \, \mathrm{d}^M x' .\qquad(3.76)$$

From (3.75) and (3.76) we obtain the integral equation

$$P(\mathbf{x}, t; u) = \int P(\mathbf{x}, t | \mathbf{x}', t') P(\mathbf{x}', t'; u) \, \mathrm{d}^M x' ,\qquad(3.77)$$

for $t \geq t' \geq t_0$. For linear families of Markov diffusion processes from (3.75) and (3.77) we obtain Fokker–Planck equations of the form [221, 498]

$$\frac{\partial}{\partial t}P(\mathbf{x},t;u) = -\sum_{i=1}^{M}\frac{\partial}{\partial x_i}D_i'(\mathbf{x},t)P(\mathbf{x},t;u)$$
$$+\sum_{i,k=1}^{M}\frac{\partial^2}{\partial x_i \partial x_k}D_{ik}'(\mathbf{x},t)P(\mathbf{x},t;u) \,,$$

$$\frac{\partial}{\partial t}P(\mathbf{x},t|\mathbf{x}',t') = -\sum_{i=1}^{M}\frac{\partial}{\partial x_i}D_i'(\mathbf{x},t)P(\mathbf{x},t|\mathbf{x}',t')$$
$$+\sum_{i,k=1}^{M}\frac{\partial^2}{\partial x_i \partial x_k}D_{ik}'(\mathbf{x},t)P(\mathbf{x},t|\mathbf{x}',t') \qquad (3.78)$$

and solutions $P(\mathbf{x},t;u)$ that satisfy the superposition principle

$$pP(\mathbf{x},t,u_1) + (1-p)P(\mathbf{x},t,u_2) = P(\mathbf{x},t,pu_1+(1-p)u_2) \,, \qquad (3.79)$$

with $p \in [0,1]$.

3.7.3 Nonlinear Families of Markov Diffusion Processes

For nonlinear families of Markov processes the Kramers–Moyal expansion [221, 498] of (3.75) gives us

$$\frac{\partial}{\partial t}P(\mathbf{x},t|\mathbf{x}',t';u) =$$
$$\sum_{n=1}^{\infty}(-1)^n \sum_{i_1=1}^{M}\cdots\sum_{i_n=1}^{M}\frac{\partial^n}{\partial x_{i_1}\cdots \partial x_{i_n}}\left[D_{i_1,\ldots,i_n}'(\mathbf{x},t;u)P(\mathbf{x},t|\mathbf{x}',t';u)\right]$$
$$(3.80)$$

with the Kramers–Moyal coefficients defined by

$$M_{i_1,\ldots,i_n}(\mathbf{x}',t,t';u) = \int_{\Omega}(y_{i_1}-x_{i_1})\cdots(y_{i_n}-x_{i_n})\,P(\mathbf{y},t|\mathbf{x}',t';u)\,\mathrm{d}^M y \,,$$
$$D_{i_1,\ldots,i_n}'(\mathbf{x},t;u) = \frac{1}{n!}\lim_{\tau\to 0}\frac{M_{i_1,\ldots,i_n}(\mathbf{x},t+\tau,t;u)}{\tau} \,,$$
$$= \frac{1}{n!}\lim_{\tau\to 0}\int_{\Omega}(y_{i_1}-x_{i_1})\cdots(y_{i_n}-x_{i_n})\,P(\mathbf{y},t+\tau|\mathbf{x},t;u)\,\mathrm{d}^M y \,. \qquad (3.81)$$

As opposed to a linear family of Markov processes, the Kramers–Moyal coefficients now depend on u. The Pawula theorem [298, 498] applies to any transition probability density and, consequently, to the transition probability density (3.73) as well. Therefore, nonlinear families of Markov diffusion processes (i.e., Markov processes with a finite number of nonvanishing Kramers–Moyal coefficients [298]) are described by

$$\frac{\partial}{\partial t}P(\mathbf{x},t;u) = -\sum_{i=1}^{M}\frac{\partial}{\partial x_i}D'_i(\mathbf{x},t;u)P(\mathbf{x},t;u)$$
$$+\sum_{i,k=1}^{M}\frac{\partial^2}{\partial x_i \partial x_k}D'_{ik}(\mathbf{x},t;u)P(\mathbf{x},t;u) \;,$$

$$\frac{\partial}{\partial t}P(\mathbf{x},t|\mathbf{x}',t';u) = -\sum_{i=1}^{M}\frac{\partial}{\partial x_i}D'_i(\mathbf{x},t;u)P(\mathbf{x},t|\mathbf{x}',t';u)$$
$$+\sum_{i,k=1}^{M}\frac{\partial^2}{\partial x_i \partial x_k}D'_{ik}(\mathbf{x},t;u)P(\mathbf{x},t|\mathbf{x}',t';u) \;. \quad (3.82)$$

The hierarchy of probability densities of a family of Markov diffusion processes defined by (3.82) is given by

$$P(\mathbf{x},t;u) = \int_\Omega P(\mathbf{x},t|\mathbf{x}',t_0;u)\,u(\mathbf{x}')\,\mathrm{d}^M x' \;,$$
$$P(\mathbf{x}_n,t_n;\ldots;\mathbf{x}_1,t_1;u)$$
$$= P(\mathbf{x}_n,t_n|\mathbf{x}_{n-1},t_{n-1};u)\cdots P(\mathbf{x}_2,t_2|\mathbf{x}_1,t_1;u)P(\mathbf{x}_1,t_1;u) \;. \quad (3.83)$$

Note that the evolution equation (3.82) for $P(\mathbf{x},t,u)$ is nonlinear with respect to u. Consequently, the superposition principle (3.79) does not necessarily hold.

Example

We give here an example of a nonlinear family of Markov processes involving a drift coefficient $D_1(x,t,u)$. First, we consider the driven stochastic system described by the Langevin equation

$$\frac{\mathrm{d}}{\mathrm{d}t}X(t) = -\gamma X - A\sin[\omega(t-t_0)] + \sqrt{Q}\Gamma(t) \;, \quad (3.84)$$

for $t \geq t_0$ and $X(t) \in \Omega = \mathbb{R}$. This Langevin equation involves a drift coefficient D_1 that depends on the parameters γ, A, ω, and t_0 and reads $D_1(x,t;\gamma,A,\omega,t_0) = \gamma x + A\sin[\omega(t-t_0)]$. Second, we assume that the driving frequency ω can assume only two possible values: $+\tilde{\omega}$ or $-\tilde{\omega}$ with $\tilde{\omega} > 0$. We assume that for initial distributions $u(x)$ with first moments larger than or equal to zero we have $\omega = \tilde{\omega}$, whereas for initial distributions $u(x)$ with first moments smaller than zero we have $\omega = -\tilde{\omega}$. Thus, we obtain a nonlinear family of Markov diffusion processes described by

$$\frac{\mathrm{d}}{\mathrm{d}t}X(t) = -\gamma X - A\sin[\tilde{\omega}\,\mathrm{sgn}\{\langle X(t_0)\rangle\}\,(t-t_0)] + \sqrt{Q}\Gamma(t) \;, \quad (3.85)$$

where the sgn-function is defined by sgn$\{x\} = 1$ for $x \geq 0$ and sgn$\{x\} = -1$ for $x < 0$. The drift coefficient now reads

$$D_1(x,t;\gamma, A, \tilde{\omega}, u, t_0) = \gamma x + A\sin[\tilde{\omega}\,\mathrm{sgn}\left\{\int_\Omega xu(x)\,\mathrm{d}x\right\}(t-t_0)]\,. \qquad (3.86)$$

3.7.4 Markov Embedding of Strongly Nonlinear Fokker–Planck Equations

We are now in the position to interpret the nonlinear Fokker–Planck equation (2.4) in the context of nonlinear families of Markov processes [188]. If for solutions $P(\mathbf{x}, t; u)$ of (2.4) the coefficients $D_i(\mathbf{x}, t, P(\mathbf{x}, t; u))$ and $D_{ik}(\mathbf{x}, t, P(\mathbf{x}, t; u))$ correspond to the first and second Kramers–Moyal coefficients $D'_i(\mathbf{x}, t, u) = D_i(\mathbf{x}, t, P(\mathbf{x}, t; u))$ and $D'_{ik}(\mathbf{x}, t, u) = D_{ik}(\mathbf{x}, t, P(\mathbf{x}, t; u))$ of a nonlinear family of Markov diffusion processes, then we will refer to (2.4) as a strongly nonlinear Fokker–Planck equation. In other words, if for solutions of (2.4) the evolution equations (3.11,...,3.13) describe a nonlinear family of Markov diffusion processes, then (2.4) corresponds to a strongly nonlinear Fokker–Planck equation and its solutions $P(\mathbf{x}, t; u)$ can be embedded into the Markov diffusion processes described by (3.11-3.13).

3.7.5 Hitchhiker Processes

The close relationship between nonlinear Fokker–Planck equations and nonlinear families of Markov diffusion processes that has been discussed in the previous section can be illustrated by means of so-called Hitchhiker processes [188]. Let us consider (2.4) for coefficients D_i and D_{ik} that do not explicitly depend on time:

$$\frac{\partial}{\partial t}P(\mathbf{x},t;u) = -\sum_{i=1}^{M}\frac{\partial}{\partial x_i}D_i(\mathbf{x};P)P + \sum_{i,k=1}^{M}\frac{\partial^2}{\partial x_i \partial x_k}D_{ik}(\mathbf{x};P)P\,. \qquad (3.87)$$

By means of the coefficients $D_i(\mathbf{x}, P)$ and $D_{ik}(\mathbf{x}, P)$ we can define the linear Fokker–Planck equation

$$\frac{\partial}{\partial t}W(\mathbf{x},t;u)_{t_0} = -\sum_{i=1}^{M}\frac{\partial}{\partial x_i}D_i(\mathbf{x};u)W(\mathbf{x},t;u)_{t_0}$$
$$+ \sum_{i,k=1}^{M}\frac{\partial^2}{\partial x_i \partial x_k}D_{ik}(\mathbf{x};u)W(\mathbf{x},t;u)_{t_0} \qquad (3.88)$$

for the probability density W. Since (3.87) is assumed to be a strongly nonlinear Fokker–Planck equation, (3.88) defines for every initial distribution $u = P$ a time-dependent probability density $W(\mathbf{x}, t; u)_{t_0}$ of a Markov process

3.7 Nonlinear Families of Markov Processes* 51

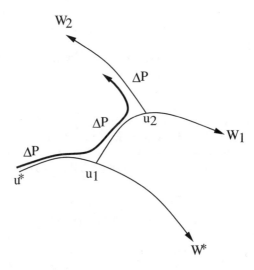

Fig. 3.1. Hitchhiker behavior of a stochastic process defined by a nonlinear Fokker–Planck equation; reprinted from [188], © 2004, with permission from Elsevier

satisfying $W(\mathbf{x}, z; u)_z = u$. In Fig. 3.1 we have plotted schematically three solutions $W^* = W(\mathbf{x}, t; u^*)_{t_0}$, $W^1 = W(\mathbf{x}, t; u_1)_{t_1}$, and $W^2(\mathbf{x}, t; u_2)_{t_2}$ of (3.88) for the initial distributions u^*, u_1, and u_2.
Comparing (3.87) and (3.88) we realize that the equivalence

$$\frac{\partial}{\partial t} P(\mathbf{x}, t; u) = \frac{\partial}{\partial t} W(\mathbf{x}, t; P)_t \qquad (3.89)$$

holds. Integrating (3.89) with respect to t, gives

$$P(\mathbf{x}, t; u) = u(\mathbf{x}) + \int_{t_0}^{t} \frac{\partial}{\partial s} W(\mathbf{x}, s; P(\mathbf{x}, s; u))_s \, ds \ . \qquad (3.90)$$

Equations (3.87-3.90) can be interpreted as follows. On the one hand, we deal with Markov processes B^i that are described by the transient probability densities $W(\mathbf{x}, t; u_i)_{t_i}$ and belong to different Fokker–Planck operators

$$\hat{F}_i = -\sum_{j=1}^{M} \frac{\partial}{\partial x_j} D_j(\mathbf{x}; u_i) + \sum_{i,k=1}^{M} \frac{\partial^2}{\partial x_j \partial x_k} D_{jk}(\mathbf{x}; u_i) \ , \qquad (3.91)$$

see (3.88). On the other hand, we have a process A described by the nonlinear Fokker–Planck equation (3.87) for a particular initial distribution u^*, see Fig. 3.1. Then, P first evolves like W^* and in the function space of probability densities approaches the "point" u_1. If P corresponds to u_1 then it will evolve like W_1 and will converge to the "point" u_2. If P corresponds to u_2 then it will evolve like W_2 and so on. Strictly speaking, this argument only holds when we consider infinitesimally small changes of P and infinitesimally small time intervals Δt. Nevertheless, Fig. 3.1 elucidates the basic mechanism that determines the solutions of nonlinear Fokker–Planck equations. The stochastic process A described by the nonlinear Fokker–Planck equation (3.87) evolves within the family of Markov processes with probability densities $W(\mathbf{x}, t; u_i)_{t_i}$ defined by the linear Fokker–Planck equation (3.88). We may also say that the process A "rides on the back" of the Markov processes B^i. Since A permanently changes its host processes B^i, the behavior of A resembles that of a hitchhiker, which is the reason why one may refer to processes defined by nonlinear Fokker–Planck equations as hitchhiker processes.

3.8 Top-Down Versus Bottom-Up Approaches*

In Sect. 3.2 we have embedded solutions of strongly nonlinear Fokker–Planck equations into Markov diffusion processes. This embedding procedure should be regarded as a top-down approach because our departure point is the nonlinear Fokker–Planck equation (2.4) and we assume that we do not have any further information at our disposal. In contrast, various studies have been concerned with the derivation of nonlinear Fokker–Planck equations by means of bottom-up approaches, see, for example, [60, 103, 135, 365, 568]. We will show next that for nonlinear Fokker–Planck equations with drift and diffusion coefficients that depend on an order parameter both approaches yield the same result. To this end, we will derive from a microscopic model the nonlinear Fokker–Planck equation

$$\frac{\partial}{\partial t} P(x, t; u) = -\frac{\partial}{\partial x} D_1(x, t, \langle A \rangle) P + \frac{\partial^2}{\partial x^2} D_2(x, t, \langle A \rangle) P \qquad (3.92)$$

involving the order parameter $\langle A(X) \rangle$.

We start off with a many-body system composed of subsystems described by the state variables $X_k(t)$ with $k = 1, \ldots, N_0$. We consider the thermodynamic limit: $N_0 \to \infty$. In order to deal with the limit $N_0 \to \infty$, in what follows, we will focus on finite subsets of the subsystem ensemble. Let $I_L = \{i_1, \ldots, i_L\}$ denote a set of L different indices with $i_l \geq 1$. Then, we assume that for all $L \geq 1$ and all possible index-sets I_L the dynamics of the subsystems can be described by the L-dimensional Ito–Langevin equation

$$\frac{d}{dt} X_{i_k}(t) = D_1 \left(X_{i_k}, t, \lim_{N_0 \to \infty} \frac{1}{N_0} \sum_{i=1}^{N_0} A(X_i) \right)$$

$$+ \sqrt{D_2 \left(X_{i_k}, t, \lim_{N_0 \to \infty} \frac{1}{N_0} \sum_{i=1}^{N_0} A(X_i) \right)} \Gamma_{i_k}(t) \qquad (3.93)$$

3.8 Top-Down Versus Bottom-Up Approaches*

for $i_k \in I_L$ and $\langle \Gamma_v(t')\Gamma_w(t)\rangle = 2\delta_{vw}\delta(t-t')$. Note that (3.93) accounts for the impact of all ensemble members on a single subsystem i_k (i.e., the sum runs from one to infinity and does not run from 1 to L). Let us describe the subpopulation of ensemble members I_L by means of the state vector $\mathbf{X}_L = (X_{i_1},\ldots,X_{i_L})$. We will prove now that if at time t^* the subsystems are statistically-independent and distributed according to the same law then they are statistically-independent for all times $t \geq t^*$. Therefore, we assume that there is a time t^* for which

$$P(\mathbf{x}_L, t^*) = \prod_{k=1}^{L} P(x, t^*)|_{x_{i_k}} \tag{3.94}$$

holds for all $L \geq 1$ and all index-sets I_L. Here, $P(\mathbf{x}_L, t^*)$ denotes the multivariable probability density $P(\mathbf{x}_L, t^*) = \langle \delta(\mathbf{x}_L - \mathbf{X}_L(t^*))\rangle$, while $P(x,t)$ denotes a single subsystem probability density. Since X_i are statistically-independent random variables and are all distributed the same way (namely, like P), they can be regarded as the realizations X^i of a random variable X that is distributed like P, which implies that $\lim_{N_0 \to \infty} N_0^{-1} \sum_{i=1}^{N_0} A(X_i(t^*)) = \langle A\rangle(t^*) = \int A(x)P(x,t^*)\,dx$ holds. To proceed further, we assume (just as in the top-down approach) that the coefficients

$$D'_1(x,t) = D_1(x,t,\langle A\rangle(t)), \quad D'_2(x,t) = D_2(x,t,\langle A\rangle(t)) \tag{3.95}$$

denote first and second order Kramers–Moyal coefficients of a Markov diffusion process at time t^*. Consequently, from (3.93) we obtain the L-dimensional Fokker–Planck equation

$$\frac{\partial}{\partial t}P(\mathbf{x}_L, t^*) = \sum_{k=1}^{L}\left[-\frac{\partial}{\partial x}D_1(x,t^*,\langle A\rangle(t^*)) + \frac{\partial^2}{\partial x^2}D_2(x,t^*,\langle A\rangle(t^*))\right]P(\mathbf{x}_L,t^*)\bigg|_{x_{i_k}}, \tag{3.96}$$

for all $L \geq 1$ and all subpopulations I_L (note that the partial derivatives also act beyond the squared bracket). From (3.94) and (3.96) we read off that P at time t^* evolves as

$$\frac{\partial}{\partial t}P(\mathbf{x}_L, t^*) = \sum_{k=1}^{L}\left\{\left(\prod_{l=1,l\neq k}^{L} P(x,t^*)|_{x_{i_l}}\right)\frac{\partial}{\partial t}P(x,t^*)|_{x_{i_k}}\right\}, \tag{3.97}$$

with

$$\frac{\partial}{\partial t}P(x,t) = \left[-\frac{\partial}{\partial x}D_1(x,t,\langle A\rangle(t)) + \frac{\partial^2}{\partial x^2}D_2(x,t,\langle A\rangle(t))\right]P(x,t). \tag{3.98}$$

As a result, for infinitesimally small time steps Δt we find that the relation

$$P(\mathbf{x}_L, t^* + \Delta t) = \prod_{k=1}^{L} P(x, t^* + \Delta t)|_{x_{i_k}} \tag{3.99}$$

holds for all $L \geq 1$ and all index-sets I_L. If we repeat our arguments for the initial time $t^* + \Delta t$, we obtain (3.99) again, where we need to replace Δt by $2\Delta t$. Proceeding further in this way, we obtain (3.99) for a finite interval Δt. For finite intervals Δt (3.99) states that the many-body system described by (3.93) evolves from a state of statistically-independent subsystems that are distributed according to the same law in such way that we again obtain a state of statistically-independent subsystems that are distributed according to the same law. Therefore, we draw the conclusion that if for solutions $P(x,t)$ of (3.98) the drift and diffusion coefficients defined by (3.95) correspond to Kramers–Moyal coefficients and if all subsystems at time t_0 are statistically-independent and distributed according to the same law, then $P(\mathbf{x}_L, t)$ is given by $P(\mathbf{x}_L, t) = \prod_{k=1}^{L} P(x_{i_k}, t)$, where $P(x,t)$ is the solution of the strongly nonlinear Fokker–Planck equation (3.98). Moreover, from (3.93) and (3.95) it follows that the transition probability density $P(\mathbf{x}_L, t|\mathbf{x}'_L, t')$ satisfies the evolution equation

$$\frac{\partial}{\partial t} P(\mathbf{x}_L, t|\mathbf{x}'_L, t') =$$
$$\sum_{k=1}^{L} \left[-\frac{\partial}{\partial x} D_1(x, t, \langle A \rangle_{P(x,t)}) + \frac{\partial^2}{\partial x^2} D_2(x, t, \langle A \rangle_{P(x,t)}) \right]_{x_{i_k}} P(\mathbf{x}_L, t|\mathbf{x}'_L, t') ,$$
$$\tag{3.100}$$

which, in turn, can be solved by $P(\mathbf{x}_L, t|\mathbf{x}'_L, t') = \prod_{k=1}^{L} P(x, t|x', t')|_{x_{i_k}}$ and

$$\frac{\partial}{\partial t} P(x, t|x', t') =$$
$$\left[-\frac{\partial}{\partial x} D_1(x, t, \langle A \rangle_{P(x,t)}) + \frac{\partial^2}{\partial x^2} D_2(x, t, \langle A \rangle_{P(x,t)}) \right] P(x, t|x', t') .$$
$$\tag{3.101}$$

Now let us write the self-consistent Ito–Langevin equation (3.36) for $D_1(x, t, P) = D_1(x, t, \langle A \rangle)$ and $D_2(x, t, P) = D_2(x, t, \langle A \rangle)$ in terms of the realizations X^k of $X(t)$:

$$\frac{\mathrm{d}}{\mathrm{d}t} X^k(t) = D_1\left(X^k, t, \lim_{N \to \infty} \frac{1}{N} \sum_{i=1}^{N} A(X^i)\right)$$
$$+ \sqrt{D_2\left(X^k, t, \lim_{N \to \infty} \frac{1}{N} \sum_{i=1}^{N} A(X^i)\right)} \Gamma^k(t) . \tag{3.102}$$

Since realizations of random variables are by definition statistically-independent quantities, by comparison of (3.93) and (3.102) we find that we may interpret the realizations of random variables defined by nonlinear Fokker–Planck equations as the state variables of the statistically-independent (but interacting) subsystems of many-body systems. In symbols, we may put $N_0 = N$ and $X_k = X^k$. In this sense, the bottom-up approach is consistent with the top-down approach discussed in Sect. 3.2.

Propagation of molecular chaos

A many-body system with statistically-independent subsystems is said to exhibit molecular chaos. The notion is that we regard the subsystems as gas molecules that move through the high-dimensional phase space of the gas in an independent and irregular fashion. If a many-body system that has been prepared in a state of molecular chaos at a particular time t^* remains in states of molecular chaos at subsequent times $t > t^*$, then we say, there is propagation of molecular chaos. Equations (3.94) and (3.99) illustrate that if the microscopic model (3.93) is related to a strongly nonlinear Fokker-Planck equation, then it describes a many-body system that exhibits propagation of molecular chaos. In fact, as shown by Bonilla, propagation of molecular chaos is a typical property of mean field systems described by nonlinear Fokker-Planck equations [60].

3.9 Transient Solutions and Transition Probability Densities

3.9.1 Nonequivalence of Transient Solutions and Transition Probability Densities

We will study now the relationship between transition probability densities and transient probability densities. Let us first elucidate this relationship for linear Fokker–Planck equations. To this end, we consider a Markov diffusion process described by

$$\frac{\partial}{\partial t}P(\mathbf{x},t;u) = -\sum_{i=1}^{M}\frac{\partial}{\partial x_i}D_i(\mathbf{x},t,t_0)P(\mathbf{x},t;u)$$

$$+ \sum_{i,k=1}^{M}\frac{\partial^2}{\partial x_i \partial x_k}D_{ik}(\mathbf{x},t,t_0)P(\mathbf{x},t;u) ,$$

$$\frac{\partial}{\partial t}P(\mathbf{x},t|\mathbf{x}',t';u) = -\sum_{i=1}^{M}\frac{\partial}{\partial x_i}D_i(\mathbf{x},t,t_0)P(\mathbf{x},t|\mathbf{x}',t';u)$$

$$+ \sum_{i,k=1}^{M}\frac{\partial^2}{\partial x_i \partial x_k}D_{ik}(\mathbf{x},t,t_0)P(\mathbf{x},t|\mathbf{x}',t';u) \quad (3.103)$$

for $t \geq t' \geq t_0$. Note that here the drift and diffusion coefficients depend explicitly on the parameter t_0 which denotes the initial time of the process. As an example, we may think of the periodically driven system described by the Langevin equation (3.84) involving the drift coefficient $D_1(x, t; \gamma, A, \omega, t_0)$. It is important to realize that the fact that the coefficients D_i and D_{ik} now depend on t_0 does not violate the requirement that we deal with a Markov process. In a similar vein, we would like to emphasize that from (3.103) it follows that $P(\mathbf{x}, t; u)$ and $P(\mathbf{x}, t|\mathbf{x}', t'; u)$ depend on the parameter t_0 such that we could also write $P(\mathbf{x}, t; u) = P(\mathbf{x}, t; u, t_0)$ and $P(\mathbf{x}, t|\mathbf{x}', t'; u) = P(\mathbf{x}, t|\mathbf{x}', t'; u, t_0)$, or alternatively $P(\mathbf{x}, t; u) = P(\mathbf{x}, t; u)_{t_0}$ and $P(\mathbf{x}, t|\mathbf{x}', t'; u) = P(\mathbf{x}, t|\mathbf{x}', t'; u)_{t_0}$. We prefer the notation $P(\mathbf{x}, t|\mathbf{x}', t'; u)_{t_0}$ (and likewise $P(\mathbf{x}, t; u)_{t_0}$) because the term $P(\mathbf{x}, t|\mathbf{x}', t'; u, t_0)$ could wrongly be interpreted as a non-Markovian conditional probability density of the form $P(\mathbf{x}, t|\mathbf{x}', t'; \mathbf{x}_{t_0}, t_0)$. Just to avoid confusion: for every initial condition u the solution $P(\mathbf{x}, t|\mathbf{x}', t'; u)_{t_0}$ of (3.103) describes a Markov transition probability density. However, $P(\mathbf{x}, t|\mathbf{x}', t'; u)_{t_0}$ depends in general on all parameters that occur in D_i and D_{ik} and, consequently, on the parameter t_0. By comparing the evolution equations for $P(\mathbf{x}, t; u)$ and $P(\mathbf{x}, t|\mathbf{x}', t'; u)$ of (3.103), we can verify that the equivalence

$$P(\mathbf{x}, t; \delta(\mathbf{x} - \mathbf{x}_0))_{t_0} = P(\mathbf{x}, t|\mathbf{x}_0, t_0; \delta(\mathbf{x} - \mathbf{x}_0))_{t_0} \quad (3.104)$$

holds. In other words, the transient probability density $P(\mathbf{x}, t; \delta(\mathbf{x}-\mathbf{x}_0))$ with initial distribution $\delta(\mathbf{x}-\mathbf{x}_0)$ can be computed from the transition probability density $P(\mathbf{x}, t|\mathbf{x}_0, t_0; \delta(\mathbf{x} - \mathbf{x}_0))_{t_0}$. However, for $t \geq t' > t_0$ we have

$$P(\mathbf{x}, t|\mathbf{x}', t'; \delta(\mathbf{x} - \mathbf{x}'))_{t_0} \neq P(\mathbf{x}, t; \delta(\mathbf{x} - \mathbf{x}_0))_{t_0}|_{\mathbf{x}_0 = \mathbf{x}', t_0 = t'} \, . \quad (3.105)$$

This relation tells us that if we determine the transient probability density $P(\mathbf{x}, t; \delta(\mathbf{x}-\mathbf{x}_0))_{t_0}$ of the stochastic process given by (3.103) and then replace $\mathbf{x}_0$ and t_0 by $\mathbf{x}'$ and t', then we do not obtain the transition probability density $P(\mathbf{x}, t|\mathbf{x}', t'; \delta(\mathbf{x} - \mathbf{x}'))_{t_0}$. In other words, transient solutions with delta-distributed initial distributions are not necessarily equivalent to transition probability densities. Roughly speaking, a transient solution involves at most two time variables, namely, t and t_0, whereas a transition probability density in general involves three time variables, t, t', and t_0. This implies that transient solutions do not contain enough information to determine transition probability densities.

This difference between transient solutions and transition probability densities becomes crucial for strongly nonlinear Fokker–Planck equations. The reason for this is that strongly nonlinear Fokker–Planck equations can be mapped to linear Fokker–Planck equations with drift and diffusion coefficients that depend on $P(\mathbf{x}, t; u)$. Since $P(\mathbf{x}, t; u)$ depends on the initial time t_0, strongly nonlinear Fokker–Planck equations can be mapped to linear Fokker–Planck equations with drift and diffusion coefficients that depend on t_0. For this reason, transient solutions of nonlinear Fokker–Planck equations cannot

3.9 Transient Solutions and Transition Probability Densities

be considered as transition probability densities of Markov diffusion processes and do not necessarily satisfy the Chapman–Kolmogorov equation (3.75). This can be explicitly illustrated for transient solutions that can be written as Gaussian distributions.

3.9.2 Gaussian Distributions*

By means of Gaussian time-dependent probability densities we can illustrate that some transition probability densities satisfy the Chapman–Kolmogorov equation and others do not.

Integral Relations

In the context of Gaussian distributions, we often deal with the integrals

$$\int_{-\infty}^{\infty} \exp\left\{-\lambda[x - x_0]^2\right\} dx = \sqrt{\frac{\pi}{\lambda}} \,, \tag{3.106}$$

$$\int_{-\infty}^{\infty} x^2 \exp\left\{-\lambda x^2\right\} dx = -\frac{d}{d\lambda} \int_{-\infty}^{\infty} \exp\{-\lambda x^2\} dx = \frac{1}{2}\sqrt{\frac{\pi}{\lambda^3}} \,, \tag{3.107}$$

$$\int_{-\infty}^{\infty} \exp\left\{-\lambda x^2\right\} \cos(bx) \, dx = \sqrt{\frac{\pi}{\lambda}} \exp\left\{-\frac{b^2}{4\lambda}\right\} \tag{3.108}$$

for $\lambda > 0$ and $x_0, b \in \mathbb{R}$ [79, 333, 627]. For a complex constant $z_0 = x_0 + iy_0$ we conclude that

$$\int_{-\infty}^{\infty} \exp\left\{-\lambda[x - (x_0 + iy_0)]^2\right\} dx = \frac{1}{\sqrt{\lambda}} \int_{-\infty}^{\infty} \exp\left\{-[x' - iy_0']^2\right\} dx'$$

$$= \frac{e^{[y_0']^2}}{\sqrt{\lambda}} \int_{-\infty}^{\infty} e^{-[x']^2} [\cos(2y_0' x') + i \sin(2y_0' x')] \, dx' = \sqrt{\frac{\pi}{\lambda}} \,, \tag{3.109}$$

where we have used the scaled variables $x' = x/\sqrt{\lambda}$ and $y_0' = y_0/\sqrt{\lambda}$ and exploited (3.108) in combination with $\int \exp\{-[x']^2\} \sin(2y_0' x') \, dx' = 0$.

Gaussian Probability Densities

The Gaussian distribution of a one-dimensional random variable $X \in \mathbb{R}$ reads

$$P(x) = \sqrt{\frac{\lambda}{\pi}} \exp\left\{-\lambda[x - x_0]^2\right\} \,, \tag{3.110}$$

for $\lambda > 0$ and $x_0 \in \mathbb{R}$. The normalization constant $\sqrt{\lambda/\pi}$ can be obtained from (3.106). Using (3.106) and (3.107), we can derive the mean value $M_1 = \langle X \rangle$ and the variance $K = \langle (X - \langle X \rangle)^2 \rangle = \langle X^2 \rangle - \langle X \rangle^2$ and find

$$M_1 = \int_{-\infty}^{\infty} x P(x) \, dx = x_0, \quad K = \int_{-\infty}^{\infty} [x-x_0]^2 P(x) \, dx = \frac{1}{2\lambda}. \quad (3.111)$$

Therefore, (3.110) can be written as

$$P(x) = \frac{1}{\sqrt{2\pi K}} \exp\left\{-\frac{(x-x_0)^2}{2K}\right\}. \quad (3.112)$$

Gaussian distributions converge to delta distributions in the limit $\lambda \to \infty$, that is, in the limit $K \to 0$ [430]:

$$\delta(x-x_0) = \lim_{\lambda \to \infty} \sqrt{\frac{\lambda}{\pi}} \exp\left\{-\lambda[x-x_0]^2\right\}, \quad (3.113)$$

see also (3.49). Gaussian distributions are also called normal distributions.

Let us consider now a stochastic process described by the Gaussian conditional probability density

$$P(x,t|x',t') = \frac{1}{\sqrt{2\pi K(t,t')}} \exp\left\{-\frac{[x-x'm(t,t')]^2}{2K(t,t')}\right\}, \quad (3.114)$$

for $t \geq t'$. We assume that the conditions $m(z,z) = 1$, $\lim_{t \to t'} K(t,t') = 0$, and $K(z,z') > 0$ for $z > z'$ hold. Obviously, we have $P(x,t|x',t') > 0$ and we can verify that $P(x,t|x',t')$ satisfies the normalization condition $\int_\Omega P(x,t|x't') \, dx = 1$. In addition, the limiting case $\lim_{t \to t'} P(x,t|x',t') = \delta(x-x')$ holds because of $\lim_{t \to t'} K(t,t') = 0$ and (3.113). In sum, the Gaussian distribution can be regarded as a transition probability density $P(x,t|x',t'; \delta(x-x'))$. Note that m measures the mean of $P(x,t|x',t')$ with respect to averages over x: $m(t,t') = \int_\Omega x P(x,t|x',t') dx$. Likewise, $K(t,t')$ describes the variance of $P(x,t|x',t')$ if we regard $P(x,t|x',t')$ as a distribution of x: $K(t,t') = \int_\Omega x^2 P(x,t|x',t') dx - [m(t,t')]^2$. At issue is now to determine under which conditions the Gaussian transition probability density (3.114) satisfies the Chapman–Kolmogorov equation (3.75) [273].

In the univariate case, the Chapman–Kolmogorov equation (3.75) becomes

$$P(x,t|x'',t'') = \int_\Omega P(x,t|x',t') P(x',t'|x'',t'') \, dx' \quad (3.115)$$

with $t \geq t' \geq t''$. If we substitute (3.114) into (3.115), then the right-hand side of (3.115) reads

$$\text{RHS} = \int_\Omega P(x,t|x',t') P(x',t'|x'',t'') \, dx' =$$
$$\frac{1}{2\pi} \frac{1}{\sqrt{K(t,t')K(t',t'')}} \int_\Omega \exp\left\{-\frac{[x-x'm(t,t')]^2}{2K(t,t')} - \frac{[x'-x''m(t',t'')]^2}{2K(t',t'')}\right\} dx'. \quad (3.116)$$

3.9 Transient Solutions and Transition Probability Densities

Introducing the new variables

$$a = m(t,t'), \quad b = 2K(t,t'), \quad c = m(t',t''), \quad d = 2K(t',t''), \tag{3.117}$$

with $a, b, c, d \in \mathbb{R}$ and $b, d > 0$, we find that (3.116) involves the integral

$$I = \int_{-\infty}^{\infty} \exp\left\{-\frac{[x - ax']^2}{b} - \frac{[x' - cx'']^2}{d}\right\} dx'. \tag{3.118}$$

By means of (3.106) we can carry out the integration with respect to x' and obtain

$$I = \sqrt{\frac{\pi b d}{a^2 d + b}} \exp\left\{-\frac{[x - acx'']^2}{a^2 d + b}\right\}. \tag{3.119}$$

Using (3.116,...,3.119), we obtain

$$\text{RHS} = \frac{1}{2\pi} \frac{1}{\sqrt{K(t,t')K(t',t'')}} \sqrt{\frac{\pi b d}{a^2 d + b}} \exp\left\{-\frac{[x - acx'']^2}{a^2 d + b}\right\}, \tag{3.120}$$

with $bd = 4K(t,t')K(t',t'')$ and $a^2 d + b = 2K(t,t') + 2m^2(t,t')K(t',t'')$. Consequently, (3.120) explicitly reads

$$\text{RHS} = \sqrt{\frac{1}{2\pi[K(t,t') + m^2(t,t')K(t',t'')]}} \exp\left\{-\frac{[x - x''m(t,t')m(t',t'')]^2}{2[K(t,t') + m^2(t,t')K(t',t'')]}\right\}. \tag{3.121}$$

For the Gaussian distribution (3.114) the left-hand side of the Chapman–Kolmogorov equation (3.115) reads

$$\text{LHS} = \sqrt{\frac{1}{2\pi K(t,t'')}} \exp\left\{-\frac{[x - x''m(t,t'')]^2}{2K(t,t'')}\right\}. \tag{3.122}$$

The Chapman–Kolmogorov equation is satisfied by $P(x,t|x',t')$ if and only if the left- and the right-hand sides are equivalent for all x, x''. Let us put $x = x'' = 0$. Then, from the requirement RHS = LHS we obtain

$$K(t,t'') = K(t,t') + m^2(t,t')K(t',t''). \tag{3.123}$$

Substituting this result into (3.121), the equivalence RHS = LHS reads

$$\exp\left\{-\frac{[x - x''m(t,t'')]^2}{2K(t,t'')}\right\} = \exp\left\{-\frac{[x - x''m(t,t')m(t',t'')]^2}{2K(t,t'')}\right\} \tag{3.124}$$

for all x, x''. Now we let us use $x'' = 1$ and $x = m(t,t'')$. Then, (3.124) becomes

$$1 = \exp\left\{-\frac{[m(t,t'') - m(t,t')m(t',t'')]^2}{2K(t,t'')}\right\}. \tag{3.125}$$

Since we have $K(t,t'') > 0$ for $t > t''$, we conclude that $[m(t,t'') - m(t,t')m(t',t'')]^2 = 0$ for $t > t''$ holds, which implies

$$m(t,t'') = m(t,t')m(t',t''). \tag{3.126}$$

So far, we have shown that (3.123) and (3.126) are necessary conditions for the Gaussian probability density (3.114) to satisfy the Chapman–Kolmogorov equation. In fact, we can show that (3.123) and (3.126) are also sufficient conditions for the Gaussian distribution (3.114) to satisfy the Chapman–Kolmogorov equation. Substituting (3.123) and (3.126) into (3.121) and (3.122) we realize that the expressions RHS and LHS are equivalent.

We consider now a larger class of Gaussian transition probability densities

$$P(x,t|x',t') = \frac{1}{\sqrt{2\pi K(t,t')}} \exp\left\{-\frac{[x - \int_{t'}^{t} m(t,z)F(z)\,dz - x'm(t,t')]^2}{2K(t,t')}\right\}, \tag{3.127}$$

where K and m satisfy (3.123) and (3.126). Then, the right-hand side of the Chapman–Kolmogorov equation is given by

$$\text{RHS} = \int_{\Omega} P(x,t|x',t')P(x',t'|x'',t'')\,dx' = \frac{1}{2\pi}\frac{I}{\sqrt{K(t,t')K(t',t'')}} \tag{3.128}$$

with

$$I = \int_{\Omega} \exp\left\{-\frac{[x - \int_{t'}^{t} m(t,z)F(z)\,dz - x'm(t,t')]^2}{2K(t,t')} -\frac{[x' - \int_{t''}^{t'} m(t',z)F(z)\,dz - x''m(t',t'')]^2}{2K(t',t'')}\right\} dx'. \tag{3.129}$$

Using $y' = x' - \int_{t''}^{t'} m(t',z)F(z)\,dz$, we obtain

$$I = \int_{\Omega} \exp\left\{-\frac{[x - \int_{t'}^{t} m(t,z)F(z)\,dz - m(t,t')\int_{t''}^{t'} m(t',z)F(z)\,dz - y'm(t,t')]^2}{2K(t,t')} -\frac{[y' - x''m(t',t'')]^2}{2K(t',t'')}\right\} dy'. \tag{3.130}$$

From (3.126) and $\int_{t''}^{t} = \int_{t'}^{t} + \int_{t''}^{t'}$ it follows that

$$\int_{t'}^{t} m(t,z)F(z)\,dz + m(t,t')\int_{t''}^{t'} m(t',z)F(z)\,dz = \int_{t''}^{t} m(t,z)F(z)\,dz. \tag{3.131}$$

3.9 Transient Solutions and Transition Probability Densities

Consequently, we have

$$I = \int_\Omega \exp\left\{-\frac{[x - \int_{t''}^t m(t,z)F(z)\,dz - y'm(t,t')]^2}{2K(t,t')} - \frac{[y' - x''m(t',t'')]^2}{2K(t',t'')}\right\} dy' . \tag{3.132}$$

By means of $y = x - \int_{t''}^t m(t,z)F(z)\,dz$, we can simplify this expression and obtain

$$I = \int_\Omega \exp\left\{-\frac{[y - y'm(t,t')]^2}{2K(t,t')} - \frac{[y' - x''m(t',t'')]^2}{2K(t',t'')}\right\} dy' . \tag{3.133}$$

Equation (3.133) exhibits the form of the integral (3.118) and, consequently, can be transformed into

$$I = \sqrt{\frac{\pi K(t,t')K(t',t'')}{m^2(t,t')K(t',t'') + K(t,t')}} \exp\left\{-\frac{[y - m(t,t')m(t',t'')x'']^2}{m^2(t,t')K(t',t'') + K(t,t')}\right\}, \tag{3.134}$$

see (3.119). Finally, if we use (3.128), (3.123), and (3.126), we get

$$\begin{aligned}
\text{RHS} &= \int_\Omega P(x,t|x',t')P(x',t'|x'',t'')\,dx' \\
&= \frac{1}{\sqrt{2\pi K(t,t'')}} \exp\left\{-\frac{[y - x''m(t,t'')]^2}{2K(t,t'')}\right\} \\
&= \frac{1}{\sqrt{2\pi K(t,t'')}} \exp\left\{-\frac{[x - \int_{t''}^t m(t,z)F(z)\,dz - x''m(t,t'')]^2}{2K(t,t'')}\right\} .
\end{aligned} \tag{3.135}$$

From (3.127) and (3.135) it follows that the expression RHS is equivalent to the left-hand side of (3.115): RHS = LHS = $P(x,t|x'',t'')$. That is, Gaussian distributions that can be cast into the form (3.127) satisfy the Chapman–Kolmogorov equation (3.115) if $K(t,t')$ and $m(t,t')$ satisfy (3.123) and (3.126). Let us next discuss some examples of Gaussian transition probability densities.

3.9.3 Purely Random Processes

In the special case

$$m(t,t') = 0 , \tag{3.136}$$
$$K(t,t') = K(t) \geq 0 , \tag{3.137}$$

the transition probability density $P(x,t|x',t')$ given by (3.114) becomes independent of t' and reads

$$P(x,t|x',t') = \frac{1}{\sqrt{2\pi K(t)}} \exp\left\{-\frac{x^2}{2K(t)}\right\} . \tag{3.138}$$

That is, we deal with a purely random process [498]. The functions $m(t,t')$ and $K(t,t')$ satisfy (3.123) and (3.126). Consequently, (3.138) satisfies the Chapman–Kolmogorov equation (3.115).

3.9.4 Wiener Processes

A stochastic process described by an initial distribution $P(x,t_0;u) = u(x)$ and the transition probability density (3.114) with

$$m(t,t') = 1 , \tag{3.139}$$
$$K(t,t') = 2Q\Delta t \tag{3.140}$$

for $\Delta t = t - t' \geq 0$ and $Q > 0$ is referred to as a Wiener process [221, 498]. There is a linear increase of the variance K with time: $K \propto \Delta t$. The fluctuation strength Q determines the slope of this increase. The limiting case $\lim_{t \to t'} K(t,t') = 0$ holds. For $t - t' \to \infty$ the variance becomes infinite. The mean value $m(t,t')$ satisfies (3.126). The variance $K(t,t')$ satisfies (3.123):

$$K(t,t') + m^2(t,t')K(t',t'') = 2Q(t - t'') = K(t,t'') . \tag{3.141}$$

Consequently, the transition probability density of a Wiener process satisfies the Chapman–Kolmogorov equation (3.115). We can verify that the transition probability density satisfies the linear Fokker–Planck equation

$$\frac{\partial}{\partial t} P(x,t|x',t') = Q \frac{\partial^2}{\partial x^2} P(x,t|x',t') . \tag{3.142}$$

In addition, the probability densities $P(x,t;u) = \int_\Omega P(x,t|x',t_0)u(x')\,dx'$ satisfy the linear Fokker–Planck equation

$$\frac{\partial}{\partial t} P(x,t;u) = Q \frac{\partial^2}{\partial x^2} P(x,t;u) . \tag{3.143}$$

3.9.5 Ornstein–Uhlenbeck Processes

If $m(t,t')$ and $K(t,t')$ are given by

$$m(t,t') = e^{-\gamma(t-t')} , \tag{3.144}$$
$$K(t,t') = \frac{Q}{\gamma}\left[1 - e^{-2\gamma(t-t')}\right] \tag{3.145}$$

for $t \geq t'$ with $\gamma > 0$, $Q > 0$, and $P(x,t_0) = u(x)$, we deal with an Ornstein–Uhlenbeck process. Note that we have $m(t,t) = 1$ and $\lim_{t \to t'} K(t,t') = 0$.

For $t - t' \to \infty$ we obtain $m = 0$ and $K = Q/\gamma$. For the mean value $m(t,t')$ the relation

$$m(t,t')m(t',t'') = e^{-\gamma(t-t')}e^{-\gamma(t'-t'')} = e^{-\gamma(t-t'')} = m(t,t'') \qquad (3.146)$$

can be found. Consequently, (3.126) is satisfied. Equation (3.123) is satisfied as well, which can be shown as follows:

$$\begin{aligned} K(t,t') &+ m^2(t,t')K(t',t'') \\ &= \frac{Q}{\gamma}\left[1 - e^{-2\gamma(t-t')}\right] + \frac{Q}{\gamma}e^{-2\gamma(t-t')}\left[1 - e^{-2\gamma(t'-t'')}\right] \\ &= \frac{Q}{\gamma}\left[1 - e^{-2\gamma(t-t')} + e^{-2\gamma(t-t')} - e^{-2\gamma(t-t')}e^{-2\gamma(t'-t'')}\right] \\ &= \frac{Q}{\gamma}\left[1 - e^{-2\gamma(t-t'')}\right] = K(t,t'') \: . \end{aligned} \qquad (3.147)$$

Therefore, we conclude that the transition probability density $P(x,t|x',t')$ of Ornstein–Uhlenbeck processes solves the Chapman–Kolmogorov equation. Ornstein–Uhlenbeck processes are completely described by an initial distribution $u(x)$ and the transition probability density $P(x,t|x',t')$ given by (3.114), (3.144), and (3.145) because from these quantities the hierarchy of joint probability densities (3.83) can be obtained. In addition, the probability densities $P(x,t|x',t')$ and $P(x,t;u)$ of Ornstein–Uhlenbeck processes satisfy the linear Fokker–Planck equations

$$\frac{\partial}{\partial t}P(x,t|x',t') = \gamma\frac{\partial}{\partial x}xP(x,t|x',t') + Q\frac{\partial^2}{\partial x^2}P(x,t|x',t') \: , \qquad (3.148)$$

and

$$\frac{\partial}{\partial t}P(x,t;u) = \gamma\frac{\partial}{\partial x}xP(x,t;u) + Q\frac{\partial^2}{\partial x^2}P(x,t;u) \: . \qquad (3.149)$$

3.9.6 Transient Solutions of Nonlinear Fokker–Planck Equations: Two Examples

We consider now the evolution of the probability densities $P(x,t;\delta(x-x_0))_{t_0}$ of two stochastic processes defined by nonlinear Fokker–Planck equations [174]. We consider again Gaussian probability densities of the form

$$P(x,t;\delta(x-x_0))_{t_0} = \frac{1}{\sqrt{2\pi K(t)}}\exp\left\{-\frac{[x - x_0 m(t)]^2}{2K(t)}\right\} \qquad (3.150)$$

for $t \geq t_0$, $m(t_0) = 1$, and $\lim_{t \to t_0} K(t) = 0$. Our objective will be to examine whether or not for these processes $P(x,t;\delta(x-x_0))_{t_0}$ can be regarded as a Markov transition probability density $P(x,t|x',t')$ that satisfies the Chapman–Kolmogorov equation (3.115). To this end, we will put $P(x,t|x',t') = P(x,t;\delta(x-x'))_{t'}$ and examine just as in the previous paragraph if (3.123) and (3.126) are satisfied.

Transient Solutions of the Shimizu–Yamada Model

The Shimizu–Yamada model will be introduced in Sect. 3.10, see (3.165). A transient solution of the Shimizu–Yamada model (3.165) is given by the Gaussian distribution (3.150) and

$$m(t) = e^{-\gamma(t-t_0)}, \tag{3.151}$$

$$K(t) = \frac{Q}{\gamma+\kappa}\left[1 - e^{-2(\gamma+\kappa)(t-t')}\right]. \tag{3.152}$$

Note that we have $\kappa \geq 0$, $\gamma > -\kappa$, and $Q > 0$. For $\kappa = 0$ (3.150-3.152) describe the Gaussian transition probability density of Ornstein–Uhlenbeck processes. If we interpret the transient solution as a transition probability density, replace in (3.151) the variable t_0 by t', and put $m(t,t') = m(t)$, then $m(t,t')$ solves (3.126):

$$m(t,t')m(t',t'') = e^{-\gamma(t-t')}e^{-\gamma(t'-t'')} = e^{-\gamma(t-t'')} = m(t,t''). \tag{3.153}$$

Likewise, if we replace in (3.152) the variable t_0 by t' and put $K(t,t') = K(t)$, then the right-hand side of (3.123) becomes

$$K(t,t') + m^2(t,t')K(t',t'')$$
$$= \frac{Q}{\gamma+\kappa}\left[1 - e^{-2(\gamma+\kappa)(t-t')}\right] + \frac{Q}{\gamma+\kappa}e^{-2\gamma(t-t')}\left[1 - e^{-2(\gamma+\kappa)(t'-t'')}\right]$$
$$= \frac{Q}{\gamma+\kappa}\left[1 - e^{-2(\gamma+\kappa)(t-t')} + e^{-2\gamma(t-t')} - e^{-2\gamma(t-t')}e^{-2(\gamma+\kappa)(t'-t'')}\right]$$
$$= \frac{Q}{\gamma+\kappa}e^{-2(\gamma+\kappa)(t-t')}\left[e^{2\kappa(t-t')} - 1\right] + \frac{Q}{\gamma+\kappa}\left[1 - e^{2\kappa(t-t')}e^{-2(\gamma+\kappa)(t-t'')}\right], \tag{3.154}$$

whereas the left-hand side of (3.123) is given by $K(t,t'') = Q[\gamma+\kappa]^{-1}[1-\exp\{-2(\gamma+\kappa)(t-t'')\}]$. Consequently, (3.123) holds if the relation

$$1 - e^{-2(\gamma+\kappa)(t-t'')} = e^{-2(\gamma+\kappa)(t-t')}\left[e^{2\kappa(t-t')} - 1\right] + \left[1 - e^{2\kappa(t-t')}e^{-2(\gamma+\kappa)(t-t'')}\right] \tag{3.155}$$

is satisfied for all $t \geq t' \geq t''$. Equation (3.155) can be written as

$$e^{-2(\gamma+\kappa)(t''-t')}\left[e^{2\kappa(t-t')} - 1\right] = \left[e^{2\kappa(t-t')} - 1\right]. \tag{3.156}$$

For $\kappa = 0$ (3.156) is satisfied for all t, t'' with $t \geq t''$. That is, we reobtain the previously derived result, which says that Ornstein–Uhlenbeck processes satisfy the Chapman–Kolmogorov equation. For $\kappa > 0$ we have $\exp\{2\kappa(t-t')\} \neq 1$. Then, (3.156) becomes

$$e^{-2(\gamma+\kappa)(t''-t')} = 1. \tag{3.157}$$

3.9 Transient Solutions and Transition Probability Densities 65

Equation (3.157) is only satisfied for $t' = t''$. Consequently, for $\kappa > 0$ the transient probability density $P(x, t; \delta(x - x_0))_{t_0}$ that is defined by (3.150-3.152) cannot be regarded as a Markov transition probability density $P(x, t|x', t')$ when replacing (x_0, t_0) by (x', t').

Transient Solutions of the Gaussian Entropy Fokker–Planck Equation

Our second example is about transient solutions of the nonlinear Fokker–Planck equation (6.217) that will be discussed in Sect. 6.5.3. Substituting (3.150) into (6.217), we get

$$m(t) = e^{-\gamma(t-t_0)}, \qquad (3.158)$$

$$K(t) = [2\pi e]^{(1-q)/(1+q)} \left[\frac{Q}{\gamma}\right]^{2/(1+q)} \left[1 - e^{-(1+q)\gamma(t-t_0)}\right]^{2/(1+q)} \qquad (3.159)$$

(for details see Sect. 6.5.3). The parameters satisfy the inequalities $q \geq 0$, $\gamma > 0$, and $Q > 0$. For $q = 1$ the nonlinear Fokker–Planck equation (6.217) becomes linear and reduces to the Fokker–Planck equation (3.149) of Ornstein–Uhlenbeck processes. Likewise, for $q = 1$ (3.150) with (3.158) and (3.159) corresponds to the transition probability density of Ornstein–Uhlenbeck processes and, consequently, in this case the Chapman–Kolmogorov equation (3.115) is satisfied. Let us examine now whether or not for $q \neq 1$ the Gaussian transition probability density (3.150) with cumulants given by (3.158) and (3.159) satisfies the Chapman–Kolmogorov equation (3.115). To this end, we replace in (3.158) and (3.159) the variable t_0 by t', put $m(t, t') = m(t)$ and $K(t, t') = K(t)$, and substitute the expressions thus obtained into (3.123) and (3.126). In particular, we find $m(t, t') = \exp\{-\gamma(t - t')\}$, which is equivalent to the function $m(t, t')$ of an Ornstein–Uhlenbeck process and implies that (3.126) is satisfied. The right-hand side of (3.123) reads

$$K(t,t') + m^2(t,t')K(t',t'') = [2\pi e]^{(1-q)/(1+q)} \left[\frac{Q}{\gamma}\right]^{2/(1+q)} B(t,t',t''), \qquad (3.160)$$

with B defined by

$$B = \left[1 - e^{-(1+q)\gamma(t-t')}\right]^{2/(1+q)} + e^{-2\gamma(t-t')}\left[1 - e^{-(1+q)\gamma(t'-t'')}\right]^{2/(1+q)}$$
$$= \left[1 - e^{-(1+q)\gamma(t-t')}\right]^{2/(1+q)} + \left[e^{-(1+q)\gamma(t-t')} - e^{-(1+q)\gamma(t'-t'')}\right]^{2/(1+q)}. \qquad (3.161)$$

Using (3.159), we can determine the left-hand side of (3.123). Comparing the left-hand side of (3.123) with the right-hand side of (3.160), we obtain

$B(t,t',t'') = [1 - \exp\{-(1+q)\gamma(t-t'')\}]^{2/(1+q)}$ for $t \geq t' \geq t''$. This relation reduces to

$$[1 - e^{-1}]^{2/(1+q)} + [e^{-1} - e^{-2}]^{2/(1+q)} = [1 - e^{-2}]^{2/(1+q)}, \qquad (3.162)$$

for $t - t' = 1/[(1+q)\gamma]$, $t' - t'' = 1/[(1+q)\gamma]$, and $t - t'' = 2/[(1+q)\gamma]$. For the sake of convenience, let us introduce the variable $\alpha = 2/(1+q) \in (0,2]$. Then, (3.162) reads

$$[1 - e^{-1}]^\alpha (1 + e^{-\alpha}) = [1 - e^{-2}]^\alpha. \qquad (3.163)$$

Using $[1 - e^{-2}]^\alpha = [(1 - e^{-1})(1 + e^{-1})]^\alpha$, we can transform (3.163) into

$$1 + e^{-\alpha} = [1 + e^{-1}]^\alpha. \qquad (3.164)$$

Equation (3.164) is satisfied for $\alpha = 1 \Rightarrow q = 1$. Let us show that $\alpha = 1$ is the only solution of (3.164). To this end, we consider the functions $f_1(\alpha) = 1 + e^{-\alpha}$ and $f_2(\alpha) = [1 + e^{-1}]^\alpha$ with the function values $f_1(0) = 2$ and $f_2(0) = 1$. For $\alpha > 0$ the derivatives of f_1 and f_2 satisfy the inequalities $df_1/d\alpha = -\alpha e^{-\alpha} < 0$ and $df_2/d\alpha = [1 + e^{-1}]^\alpha \ln(1 + e^{-1}) > 0$, respectively. Since f_2 is a strictly monotonically increasing function, f_1 is a strictly monotonically decreasing function, and the inequality $f_2(0) < f_1(0)$ holds, we conclude that there is one and only one intersection point α given by $f_1(\alpha) = f_2(\alpha)$ and this intersection point is at $\alpha = 1$. Therefore, we conclude that for $\alpha \neq 1$, that is, for $q \neq 1$ the function $P(x,t|x',t') = P(x,t;\delta(x-x'))_{t'}$ described by (3.150), (3.158), and (3.159) does not satisfy the Chapman–Kolmogorov equation (3.115).

3.10 Shimizu–Yamada Model – Transition Probabilities and Transient Solutions

In line with work by Shimizu and Yamada [528, 529] and others [648], we consider the nonlinear Fokker–Planck equation

$$\frac{\partial}{\partial t} P(x,t;u) = \frac{\partial}{\partial x}\left[\gamma x + \kappa(x - \langle X \rangle)\right] P + Q \frac{\partial^2}{\partial x^2} P, \qquad (3.165)$$

for $\kappa \geq 0$, $\gamma > 0$, $t \geq t_0$, and $x \in \mathbb{R}$. We will refer to this Fokker–Planck equation as Shimizu–Yamada model and return to the model in Sects. 5.2.1 and 5.5.4. For $\kappa = 0$, (3.165) reduces to the linear Fokker–Planck equation of Ornstein–Uhlenbeck processes. In what follows, we will consider the case $\kappa > 0$. For (3.165) an exact time-dependent solution can be found. Substituting

$$P(x,t;\delta(x-x_0)) = \frac{1}{\sqrt{2\pi K(t)}} \exp\left\{-\frac{[x - x_0 m(t)]^2}{2K(t)}\right\} \qquad (3.166)$$

3.10 Shimizu–Yamada Model – Transient Solutions

into (3.165) gives us [174, 648]

$$m(t) = \exp\{-\gamma(t-t_0)\}, \tag{3.167}$$

$$K(t) = \frac{Q}{\gamma+\kappa}[1 - e^{-2(\gamma+\kappa)(t-t_0)}]. \tag{3.168}$$

For $\gamma > 0$ the time-dependent Gaussian distribution converges to a stationary one in the long time limit. As shown in Sect. 3.2 we can assign to the nonlinear Fokker–Planck equation (3.165) a linear evolution equation for the transition probability density

$$\frac{\partial}{\partial t}P(x,t|x',t';u) = \frac{\partial}{\partial x}\left[\gamma x + \kappa(x - \langle X \rangle_{P(x,t;u)})\right]P(x,t|x',t';u)$$
$$+ Q\frac{\partial^2}{\partial x^2}P(x,t|x',t';u). \tag{3.169}$$

The evolution of $M_1(t) = \langle X \rangle$ can be computed from (3.165) and satisfies $dM_1(t)/dt = -\gamma M_1(t)$ and depends on the mean value of the initial distribution $u(x)$. We have

$$M_1(t;u) = \exp\{-\gamma(t-t_0)\}\int_\Omega xu(x)\,dx. \tag{3.170}$$

This implies that the drift coefficient D_1' of (3.169) reads

$$D_1'(x,t,u,t_0) = D_1(x,t,\langle X \rangle) = -(\gamma+\kappa)x + \kappa\exp\{-\gamma(t-t_0)\}\int_\Omega xu(x)\,dx. \tag{3.171}$$

Consequently, D_1' can be regarded as the first Kramers–Moyal coefficient of a Markov diffusion process and solutions of (3.169) can be regarded as solutions of the linear Fokker–Planck equation

$$\frac{\partial}{\partial t}P(x,t|x',t';u) = -\frac{\partial}{\partial x}D_1'(x,t,u,t_0)P(x,t|x',t';u) + Q\frac{\partial^2}{\partial x^2}P(x,t|x',t';u). \tag{3.172}$$

This, in turn, implies that (3.165) is a strongly nonlinear Fokker–Planck equation. For the time-dependent probability density (3.166) we obtain $M_1(t) = x_0\exp\{-\gamma(t-t_0)\}$, which leads to

$$\frac{\partial}{\partial t}P(x,t|x',t';\delta(x-x_0)) = \frac{\partial}{\partial x}\left[(\gamma+\kappa)x - \kappa x_0 e^{-\gamma(t-t_0)}\right]P + Q\frac{\partial^2}{\partial x^2}P. \tag{3.173}$$

Equation (3.173) is a linear Fokker–Planck equation describing an Ornstein–Uhlenbeck process subjected to a time-dependent driving force $f(t-t_0) = \kappa x_0\exp\{-\gamma(t-t_0)\}$. By means of a moving frame transformation (similar to the one that will be discussed in Sect. 7.2.1), we can show that (3.173) is solved by

$$P(x,t|x',t';\delta(x-x_0)) =$$
$$\sqrt{\frac{1}{2\pi K(t,t')}} \exp\left\{-\frac{[x-g(t,t',t_0)-x'm(t,t')]^2}{2K(t,t')}\right\} , \quad (3.174)$$

with

$$m(t,t') = \exp\{-(\gamma+\kappa)(t-t')\} , \quad (3.175)$$
$$K(t,t') = \frac{Q}{\gamma+\kappa}[1-e^{-2(\gamma+\kappa)(t-t')}] , \quad (3.176)$$
$$g(t,t',t_0) = x_0\left[\exp\{-\gamma(t-t_0)\} - \exp\{-(\gamma+\kappa)t+\gamma t_0+\kappa t'\}\right] . \quad (3.177)$$

Comparing (3.166-3.168) with (3.174-3.177), we realize that

$$P(x,t|x',t';\delta(x-x_0)) \neq P(x,t;\delta(x-x_0))_{t_0}|_{x_0=x',t_0=t'} , \quad (3.178)$$

for $t' > t_0$. That is, for $t' > t_0$ transition and transient probability densities differ from each other (see also Sect. 3.9.1). However, for $t' = t_0$ the expression $g(t,t',t_0) + x_0 m(t,t')$ becomes

$$g(t,t_0,t_0) + x_0 m(t,t_0)$$
$$= x_0\left[\exp\{-\gamma(t-t_0)\} - \exp\{-(\gamma+\kappa)(t-t_0)\}\right] + x_0\exp\{-(\gamma+\kappa)(t-t_0)\}$$
$$= x_0\exp\{-\gamma(t-t_0)\} = x_0 m(t) , \quad (3.179)$$

leading to

$$P(x,t|x_0,t_0;\delta(x-x_0)) = P(x,t;\delta(x-x_0))_{t_0} . \quad (3.180)$$

In Sect. 3.9.2 we have shown that the transient solution given by (3.166-3.168) violates the Chapman–Kolmogorov equation if it is interpreted as a transition probability density. In contrast, the transition probability density given by (3.174-3.177) satisfies the Chapman–Kolmogorov equation because it can be cast into the form of the Gaussian transition probability density (3.127). With (3.166) and (3.174) at hand, we can explicitly describe Markov diffusion processes defined by the Shimizu–Yamada model for initial distributions $u(x_1) = \delta(x_1 - x_0)$:

$$P(x_n,t_n;\ldots;x_1,t_1;\delta(x_1-x_0)) =$$
$$P(x_n,t_n|x_{n-1},t_{n-1};\delta(x_1-x_0))\cdots P(x_2,t_2|x_1,t_1;\delta(x_1-x_0))P(x_1,t_1;\delta(x_1-x_0)) . \quad (3.181)$$

For arbitrary initial distributions $u(x)$ transient solutions $P(x,t;u)$ are defined by (3.165) and $M_1(t) = \langle X \rangle$ satisfies (3.170), while the transition probability density $P(x,t|x',t';u)$ satisfies (3.172). If we proceed just as in the aforementioned special case $u(x) = \delta(x-x_0)$, we find that $P(x,t|x',t';u)$ is given by

3.10 Shimizu–Yamada Model – Transient Solutions

$$P(x,t|x',t';u) = \sqrt{\frac{1}{2\pi K(t,t')}} \exp\left\{-\frac{[x - g(t,t',t_0,u) - x'm(t,t')]^2}{2K(t,t')}\right\} , \quad (3.182)$$

with

$$m(t,t') = \exp\{-(\gamma + \kappa)(t - t')\} , \quad (3.183)$$

$$K(t,t') = \frac{Q}{\gamma + \kappa}[1 - e^{-2(\gamma+\kappa)(t-t')}] , \quad (3.184)$$

$$g(t,t',t_0,u) = [\exp\{-\gamma(t - t_0)\} - \exp\{-(\gamma + \kappa)t + \gamma t_0 + \kappa t'\}] \int_\Omega x\, u(x)\, dx , \quad (3.185)$$

see (3.174,...,3.177). In sum, the strongly nonlinear Fokker–Planck equation described by (3.165) and (3.169) defines Markov diffusion processes with joint probability densities

$$P(x_n,t_n;\ldots;x_1,t_1;u) = P(x_n,t_n|x_{n-1},t_{n-1};u)\cdots P(x_2,t_2|x_1,t_1;u)P(x_1,t_1;u) , \quad (3.186)$$

where $P(x_1,t_1;u)$ is a solution of (3.165) and $P(x,t|x',t';u)$ is given by (3.182). In the stationary case, that is, for $t \to \infty$, $t' \to \infty$, and $t-t' = z > 0$ we have $g(t,t',t_0,u) = 0$ (hint: $\exp\{-(\gamma + \kappa)t + \kappa t'\} = 0$ because of $\gamma > 0$), which implies that $P(x,t|x',t';u)$ converges to the stationary transition probability density

$$P_{st}(x,t|x',t') = \sqrt{\frac{1}{2\pi K(t,t')}} \exp\left\{-\frac{[x - x'm(t,t')]^2}{2K(t,t')}\right\} , \quad (3.187)$$

with $m(t,t')$ and $K(t,t')$ defined above. This transition probability density only depends on $t - t'$ and corresponds to the transition probability density of an Ornstein–Uhlenbeck process with a damping constant $\gamma + \kappa$.

While (3.186) describes the stochastic processes under consideration in terms of distribution functions, the Langevin equation

$$\frac{d}{dt}X(t) = -(\gamma + \kappa)X(t) + \kappa \langle X(t)\rangle + \sqrt{Q}\Gamma(t) , \quad (3.188)$$

for $t \geq t_0$ and $X(t_0)$ distributed like $u(x)$ describes the very same processes in terms of stochastic trajectories. For $X(t_0) = x_0$ (i.e., for $u(x) = \delta(x - x_0)$) from (3.166) and (3.174) we can compute the probability density $P(x,t;x',t';\delta(x - x_0))$ and the correlation function $\langle X(t)X(t')\rangle$. Thus, we obtain

$$\langle X(t)X(t')\rangle = g(t,t',t_0)M_1(t') + e^{-(\gamma+\kappa)(t-t')}[K(t') + M_1(t')^2] , \quad (3.189)$$

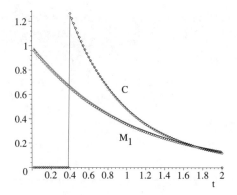

Fig. 3.2. $M_1(t)$ and $C(t,t') = \langle X(t)X(t') \rangle$ as functions of t. Solid lines represent analytical results obtained from (3.167) and (3.189). Diamonds represent results obtained from a simulation of (3.188) using the averaging method for self-consistent Langevin equations. The autocorrelation function $\langle X(t)X(t') \rangle$ is given for $t \geq t'$. For $t < t'$ we have put $\langle X(t)X(t') \rangle = 0$. Parameters: $Q = 2$, $x_0 = 1$, $\gamma = \kappa = 1$, and $t' = 0.4$ ($L = 10^6$, $\Delta t = 0.2$, w_n via Box–Muller); reprinted from [188], © 2004, with permission from Elsevier

with $M_1(t') = x_0 \exp\{-\gamma(t - t_0)\}$ and K and g, respectively, defined by (3.168) and (3.177). We solve numerically (3.188) and compute $\langle X(t)X(t') \rangle$ by means of $\langle X(t)X(t') \rangle = L^{-1} \sum_{k=1}^{L} X^l(t) X^l(t')$ for large L, where $X^l(t)$ are realizations of $X(t)$, and compare the result with the analytical expression (3.189). Figure 3.2 shows $M_1(t)$ and $\langle X(t)X(t') \rangle$ as predicted by the theory of nonlinear Fokker–Planck equations and the corresponding quantities as obtained from a simulation of the Langevin equation (3.188). Finally, we would like to note that in the stationary case from (3.189) and $g = 0$ it follows that

$$\lim_{t,t' \to \infty} \langle X(t)X(t') \rangle = \langle X(t+z)X(t) \rangle_{\text{st}} = K_{\text{st}} \exp\{-(\gamma + \kappa)z\} \,. \quad (3.190)$$

That is, the stationary autocorrelation function $\langle X(t+z)X(t) \rangle_{\text{st}}$ decreases exponentially from $K_{\text{st}} = Q/(\gamma + \kappa)$ to zero as a function of z.

3.11 Fluctuation–Dissipation Theorem

In Sect. 1.1.1 we have argued that fluctuating forces arise from the interactions between systems and their environments. In this context, we have stated that these interactions can also lead to the emergence of damping forces. When fluctuating forces and damping forces emerge due to the very same mechanism, then we may speculate that they are correlated to each other. In fact, the relationship between fluctuating forces and damping forces

3.11 Fluctuation–Dissipation Theorem

can often be described in terms of so-called fluctuation–dissipation theorems. Fluctuation–dissipation theorems relate dissipative properties of systems as described by response functions, damping constants and transport coefficients to the properties of fluctuating forces as described by autocorrelation functions and noise amplitudes and can be found for equilibrium systems (Green–Kubo relations) [89, 240, 360] and stochastic systems described by Fokker–Planck equations [10, 221, 254, 498]. In addition, there are several approaches to derive fluctuation–dissipation theorems for stochastic systems described by nonlinear Fokker–Planck equations [183, 301, 523]. In line with these studies, we first consider the Shimizu–Yamada model (3.165) supplemented with an external driving force $f(t)$:

$$\frac{\partial}{\partial t} P(x,t;u) = \frac{\partial}{\partial x}\left[\gamma x + \kappa(x - \langle X \rangle) - f(t)\right]P + Q\frac{\partial^2}{\partial x^2}P . \quad (3.191)$$

The driving force is assume to be a small quantity such that linear response theory can be applied. That is, we are looking for solutions of (3.191) of the form $P(x,t; P_{0,\text{st}}) = P_{0,\text{st}}(x) + \epsilon(x,t)$, where $P_{0,\text{st}}$ is a stationary solution of (3.191) for $f = 0$ (unperturbed system) and $\epsilon(x,t)$ is a small deviation from $P_{0,\text{st}}$. It is clear that the mean value $\langle X(t) \rangle$ of (3.191) evolves like $\mathrm{d}\langle X \rangle/\mathrm{d}t = -\gamma \langle X \rangle + f(t)$. We focus now on the evolution of the deviation $\langle X \rangle_\epsilon$ of the mean value from the stationary mean value $\langle X \rangle_{0,\text{st}}$ of the reference state $P_{0,\text{st}}$. Here, $\langle X \rangle_\epsilon$ is defined by by $\langle X(t) \rangle_\epsilon = \int_\Omega x\epsilon(x,t)\,\mathrm{d}x$. Due to the linearity of the problem, we conclude that $\langle X \rangle_\epsilon$ evolves just like $\langle X(t) \rangle$. That is, we have

$$\frac{\mathrm{d}}{\mathrm{d}t}\langle X \rangle_\epsilon = -\gamma \langle X \rangle_\epsilon + f(t) . \quad (3.192)$$

This evolution equation determines the response of the order parameter $\langle X \rangle(t) = \langle X \rangle_{0,\text{st}} + \langle X(t) \rangle_\epsilon$ to a small driving force $f(t)$ and may be solved explicitly like

$$\langle X(t) \rangle_\epsilon = \int_0^t \exp\{-\gamma(t-t')\}f(t')\,\mathrm{d}x . \quad (3.193)$$

Equation (3.193) can be written in terms of the stationary autocorrelation function $\langle X(t)X(t') \rangle_{0,\text{st}}$ of the unperturbed systems, which is explicitly given by (3.190). We get

$$\langle X(t) \rangle_\epsilon = \int_0^t [f(t') + \kappa \langle X(t') \rangle_\epsilon] \underbrace{\left(-\frac{1}{Q}\frac{\partial}{\partial t}\langle X(t)X(t') \rangle_{0,\text{st}}\right)}_{G(t-t')} \,\mathrm{d}t' . \quad (3.194)$$

In fact, substituting (3.190) into (3.194) and differentiating the left- and right-hand sides of (3.194) with respect to t we obtain (3.192). The response function $G(t-t')$ has two faces. On the one hand, it reflects dissipative properties of our system [89], which is also indicated by the fact that it involves

3 Strongly Nonlinear Fokker–Planck Equations

the damping constant γ (see (3.190)). On the other hand, it is related to the second-order fluctuations of the unperturbed system as expressed by the autocorrelation function $\langle X(t)X(t')\rangle_{0,\text{st}}$. Therefore, (3.194) relates dissipative properties of the Shimizu–Yamada model to the second-order fluctuations of the model and is an example of a fluctuation–dissipation theorem for a nonlinear Fokker–Planck equation. We would like to emphasize that (3.194) not only holds for the Shimizu–Yamada model. In fact, it can be shown that the fluctuation–dissipation theorem (3.194) holds for all strongly nonlinear Fokker-Planck equations that can be cast into the form

$$\frac{\partial}{\partial t}P(x,t;u) = -\frac{\partial}{\partial x}\left[h(x) - \kappa(x - \langle X \rangle) + f(t)\right]P + Q\frac{\partial^2}{\partial x^2}P, \qquad (3.195)$$

where $h(x)$ may correspond to a nonlinear function [183].

4 Free Energy Fokker–Planck Equations

Free Energy Fokker–Planck Equations for Probability Measures

We consider free energy measures that can be cast into the form

$$F[P] = U[P] - QS[P] \,, \tag{4.1}$$

where S denotes an entropy or information measure, $U[P]$ describes the internal energy of a system, and Q is the noise amplitude of the system. The noise amplitude Q can often be interpreted as a measure for the overall strength of the fluctuating forces that act on the system. Note that Q can also be regarded as some kind of temperature, see Sect. 4.2. For the sake of convenience, we may decompose $U[P]$ into a linear functional $\int_\Omega U_0(\mathbf{x}) P(\mathbf{x}) \, \mathrm{d}^M x$ and a nonlinear functional $U_{\mathrm{NL}} \propto O(P^2)$ as follows:

$$U[P] = \int_\Omega U_0(\mathbf{x}) P(\mathbf{x}) \, \mathrm{d}^M x + U_{\mathrm{NL}}[P] \,. \tag{4.2}$$

We assume that F is bounded from below:

$$F[P] \geq F_{\min} \,. \tag{4.3}$$

We further assume that the evolution of P is given by the multivariate nonlinear Fokker–Planck equation

$$\frac{\partial}{\partial t} P(\mathbf{x}, t; u) = \nabla \cdot \left\{ P \nabla \frac{\delta F}{\delta P} \right\} \,. \tag{4.4}$$

We refer to (4.4) as a free energy Fokker–Planck equation, see Sect. 2.6. In many cases, S can be written as

$$S = B\left(\int_\Omega s(P) \, \mathrm{d}^M x \right) \,. \tag{4.5}$$

We refer to $B(z)$ as outer function and to $s(z)$ as entropy kernel or kernel of an information measure. The outer function $B(z)$ may be interpreted as an entropy or information scale [190, 193]. Accordingly, the expression $\int_\Omega s(P) \, \mathrm{d}^M x$ measures the scale-free amount of entropy and information. This interpretation applies in particular for monotonically increasing functions B (i.e., for functions with $\mathrm{d}B/\mathrm{d}z > 0$). In the linear case, that is, for $B(z) = z$, the entropy and information measure (4.5) reduces to $S = \int_\Omega s(P) \, \mathrm{d}^M x$.

Next, let us introduce the operator

$$\hat{L}_f(z) = f(z) - z\frac{\mathrm{d}f(z)}{\mathrm{d}z} \tag{4.6}$$

that acts on a function $f(z)$ [190, 193]. Using (4.5) and (4.6), the free energy Fokker–Planck equation (4.4) becomes

$$\frac{\partial}{\partial t}P(\mathbf{x},t;u) = \nabla \cdot \left[\left\{\nabla U_0(\mathbf{x}) + \nabla \frac{\delta U_{\mathrm{NL}}[P]}{\delta P}\right\} P\right] + Q\Delta \hat{L}_s(P) \tag{4.7}$$

and involves the drift and diffusion coefficients

$$D_i(\mathbf{x},P) = -\frac{\partial}{\partial x_i}U_0(\mathbf{x}) - \frac{\partial}{\partial x_i}\frac{\delta U_{\mathrm{NL}}[P]}{\delta P}, \tag{4.8}$$

$$D_{ik}(\mathbf{x},P) = Q\delta_{ik}\frac{L_s(P)}{P}. \tag{4.9}$$

If (4.7) corresponds to a strongly nonlinear Fokker–Planck equation (i.e., if for solutions of (4.7) the coefficients $D'_i(\mathbf{x},t,u) = D_i$ and $D'_{ik}(\mathbf{x},t,u) = D_{ik}$ correspond to Kramers–Moyal coefficients of Markov diffusion processes), then Markov diffusion processes can be defined by (4.7) and

$$\frac{\partial}{\partial t}P(\mathbf{x},t|\mathbf{x}',t';u) = \nabla \cdot \left[\left\{\nabla U_0(\mathbf{x}) + \nabla \frac{\delta U_{\mathrm{NL}}[P]}{\delta P(\mathbf{x},t;u)}\right\} P(\mathbf{x},t|\mathbf{x}',t';u)\right]$$
$$+ Q\Delta \frac{\hat{L}_s(P(\mathbf{x},t;u))}{P(\mathbf{x},t;u)} P(\mathbf{x},t|\mathbf{x}',t';u). \tag{4.10}$$

These processes can alternatively be defined by the Ito–Langevin equation

$$\frac{\mathrm{d}}{\mathrm{d}t}\mathbf{X}(t) = -\nabla U_0(\mathbf{X}) - \nabla \frac{\delta U_{\mathrm{NL}}[P]}{\delta P(\mathbf{x},t;u)}\bigg|_{\mathbf{x}=\mathbf{X}(t)} + \sqrt{Q\frac{\hat{L}_s(P(\mathbf{x},t;u))}{P(\mathbf{x},t;u)}}\bigg|_{\mathbf{x}=\mathbf{X}(t)} \mathbf{\Gamma}(t), \tag{4.11}$$

where $\mathbf{\Gamma}(t)$ is an M-dimensional Langevin force with components $\Gamma_i(t)$ satisfying $\langle \Gamma_i(t) \rangle = 0$ and $\langle \Gamma_i(t), \Gamma_k(t') \rangle = 2\delta_{ik}\delta(t - t')$. See Chap. 3 for details.

Free Energy Fokker–Planck Equations for Density Functions

Evolution equations for density functions can be defined analogously to the evolution equations for probability measures. Let ρ denote a density function (e.g., mass density, particle density) and

$$F[\rho] = U[\rho] - QS[\rho] \tag{4.12}$$

describe a free energy measure of ρ. Then, $\rho(\mathbf{x}, t)$ may evolve according to the nonlinear evolution equation

$$\frac{\partial}{\partial t}\rho(\mathbf{x}, t) = \nabla \cdot \left\{ \rho \nabla \frac{\delta F}{\delta \rho} \right\} . \tag{4.13}$$

If we restrict ourselves to systems for which the total mass or the number of particles is constant, that is, if we have $\int_\Omega \rho(\mathbf{x}, t) \, \mathrm{d}^M x = M_0 = \mathrm{const.}$, then by means of

$$P = \frac{\rho}{M_0} \tag{4.14}$$

we can transform (4.13) into

$$\frac{\partial}{\partial t}P(\mathbf{x}, t; u) = \nabla \cdot \left\{ P \nabla \left. \frac{\delta F}{\delta \rho} \right|_{M_0 P} \right\} , \tag{4.15}$$

see Sect. 2.2. Langevin equations for systems described by (4.13) can be defined on the basis of (4.11) and (4.15) and read

$$\frac{\mathrm{d}}{\mathrm{d}t}\mathbf{X}(t) = -\left. \nabla \frac{\delta U[\rho]}{\delta \rho(\mathbf{x}, t)} \right|_{\mathbf{x} = \mathbf{X}(t)} + \sqrt{Q \left. \frac{\hat{L}_s(\rho(\mathbf{x}, t))}{\rho(\mathbf{x}, t)} \right|_{\mathbf{x} = \mathbf{X}(t)}} \mathbf{\Gamma}(t) , \tag{4.16}$$

with $\rho(\mathbf{x}, t) = M_0 \langle \delta(\mathbf{x} - \mathbf{X}(t)) \rangle$, see also Sect. 3.4.

4.1 Free Energy Principle

If solutions P_{st} of

$$\frac{\delta F[P_{\mathrm{st}}]}{\delta P} = \mu \tag{4.17}$$

satisfy the boundary conditions at hand, then they correspond to stationary solutions of (4.4). In addition, the constant μ may be regarded as a chemical potential for systems with mass equal to one. Equation (4.17) states that stationary states of the systems under consideration correspond to stationary points (or critical points) of free energy measures (free energy principle). Substituting (4.5) into (4.17), we obtain

$$\frac{\delta U[P_{\mathrm{st}}]}{\delta P} - Q \left. \frac{\mathrm{d}B(z)}{\mathrm{d}z} \right|_{z_{\mathrm{st}}} \left. \frac{\mathrm{d}s(z)}{\mathrm{d}z} \right|_{P_{\mathrm{st}}} = \mu , \tag{4.18}$$

with $z_{\mathrm{st}} = \int_\Omega s(P_{\mathrm{st}}) \, \mathrm{d}^M x$. Consequently, for entropy and information measures (4.5) the free energy principle (4.17) can be written in form of the implicit equation

$$P_{\text{st}}(\mathbf{x}) = \left[\frac{ds}{dz}\right]^{-1} \left(\frac{\frac{\delta U[P_{\text{st}}]}{\delta P} - \mu}{Q\frac{dB(z_{\text{st}})}{dz}}\right) \tag{4.19}$$

and with $U[P] = \langle U_0(\mathbf{X})\rangle + U_{\text{NL}}[P]$ as

$$P_{\text{st}}(\mathbf{x}) = \left[\frac{ds}{dz}\right]^{-1} \left(\frac{U_0(\mathbf{x}) + \frac{\delta U_{\text{NL}}[P_{\text{st}}]}{\delta P} - \mu}{Q\frac{dB(z_{\text{st}})}{dz}}\right), \tag{4.20}$$

where $[ds/dz]^{-1}$ is the inverse of ds/dz [168, 593]. Note that in some cases (4.20) may fail. For example, (4.20) cannot be applied if the inverse of ds/dz does not exist or leads to negative valued probability densities. In these cases, however, one can still try to derive stationary probability densities by means of (4.17).

If we deal with a **T**-periodic random variable for which $\mathbf{X}$ and $\mathbf{X} + \mathbf{T}$ describe the same states, we require that $f = \delta U_{\text{NL}}[P]/\delta P$ as a function of $\mathbf{x}$ and $U_0(\mathbf{x})$ are **T**-periodic functions (i.e., we have $f(\ldots, x_i + T_i, \ldots) = f(\ldots, x_i, \ldots)$ and $U_0(\ldots, x_i + T_i, \ldots) = U_0(\ldots, x_i, \ldots)$). Then, for entropy measures (4.5) the variational derivative $\delta F[P_{\text{st}}]/\delta P$ is a **T**-periodic function and (4.17) and (4.20) define stationary solutions of the free energy Fokker–Planck equation (4.4) that satisfy the required boundary conditions.

In the case of a system satisfying the density Fokker–Planck equation (4.13), the free energy principle reads

$$\frac{\delta F[\rho_{\text{st}}]}{\delta \rho} = \mu \tag{4.21}$$

and (4.19) becomes

$$\rho_{\text{st}}(\mathbf{x}) = \left[\frac{ds}{dz}\right]^{-1} \left(\frac{\frac{\delta U[\rho_{\text{st}}]}{\delta \rho} - \mu}{Q\frac{dB(z_{\text{st}})}{dz}}\right), \tag{4.22}$$

with $z_{\text{st}} = \int_\Omega s(\rho_{\text{st}}) \, d^M x$.

4.2 Maximum Entropy Principle and Relationship between Noise Amplitude and Temperature

The stationary solutions defined by (4.18) can also be obtained from the maximum entropy principle used in the theory of canonical ensembles [490] and information theory [254, 309]. According to the maximum entropy principle, systems with entropy functionals $S[P]$ and internal energy functionals $U[P]$ that exhibit a particular internal energy E have stationary distributions P_{st} that make S maximal under the constraint $U = E$. For the sake of convenience, we consider now only concave entropy measures with negative functional derivatives of second order $\delta^2 S[P](\epsilon) < 0$ (see Sect. 6.4 in general and (6.81-6.83) in particular). Then, every distribution P that makes S stationary under the constraint $U = E$ (i.e., for which $\delta S[P](\epsilon) = 0$ holds for $U = E$) corresponds to a maximum of S. Introducing the Lagrange multipliers α and β [86, 89] and the functional $I[P] = S[P] - \alpha(E - U[P]) - \beta(1 - \int_\Omega P \, \mathrm{d}^M x)$, stationary solutions P_{st} yield $\delta I[P_{\text{st}}](\epsilon) = 0$. From $\delta I[P_{\text{st}}](\epsilon) = 0$ it follows that

$$\frac{\delta S[P_{\text{st}}]}{\delta P} - \alpha \frac{\delta U[P_{\text{st}}]}{\delta P} - \beta = 0 \ . \tag{4.23}$$

If we put $\alpha = 1/Q$ and $\beta = -\mu/Q$ and compare (4.23) with (4.17) we see that stationary solutions defined by the maximum entropy principle correspond to stationary solutions defined by the free energy principle provided that the Lagrange multipliers are appropriately chosen. For a given Lagrange multiplier α and $\alpha = 1/Q$, a stationary solution can be regarded as a function of the noise amplitude Q: $P_{\text{st}}(Q)$. According to the maximum entropy principle, the parameter Q (more precisely the Lagrange multiplier $\alpha = 1/Q$) is determined by the constraint $E = U[P_{\text{st}}]$. Consequently, we can eliminate Q and thus obtain $P_{\text{st}} = P_{\text{st}}(E)$, which implies that the entropy functional $S[P_{\text{st}}]$ becomes a function of E: $S[P_{\text{st}}] = S(E)$. We can then compute the derivative of S with respect to E. In several studies it has been shown that irrespective of the explicit form of the functionals $S[P]$ and $U[P]$ the relation $\partial S/\partial E = 1/\alpha$ holds [106, 169, 204, 389, 416, 480, 482, 592, 641]. Using $\alpha = 1/Q$, we obtain

$$\frac{\partial S}{\partial E} = Q \ . \tag{4.24}$$

That is, the parameter Q determines the relationship between entropy and energy changes in the stationary case. In line with equilibrium thermodynamics [254, 490], we can therefore interpret the noise amplitude Q as some kind of temperature measure.

4.3 H-Theorem for Free Energy Fokker–Planck Equations

Before deriving the H-theorem for free energy Fokker–Planck equations, let us briefly elucidate the key idea of H-theorems. Let us consider a continuously differentiable, time-dependent function $f(t)$ defined on $[t_0, \infty)$. Let us assume that f satisfies

$$f(t) \geq f_{\min} \ , \quad \frac{\mathrm{d}}{\mathrm{d}t} f \leq 0 \ , \tag{4.25}$$

for $t \geq t_0$. That is, $f(t)$ is bounded from below and corresponds to a monotonically decreasing function. Then, $f(t)$ has a finite limit value for $t \to \infty$ [74]:

$$\lim_{t \to \infty} f(t) = f_\infty . \qquad (4.26)$$

Equation (4.26) implies that the limiting case

$$\lim_{t \to \infty} \frac{\mathrm{d}}{\mathrm{d}t} f(t) = 0 \qquad (4.27)$$

holds. In sum, the following proposition can be made: every monotonically decreasing function $f(t)$ that is bounded from below becomes stationary in the limit $t \to \infty$. With this result at hand, an H-theorem for free energy Fokker–Planck equations can be derived. Using partial integration, from (4.4) we obtain the inequality

$$\frac{\mathrm{d}}{\mathrm{d}t} F = -\int_\Omega P \left[\nabla \frac{\delta F}{\delta P} \right]^2 \mathrm{d}^M x \leq 0 . \qquad (4.28)$$

Since we assume that F does not depend explicitly on time, the implication $\partial P/\partial t = 0 \Rightarrow \mathrm{d}F/\mathrm{d}t = 0$ holds. From (4.4) and (4.28) we conclude that $\mathrm{d}F/\mathrm{d}t = 0 \Rightarrow \delta F/\delta P =$ constant $\Rightarrow \partial P/\partial t = 0$. In sum, F satisfies the following properties:

$$F \geq F_{\min} , \quad \frac{\mathrm{d}}{\mathrm{d}t} F \leq 0 , \quad \frac{\mathrm{d}}{\mathrm{d}t} F = 0 \Leftrightarrow \frac{\partial}{\partial t} P = 0 . \qquad (4.29)$$

In line with our preceding consideration, from (4.29) it follows that F becomes stationary in the limit $t \to \infty$. Using the implication $\mathrm{d}F/\mathrm{d}t = 0 \Leftrightarrow \partial P/\partial t = 0$, we further conclude that the limiting case

$$\lim_{t \to \infty} \frac{\partial P}{\partial t} = 0 \qquad (4.30)$$

holds. A functional that satisfies the properties (4.29) is called a Lyapunov functional. By means of the Lyapunov functional F, we have shown that transient solutions converge to stationary ones in the long time limit. In the context of Fokker–Planck equations, this convergence theorem is called an H-theorem. H-theorems for the linear Fokker–Planck equation have been discussed, for example, in [69, 148, 221, 236, 241, 498, 524]. H-theorems for nonlinear Fokker–Planck equations have been studied, for example, in [30, 62, 192, 199, 322, 323, 394, 522, 523, 525]. The key features of these H-theorems also become apparent in studies of H-theorems for the Boltzmann equation and the master equation [386, 607].

One caveat: in order to derive (4.28) we have assumed that the surface integral

$$I = \oint_{\partial \Omega} P \frac{\delta F}{\delta P} \nabla \frac{\delta F}{\delta P} \cdot \mathbf{n}(\mathbf{x}) \, \mathrm{d}A \qquad (4.31)$$

vanishes, where $\partial\Omega$ denotes the boundary of the phase space Ω, $\mathbf{n}$ is a normal vector on the surface $\partial\Omega$, and $\mathrm{d}A$ is an $M{-}1$-dimensional integration element. This integral does not necessarily vanish. For natural boundary conditions, we usually consider only such processes for which P decreases sufficiently fast for $|\mathbf{x}| \to \infty$ such that the vector $P\delta F/\delta P \nabla \delta F/\delta P$ vanishes in the limit $|\mathbf{x}| \to \infty$, which implies that the surface integral I vanishes. For periodic boundary conditions, by means of symmetry arguments one can prove that I vanishes if $\delta F/\delta P$ as a function of $\mathbf{x}$ describes a periodic function.

4.4 Boltzmann Statistics

Let us consider the free energy (4.1) for the Boltzmann entropy $^{\mathrm{B}}S = -\int_\Omega P \ln P \, \mathrm{d}^M x$ (see also Sect. 6.4.2) and a vanishing nonlinear internal energy functional (i.e, we put $U_{\mathrm{NL}} = 0$):

$$F[P] = \int_\Omega U_0(\mathbf{x})P(\mathbf{x})\,\mathrm{d}^M x + Q \int_\Omega P(\mathbf{x}) \ln P(\mathbf{x})\,\mathrm{d}^M x \ . \tag{4.32}$$

Since for $f(z) = -z \ln z$ the operator $\hat{L}_f(z)$ given by (4.6) reduces to an identical mapping, that is, we have

$$f(z) = -z \ln z \Rightarrow \hat{L}_f(z) = z \ , \tag{4.33}$$

the nonlinear Fokker–Planck equation (4.7) becomes linear with respect to P and reads

$$\frac{\partial}{\partial t}P(\mathbf{x}, t; u) = \nabla \cdot [P \nabla U_0(\mathbf{x})] + Q\Delta P \ . \tag{4.34}$$

The stationary distribution is given by

$$P_{\mathrm{st}}(\mathbf{x}) = \frac{\exp\{-U_0(\mathbf{x})/Q\}}{\displaystyle\int_\Omega \exp\{-U_0(\mathbf{x})/Q\}\,\mathrm{d}^M x} \ , \tag{4.35}$$

which can be proven by substituting (4.35) into (4.34). The distribution function also satisfies the free energy principle and can be derived from (4.20). To see this, we use $B(z) = z$ and $S = \int s(P)\,\mathrm{d}^M x$ with $s(z) = -z \ln z$ which implies $\mathrm{d}s/\mathrm{d}z = -1 - \ln z$ and $[\mathrm{d}s/\mathrm{d}z]^{-1} = \exp\{-1-z\}$. Then, from (4.20) we get

$$P_{\mathrm{st}}(\mathbf{x}) = \exp\{-[U_0(\mathbf{x}) - \mu]/Q - 1\} = \frac{1}{Z}\exp\{-U_0(\mathbf{x})/Q\} \ , \tag{4.36}$$

with the partition function Z given by $Z = \int_\Omega \exp\{-U_0(\mathbf{x})/Q\}\,\mathrm{d}^M x$ and $\ln Z = 1 - \mu/Q$.

4.5 Linear Nonequilibrium Thermodynamics

4.5.1 Derivation of Free Energy Fokker–Planck Equations

Thermodynamic perspective

In this section, nonlinear Fokker–Planck equations are studied from the perspective of linear nonequilibrium thermodynamics [130, 228, 332, 353, 449]. To this end, we consider systems with state variables $\mathbf{x} \in \Omega$ and describe them by means of probability densities P, entropy and information measures $S[P]$ and internal energy measures $U[P]$. Our objective is to derive from the principles of linear nonequilibrium thermodynamics that determine the behavior of S and U an evolution equation for P. We will proceed as illustrated in Fig. 4.1 [99, 167, 168].

$$\mathrm{d}S[P]$$
$$\downarrow$$
$$\mathrm{d}S = \mathrm{d}_\mathrm{i}S + \mathrm{d}_\mathrm{e}S$$

$$\mathrm{d}_\mathrm{i}S = \int_\Omega \mathbf{X}^\mathrm{th}\mathbf{J}^\mathrm{th}\,\mathrm{d}^M x \qquad Q\mathrm{d}_\mathrm{e}S = \mathrm{d}U$$

$$\downarrow$$
$$\mathbf{J}^\mathrm{th} = \underbrace{P(\mathbf{I} + \mathcal{M}\mathbf{X}^\mathrm{th})}$$

$$\int_\Omega P\mathbf{I}\cdot\mathbf{X}^\mathrm{th}\,\mathrm{d}^M x = 0$$
$$\mathbf{X}^\mathrm{th} = -\nabla\frac{\delta F}{\delta P}$$

$$\mathbf{J} \propto \mathbf{J}^\mathrm{th}$$

$$\frac{\partial P}{\partial t} = -\mathrm{div}\mathbf{J}$$

Fig. 4.1. Basic steps in the derivation of the free energy Fokker–Planck equation (4.49). See text for details

Our departure point is a state change of a system that involves a stochastic process described by $P(\mathbf{x}, t; u)$, on the one hand, and may involve a change of entropy $\mathrm{d}S$ and a change of internal energy $\mathrm{d}U$, on the other hand. We assume that the change of energy results from the contact of the system with its environment, see Fig. 4.2 (upper panel). According to the principles of linear nonequilibrium thermodynamics we assume that the change $\mathrm{d}S$ is due

4.5 Linear Nonequilibrium Thermodynamics

to two effects. First, S may change due to the exchange of energy of the system with its environment, that is, there is an entropy change $d_e S(dU)$. Here the index e stands for "external". Second, S may change due to processes that are also present if we isolate the system from its environment. These processes are called intrinsic and the change of S induced by intrinsic processes is denoted by $d_i S$, see Fig. 4.2 (lower panel). Since in isolated systems entropy cannot decrease, we require that $d_i S$ satisfies the inequality $d_i S \geq 0$. In sum, the change dS related to a state change of the system under consideration can be written as

$$dS = d_i S + d_e S(dU) . \tag{4.37}$$

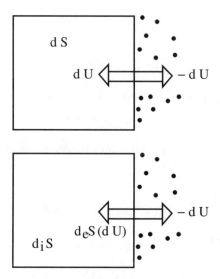

Fig. 4.2. Two kinds of entropy changes: $d_i S$ and $d_e S$. The environment is considered as an ensemble of heat bath particles (which are represented by full dots)

We need to determine now $d_i S$ and $d_e S$. For reversible processes of equilibrium systems the heat change $dH_T = TdS$ is equal to the energy change dU. We assume that for the systems under consideration a similar relation holds, where the noise amplitude Q plays the role of the temperature T. That is, we put

$$Q\, d_e S = dU . \tag{4.38}$$

According to linear nonequilibrium thermodynamics, $d_i S$ can be expressed by means of the local entropy production $\sigma(\mathbf{x}, t)$ as $d_i S/dt = \int_\Omega \sigma(\mathbf{x}, t)\, d^M x$. The local entropy production in turn is given by the product of the thermodynamic force $\mathbf{X}^{\mathrm{th}}$ and the thermodynamic flux $\mathbf{J}^{\mathrm{th}}$ as $\sigma(\mathbf{x}, t) \propto \mathbf{X}^{\mathrm{th}} \cdot \mathbf{J}^{\mathrm{th}}$

with $\mathbf{X}^{\text{th}} = (X_1^{\text{th}}, \ldots, X_M^{\text{th}})$ and $\mathbf{J}^{\text{th}} = (J_1^{\text{th}}, \ldots, J_M^{\text{th}})$:

$$Q\frac{d_i S}{dt} = \int_\Omega \mathbf{X}^{\text{th}} \cdot \mathbf{J}^{\text{th}} \, d^M x \, . \tag{4.39}$$

Next, we need to determine $\mathbf{X}^{\text{th}}$ and $\mathbf{J}^{\text{th}}$. To this end, we consider state changes for which the free energy variation $\delta F/\delta P$ describes a potential for the thermodynamic force $\mathbf{X}^{\text{th}}$. Then $\mathbf{X}^{\text{th}}$ is given by

$$\mathbf{X}^{\text{th}} = -\nabla \frac{\delta F}{\delta P} \, . \tag{4.40}$$

Alternatively, we may regard the variational derivative $\delta F[P]/\delta P$ as a functional depending on P and $\mathbf{x}$ and thus introduce the chemical potential functional $\mu(\mathbf{x}, P)$ defined by

$$\mu(\mathbf{x}, P) = \frac{\delta F}{\delta P} \, . \tag{4.41}$$

Then, (4.40) becomes

$$\mathbf{X}^{\text{th}} = -\nabla \mu(\mathbf{x}, P) \, . \tag{4.42}$$

In addition, we assume that there is a linear relationship between $\mathbf{X}^{\text{th}}$ and $\mathbf{J}^{\text{th}}$:

$$\mathbf{J}^{\text{th}} \propto \mathbf{X}^{\text{th}} \tag{4.43}$$

(whence the name "linear nonequilibrium thermodynamics"). We introduce a conservative force vector $\mathbf{I}$ and an $M \times M$-matrix $\mathcal{M}$ with coefficients M_{ik} and write (4.43) as

$$\mathbf{J}^{\text{th}} = P(\mathbf{x}, t; u) \left[\mathbf{I} + \mathcal{M} \cdot \mathbf{X}^{\text{th}} \right] \, , \tag{4.44}$$

which reads in components like

$$J_i^{\text{th}} = P(\mathbf{x}, t; u) \left[I_i + \sum_{k=1}^M M_{ik} X_k^{\text{th}} \right] \, . \tag{4.45}$$

Let us illustrate the meaning of (4.44). To this end, we regard the thermodynamic flux as a particle current and consider the univariate case $J^{\text{th}} = J^{\text{th}}(x, t)$, where x denotes particle positions on the real line $\mathbb{R}^1$. Then, the particle current at position x and time t is defined by the number of particles ΔM_0 that pass the position x at time t in a small time interval Δt: $J(x, t) = \Delta M_0/\Delta t$. If $v(x, t)$ denotes the velocity of the particles, then $J(x, t)$ can equivalently be expressed as $J(x, t) = v(x, t) P(x, t)$. The reason for this is that ΔM_0 can be computed from $\Delta M_0 = P(x, t) \Delta x$ and $v(x, t)$ may be related to the ratio $\Delta x/\Delta t$, where Δx denotes a small interval. Substituting $\Delta M_0 = P(x, t) \Delta x$ and $v(x, t) = "\Delta x/\Delta t"$ into $J(x, t) = \Delta M_0/\Delta t$, we obtain $J(x, t) = v(x, t) P(x, t)$. Next, we assume that the velocity is proportional to the thermodynamic force X^{th}, where the proportional factor is given by the

4.5 Linear Nonequilibrium Thermodynamics

mobility coefficient M. That is, we put $v(x,t) = MX(x,t)$, which gives us $J = vP = MPX$. In the multivariate case we can proceed in a similar fashion but it is helpful to distinguish between two kinds of forces as shown in (4.44). From our pervious discussion it is clear that in general $\mathcal{M}$ can interpreted as a mobility matrix. Note also that $\mathcal{M}$ can be put in front of the brackets in (4.44) such that we have $\mathbf{J}^{\text{th}} = P(\mathbf{x},t;u)\mathcal{M}\cdot[\mathbf{I}'+\mathbf{X}^{\text{th}}]$ with $\mathbf{I} = \mathcal{M}\cdot\mathbf{I}'$. In general, $\mathcal{M}$ and $\mathbf{I}$ may depend on $\mathbf{x}$, t and P: $\mathcal{M} = \mathcal{M}(\mathbf{x},t,P)$ and $\mathbf{I} = \mathbf{I}(\mathbf{x},t,P)$. We assume that the probability density P evolves according to a differential equation which is of first order in time. More explicitly, we consider systems for which the continuity equation

$$\frac{\partial}{\partial t}P(\mathbf{x},t;u) = -\text{div}\,\mathbf{J}(\mathbf{x},t;u) \tag{4.46}$$

holds. Finally, we assume that the probability current $\mathbf{J}(\mathbf{x},t;u)$ is proportional to the thermodynamic flux $\mathbf{J}^{\text{th}}(\mathbf{x},t;u)$:

$$\mathbf{J}(\mathbf{x},t;u) \propto \mathbf{J}^{\text{th}}(\mathbf{x},t;u) \ . \tag{4.47}$$

For the sake of convenience, we put the proportional constant equal to one and thus obtain

$$\mathbf{J}(\mathbf{x},t;u) = \mathbf{J}^{\text{th}}(\mathbf{x},t;u) \ . \tag{4.48}$$

Taking (4.40-4.48) together, we obtain the nonlinear Fokker–Planck equation

$$\frac{\partial}{\partial t}P(\mathbf{x},t;u) = -\nabla\cdot[\mathbf{I}(\mathbf{x},t,P)\,P] + \nabla\cdot\left[\mathcal{M}(\mathbf{x},t,P)P\cdot\nabla\frac{\delta F}{\delta P}\right] \ , \tag{4.49}$$

which reads in components

$$\frac{\partial}{\partial t}P(\mathbf{x},t;u) = -\sum_{i=1}^{M}\frac{\partial}{\partial x_i}[I_i(\mathbf{x},t,P)\,P]$$
$$+ \sum_{i,k=1}^{M}\frac{\partial}{\partial x_i}\left[M_{ik}(\mathbf{x},t,P)P\frac{\partial}{\partial x_k}\frac{\delta F}{\delta P}\right] \ . \tag{4.50}$$

If S corresponds to the Boltzmann entropy $^{\text{B}}S = -\int_\Omega P\ln P\,\mathrm{d}^M x$ and U_{NL} vanishes and if we assume that $\mathcal{M}$ and $\mathbf{I}$ are independent of P, then (4.49) reduces to the linear Fokker–Planck equation

$$\frac{\partial}{\partial t}P(\mathbf{x},t;u) = -\nabla\cdot(P[\mathbf{I}(\mathbf{x},t) - \mathcal{M}(\mathbf{x},t)\cdot\nabla U_0(\mathbf{x})]) + Q\nabla\cdot\mathcal{M}(\mathbf{x},t)\cdot\nabla P \ . \tag{4.51}$$

In the linear case, the mobility matrix $\mathcal{M}$ is also called a friction matrix and QM_{ik} is referred to as a (thermodynamic) diffusion matrix. By analogy to the linear case, we will in general regard $\mathcal{M}$ as a friction or mobility matrix and $Q\mathcal{M}$ as a diffusion matrix. Comparing (2.16) and (4.49), we realize that

84   4 Free Energy Fokker–Planck Equations

the (thermodynamic) diffusion matrix QM_{ik} is equivalent to the (stochastic) diffusion matrix D_{ik}. We therefore require that M_{ik} satisfies the same properties as the diffusion coefficients of linear Fokker–Planck equations, which means that we assume that M_{ik} is a symmetric and semi-positive definite matrix. The free energy Fokker–Planck equation (4.49) can be regarded as a generalization of (4.4).

Fokker–Planck perspective

In the preceding paragraph, the focus was on thermodynamic relations. Equation (4.49) can be derived in an alternative way that emphasizes the continuity equation and the Fokker–Planck perspective. For the sake of convenience, in what follows we neglect the conservative force $\mathbf{I}$. Then, we consider a stochastic process for which the continuity equation describes the relationship between P and $\mathbf{J}^{\text{th}}$ and linear irreversible thermodynamics describes the relationship between $\mathbf{J}^{\text{th}}$ and $\mathbf{X}^{\text{th}}$ such that we get

$$\frac{\partial}{\partial t}P = -\nabla \mathbf{J}^{\text{th}} \ , \ \ \mathbf{J}^{\text{th}} = \mathbf{v}P = P\mathcal{M} \cdot \mathbf{X}^{\text{th}} \ , \tag{4.52}$$

where $\mathbf{v}$ denotes some kind of particle (or subsystem) velocity. We assume that the thermodynamic force $\mathbf{X}^{\text{th}}$ can be computed from a chemical potential μ that in turn is determined by a free energy F such that we have

$$\mathbf{X}^{\text{th}} = -\nabla \mu \ , \ \ \mu(\mathbf{x}, P) = \frac{\delta F}{\delta P} \ . \tag{4.53}$$

Taking (4.52) and (4.53) together we obtain

$$\frac{\partial}{\partial t}P(\mathbf{x}, t; u) = \nabla \cdot \mathcal{M} P \cdot \nabla \frac{\delta F}{\delta P} \ , \tag{4.54}$$

which corresponds to (4.49) for $\mathbf{I} = 0$.

Notations

In deriving (4.49) and (4.54) we have used several quantities, which are listed in Table 4.1.

4.5.2 Drift and Diffusion Coefficients

Using the entropy measure (4.5) and the operator $\hat{L}_f(z)$ defined by (4.6), we obtain the intermediate result

$$P\frac{\partial}{\partial x_k}\frac{\delta S}{\delta P} = P\left.\frac{\mathrm{d}B(z)}{\mathrm{d}z}\right|_{z_0}\frac{\partial}{\partial x_k}\frac{\mathrm{d}s}{\mathrm{d}P} = -\left.\frac{\mathrm{d}B(z)}{\mathrm{d}z}\right|_{z_0}\frac{\partial}{\partial x_k}\hat{L}_s(P) \ , \tag{4.55}$$

4.5 Linear Nonequilibrium Thermodynamics

Table 4.1. Summary of some variables used in the text

Symbol	Description	Reference
F	free energy	–
U	internal energy functional	–
S	entropy or information measure	–
μ	chemical potential	–
$d_e S$	external entropy change (entropy flux)	[130, 449]
$d_i S$	intrinsic entropy change (entropy production)	″
σ	local entropy production	″
J_i^{th}	thermodynamic flux	″
X_i^{th}	thermodynamic force	″
I_i	conservative force (reversible drift force)	[498]
Q	noise amplitude, fluctuation strength, temperature	–
$M_{i,k}$	mobility matrix	[19, 458, 432, 501]
$QM_{i,k}$	diffusion matrix	[19, 100, 101, 458]

with $z_0 = \int_\Omega s(P) \, \mathrm{d}^M x$. Exploiting the equivalence $\delta F/\delta P = \delta U/\delta P - Q \delta S/\delta P$, we can write (4.49) as

$$\frac{\partial}{\partial t} P(\mathbf{x}, t; u) = -\nabla \cdot P(\mathbf{x}, t; u) \left[\mathbf{I}(\mathbf{x}, t, P) - \mathcal{M}(\mathbf{x}, t, P) \cdot \nabla \frac{\delta U}{\delta P} \right]$$
$$+ Q \left. \frac{\mathrm{d}B(z)}{\mathrm{d}z} \right|_{z_0} \nabla \cdot \left[\mathcal{M}(\mathbf{x}, t, P) \cdot \nabla \hat{L}_s(P) \right] . \quad (4.56)$$

In components, (4.56) reads

$$\frac{\partial}{\partial t} P(\mathbf{x}, t; u) =$$
$$- \sum_{i=1}^{M} \frac{\partial}{\partial x_i} \left\{ P \left[I_i - \sum_{k=1}^{M} M_{ik} \frac{\partial}{\partial x_k} \frac{\delta U}{\delta P} \right] + Q \left. \frac{\mathrm{d}B(z)}{\mathrm{d}z} \right|_{z_0} \hat{L}_s(P) \sum_{k=1}^{M} \frac{\partial M_{ik}}{\partial x_k} \right\}$$
$$+ Q \left. \frac{\mathrm{d}B(z)}{\mathrm{d}z} \right|_{z_0} \frac{\partial^2}{\partial x_i \partial x_k} M_{ik} \hat{L}_s(P) . \quad (4.57)$$

Note that in (4.57) we have changed the position of the mobility matrix in the diffusion term which yields an additional expression in the drift term. If we define D_i and D_{ik} like

$$P D_i = P \left[I_i - \sum_{k=1}^{M} M_{ik} \frac{\partial}{\partial x_k} \frac{\delta U}{\delta P} \right] + Q \left. \frac{\mathrm{d}B(z)}{\mathrm{d}z} \right|_{z_0} \hat{L}_s(P) \sum_{k=1}^{M} \frac{\partial M_{ik}}{\partial x_k} ,$$

$$PD_{ik} = Q \left.\frac{\mathrm{d}B(z)}{\mathrm{d}z}\right|_{z_0} M_{ik}\hat{L}_s(P) \tag{4.58}$$

then the free energy Fokker–Planck equation (4.57) can be regarded as a special case of the general nonlinear Fokker–Planck equation (2.4).

4.5.3 Transition Probability Densities and Langevin Equations

If (4.56) corresponds to a strongly nonlinear Fokker–Planck equation, then nonlinear families of Markov diffusion processes with transition probability densities $P(\mathbf{x},t|\mathbf{x}',t';u)$ can be defined by (4.56) and

$$\frac{\partial}{\partial t}P(\mathbf{x},t|\mathbf{x}',t';u) =$$
$$-\nabla \cdot P(\mathbf{x},t|\mathbf{x}',t';u)\left[\mathbf{I}(\mathbf{x},t,P(\mathbf{x},t;u)) - \mathcal{M}(\mathbf{x},t,P(\mathbf{x},t;u)) \cdot \nabla\frac{\delta U}{\delta P(\mathbf{x},t;u)}\right]$$
$$+ Q\left.\frac{\mathrm{d}B(z)}{\mathrm{d}z}\right|_{z_0} \nabla \cdot \left[\mathcal{M}(\mathbf{x},t,P(\mathbf{x},t;u)) \cdot \nabla\frac{\hat{L}_s(P(\mathbf{x},t;u))}{P(\mathbf{x},t;u)}P(\mathbf{x},t|\mathbf{x}',t';u)\right]$$
$$\tag{4.59}$$

and $z_0 = \int_\Omega s(P(\mathbf{x},t;u))\,\mathrm{d}^M x$. This result can be verified using the drift and diffusion coefficients (4.58). By means of the coefficients (4.58) we can also define the Ito–Langevin equation that corresponds to (4.56) and (4.59). It reads

$$\frac{\mathrm{d}}{\mathrm{d}t}\mathbf{X}(t) = \left[\mathbf{I}(\mathbf{x},t,P(\mathbf{x},t;u)) - \mathcal{M}(\mathbf{x},t,P(\mathbf{x},t;u)) \cdot \nabla\frac{\delta U}{\delta P(\mathbf{x},t;u)}\right]_{\mathbf{x}=\mathbf{X}(t)}$$
$$+ \underbrace{Q\left.\frac{\mathrm{d}B(z)}{\mathrm{d}z}\right|_{z_0}\left[\frac{\hat{L}_s(P(\mathbf{x},t;u))}{P(\mathbf{x},t;u)}\nabla \cdot \mathcal{M}(\mathbf{x},t,P(\mathbf{x},t;u))\right]_{\mathbf{x}=\mathbf{X}(t)}}_{Y}$$
$$+ G(\mathbf{x},t,P(\mathbf{x},t;u))|_{\mathbf{x}=\mathbf{X}(t)} \cdot \mathbf{\Gamma}(t) , \tag{4.60}$$

where G denotes an $M \times M$ matrix with coefficients G_{ik} that satisfy

$$\sum_{l=1}^{M} G_{il}G_{lk} = QM_{ik}\frac{\hat{L}_s(P(\mathbf{x},t;u))}{P(\mathbf{x},t;u)} . \tag{4.61}$$

The Ito–Langevin equation (4.60) involves a spurious drift term Y that vanishes if we use the Stratonovich calculus instead of the Ito calculus (see Stratonovich-Langevin equations in Sects. 7.2.3 and 7.4.4).

4.5.4 Density Functions

If our departure point is the free energy functional (4.12) for a density measure ρ with $M_0 = \int_\Omega \rho(\mathbf{x},t)\,\mathrm{d}^M x = \text{const.}$, then (4.49) may be generalized to

$$\frac{\partial}{\partial t}\rho(\mathbf{x},t) = -\nabla \cdot [\mathbf{I}(\mathbf{x},t,\rho)\,\rho] + \nabla \cdot \left[\mathcal{M}(\mathbf{x},t,\rho)\rho \cdot \nabla \frac{\delta F}{\delta \rho}\right]. \quad (4.62)$$

This evolution equation, in turn, can be written as a free energy Fokker–Planck equation for a probability density $P = \rho/M_0$. Substituting $P = \rho/M_0$ into (4.62), we get

$$\frac{\partial}{\partial t}P(\mathbf{x},t;u) = -\nabla \cdot [\mathbf{I}(\mathbf{x},t,M_0 P)\,P] + \nabla \cdot \left[\mathcal{M}(\mathbf{x},t,M_0 P) P \cdot \nabla \left.\frac{\delta F}{\delta \rho}\right|_{M_0 P}\right]. \quad (4.63)$$

4.5.5 Entropy Production and Conservative Force

Substituting (4.44) into (4.39), the intrinsic entropy production is given by $\mathrm{d}_i S/\mathrm{d}t = Q^{-1}[\int_\Omega P\mathbf{X}^{\text{th}} \cdot \mathcal{M} \cdot \mathbf{X}^{\text{th}}\,\mathrm{d}^M x + \int_\Omega P\mathbf{I} \cdot \mathbf{X}^{\text{th}}\,\mathrm{d}^M x]$. As stated earlier, the drift vector $\mathbf{I}$ is regarded as a conservative force which means that $\mathbf{I}$ does not contribute to the entropy production $\mathrm{d}_i S$. That is, we require that the integral relation

$$\int_\Omega P\mathbf{I}(\mathbf{x},t,P) \cdot \mathbf{X}^{\text{th}}(\mathbf{x},t,P)\,\mathrm{d}^M x = 0 \quad (4.64)$$

holds, which is reminiscent to the null space hypothesis used in the GENERIC approach to the thermodynamics of complex fluids [153, 243, 459]. From (4.64) and the assumption that $\mathcal{M}$ is semi-positive definite, we find that $\mathrm{d}_i S$ can only increase as a function of time or is constant:

$$Q\frac{\mathrm{d}_i S}{\mathrm{d}t} = \langle \mathbf{X}^{\text{th}} \cdot \mathcal{M} \cdot \mathbf{X}^{\text{th}} \rangle \geq 0. \quad (4.65)$$

Let us derive now a sufficient condition for which (4.64) is satisfied. Substituting (4.40) and (4.55) into (4.64), we obtain

$$\int_\Omega P\mathbf{I} \cdot \mathbf{X}^{\text{th}}\,\mathrm{d}^M x = -\int_\Omega P\mathbf{I} \cdot \nabla \frac{\delta U}{\delta P}\,\mathrm{d}^M x + Q\int_\Omega P\mathbf{I} \cdot \nabla \frac{\delta S}{\delta P}\,\mathrm{d}^M x$$

$$= -\int_\Omega P\mathbf{I} \cdot \nabla \frac{\delta U}{\delta P}\,\mathrm{d}^M x - Q\left.\frac{\mathrm{d}B(z)}{\mathrm{d}z}\right|_{z_0} \int_\Omega \mathbf{I} \cdot \nabla L_s(P)\,\mathrm{d}^M x. \quad (4.66)$$

Integration by part and assuming that surface terms vanish, gives us

$$\int_\Omega P\mathbf{I} \cdot \mathbf{X}^{\text{th}} \, d^M x = -\int_\Omega P\mathbf{I} \cdot \nabla \frac{\delta U}{\delta P} d^M x + Q \left. \frac{dB(z)}{dz} \right|_{z_0} \int_\Omega L_s(P) \nabla \cdot \mathbf{I} \, d^M x \,. \tag{4.67}$$

It is clear that (4.64) is satisfied for

$$Q \left. \frac{dB(z)}{dz} \right|_{z_0} L_s(P) \nabla \cdot \mathbf{I} - P\mathbf{I} \cdot \nabla \frac{\delta U}{\delta P} = 0 \,. \tag{4.68}$$

Equation (4.68), in turn, is satisfied for

$$\nabla \cdot \mathbf{I} = 0 \,, \quad \mathbf{I} \cdot \nabla \frac{\delta U}{\delta P} = 0 \,. \tag{4.69}$$

In fact, as we will show next, in order to derive stationary solutions it is convenient to require that the conditions in (4.69) hold. That is, we may focus on systems that involve conservative forces $\mathbf{I}$ with vanishing divergence and energy functionals U that satisfy $\mathbf{I} \cdot \nabla \delta U/\delta P = 0$. Finally, note that in the linear case given by $\delta U/\delta P = U_0$, $B(z) = z$ and $L_s(z) = z$, (4.68) yields the relation $\nabla \cdot \mathbf{I} - \mathbf{I} \cdot \nabla(U_0/Q) = 0$ known from the theory of linear Fokker–Planck equations [237, 498].

4.5.6 Stationary Solutions

Let us show now that (4.20) defines stationary probability densities of the multivariate free energy Fokker–Planck equation (4.49). Substituting (4.20) in form of (4.17) into (4.49), gives us

$$\frac{\partial}{\partial t} P(\mathbf{x}, t; u = P_{\text{st}}) = -\nabla \cdot [P_{\text{st}}(\mathbf{x}) \mathbf{I}(\mathbf{x}, t, P_{\text{st}})] = -P_{\text{st}} \nabla \cdot \mathbf{I} - \mathbf{I} \cdot \nabla P_{\text{st}} \,. \tag{4.70}$$

In view of (4.20), we write P_{st} like $P_{\text{st}} = f(\delta U/\delta P)$, which leads to

$$\frac{\partial}{\partial t} P(\mathbf{x}, t; u = P_{\text{st}}) = -P_{\text{st}} \nabla \cdot \mathbf{I} - \left. \frac{df}{dz} \right|_{\delta U/\delta P} \mathbf{I} \cdot \nabla \frac{\delta U[P_{\text{st}}]}{\delta P} \,. \tag{4.71}$$

By means of (4.69), we finally obtain

$$\frac{\partial}{\partial t} P(\mathbf{x}, t; u = P_{\text{st}}) = 0 \,. \tag{4.72}$$

We arrive at the conclusion that stationary points of free energy measures describe stationary probability densities of the free energy Fokker–Planck equation (4.49). If we take the gradient of (4.17), we obtain $\nabla \delta F[P_{\text{st}}]/\delta P = 0$. Consequently, in the context of linear nonequilibrium thermodynamics the free energy principle (4.17) means that the thermodynamics force (4.40) vanishes in the stationary case:

$$\mathbf{X}_{\text{st}}^{\text{th}} = -\nabla \frac{\delta F[P_{\text{st}}]}{\delta P} = 0 \,. \tag{4.73}$$

We can also conclude in the opposite direction. Accordingly, we find that the vanishing of $\mathbf{X}^{\text{th}}$ implies that we have a stationary solution of (4.49): $\mathbf{X}^{\text{th}} = 0 \Rightarrow P = P_{\text{st}}$. Finally, note that this result can easily be obtained using the relationship between the thermodynamic force $\mathbf{X}^{\text{th}}$ and the chemical potential functional $\mu(\mathbf{x}, P)$, see (4.42). In line with the free energy principle (4.17) we require that $\mu(\mathbf{x}, P)$ corresponds in the stationary case to a constant μ. From (4.42) and $\mu(\mathbf{x}, P) = \mu$ it then follows that $\mathbf{X}^{\text{th}} = 0$.

4.5.7 H-Theorem for Systems with Conservative Forces and Nontrivial Mobility Coefficients

For solutions of (4.49) the free energy F evolves like

$$\frac{\mathrm{d}}{\mathrm{d}t} F[P] = - \int_\Omega \frac{\delta F}{\delta P} \nabla (\mathbf{I} P) \, \mathrm{d}^M x + \int_\Omega \frac{\delta F}{\delta P} \nabla \left[\mathcal{M}(\mathbf{x}, t, P) P(\mathbf{x}, t; u) \nabla \frac{\delta F}{\delta P} \right] \mathrm{d}^M x \, . \tag{4.74}$$

Integrating by parts, assuming that surface terms arising due to partial integration vanish, and taking (4.69) into account, we obtain

$$\frac{\mathrm{d}}{\mathrm{d}t} F[P] = \underbrace{\int_\Omega \mathbf{X}^{\text{th}} \cdot \mathbf{I} P \, \mathrm{d}^M x}_{Y} + \int_\Omega P \mathcal{M}(\mathbf{x}, t, P) : \left(\nabla \frac{\delta F}{\delta P} \right) \left(\nabla \frac{\delta F}{\delta P} \right) \mathrm{d}^M x \, . \tag{4.75}$$

The integral Y describes the entropy production by the force $\mathbf{I}$ and vanishes (see (4.64)) because we assume that (4.69) holds. This leads us to

$$\frac{\mathrm{d}}{\mathrm{d}t} F[P] = - \underbrace{\int_\Omega P \sum_{i,k=1}^M M_{ik}(\mathbf{x}, t, P) \left(\frac{\partial}{\partial x_i} \frac{\delta F}{\delta P} \right) \left(\frac{\partial}{\partial x_k} \frac{\delta F}{\delta P} \right) \mathrm{d}^M x}_{Y' \geq 0} \leq 0 \, . \tag{4.76}$$

The expression Y' is larger than or equal to zero because $\mathcal{M}$ describes a semi-positive matrix. Taking (4.3) into account, we find

$$F \geq F_{\min} \, , \quad \frac{\mathrm{d}}{\mathrm{d}t} F \leq 0 \, , \tag{4.77}$$

which implies that the limiting case

$$\lim_{t \to \infty} \frac{\mathrm{d}}{\mathrm{d}t} F = 0 \tag{4.78}$$

holds. Furthermore, we conclude that $\partial P / \partial t = 0 \Rightarrow \mathrm{d}F/\mathrm{d}t = 0$ holds because F does not depend explicitly on time (which comes from the fact that we assume that U and S do not depend explicitly on time). If $\mathcal{M}$ is positive definite, the implication $\mathrm{d}F/\mathrm{d}t = 0 \Rightarrow \delta F/\delta P = \text{constant}$ holds. As shown earlier, for $\delta F/\delta P = $ constant we obtain $\partial P/\partial t = 0$. Consequently, for the free energy Fokker–Planck equation (4.49) with a positive-definite friction matrix the relations

$$F \geq F_{\min} \;, \quad \frac{d}{dt}F \leq 0 \;, \quad \frac{d}{dt}F = 0 \Leftrightarrow \frac{\partial}{\partial t}P = 0 \qquad (4.79)$$

hold. From (4.79) it is clear that transient solutions converge to stationary ones (H-theorem, see Sec. 4.3):

$$\lim_{t\to\infty} \frac{\partial}{\partial t} P = 0 \;. \qquad (4.80)$$

In sum, if $\mathcal{M}$ is semi-positive definite then systems become stationary in the sense of (4.78), whereas if $\mathcal{M}$ is positive definite then systems become stationary in the sense of (4.80).

4.6 Canonical-Dissipative Systems

4.6.1 Linear Case

The derivation of Fokker–Planck equations by means of linear nonequilibrium thermodynamics is tailored to address canonical-dissipative systems and the Kramers equation. Canonical-dissipative systems are systems that involve a deterministic part satisfying a Hamiltonian dynamics and a dissipative stochastic part [146, 147, 149, 193, 251, 296, 515, 516]. We consider a $M = 2M'$-dimensional phase space Ω spanned by the generalized coordinates $\mathbf{q} = (q_1, \ldots, q_{M'})$ and momentum variables $\mathbf{p} = (p_1, \ldots, p_{M'})$. That is, we have $\mathbf{x} = (\mathbf{p}, \mathbf{q}) \in \Omega$. In addition, we decompose $\mathbf{I}$ like $\mathbf{I} = (\mathbf{I}_p, \mathbf{I}_q)$. Then, the deterministic part of the canonical-dissipative system is given by

$$\frac{d}{dt}\mathbf{q}(t) = \mathbf{I}_q(\mathbf{p}, \mathbf{q}) = \frac{\partial}{\partial \mathbf{p}} H(\mathbf{p}, \mathbf{q}) \;, \qquad (4.81)$$

$$\frac{d}{dt}\mathbf{p}(t) = \mathbf{I}_p(\mathbf{p}, \mathbf{q}) = -\frac{\partial}{\partial \mathbf{q}} H(\mathbf{p}, \mathbf{q}) \qquad (4.82)$$

and a Hamilton function $H(\mathbf{p}, \mathbf{q})$. The Hamilton function is the counterpart of the energy potential $U_0(\mathbf{x})$ occurring in the free energy equation (4.32):

$$U_0(\mathbf{x}) \to H(\mathbf{p}, \mathbf{q}) \;. \qquad (4.83)$$

Let us regard $\mathcal{M}$ as a $2M' \times 2M'$ matrix. Then, substituting (4.81-4.83) into (4.51) we obtain

$$\frac{\partial}{\partial t} P(\mathbf{x}, t; u) = \underbrace{-\sum_{i=1}^{M'} \frac{\partial}{\partial q_i}\left[\frac{\partial H}{\partial p_i} P\right] + \sum_{i=1}^{M'} \frac{\partial}{\partial p_i}\left[\frac{\partial H}{\partial q_i} P\right]}_{Y_1}$$

$$+ \underbrace{\sum_{i,k=1}^{M} \frac{\partial}{\partial x_i}\left[PM_{ik}(\mathbf{p},\mathbf{q})\frac{\partial}{\partial x_k}(H+Q\ln P)\right]}_{Y_2}, \quad (4.84)$$

which can equivalently be expressed as

$$\frac{\partial}{\partial t}P(\mathbf{x},t;u) = \underbrace{-\sum_{i=1}^{M'}\frac{\partial}{\partial q_i}\left[\frac{\partial H}{\partial p_i}P\right] + \sum_{i=1}^{M'}\frac{\partial}{\partial p_i}\left[\frac{\partial H}{\partial q_i}P\right]}_{Y_1}$$

$$+\underbrace{\sum_{i,k=1}^{M}\frac{\partial}{\partial x_i}\left[M_{ik}(\mathbf{p},\mathbf{q})\frac{\partial H}{\partial x_k}P\right] + Q\sum_{i,k=1}^{M}\frac{\partial}{\partial x_i}\left[M_{ik}(\mathbf{p},\mathbf{q})\frac{\partial}{\partial x_k}P\right]}_{Y_2}.$$

$$(4.85)$$

The dynamics described by these evolution equations involves a conservative and a dissipative part. The expression Y_1 describes the canonical conservative part related to the conservative drift force $\mathbf{I}$. The dissipative stochastic part is described by the expression Y_2 and vanishes for $\mathcal{M} = 0$. From (4.85) we read off that the dissipative stochastic part Y_2 is composed of a drift term and a diffusion term. The latter term vanishes for $Q = 0$. The drift term usually describes damping forces (friction forces). The fact that in our derivation there occurs simultaneously a damping force and a diffusion force indicates that there is a close relationship between the emergence of fluctuations and damping forces. This relationship has been studied, for example, in [89, 254, 308, 317, 517] (see also Sect. 1.1.1) and involves the so-called fluctuation-dissipation theorem [358, 254, 490, 498]. Note that we may decompose the mobility matrix $\mathcal{M}$ into four submatrices $\mathcal{M}_{pp}$, $\mathcal{M}_{pq}$, $\mathcal{M}_{qp}$, and $\mathcal{M}_{qq}$. In doing so, we can eliminate the vector $\mathbf{x}$ and (4.84) and (4.85) can be written exclusively in terms of $\mathbf{p}$ and $\mathbf{q}$ coordinates (see Sect. 4.6.2 below). Finally, note that canonical-dissipative systems described by linear Fokker–Planck equations such as (4.85) have in particular been studied in the context of odd and even variables [237, 498, 604, 605].

Let us illustrate now that (4.69) with $\delta U/\delta P = H$ is satisfied for canonical-dissipative systems. First, the divergence of the force $\mathbf{I}$ vanishes:

$$\nabla \cdot \mathbf{I} = \frac{\partial}{\partial \mathbf{p}}\cdot \mathbf{I}_p + \frac{\partial}{\partial \mathbf{q}}\cdot \mathbf{I}_q = -\frac{\partial}{\partial \mathbf{p}}\cdot\frac{\partial H}{\partial \mathbf{q}} + \frac{\partial}{\partial \mathbf{q}}\cdot\frac{\partial H}{\partial \mathbf{p}} = 0, \quad (4.86)$$

with $\nabla = (\partial/\partial p_1, \ldots, \partial/\partial p_M, \partial/\partial q_1, \ldots, \partial/\partial q_M)$. Second, $\mathbf{I}$ is orthogonal to the gradient of the variational derivative $\delta U/\delta P$:

$$\mathbf{I}\cdot \nabla \frac{\delta U}{\delta P} = \mathbf{I}_p \cdot \frac{\partial}{\partial \mathbf{p}}\frac{\delta U}{\delta P} + \mathbf{I}_q \cdot \frac{\partial}{\partial \mathbf{q}}\frac{\delta U}{\delta P} = -\frac{\partial H}{\partial \mathbf{q}}\cdot \frac{\partial H}{\partial \mathbf{p}} + \frac{\partial H}{\partial \mathbf{p}}\cdot \frac{\partial H}{\partial \mathbf{q}} = 0. \quad (4.87)$$

Consequently, the force **I** does not induce an intrinsic entropy change dS_i.

As an example of a canonical-dissipative system we consider a single particle with mass $m = 1$, position $\mathbf{q} = (q_1, q_2, q_3)$, and momentum $\mathbf{p} = (p_1, p_2, p_3)$ that moves under the impact of a force $\mathbf{h}(\mathbf{q})$ derived from a potential $V(\mathbf{q})$. The deterministic system would read $\dot{\mathbf{q}} = \mathbf{p}$ and $\dot{\mathbf{p}} = \mathbf{h}(\mathbf{q}) = -\partial V(\mathbf{q})/\partial \mathbf{q}$. Consequently, the Hamiltonian of the system reads

$$H = \frac{p^2}{2} + V(\mathbf{q}) , \qquad (4.88)$$

with $p^2 = |\mathbf{p}|^2$. We assume that the particle is subjected to fluctuations and a linear damping force. This dissipative impact can be taken into account by means of the mobility matrix

$$\mathcal{M} = \gamma \begin{pmatrix} \begin{array}{c|c} \begin{matrix} 1 & 0 & 0 \\ 0 & 1 & 0 \\ 0 & 0 & 1 \end{matrix} & \\ \hline 0 & 0 \end{array} \end{pmatrix} \qquad (4.89)$$

involving the damping constant $\gamma > 0$. Then, (4.85) reads

$$\frac{\partial}{\partial t} P(\mathbf{p},\mathbf{q},t;u) = -\sum_{i=1}^{3} \frac{\partial}{\partial q_i} p_i P + \sum_{i=1}^{3} \frac{\partial}{\partial p_i}[\gamma p_i - h_i(\mathbf{q})] P + \gamma Q \sum_{i=1}^{3} \frac{\partial^2}{\partial p_i^2} P . \qquad (4.90)$$

From (4.90) we can see that the fluctuation strength is given by $Q' = \gamma Q$. Equation (4.90) is referred to as Kramers equation. Using $\Delta_p = \sum_{i=1}^{3}(\partial^2/\partial p_i^2)$, (4.90) can be written as

$$\frac{\partial}{\partial t} P(\mathbf{p},\mathbf{q},t;u) = -\mathbf{p}\frac{\partial}{\partial \mathbf{q}} P + \frac{\partial}{\partial \mathbf{p}}[\gamma \mathbf{p} - \mathbf{h}(\mathbf{q})] P + Q' \Delta_p P . \qquad (4.91)$$

4.6.2 Nonlinear Case

We now generalize our preceding considerations to the nonlinear case. Again, we describe a single particle defined on an $M = 2M'$-dimensional phase space Ω spanned by the generalized coordinates $\mathbf{q} = (q_1, \ldots, q_{M'})$ and momentum variables $\mathbf{p} = (p_1, \ldots, p_{M'})$. That is, we have $\mathbf{x} = (\mathbf{p},\mathbf{q}) \in \Omega$ and $P(\mathbf{x},t;u) \to P(\mathbf{p},\mathbf{q},t;u)$. We write **I** as $\mathbf{I} = (\mathbf{I}_p, \mathbf{I}_q)$ with $\mathbf{I}_p = \mathbf{I}_p(\mathbf{p},\mathbf{q},t,P)$ and $\mathbf{I}_q = \mathbf{I}_q(\mathbf{p},\mathbf{q},t,P)$. The mobility matrix $\mathcal{M}$ corresponds again to a $2M' \times 2M'$ matrix $\mathcal{M}(\mathbf{p},\mathbf{q},t,P)$ but now depends on P as indicated. Then, the free energy Fokker–Planck equation (4.49) becomes

$$\frac{\partial}{\partial t} P(\mathbf{p},\mathbf{q},t;u) = -\frac{\partial}{\partial \mathbf{p}} \cdot [\mathbf{I}_p P] - \frac{\partial}{\partial \mathbf{q}} \cdot [\mathbf{I}_q P] + \nabla \cdot \left[\mathcal{M} P \cdot \nabla \frac{\delta F}{\delta P}\right] . \qquad (4.92)$$

4.6 Canonical-Dissipative Systems

Alternatively, if S is given by $S = B[\int_\Omega s(P)\,\mathrm{d}^{M'}p\,\mathrm{d}^{M'}q]$ (see (4.5)), we have [168]

$$\frac{\partial}{\partial t}P(\mathbf{p},\mathbf{q},t;u) = -\frac{\partial}{\partial \mathbf{p}}\cdot[\mathbf{I}_p P] - \frac{\partial}{\partial \mathbf{q}}\cdot[\mathbf{I}_q P] + \nabla\cdot\left[\mathcal{M}P\cdot\nabla\frac{\delta U}{\delta P}\right]$$
$$+ Q\left.\frac{\mathrm{d}B}{\mathrm{d}z}\right|_{z_0}\nabla\cdot\mathcal{M}\cdot\nabla\hat{L}_s(P) \qquad (4.93)$$

with $z_0 = \int_\Omega s(P)\,\mathrm{d}^{M'}p\,\mathrm{d}^{M'}q$. The expression $\nabla\cdot\mathcal{M}\cdot\nabla$ can be expressed using submatrices of $\mathcal{M}$ defined by M_{lm}^{pp}, M_{lm}^{pq}, M_{lm}^{qp}, M_{lm}^{qq} of M_{ik} (with $l,m \in \{1,\ldots,M'\}$), which gives us [168, 193]

$$\frac{\partial}{\partial x_i}M_{ik}[\mathbf{x},P]\frac{\partial}{\partial x_k} = \frac{\partial}{\partial p_l}M_{lm}^{pp}[\mathbf{p},\mathbf{q},P]\frac{\partial}{\partial p_m} + \frac{\partial}{\partial p_l}M_{lm}^{pq}[\mathbf{p},\mathbf{q},P]\frac{\partial}{\partial q_m}$$
$$+ \frac{\partial}{\partial q_l}M_{lm}^{qp}[\mathbf{p},\mathbf{q},P]\frac{\partial}{\partial p_m} + \frac{\partial}{\partial q_l}M_{lm}^{qq}[\mathbf{p},\mathbf{q},P]\frac{\partial}{\partial q_m}.$$
$$(4.94)$$

Analogously, the term $\nabla\cdot\mathcal{M}P\cdot\nabla$ can be treated. Next, we discuss two cases in which $\mathbf{I}$ and U are chosen such that (4.69) is satisfied.

Case A

In line with (4.81) and (4.82), we assume that the conservative drift forces are given by

$$I_l^p[\mathbf{p},\mathbf{q},P] = -\frac{\partial}{\partial q_l}\frac{\delta U[P]}{\delta P}, \quad I_l^q[\mathbf{p},\mathbf{q},P] = \frac{\partial}{\partial p_l}\frac{\delta U[P]}{\delta P}. \qquad (4.95)$$

Then, we have

$$\nabla\cdot\mathbf{I} = \frac{\partial}{\partial \mathbf{p}}\cdot\mathbf{I}_p + \frac{\partial}{\partial \mathbf{q}}\cdot\mathbf{I}_q = -\frac{\partial}{\partial \mathbf{p}}\cdot\frac{\partial}{\partial \mathbf{q}}\frac{\delta U}{\delta P} + \frac{\partial}{\partial \mathbf{q}}\cdot\frac{\partial}{\partial \mathbf{p}}\frac{\delta U}{\delta P} = 0. \qquad (4.96)$$

In addition, we have

$$\mathbf{I}\cdot\nabla\frac{\delta U}{\delta P} = \mathbf{I}_p\cdot\frac{\partial}{\partial \mathbf{p}}\frac{\delta U}{\delta P} + \mathbf{I}_q\cdot\frac{\partial}{\partial \mathbf{q}}\frac{\delta U}{\delta P}$$
$$= -\left(\frac{\partial}{\partial \mathbf{q}}\frac{\delta U}{\delta P}\right)\cdot\left(\frac{\partial}{\partial \mathbf{p}}\frac{\delta U}{\delta P}\right) + \left(\frac{\partial}{\partial \mathbf{p}}\frac{\delta U}{\delta P}\right)\cdot\left(\frac{\partial}{\partial \mathbf{q}}\frac{\delta U}{\delta P}\right) = 0. \quad (4.97)$$

That is, (4.69) is satisfied. Furthermore, the Fokker–Planck equation (4.93) reads

$$\frac{\partial}{\partial t}P(\mathbf{p},\mathbf{q},t;u) = \frac{\partial}{\partial \mathbf{p}}\cdot\left[P\frac{\partial}{\partial \mathbf{q}}\frac{\delta U[P]}{\delta P}\right] - \frac{\partial}{\partial \mathbf{q}}\cdot\left[P\frac{\partial}{\partial \mathbf{p}}\frac{\delta U[P]}{\delta P}\right]$$
$$+ \nabla\cdot\left[\mathcal{M}P\cdot\nabla\frac{\delta U}{\delta P}\right] + Q\left.\frac{\mathrm{d}B}{\mathrm{d}z}\right|_{z_0}\nabla\cdot\mathcal{M}\cdot\nabla\hat{L}_s(P) \qquad (4.98)$$

and the stationary distributions of (4.98) are given by (4.20).

Case B

Here, let us put

$$U[P] = \int_\Omega \tilde{g}(H) P \, \mathrm{d}^{M'} p \, \mathrm{d}^{M'} q \,,$$

$$I_l^p[\mathbf{p}, \mathbf{q}] = -\frac{\partial}{\partial q_l} H \,,$$

$$I_l^q[\mathbf{p}, \mathbf{q}] = \frac{\partial}{\partial p_l} H \,, \qquad (4.99)$$

where $\tilde{g}(z)$ is a function of z and H denotes a Hamiltonian $H = H(\mathbf{p}, \mathbf{q})$. Then, one finds that the relation $\nabla \cdot \mathbf{I} = 0$ is satisfied because of (4.86). In addition, the condition $\mathbf{I} \cdot \nabla \delta U / \delta P = 0$ holds provided that $\mathrm{d}\tilde{g}/\mathrm{d}z$ exists because we have

$$\mathbf{I} \cdot \nabla \frac{\delta U}{\delta P} = \frac{\mathrm{d}\tilde{g}}{\mathrm{d}H} [\mathbf{I} \cdot \nabla H] \qquad (4.100)$$

and $\mathbf{I} \cdot \nabla H = 0$. Now, the Fokker–Planck equation (4.93) reads

$$\frac{\partial}{\partial t} P(\mathbf{p}, \mathbf{q}, t; u) = \frac{\partial}{\partial \mathbf{p}} \cdot \left[\frac{\partial H}{\partial \mathbf{q}} P \right] - \frac{\partial}{\partial \mathbf{q}} \cdot \left[\frac{\partial H}{\partial \mathbf{p}} P \right]$$
$$+ \nabla \cdot [\mathcal{M} P \cdot \nabla \tilde{g}(H)] + Q \left. \frac{\mathrm{d}B}{\mathrm{d}z} \right|_{z_0} \nabla \cdot \mathcal{M} \cdot \nabla \hat{L}_s(P) \qquad (4.101)$$

and the stationary distributions of (4.101) can be obtained from (4.20) and satisfy

$$P_{\mathrm{st}}(\mathbf{p}, \mathbf{q}) = \left[\frac{\mathrm{d}s}{\mathrm{d}z} \right]^{-1} \left\{ \frac{\tilde{g}(H(\mathbf{p}, \mathbf{q})) - \mu}{Q \, \mathrm{d}B(z_{0,\mathrm{st}})/\mathrm{d}z} \right\} \qquad (4.102)$$

with $z_{0,\mathrm{st}} = \int_\Omega s(P_{\mathrm{st}}(\mathbf{p}, \mathbf{q})) \, \mathrm{d}^{M'} p \, \mathrm{d}^{M'} q$.

Example I

Let us consider the energy functional

$$U[P(\mathbf{p}, \mathbf{q})] = \int_\Omega H_0(\mathbf{p}, \mathbf{q}) P(\mathbf{p}, \mathbf{q}) \, \mathrm{d}^{M'} p \, \mathrm{d}^{M'} q$$
$$+ \frac{1}{2} \int_\Omega \int_\Omega H_{\mathrm{MF}}(\mathbf{q}, \mathbf{q}') P(\mathbf{p}, \mathbf{q}) P(\mathbf{p}', \mathbf{q}') \, \mathrm{d}^{M'} p \, \mathrm{d}^{M'} q \, \mathrm{d}^{M'} p' \, \mathrm{d}^{M'} q' \quad (4.103)$$

for a symmetric interaction Hamiltonian H_{MF}. That is, we require $H_{\mathrm{MF}}(\mathbf{q}, \mathbf{q}') = H_{\mathrm{MF}}(\mathbf{q}', \mathbf{q})$. Using (4.95) and (4.103), the Fokker–Planck equation (4.93) reads

$$\frac{\partial}{\partial t}P(\mathbf{p},\mathbf{q},t;u) = \frac{\partial}{\partial \mathbf{p}} \cdot \left[\frac{\partial H_0}{\partial \mathbf{q}} P\right] - \frac{\partial}{\partial \mathbf{q}}\left[\frac{\partial H_0}{\partial \mathbf{p}} P\right]$$
$$+ \nabla \cdot \left[\mathcal{M} P \cdot \left\{\nabla H_0 + \frac{\partial}{\partial \mathbf{q}} \int_\Omega H_{\mathrm{MF}}(\mathbf{q},\mathbf{q}')P(\mathbf{p}',\mathbf{q}',t;u)\,\mathrm{d}^{M'}p\,\mathrm{d}^{M'}q\,\mathrm{d}^{M'}p'\,\mathrm{d}^{M'}q'\right\}\right]$$
$$+ Q\left.\frac{\mathrm{d}B}{\mathrm{d}z}\right|_{z_0} \nabla \cdot \mathcal{M} \cdot \nabla \hat{L}_s(P) . \tag{4.104}$$

Using (4.20) and (4.103), we conclude that the stationary distribution of (4.104) is given by

$$P_{\mathrm{st}}(\mathbf{p},\mathbf{q}) = \left[\frac{\mathrm{d}s}{\mathrm{d}z}\right]^{-1} \left\{\frac{H_0(\mathbf{p},\mathbf{q}) + \int_\Omega H_{\mathrm{MF}}(\mathbf{q}-\mathbf{q}')P_{\mathrm{st}}(\mathbf{p}',\mathbf{q}')\,\mathrm{d}^{M'}p'\,\mathrm{d}^{M'}q' - \mu}{Q\,\mathrm{d}B(z_{0,\mathrm{st}})/\mathrm{d}z}\right\} \tag{4.105}$$

with $z_{0,\mathrm{st}} = \int_\Omega s(P_{\mathrm{st}}(\mathbf{p},\mathbf{q}))\,\mathrm{d}^{M'}p\,\mathrm{d}^{M'}q$.

Example II: Generalized Kramers Equation

Now, we require $M^{pp}_{lm} = \gamma \delta_{lm}$, $M^{pq}_{lm} = M^{qp}_{lm} = M^{qq}_{lm} \equiv 0$, and $H_0 = p^2/(2m) + V(\mathbf{q})$, where $p^2 = |\mathbf{p}|^2$ and $m > 0$ describes the mass of the particle under consideration. The states of the particle are described by $\mathbf{q} = (q_1, q_2, q_3)$ and $\mathbf{p} = (p_1, p_2, p_3)$. Then, (4.104) becomes

$$\frac{\partial}{\partial t}P(\mathbf{p},\mathbf{q},t) =$$
$$\sum_{l=1}^{3} \frac{\partial}{\partial p_l}\left[\left(\gamma \frac{p_l}{m} + \frac{\partial V(\mathbf{q})}{\partial q_l} + \frac{\partial}{\partial q_l}\int_\Omega H_{\mathrm{MF}}(\mathbf{q}-\mathbf{q}')P(\mathbf{p}',\mathbf{q}',t)\,\mathrm{d}^3 p'\,\mathrm{d}^3 q'\right)P\right]$$
$$-\sum_{l=1}^{3} \frac{\partial}{\partial q_l}\frac{p_l}{m}P + \gamma Q \left.\frac{\mathrm{d}B}{\mathrm{d}z}\right|_{\int_\Omega s(P)\,\mathrm{d}^3 p\,\mathrm{d}^3 q} \sum_{l=1}^{3}\frac{\partial^2}{\partial p_l^2}\hat{L}_s(P) \tag{4.106}$$

or

$$\frac{\partial}{\partial t}P(\mathbf{p},\mathbf{q},t) =$$
$$\frac{\partial}{\partial \mathbf{p}}\cdot\left[\left(\gamma \frac{\mathbf{p}}{m} + \frac{\mathrm{d}V(\mathbf{q})}{\mathrm{d}\mathbf{q}} + \frac{\partial}{\partial \mathbf{q}}\int_\Omega H_{\mathrm{MF}}(\mathbf{q}-\mathbf{q}')P(\mathbf{p}',\mathbf{q}',t)\,\mathrm{d}^3 p'\,\mathrm{d}^3 q'\right)P\right]$$
$$-\frac{\mathbf{p}}{m}\cdot\frac{\partial}{\partial \mathbf{q}}P(\mathbf{p},\mathbf{q},t) + \gamma Q\left.\frac{\mathrm{d}B}{\mathrm{d}z}\right|_{\int_\Omega s(P)\,\mathrm{d}^3 p'\,\mathrm{d}^3 q'} \frac{\partial}{\partial \mathbf{p}}\cdot\frac{\partial}{\partial \mathbf{p}}\hat{L}_s(P) \tag{4.107}$$

with

$$P_{\mathrm{st}}(\mathbf{p},\mathbf{q}) =$$
$$\left[\frac{\mathrm{d}s}{\mathrm{d}z}\right]^{-1}\left\{\frac{\mathbf{p}^2/(2m) + V(\mathbf{q}) + \int_\Omega H_{\mathrm{MF}}(\mathbf{q}-\mathbf{q}')P_{\mathrm{st}}(\mathbf{p}',\mathbf{q}')\,\mathrm{d}^3 p'\,\mathrm{d}^3 q' - \mu}{Q\mathrm{d}B(z_{0,\mathrm{st}})/\mathrm{d}z}\right\} . \tag{4.108}$$

For $S = {}^\mathrm{B}S[P] = -\int P \ln P \, \mathrm{d}^3 p \, \mathrm{d}^3 q$ and $V(\mathbf{q}) = \mathbf{q}^2/2$ (4.106) can be used to describe particle bunches in particle accelerator. We will return to this issue in Sect. 7.4.2.

Example III

We consider (4.101) for particles with position and momentum vectors $\mathbf{q} = (q_1, q_2, q_3)$ and $\mathbf{p} = (p_1, p_2, p_3)$. In addition, we put $M^{pp}_{lm} = \gamma \delta_{lm}$, $\gamma > 0$, and $M^{pq}_{lm} = M^{qp}_{lm} = M^{qq}_{lm} = 0$, where γ can be seen as a friction coefficient, see (4.109) below. In order to make contact with other studies (e.g. [149]), we introduce the function $g(z) = \mathrm{d}\tilde{g}/\mathrm{d}z \Rightarrow \tilde{g}(z) = \int_0^z g(z') \, \mathrm{d}z'$. Then, from (4.94) and (4.101) we obtain

$$\frac{\partial}{\partial t} P(\mathbf{p}, \mathbf{q}, t) = \sum_{l=1}^{3} \frac{\partial}{\partial p_l} \left[\left(\frac{\partial H_0}{\partial q_l} + \gamma g(H_0) \frac{\partial H_0}{\partial p_l} \right) P \right] - \sum_{l=1}^{3} \frac{\partial}{\partial q_l} \left[\frac{\partial H_0}{\partial p_l} P \right]$$

$$+ \gamma Q \left. \frac{\mathrm{d}B}{\mathrm{d}z} \right|_{\int_\Omega s(P) \mathrm{d}^3 p \, \mathrm{d}^3 q} \sum_{l=1}^{3} \frac{\partial^2}{\partial p_l^2} \hat{L}_s(P) \,, \qquad (4.109)$$

and stationary solutions are determined by

$$P_\mathrm{st}(\mathbf{p}, \mathbf{q}) = \left[\frac{\mathrm{d}s}{\mathrm{d}z} \right]^{-1} \left\{ \frac{\int^{H_0} g(z) \, \mathrm{d}z - \mu}{Q \, \mathrm{d}B/\mathrm{d}z} \right\} . \qquad (4.110)$$

Equation (4.109) can be regarded as a generalization of the linear Fokker–Planck equations of canonical-dissipative systems as discussed in [149, 516] and may be written as

$$\frac{\partial}{\partial t} P(\mathbf{p}, \mathbf{q}, t) = \frac{\partial}{\partial \mathbf{p}} \cdot \left[\left(\frac{\partial H_0}{\partial \mathbf{q}} + \gamma g(H_0) \frac{\partial H_0}{\partial \mathbf{p}} \right) P \right] - \frac{\partial}{\partial \mathbf{q}} \cdot \left[\frac{\partial H_0}{\partial \mathbf{p}} P \right]$$

$$+ Q' \left. \frac{\mathrm{d}B}{\mathrm{d}z} \right|_{\int_\Omega s(P) \mathrm{d}^3 p \, \mathrm{d}^3 q} \frac{\partial}{\partial \mathbf{p}} \cdot \frac{\partial}{\partial \mathbf{p}} \hat{L}_s(P) \,, \qquad (4.111)$$

with $Q' = \gamma Q$.

4.7 Boundedness of Free Energy Functionals*

So far, we have assumed that the free energy measures of interest are bounded from below, see (4.3). Our objective now is to derive sufficient conditions for the boundedness of free energy functionals [175].

4.7.1 Distortion Functionals

Distortion functionals relate stationary solutions of nonlinear Fokker–Planck equations in general and free energy Fokker–Planck equations in particular to Boltzmann distributions and are helpful tools to investigate the boundedness of free energy measures. We will consider here distortion functionals related to free energy measures, whereas in Chap. 7 we will be concerned with distortion functionals of a more general type.

The distortion functional G related to free energy measures F given by (4.1) and (4.2) is defined by [166, 190, 192, 193, 199]

$$G[u] = \exp\left\{-\frac{\delta S}{\delta u} - 1 + \frac{1}{Q}\frac{\delta U_{\rm NL}}{\delta u}\right\} . \qquad (4.112)$$

Let us express the variational principle $\delta F/\delta P = \mu$ as

$$\frac{\delta S}{\delta P_{\rm st}} = \frac{1}{Q}\left[\frac{\delta U}{\delta P_{\rm st}} - \mu\right] = \frac{1}{Q}\left[U_0(\mathbf{x}) + \frac{\delta U_{\rm NL}}{\delta P_{\rm st}} - \mu\right] . \qquad (4.113)$$

Then, (4.113) can be written as

$$G[P_{\rm st}] = \exp\left\{-\frac{\delta S}{\delta P_{\rm st}} - 1 + \frac{1}{Q}\frac{\delta U_{\rm NL}}{\delta P_{\rm st}}\right\} = \exp\left\{-\frac{[U_0(\mathbf{x}) - \mu]}{Q} - 1\right\} . \qquad (4.114)$$

Using the Boltzmann distribution

$$W(\mathbf{x}) = \frac{\exp\{-U_0(\mathbf{x})/Q\}}{\int_\Omega \exp\{-U_0(\mathbf{x})/Q\}\, d^M x} \qquad (4.115)$$

(see also Sect. 4.4), we obtain the mapping $P_{\rm st} \to W$ defined by

$$G[P_{\rm st}] = \frac{1}{Z} W(\mathbf{x}) . \qquad (4.116)$$

Here, Z is a normalization constant that is implicitly determined by the normalization condition $\int_\Omega P_{\rm st}(\mathbf{x})\, d^M x = 1$.

4.7.2 Kullback Measure and Entropy Inequality

Let us consider an entropy or information measure $S[P]$ that satisfies the concavity inequality

$$S[P] \leq S[P_0] + \delta S[P_0](P - P_0) , \qquad (4.117)$$

for $P, P_0 > 0$, where the equal sign holds only for $P = P_0$ (for properties of S such as concavity see Sect. 6.4.1). Equation (4.117) can be written as

$$S[P_0] - S[P] + \delta S[P_0](P - P_0) \geq 0 \ . \tag{4.118}$$

On the basis of the inequality (4.118), we define the semi-positive definite measure [167, 168, 192]

$$K[P, P_0] = S[P_0] - S[P] + \delta S[P_0](P - P_0) \geq 0 \ . \tag{4.119}$$

The functional K takes two arguments: P and P_0. It can be regarded as a distance measure. That is, it measures the difference between the probability densities P and P_0. If P equals P_0 we have $K = 0$. If P differs from P_0 we have $K > 0$:

$$K > 0 \Leftrightarrow P \neq P_0 \ , \quad K = 0 \Leftrightarrow P = P_0 \ . \tag{4.120}$$

We refer to K as a general Kullback measure because it recovers the Kullback distance measure in the case of the Boltzmann entropy (see below). Our considerations reveal that the general Kullback measure is nothing but the concavity inequality of entropy and information measures:

Kullback measure $\Leftrightarrow$ concavity of entropies and information measures.
$$\tag{4.121}$$

Note that the Kullback measure (4.119) can also be defined for two discrete probability distributions $\{p_i\}$ and $\{p_i^{(0)}\}$ and entropy measures of the form $S = S(p_1, \ldots, p_N)$:

$$K\left(\{p_i\}, \{p_i^{(0)}\}\right) = S(\{p_i^{(0)}\}) - S(\{p_i\}) + \sum_{i=1}^{N} \left.\frac{\partial S}{\partial p_i}\right|_{\{p_i\}=\{p_i^{(0)}\}} (p_i - p_i^{(0)}) \ , \tag{4.122}$$

see also [185] for the special case $S = \sum_{i=1}^{N} s(p_i)$.

Boltzmann–Kullback Measure

For the Boltzmann entropy ${}^BS[P] = -\int_\Omega P \ln P \, d^M x$, the Kullback measure (4.119) reads

$$\begin{aligned}{}^BK[P, P_0] &= -\int_\Omega P_0 \ln P_0 \, d^M x \\ &\quad + \int_\Omega P \ln P \, d^M x - \int_\Omega [1 + \ln P_0](P - P_0) \, d^M x \\ &= \int_\Omega P \ln\left[\frac{P}{P_0}\right] d^M x \ . \end{aligned} \tag{4.123}$$

This measure is known in the literature as the Kullback measure or the Kullback distance measure [257, 361, 362]. In what follows, we will refer to ${}^BK[P, P_0]$ as the Boltzmann–Kullback measure. Since the Boltzmann entropy satisfies the concavity inequality (4.117), the measure ${}^BK$ is semi-positive definite. We would like to present here also an alternative proof of this property.

4.7 Boundedness of Free Energy Functionals*

We start off with the logarithm $\ln(x)$. The logarithm satisfies $\ln(1) = 0$. Furthermore, if we plot $\ln(x)$ and $x - 1$ in one diagram, we see that the inequality

$$\ln(x) \leq x - 1 \tag{4.124}$$

holds for $x > 0$, where the equal sign holds for $x = 1$ only. Now, let us replace x by $P_0(\mathbf{x})/P(\mathbf{x})$ for $P, P_0 > 0$. Then, we conclude that

$$\ln\left[\frac{P_0}{P}\right] \leq \frac{P_0}{P} - 1$$
$$\Rightarrow P \ln\left[\frac{P_0}{P}\right] \leq P_0 - P$$
$$\Rightarrow \int_\Omega P \ln\left[\frac{P_0}{P}\right] d^M x \leq 0$$
$$\Rightarrow {}^B K[P, P_0] = \int_\Omega P \ln\left[\frac{P}{P_0}\right] d^M x \geq 0 \;. \tag{4.125}$$

Note that in the derivation above we used the normalization condition of probability densities in terms of $\int_\Omega [P_0 - P]\, d^M x = 0$.

General Kullback Measure for Stationary Probability Densities

In particular, the general Kullback measure can be used to compare arbitrary probability densities P with stationary probability densities P_{st} obtained from the free energy principle. In this case, we put $P_0 = P_{st}$ and (4.119) becomes

$$K[P, P_{st}] = S[P_{st}] - S[P] + \delta S[P_{st}](P - P_{st}) \;. \tag{4.126}$$

If we multiply (4.113) with a function $f(\mathbf{x})$ and integrate with respect to $\mathbf{x}$, we obtain $\delta S[P_{st}](f) = \delta U[P_{st}](f)/Q - \mu \int_\Omega f(\mathbf{x})\, d^M x$. In particular, for $f = P - P_{st}$ we obtain

$$\delta S[P_{st}](P - P_{st}) = \frac{1}{Q}\delta U[P_{st}](P - P_{st}) - \frac{\mu}{Q}\underbrace{\int_\Omega [P - P_{st}]\, d^M x}_{= 0} \;. \tag{4.127}$$

Consequently, (4.126) becomes

$$K[P, P_{st}] = \frac{1}{Q}\delta U[P_{st}](P - P_{st}) - S[P] + S[P_{st}] \geq 0 \;. \tag{4.128}$$

Using (4.2) and $\delta U[P_{st}](P - P_{st}) = \delta U_{NL}[P_{st}](P - P_{st}) + U_L[P] - U_L[P_{st}]$, which implies

$$\delta U[P_{\mathrm{st}}](P-P_{\mathrm{st}}) = \delta U_{\mathrm{NL}}[P_{\mathrm{st}}](P-P_{\mathrm{st}}) + U[P] - U[P_{\mathrm{st}}] - (U_{\mathrm{NL}}[P] - U_{\mathrm{NL}}[P_{\mathrm{st}}]) ,\tag{4.129}$$

we can write (4.128) as

$$Q K[P, P_{\mathrm{st}}] = F[P] - F[P_{\mathrm{st}}] + \delta U_{\mathrm{NL}}[P_{\mathrm{st}}](P - P_{\mathrm{st}}) - (U_{\mathrm{NL}}[P] - U_{\mathrm{NL}}[P_{\mathrm{st}}]) .\tag{4.130}$$

4.7.3 Generic Cases and Schlögl's Decomposition of Kullback Measures

Linear Energy Functionals

The result (4.130) suggests that we examine in more detail the case in which energy functionals can be written as $U = U_{\mathrm{L}} = \int_{\Omega} U_0(\mathbf{x}) P(\mathbf{x}) \, \mathrm{d}^M x$. Then, (4.130) becomes

$$Q K[P, P_{\mathrm{st}}] = F[P] - F[P_{\mathrm{st}}] \tag{4.131}$$

or

$$K[P, P_{\mathrm{st}}] = -(S[P] - S[P_{\mathrm{st}}]) + \frac{1}{Q}(U[P] - U[P_{\mathrm{st}}]) .\tag{4.132}$$

For the special case of the Boltzmann entropy, this relation between the Kullback measure, on the one hand, and the energy and entropy terms, on the other hand, has been discussed, for example, by Schlögl [508, 509]. Equation (4.132) demonstrates that Schlögl's relationship is not a peculiarity of the Boltzmann entropy but holds for all kinds of Kullback measures K, entropy and information measures S, and linear internal energy functionals U [167]. From the semi-positivity of K it follows that

$$F[P] \geq F[P_{\mathrm{st}}] .\tag{4.133}$$

If $F[P_{\mathrm{st}}]$ is finite, we can draw the conclusion that $F[P]$ is bounded from below. Furthermore, it is clear from (4.133) that for linear functionals U and concave measures S stationary distributions given by the free energy principle $\delta F/\delta P = \mu$ correspond to minima of F. In terms of the so-called Massieu potential

$$J[P] = S[P] - \frac{U[P]}{Q} ,\tag{4.134}$$

K can also be written as the difference

$$K[P, P_{\mathrm{st}}] = -(J[P] - J[P_{\mathrm{st}}]) .\tag{4.135}$$

Negative Concave Energy Functionals

We consider now a negative concave (or convex) energy functional U. Then, U satisfies the convexity inequality

$$U[P] \geq U[P_0] + \delta U[P_0](P - P_0) \,, \tag{4.136}$$

see also (4.117). In particular, for $P_0 = P_{\mathrm{st}}$ we have

$$U[P] \geq U[P_{\mathrm{st}}] + \delta U[P_{\mathrm{st}}](P - P_{\mathrm{st}})$$
$$\Rightarrow \delta U[P_{\mathrm{st}}](P - P_{\mathrm{st}}) \leq U[P] - U[P_{\mathrm{st}}] \,. \tag{4.137}$$

Consequently, from (4.128) we obtain

$$0 \leq K[P, P_{\mathrm{st}}]$$
$$= \frac{1}{Q} \delta U[P_{\mathrm{st}}](P - P_{\mathrm{st}}) - S[P] + S[P_{\mathrm{st}}]$$
$$\leq \frac{1}{Q} \left(U[P] - U[P_{\mathrm{st}}] \right) - S[P] + S[P_{\mathrm{st}}] \,. \tag{4.138}$$

Just as in the case of linear energy functionals, we arrive at the result

$$F[P] \geq F[P_{\mathrm{st}}] \,, \tag{4.139}$$

which implies that F is bounded from below provided that $F[P_{\mathrm{st}}]$ is finite.

Nonlinear Energy Functionals in Finite Phase Spaces

We consider systems with random variables defined on finite phase spaces Ω. We assume that the kernel of the functional U and the kernel of the functional $\delta U[P](P')$ are continuous functions. Then, these kernels are bounded from below and from above on the phase space Ω. Usually, we will then deal with functionals $U[P]$ and $\delta U[P](P')$ that are bounded like

$$|U[P]| < C_1, \quad |\delta U[P](P')| < C_2 \,. \tag{4.140}$$

For example, for $U_0 \in C^0(\Omega)$ and $P \in C^0(\Omega)$ the integral $\int_\Omega U_0(\mathbf{x}) P(\mathbf{x}) \, \mathrm{d}^M x$ is larger than or equal to $\min_{\mathbf{x} \in \Omega} \{U_0(\mathbf{x})\}$ and smaller than or equal to $\max_{\mathbf{x} \in \Omega} \{U_0(\mathbf{x})\}$. Assuming that (4.140) holds, from (4.128) we obtain

$$0 \leq K[P, P_{\mathrm{st}}]$$
$$= S[P_{\mathrm{st}}] - S[P] + \frac{1}{Q} \delta U[P_{\mathrm{st}}](P) - \delta U[P_{\mathrm{st}}](P_{\mathrm{st}})$$
$$\leq S[P_{\mathrm{st}}] - S[P] + 2 \frac{C_2}{Q}$$
$$= S[P_{\mathrm{st}}] - \frac{1}{Q} U[P_{\mathrm{st}}] - S[P] + \frac{1}{Q} U[P] + 2 \frac{C_2}{Q} + \frac{1}{Q}(U[P_{\mathrm{st}}] - U[P])$$
$$\leq \frac{1}{Q}(F[P] - F[P_{\mathrm{st}}]) + 2 \frac{C_1 + C_2}{Q} \,. \tag{4.141}$$

Consequently, if the integral $F[P_{\mathrm{st}}]$ exists (which is usually the case due to the finiteness of Ω), then $F[P]$ is bounded from below:

$$F[P] \geq F[P_{\text{st}}] - 2(C_1 + C_2) \ . \tag{4.142}$$

The boundedness of free energy measures of systems with finite phase spaces can also be shown in an alternative fashion [172]. Let us consider a phase space Ω with $\int_\Omega \mathrm{d}^M x = V < \infty$ (e.g., $\Omega = \prod_{i=1}^M [a_i, b_i] \Rightarrow V = \prod_{i=1}^M (b_i - a_i)$). Let $U[P]$ be bounded from below by $U[P] \geq U_{\min}$. Furthermore, we assume that S is maximal for the uniform distribution: $S_{\max} = S(P = 1/\int_\Omega \mathrm{d}^M x = 1/V)$. Then, we obtain

$$F[P] = U[P] - QS[P] \geq U_{\min} - Q\, S_{\max} \ . \tag{4.143}$$

Classical Mean Field Energy Functionals and Boltzmann Statistics

Of particular interest are systems that exhibit, on the one hand, Boltzmann statistics (i.e., Boltzmann distributions) and, on the other hand, involve energy functionals that are of second order with respect to probability densities (such functionals are typically found in systems with mean field interactions). That is, we assume now that $S = {}^{\text{B}}S = -\int_\Omega P \ln P \, \mathrm{d}^M x$ and

$$U = \int_\Omega U_0(\mathbf{x}) P(\mathbf{x}) \, \mathrm{d}^M x + \frac{1}{2} \int_\Omega \int_\Omega U_{\text{MF}}(\mathbf{x}, \mathbf{y}) P(\mathbf{x}) P(\mathbf{y}) \, \mathrm{d}^M x \, \mathrm{d}^M y \ . \tag{4.144}$$

We assume that the interaction potential U_{MF} satisfies the symmetry condition $U_{\text{MF}}(\mathbf{x}, \mathbf{y}) = U_{\text{MF}}(\mathbf{y}, \mathbf{x})$. The distortion functional (4.112) reads

$$G[u] = u \exp\left\{ \frac{1}{Q} \int_\Omega U_{\text{MF}}(\mathbf{x}, \mathbf{y})\, u(\mathbf{y}) \, \mathrm{d}^M y \right\} \ . \tag{4.145}$$

We define now the functional I given by

$$\begin{aligned} I[P, P_0] = &-\frac{1}{2Q} \int_\Omega \int_\Omega U_{\text{MF}}(\mathbf{x}, \mathbf{y}) P(\mathbf{x}) P(\mathbf{y}) \, \mathrm{d}^M x \, \mathrm{d}^M y \\ &+ \int_\Omega P(\mathbf{x}) \ln\left[\frac{G(P)}{G(P_0)} \right] \mathrm{d}^M x \end{aligned} \tag{4.146}$$

that involves two probability densities P and P_0 [166, 199]. For $U_{\text{MF}} = 0$ we have $G[u] = u$ and, consequently, $I[P, P_0]$ reduces to the Boltzmann–Kullback measure (4.123). Equation (4.146) can alternatively be expressed as

$$\begin{aligned} I[P, P_0] = &\frac{1}{2Q} \int_\Omega \int_\Omega U_{\text{MF}}(\mathbf{x}, \mathbf{y}) P(\mathbf{x}) P(\mathbf{y}) \, \mathrm{d}^M x \, \mathrm{d}^M y \\ &+ \int_\Omega P(\mathbf{x}) \ln\left[\frac{P}{G(P_0)} \right] \mathrm{d}^M x \ . \end{aligned} \tag{4.147}$$

The integral (4.147) can be used to compare stationary probability densities obtained from the free energy principle with arbitrary probability densities

4.7 Boundedness of Free Energy Functionals*

P. To this end, we put $P_0 = P_{\text{st}}$, where P_{st} satisfies (4.116). That is, we replace $G(P_0)$ by $G(P_{\text{st}}) = W/Z$ and obtain

$$I[P, P_{\text{st}}] = \ln Z + \frac{1}{2Q} \int_\Omega \int_\Omega U_{\text{MF}}(\mathbf{x},\mathbf{y}) P(\mathbf{x}) P(\mathbf{y}) \, d^M x \, d^M y$$
$$+ \underbrace{\int_\Omega P(\mathbf{x}) \ln\left[\frac{P}{W}\right] d^M x}_{{}^B K[P,W] \geq 0} \ . \tag{4.148}$$

Note that Z is the normalization constant of P_{st}. We assume that the mean field potential U_{MF} is bounded from below by $U_{\text{MF}}(\mathbf{x},\mathbf{y}) \geq C$ which implies that the ensemble average $f(\mathbf{x}) = \int_\Omega U_{\text{MF}}(\mathbf{x},\mathbf{y}) P(\mathbf{y}) d^M y$ is bounded from below by C. This result, in turn, implies that the ensemble average $\int_\Omega f(\mathbf{x}) P(\mathbf{x}) \, d^M x$ is bounded from below by C. Consequently, the double integral occurring in (4.148) is bounded from below by $C/(2Q)$ and from (4.148) we obtain

$$I[P, P_{\text{st}}] \geq \ln Z + \frac{C}{2Q} \ . \tag{4.149}$$

That is, $I[P, P_{\text{st}}]$ is bounded from below. In addition, the functional $I[P, P_{\text{st}}]$ is, up to a constant, equivalent to $F[P]/Q$. The reason for this is that ${}^B K[P,W]$ can be written as ${}^B K[P,W] = -{}^B S[P] - \int_\Omega P(\mathbf{x}) \ln W(\mathbf{x}) \, d^M x$. Using (4.115) for $U(\mathbf{x}) = U_0(\mathbf{x})$ and $Z_B = \int_\Omega \exp\{-U_0(\mathbf{x})/Q\} \, d^M x$, we can write ${}^B K[P,W]$ as ${}^B K[P,W] = -{}^B S[P] + Q^{-1} \int_\Omega P(\mathbf{x}) U_0(\mathbf{x}) \, d^M x + \ln Z_B$. Substituting this result into (4.148) gives us

$$I[P, P_{\text{st}}] = \ln(Z_B Z) - {}^B S[P] + \frac{U[P]}{Q} \tag{4.150}$$

or, alternatively, the final result:

$$I[P, P_{\text{st}}] = \frac{1}{Q} F[P] + \ln[Z_B Z] \ . \tag{4.151}$$

Comparing (4.149) and (4.151), we obtain

$$F[P] \geq -Q \ln(Z_B) + \frac{C}{2} \ . \tag{4.152}$$

The inequality (4.152) states that if we deal with systems that exhibit Boltzmann statistics and involve classical mean field interaction functionals that are bounded from below, then we have free energy measures that are bounded from below as well. This result holds for mean field systems defined on all kinds of phase spaces. While for systems defined on finite phase spaces we have derived the same result under much weaker conditions in the previous paragraph, the estimate (4.152) can in particular be used for systems defined on infinite phase spaces (e.g., $\Omega = \mathbb{R}^M$) and mixed phase spaces (e.g. $\Omega = \mathbb{R}^{M'} \times \prod_{i=1}^{M''}[a_i, b_i]$ with $M' + M'' = M$ such as phase oscillator systems with inertia terms).

Nonlinear Energy Functionals and Boltzmann Statistics

Here, we assume that we deal with the Boltzmann statistics and a general nonlinear energy functional U composed of a linear part U_L and a nonlinear part U_{NL}, see (4.2). Then, G is given by

$$G[u] = u \exp\left\{\frac{1}{Q}\frac{\delta U_{NL}}{\delta u}\right\}, \qquad (4.153)$$

see (4.112). With (4.153) at hand, we can generalize the functional (4.146) to

$$I[P, P_0] = \int_\Omega d^M x \int dP \ln\left[\frac{G(P)}{G(P_0)}\right]. \qquad (4.154)$$

The expression $\int dP \ldots$ denotes a functional integration, that is, an integration with respect to a function. We interpret the functional I as a functional that satisfies a particular ordinary first order differential equation. To this end, we introduce a real parameter κ and consider a probability density $\tilde{P}(\mathbf{x}, \kappa)$ depending on κ. Then, the integral (4.154) is defined as the integral $I[\tilde{P}, P_0]$ that satisfies

$$\frac{d}{d\kappa} I[\tilde{P}(\mathbf{x}; \kappa), P_0] = \int_\Omega \frac{d\tilde{P}}{d\kappa} \ln\left[\frac{G(\tilde{P})}{G(P_0)}\right] d^M x. \qquad (4.155)$$

For $P_0 = P_{st}$ the term $G(P_0)$ can be replaced by $G(P_{st}) = W/Z$. Analogously to the derivation of (4.151), we then conclude that

$$\frac{d}{d\kappa} I[\tilde{P}, P_{st}] = \int_\Omega \frac{d\tilde{P}}{d\kappa} \ln[G(\tilde{P})] d^M x + \int_\Omega \frac{d\tilde{P}}{d\kappa} \ln\left(Z_B\, Z \exp\{U_0/Q\}\right) d^M x$$

$$= \frac{d}{d\kappa}\left[-{}^B S[\tilde{P}] + \frac{U_{NL}[\tilde{P}]}{Q}\right] + \frac{1}{Q}\frac{d}{d\kappa} \int_\Omega U_0(\mathbf{x})\, \tilde{P} d^M x. \qquad (4.156)$$

Next, we integrate both sides of (4.156) with respect to κ. Subsequently, we drop the parameter κ (that has merely been used to carry out the functional integration using standard techniques). The result reads:

$$I[P, P_{st}] = \int_\Omega d^M x \int dP \ln\left[\frac{G(P)}{G(P_{st})}\right] = \frac{1}{Q} F[P] + I_0, \qquad (4.157)$$

where I_0 denotes an integration constant. Since $F[P]$ can be decomposed into $F = U_{NL} - Q\,{}^B S + U_L$ and U_L can be written as

$$U_L = \int_\Omega U_0(\mathbf{x}) P(\mathbf{x}) d^M x = -Q \int_\Omega P(\mathbf{x}) \ln[Z_B W(\mathbf{x})] d^M x, \qquad (4.158)$$

we can write F as $F = U_{NL} - Q \ln Z_B + Q \int_\Omega P \ln[P/W] d^M x$. Due to the semi-positivity of the Boltzmann–Kullback measure (i.e., because of

$\int_\Omega P\ln[P/W]\,\mathrm{d}^M x\geq 0$), we conclude that if the nonlinear energy functional is bounded from below by $U_{\text{NL,min}}$, then the free energy functional F and the functional I are bounded from below as well:

$$F[P]\geq -Q\ln Z_B+U_{\text{NL,min}}\ ,\tag{4.159}$$

$$I[P,P_{\text{st}}]\geq \ln Z_B+\frac{U_{\text{NL,min}}}{Q}+I_0\ .\tag{4.160}$$

Finally, we would like to mention that from (4.153) and (4.154) and the relation $\int_\Omega \mathrm{d}^M x[\int \mathrm{d}P\delta U_{\text{NL}}/\delta P]=U_{\text{NL}}[P]$ (see below), it follows that $I[P,P_0]$ can equivalently be expressed as

$$I[P,P_0]=\frac{1}{Q}U_{\text{NL}}[P]+\int_\Omega P(\mathbf{x})\ln\left[\frac{P}{G(P_0)}\right]\mathrm{d}^M x+I_0\ .\tag{4.161}$$

The functional (4.161) generalizes the functional (4.147). Likewise, as stated earlier, (4.154) contains (4.146) as a special case. In order to see this, note that the functional $Y[P]=0.5\int_\Omega\int_\Omega U_{\text{MF}}(\mathbf{x},\mathbf{y})P(\mathbf{x})P(\mathbf{y})\,\mathrm{d}^M x\,\mathrm{d}^M y$ with $U_{\text{MF}}(\mathbf{x},\mathbf{y})=U_{\text{MF}}(\mathbf{y},\mathbf{x})$ satisfies $\int_\Omega\int_\Omega PY[\delta Y/\delta P]\mathrm{d}^M x\,\mathrm{d}^M y=\delta Y[P](P)=2Y[P]$ with $\delta Y/\delta P=\int_\Omega U_{\text{MF}}(\mathbf{x},\mathbf{y})P(\mathbf{y})\,\mathrm{d}^M y$. Consequently, for $U_{\text{NL}}=Y$ we can add on the right-hand side of (4.161) a zero in the form of $0=-2Q^{-1}Y+Q^{-1}\delta Y[P](P)$ and write $Q^{-1}Y[P](P)$ as $\int_\Omega P\ln[\exp\{Q^{-1}\delta Y/\delta P\}]\,\mathrm{d}^M x$ which yields

$$I[P,P_0]=-\frac{1}{Q}Y+\int_\Omega P(\mathbf{x})\ln\left[\frac{P\exp\{Q^{-1}\delta Y/\delta P\}}{G(P_0)}\right]\mathrm{d}^M x+I_0\ .\tag{4.162}$$

and corresponds to (4.146) for $I_0=0$.

Functional Integration

Let $f(x)$ by a function defined on a one-dimensional domain Ω. Then, $\int_\Omega \mathrm{d}x[\int \mathrm{d}f\, g(f(x))]$ can be written as $\int_\Omega \int^{z=f(x)} g(z)\mathrm{d}z\,\mathrm{d}x$. That is, we integrate with respect to z, replace in the result thus obtained the variable z by $f(x)$, and finally integrate with respect to x. It is clear from the notion of a functional integration that the functional integration and the functional derivative can be regarded as inverse operations. For example, the functional $Y[f]=\int_\Omega \mathrm{d}x[\int \mathrm{d}f\,g(f(x))]$ can be written as $Y[f]=\int \mathrm{d}x H(f)$ with $H(f)=\int^f \mathrm{d}f'g(f')$, which implies that $\delta Y=\int \mathrm{d}x[\mathrm{d}H/\mathrm{d}f]\delta f$ and $\delta Y/\delta f=\mathrm{d}H/\mathrm{d}f=g(f)$. This finally leads to $Y[f]=\int_\Omega \mathrm{d}x[\int \mathrm{d}f\delta Y/\delta f]$.

Nonlinear Free Energy Functionals with Matching Condition

We consider a system with a nonlinear energy functional U given by (4.2) and an arbitrary entropy and information measure S. We assume that the system has at least one stationary state for which the free energy $F = U - QS$ and the nonlinear part U_{NL} of the energy measure are stationary: $\delta F[P_{\text{st}}] = 0$ and $\delta U_{\text{NL}}[P_{\text{st}}] = 0$. In this case, from (4.130) and $K \geq 0$ it follows that

$$F[P] \geq F[P_{\text{st}}] - U_{\text{NL}}[P_{\text{st}}] + U_{\text{NL}}[P] \ . \tag{4.163}$$

Assuming that the nonlinear part U_{NL} of the internal energy is bounded from below like $U_{\text{NL}}[P] \geq U_{\text{NL,min}}$, we obtain

$$F[P] \geq F[P_{\text{st}}] - U_{\text{NL}}[P_{\text{st}}] + U_{\text{NL,min}} \ . \tag{4.164}$$

That is, if the integrals $F[P_{\text{st}}]$ and $U_{\text{NL}}[P_{\text{st}}]$ exist (i.e., are finite), then the free energy F is bounded from below.

Summary

Let us summarize the results obtained so far. We have examined the boundedness of free energy measure for several generic cases. The conditions involved in these cases as well as the results are listed in Table 4.2. We conclude that there are a few essential ingredients that lead to the boundedness of free energy functionals. For systems described by finite phase spaces the boundedness results from the following assumptions: (i) energy measures are continuous and, therefore, bounded from below and above and (ii) stationary states exist with finite free energies. Alternatively, free energies are bounded if the following assumptions hold: (i) internal energy measures are bounded from below and (ii) entropy and information measures exhibit a global maximum (which is usually assumed for the uniform distribution). For systems defined on infinite phase spaces the boundedness of free energy measures is, roughly speaking, a consequence of the following properties: (i) energy measures are bounded from below, (ii) entropy and information measures are concave measures, and (iii) there exists at least one stationary state with finite values for internal energy, entropy or information, and free energy.

Table 4.2. Boundedness of several generic free energy functionals of the form $F = U - QS$ (f.b. = from below, f.a. = from above, ∞ = infinite phase space, U_L = linear energy functional)

Generic type	S	U	Ω	K, I	$F[P]$	$F[P_{\text{st}}]$
A	general	$U = U_L$	finite/∞	K	bounded f.b.	minimal
B	general	negative concave	finite/∞	K	bounded f.b.	minimal
C	general	bounded f.b. & f.a.	finite	K	bounded f.b.	–
D	general	bounded f.b.	finite	–	bounded f.b.	–
E	^BS	$U = U_L + U_{\text{MF}}$, U_{MF} bounded f.b.	finite/∞	I	bounded f.b.	–
F	^BS	$U = U_L + U_{\text{NL}}$, U_{NL} bounded f.b	finite/∞	I	bounded f.b.	–
G	general	$U = U_L + U_{\text{NL}}$, matching condition	finite/∞	K	bounded f.b.	–

4.8 First, Second, and Third Choice Thermostatistics

In the preceding section we have considered internal energy functionals such as

$$U[P] = \int_\Omega U_0(\mathbf{x}) P(\mathbf{x}) \, d^M x \, ,$$
$$U[P] = \int_\Omega U_0(\mathbf{x}) P(\mathbf{x}) \, d^M x + \frac{1}{2} \int_\Omega \int_\Omega U_{\text{MF}}(\mathbf{x}, \mathbf{y}) P(\mathbf{x}) P(\mathbf{y}) \, d^M x \, d^M y \, . \tag{4.165}$$

Functionals of this kind can be generalized by replacing P with a function of P, say, $f(P)$:

$$U[P] = \int_\Omega U_0(\mathbf{x}) f[P(\mathbf{x})] \, d^M x \, ,$$
$$U[P] = \int_\Omega U_0(\mathbf{x}) f[P(\mathbf{x})] \, d^M x + \frac{1}{2} \int_\Omega \int_\Omega U_{\text{MF}}(\mathbf{x}, \mathbf{y}) f[P(\mathbf{x})] f[P(\mathbf{y})] \, d^M x \, d^M y \, . \tag{4.166}$$

Alternatively, we may replace P by $f(P) / \int_\Omega f(P) \, d^M x$, which gives us internal energy functionals that read as

$$U[P] = \frac{\int_\Omega U_0(\mathbf{x}) f[P(\mathbf{x})] \, d^M x}{\int_\Omega f(P) \, d^M x} \, ,$$

$$U[P] = \frac{\int_\Omega U_0(\mathbf{x}) f[P(\mathbf{x})] \, \mathrm{d}^M x}{\int_\Omega f(P) \, \mathrm{d}^M x} + \frac{\int_\Omega \int_\Omega U_{\mathrm{MF}}(\mathbf{x},\mathbf{y}) f[P(\mathbf{x})] f[P(\mathbf{y})] \, \mathrm{d}^M x \, \mathrm{d}^M y}{2 [\int_\Omega f(P) \, \mathrm{d}^M x]^2}.$$
(4.167)

In particular, if we deal with entropy and information measures that involve power laws it has been suggested to consider internal energy functionals with $f(z) = z^q$. In this context, thermostatistics based on internal energy measures of the form (4.165), (4.166), and (4.167) have been called first, second, and third choice thermostatistics, respectively [106, 592], and have attracted considerable interest [2, 136, 202, 613, 641]. Moreover, the expression $P' = P^q / \int_\Omega P^q \mathrm{d}^M x$ that occurs in the third choice thermostatistics for $f(z) = z^q$ is known as the escort distribution of P [3, 48].

5 Free Energy Fokker–Planck Equations with Boltzmann Statistics

In Sect. 4.4 we have considered systems with linear internal energy functionals that exhibit Boltzmann statistics. In this chapter we will again consider systems that exhibit Boltzmann statistics but as opposed to Sect. 4.4 we will discuss nonlinear internal energy functionals. The nonlinear energy terms are assumed to reflect subsystem-subsystem interactions (see Sect. 1.1.2). That is, we are now concerned with many-body systems that involve interacting subsystems and exhibit Boltzmann statistics. Free energy measures of this kind of system read

$$F[P] = \int_\Omega U_0(\mathbf{x}) P(\mathbf{x}) \, \mathrm{d}^M x + U_{\mathrm{NL}}[P] - Q \,^{\mathrm{B}}S[P] \,, \tag{5.1}$$

where $^{\mathrm{B}}S$ denotes the Boltzmann entropy

$$^{\mathrm{B}}S[P] = -\int P \ln P \, \mathrm{d}^M x \tag{5.2}$$

(see Sect. 6.4.2). We assume that $U_{\mathrm{NL}}[P]$ is bounded from below like $U_{\mathrm{NL}}[P] \geq U_{\mathrm{NL,min}}$. We further assume that the evolution of $P(\mathbf{x}, t; u)$ for $t \geq t_0$ is given by the free energy Fokker–Planck equation

$$\frac{\partial}{\partial t} P(\mathbf{x}, t; u) = \nabla \cdot \left\{ P \nabla \frac{\delta F}{\delta P} \right\} \,, \tag{5.3}$$

see (4.4), which can equivalently be expressed as

$$\frac{\partial}{\partial t} P(\mathbf{x}, t; u) = \nabla \cdot \left[\left\{ \nabla U_0(\mathbf{x}) + \nabla \frac{\delta U_{\mathrm{NL}}[P]}{\delta P} \right\} P \right] + Q \, \Delta P \,. \tag{5.4}$$

We restrict ourselves to considering systems for which stationary solutions can be obtained from the free energy principle (4.17). Since the entropy kernel of $^{\mathrm{B}}S$ is given by $s(z) = -z \ln z$, we have $\mathrm{d}s/\mathrm{d}z = -\ln z - 1$ and $[\mathrm{d}s/\mathrm{d}z]^{-1} = \exp\{-z-1\}$. In addition, the outer function $B(z)$ of $^{\mathrm{B}}S$ simply reads $B(z) = z$. Consequently, (4.20) becomes

$$P_{\mathrm{st}}(\mathbf{x}) = \frac{1}{Z} \exp\left\{ -\frac{U_0(\mathbf{x}) + \delta U_{\mathrm{NL}}[P_{\mathrm{st}}]/\delta P}{Q} \right\} , \tag{5.5}$$

where Z denotes a normalization constant given by $\ln Z = 1 - \mu/Q$ (see also Sect. 4.4).

Finite Dimensional Nonlinearities

In order to evaluate the implicit equation (5.5), we consider nonlinear Fokker–Planck equations with finite dimensional nonlinearities (i.e., we have $d < \infty$, see Sect. 2.6). In this case, the functional derivative $\delta U_{\rm NL}[P]/\delta P$ can be regarded as a function of $\mathbf{x}$ and a set of expectation values $\langle \mathbf{A} \rangle = \int_\Omega \mathbf{A}(\mathbf{x}) P(\mathbf{x}) \, \mathrm{d}^M x$ with $\mathbf{A}(\mathbf{x}) = (A_1(\mathbf{x}), \ldots, A_d(\mathbf{x}))$:

$$\frac{\delta U_{\rm NL}[P]}{\delta P} = \frac{\delta U_{\rm NL}}{\delta P}(\mathbf{x}, \langle \mathbf{A} \rangle) \,. \tag{5.6}$$

The expectation values may be regarded as order parameters (see Sect. 1.2). Equations (5.4) and (5.5) can now be written as

$$\frac{\partial}{\partial t} P(\mathbf{x}, t; u) = \nabla \cdot \left[\left\{ \nabla U_0(\mathbf{x}) + \nabla \frac{\delta U_{\rm NL}}{\delta P}(\mathbf{x}, \langle \mathbf{A} \rangle) \right\} P \right] + Q \, \Delta P \tag{5.7}$$

and

$$P_{\rm st}(\mathbf{x}) = \frac{1}{Z} \exp \left\{ -\frac{U_0(\mathbf{x}) + \frac{\delta U_{\rm NL}}{\delta P}(\mathbf{x}, \langle \mathbf{A} \rangle_{\rm st})}{Q} \right\} \,. \tag{5.8}$$

Equation (2.25) reads

$$P(\mathbf{x}, \mathbf{m}) = \frac{1}{Z} \exp \left\{ -\frac{U_0(\mathbf{x}) + \frac{\delta U_{\rm NL}}{\delta P}(\mathbf{x}, \mathbf{m})}{Q} \right\} , \tag{5.9}$$

with Z given by $Z(\mathbf{m}) = \int_\Omega \exp \{-\{U_0(\mathbf{x}) + \delta U_{\rm NL}/\delta P(\mathbf{x}, \mathbf{m})\}/Q\} \, \mathrm{d}^M x$. Using (2.26) in the form of

$$\mathbf{R}(\mathbf{m}) = \int_\Omega \mathbf{A}(\mathbf{x}) P(\mathbf{x}, \mathbf{m}) \, \mathrm{d}^M x \,, \tag{5.10}$$

we define the self-consistency equation

$$\mathbf{m} = \mathbf{R}(\mathbf{m}) \,. \tag{5.11}$$

As shown in Sect. 2.3, solutions of the self-consistency equation (5.11) determine the order parameter values $\langle \mathbf{A} \rangle_{\rm st}$. That is, solving (5.11) and substituting $\langle \mathbf{A} \rangle_{\rm st} = \mathbf{m}$ into (5.8) gives us the stationary solutions of the free energy Fokker–Planck equation (5.7).

5.1 Stability Analysis

In the context of nonlinear Fokker–Planck equations it is useful to distinguish between different kinds of stationary distributions: unstable, stable and

5.1 Stability Analysis

asymptotically stable ones. The basic idea is to adopt the stability analysis of deterministic systems [379, 610] in order to determine the stability of stationary distributions of nonlinear Fokker–Planck equations. In line with this idea we will determine the stability of stationary distributions by studying the evolution of transient solutions that correspond to perturbed stationary distributions. In what follows, we will restrict our attention to Fokker–Planck equations with coefficients that do not explicitly depend on time.

Let $P_{\text{st}}(\mathbf{x})$ denote a stationary probability density. Let $P(\mathbf{x},t;u)$ denote a time-dependent probability density that is at $t = t_0$ in some kind of neighborhood of P_{st}: $u \approx P_{\text{st}}$. Then, we denote by $\epsilon(\mathbf{x})$ the perturbation given by $\epsilon = P(\mathbf{x},t_0) - P_{\text{st}}$. The function ϵ is assumed to be a small quantity. We refer to a stationary probability density as asymptotically stable if all perturbations of the probability density vanish:

$$\forall \, \epsilon : \lim_{t \to \infty} P(\mathbf{x}, t, P_{\text{st}} + \epsilon) = P_{\text{st}}(\mathbf{x}) \Rightarrow P_{\text{st}} = \text{asymptotically stable} \quad (5.12)$$

(see also Fig. 5.1a). We refer to a stationary probability density as stable if all perturbations remain in a neighborhood of the probability density. A special case of a stable stationary probability density is a distribution P_{st}, for which perturbations vanish or give rise to further stationary distributions P'_{st}:

$$\forall \, \epsilon : \left[\lim_{t \to \infty} P(\mathbf{x}, t; P_{\text{st}} + \epsilon) = P_{\text{st}}(\mathbf{x}) \right] \text{ or } [P'_{\text{st}}(\mathbf{x}) = P_{\text{st}}(\mathbf{x}) + \epsilon]$$
$$\Rightarrow P_{\text{st}} = \text{stable} \quad (5.13)$$

(see also Fig. 5.1b). Finally, we refer to a stationary probability density as unstable if the probability density is not stable. In other words, a stationary probability density P_{st} is unstable if there exist perturbed distributions $P = P_{\text{st}} + \epsilon$ that leave the neighborhood of P_{st} no matter how small we choose the perturbation ϵ. In line with this notion, a sufficient condition for P_{st} to be an unstable distribution is given by

$$\exists \, \epsilon : \lim_{t \to \infty} P(\mathbf{x}, t; P_{\text{st}} + \epsilon) = P'_{\text{st}}(\mathbf{x}) \, , \; P'_{\text{st}}(\mathbf{x}) \neq P_{\text{st}}(\mathbf{x}) + \mathrm{d}\chi(\mathbf{x})$$
$$\Rightarrow P_{\text{st}} = \text{unstable}. \quad (5.14)$$

Equation (5.14) states that P_{st} is unstable if there are perturbations ϵ of P_{st} for which $P(\mathbf{x}, t; P_{\text{st}} + \epsilon)$ converges to a stationary probability density P'_{st} that is significantly different from P_{st} although ϵ can be chosen arbitrarily small. By significantly different we mean that P'_{st} is not in some kind of neighborhood of P_{st} and cannot be derived from P_{st} by means of a perturbation $\mathrm{d}\chi$ of P_{st} (see also Fig. 5.1c).

In some cases, a norm $\|\cdot\|$ for the perturbations of stationary probability densities can be found. By means of the evolution of this norm, one can then determine the stability of stationary distributions as well. For example, if for all (nonvanishing) small deviations ϵ the norm $\|\epsilon\|$ decreases with time, then

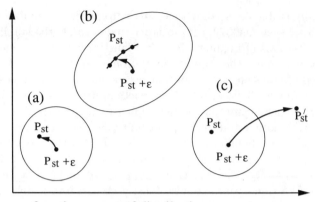

Fig. 5.1. Examples of stationary distributions P_{st}: an asymptotically stable distribution (*panel* (**a**)), a family of stable distributions (*panel* (**b**)), and an unstable distribution (*panel* (**c**)). The circles define neighborhoods of the stationary distributions. In panel (**b**) the line with the dots describes a continuous family of stationary distributions (the dots represent the stationary distributions of the family and actually are located infinitely close to each other)

for $t \to \infty$ we have $\epsilon = 0 \Rightarrow P = P_{st}$ and, consequently, P_{st} is asymptotically stable:

$$\forall \epsilon : \frac{d}{dt} \|\epsilon\| < 0 \Rightarrow P_{st} = \text{asymptotically stable.} \quad (5.15)$$

In contrast, if for all perturbations the norm increases with time, then we conclude that P_{st} is unstable;

$$\forall \epsilon : \frac{d}{dt} \|\epsilon\| > 0 \Rightarrow P_{st} = \text{unstable.} \quad (5.16)$$

Let us determine next the stability of the stationary distributions by means of Lyapunov's direct method, linear stability analysis, and self-consistency equation analysis [135, 172, 176, 187, 198, 199, 365, 522, 523, 526, 527].

5.1.1 Lyapunov's Direct Method

Lyapunov's direct method for deterministic systems [379, 610] can be adopted to examine evolution equations for probability densities P [172, 176, 187, 199, 249, 522, 523, 526, 527]. Accordingly, Lyapunov's direct method can be applied to evolution equations for P for which a functional $F[P]$ exists that satisfies

$$F \geq F_{min}, \quad \frac{d}{dt}F \leq 0, \quad \frac{d}{dt}F = 0 \Leftrightarrow \frac{\partial}{\partial t}P = 0. \quad (5.17)$$

In this case, F is referred to as the Lyapunov functional, see Sect. 4.3. In the case of free energy Fokker–Planck equations this requirement is satisfied (see (4.29)) and the Lyapunov functional corresponds to a free energy measure.

Let us consider perturbations $\epsilon(\mathbf{x})$ of a stationary distribution $P_{\text{st}}(\mathbf{x})$. If $\delta^2 F[P_{\text{st}}](\epsilon) > 0$ for all $\epsilon \neq 0$, then P_{st} corresponds to a minimum of F. From $\delta F[P_{\text{st}}]/\delta P = \mu \Rightarrow \delta F[P_{\text{st}}](\epsilon) = 0$ and $\delta^2 F[P_{\text{st}}](\epsilon) > 0$, we conclude that for small perturbations ϵ and ϵ' we have $\delta F[P_{\text{st}} + \epsilon](\epsilon') \neq 0$. That is, for small ϵ there does not exist a probability density $P = P_{\text{st}} + \epsilon$ that corresponds to another stationary point of F and describes a further stationary solution of the Fokker–Planck equation. Since from (5.17) we have $\partial P/\partial t \neq 0 \Rightarrow \mathrm{d}F/\mathrm{d}t < 0$ we conclude that every perturbed stationary solution $P = P_{\text{st}} + \epsilon$ evolves in such a way that the inequality $\mathrm{d}F/\mathrm{d}t < 0$ is satisfied as long as $\epsilon \neq 0$. Since $P = P_{\text{st}} + \epsilon$ is located in a neighborhood of a minimum of F, the time-dependent solution P cannot leave this neighborhood (because this would imply an increase of F). The only behavior of the solution $P = P_{\text{st}} + \epsilon$ that is consistent with the constraint $\mathrm{d}F/\mathrm{d}t < 0$ for $\epsilon \neq 0$ and the fact that P is located (in the function space of probability densities) in a neighborhood of a minimum of F is the relaxation to the unperturbed state described by P_{st}. Therefore, we get

$$\forall \epsilon \neq 0 : \delta^2 F[P_{\text{st}}](\epsilon) > 0 \Rightarrow P_{\text{st}} = \text{asymptotically stable}. \quad (5.18)$$

Let P_{st} correspond to a maximum or a saddle point of F such that there is at least one small perturbation ϵ which yields $\delta^2 F[P_{\text{st}}](\epsilon) < 0$. Then, the inequality $F[P_{\text{st}} + \epsilon] < F[P_{\text{st}}]$ holds and the time-dependent probability density $P(\mathbf{x}, t; u)$ with $u = P_{\text{st}} + \epsilon$ at $t = t_0$ cannot return to $P_{\text{st}}(\mathbf{x})$ for any $t \geq t_0$ (because F cannot increase and $F(t_0) < F[P_{\text{st}}]$). Since the system under consideration has a Lyapunov functional the H-theorem discussed in Sect. 4.3 applies and the system converges to a stationary distribution P'_{st}. If P_{st} corresponds to a maximum of F, then there cannot exist another stationary distribution in a neighborhood of P_{st}, which implies that P'_{st} must be significantly different from P_{st}. If P_{st} corresponds to a saddle point of F, then one often can show again that P'_{st} must be significantly different from P_{st}. Therefore, usually, the implication (5.14) holds and we obtain:

$$\exists \epsilon : \delta^2 F[P_{\text{st}}](\epsilon) < 0 \Rightarrow P_{\text{st}} = \text{unstable}. \quad (5.19)$$

In sum, the sign of the second variation of the free energy F determines the stability of stationary distributions. For F given by (5.1) the second variation reads

$$\delta^2 F[P_{\text{st}}](\epsilon) = \int_\Omega \int_\Omega \frac{\delta^2 U_{\text{NL}}[P_{\text{st}}]}{\delta P(\mathbf{x}) \delta P(\mathbf{y})} \epsilon(\mathbf{x}) \epsilon(\mathbf{y}) \, \mathrm{d}^M x \, \mathrm{d}^M y + Q \int_\Omega \frac{\epsilon^2}{P_{\text{st}}} \, \mathrm{d}^M x \ . \quad (5.20)$$

5.1.2 Linear Stability Analysis

In order to determine the stability of stationary solutions of (5.4) by means of linear stability analysis it is helpful to require that $\delta^2 U_{\text{NL}}/\delta P(\mathbf{x})\delta P(\mathbf{y}) =$

$\chi'(\mathbf{x},\mathbf{y})$ satisfies the symmetry relation $\chi'(\mathbf{x},\mathbf{y}) = \chi'(\mathbf{y},\mathbf{x})$, which implies that the symmetry relation $\chi(\mathbf{x},\mathbf{y}) = \chi(\mathbf{y},\mathbf{x})$ holds for $\delta^2 F/\delta P(\mathbf{x})\delta P(\mathbf{y}) = \chi(\mathbf{x},\mathbf{y})$. Next, we linearize (5.3) at P_{st}. That is, we put $P(\mathbf{x},t;u) = P_{\text{st}}(\mathbf{x}) + \epsilon(\mathbf{x},t)$. Using (4.17) and taking only ϵ-terms of first order into account, we obtain

$$\frac{\partial}{\partial t}\epsilon(\mathbf{x},t) = \nabla \cdot \left\{ P_{\text{st}} \nabla \int_\Omega \frac{\delta^2 F[P_{\text{st}}]}{\delta P(\mathbf{x})\delta P(\mathbf{y})} \epsilon(\mathbf{y},t)\,\mathrm{d}^M y \right\} . \tag{5.21}$$

Let us consider the functional L defined by the second variation of F as

$$L[P, P_{\text{st}}] = \frac{1}{2}\delta^2 F[P_{\text{st}}](\epsilon) , \tag{5.22}$$

with $\epsilon = P - P_{\text{st}}$. Differentiating L with respect to t, exploiting the symmetry of $\delta^2 F/\delta P(\mathbf{x})\delta P(\mathbf{y})$, integrating by parts and assuming that the surface terms vanish, from (5.21) we obtain

$$\frac{\mathrm{d}}{\mathrm{d}t}L = \frac{\mathrm{d}}{\mathrm{d}t}\frac{1}{2}\delta^2 F[P_{\text{st}}](\epsilon)$$

$$= -\int_\Omega P_{\text{st}} \left[\nabla \int_\Omega \frac{\delta^2 F[P_{\text{st}}]}{\delta P(\mathbf{x})\delta P(\mathbf{y})} \epsilon(\mathbf{y},t)\,\mathrm{d}^M y \right]^2 \mathrm{d}^M x \leq 0 . \tag{5.23}$$

It is clear from this result that $\mathrm{d}L/\mathrm{d}t = 0$ implies

$$\int_\Omega \frac{\delta^2 F[P_{\text{st}}]}{\delta P(\mathbf{x})\delta P(\mathbf{y})} \epsilon(\mathbf{y},t)\,\mathrm{d}^M y = C , \tag{5.24}$$

where C is a constant. Multiplying (5.24) by $\epsilon(\mathbf{x},t)$, integrating the result with respect to $\mathbf{x}$, and taking the normalization constraint $\int_\Omega \epsilon(\mathbf{x},t)\,\mathrm{d}^M x = 0$ into account, we get

$$\frac{\mathrm{d}}{\mathrm{d}t}L = 0 \Rightarrow \delta^2 F[P_{\text{st}}](\epsilon) = 0 . \tag{5.25}$$

In sum, using linear stability analysis, two fundamental results can be found [187]: the inequality (5.23) and the implication (5.25). These results can now be used to analyze the stability of stationary probability densities related to stationary (or critical) points of F.

Minima

Let us consider a distribution function P_{st} with $\delta^2 F[P_{\text{st}}](\epsilon) > 0$ for all $\epsilon(\mathbf{x}) \neq 0$ and $\delta^2 F[P_{\text{st}}](\epsilon) = 0 \Leftrightarrow \epsilon = 0$. Then, from (5.22), (5.23), and (5.25) it follows that

$$L \geq 0 , \quad \frac{\mathrm{d}}{\mathrm{d}t}L \leq 0 , \quad \frac{\mathrm{d}}{\mathrm{d}t}L = 0 \Leftrightarrow \epsilon = 0 . \tag{5.26}$$

Consequently, the limiting cases $\lim_{t\to\infty} \epsilon(\mathbf{x},t) = 0$ and $\lim_{t\to\infty} P(\mathbf{x},t;u) = P_{st}$ hold for all initial distributions $u \approx P_{st}$. In other words, by means of linear stability analysis related to the second variation of free energies we reobtain the proposition (5.18). Next, we define the norm $\|\cdot\|$ for positive definite $\delta^2 F$ and functions $\epsilon(\mathbf{x})$ with $\int_\Omega \epsilon(\mathbf{x}) \, \mathrm{d}^M x = 0$ by

$$\|\epsilon\| = \sqrt{\delta^2 F[P_{st}](\epsilon)} \ . \tag{5.27}$$

For details, see Sect. 5.1.8 below. Using $2L(\epsilon) = \delta^2 F[P_{st}](\epsilon) = \|\epsilon\|^2 \Rightarrow \mathrm{d}L/\mathrm{d}t = \|\epsilon\| \, \mathrm{d}\|\epsilon\|/\mathrm{d}t$ and (5.26), we obtain

$$\forall \epsilon \neq 0 \Rightarrow \frac{\mathrm{d}}{\mathrm{d}t} \|\epsilon\| < 0 \ . \tag{5.28}$$

Equation (5.28) tells us that the norm of every small perturbation of P_{st} decreases as a function of time if P_{st} corresponds to a free energy minimum with $\delta^2 F > 0$, which means once again that P_{st} is asymptotically stable.

Maxima

If P_{st} describes a maximum of F with $\delta^2 F[P_{st}](\epsilon) < 0$ for all $\epsilon(\mathbf{x}) \neq 0$, we can introduce a norm defined by

$$\|\epsilon\| = \sqrt{-\delta^2 F[P_{st}](\epsilon)} \ , \tag{5.29}$$

see Sect. 5.1.8. Using $-2L(\epsilon) = -\delta^2 F[P_{st}](\epsilon) = \|\epsilon\|^2 \Rightarrow -\mathrm{d}L/\mathrm{d}t = \|\epsilon\| \, \mathrm{d}\|\epsilon\|/\mathrm{d}t$ and (5.26), we obtain

$$\forall \epsilon \neq 0 \Rightarrow \frac{\mathrm{d}}{\mathrm{d}t} \|\epsilon\| > 0 \ . \tag{5.30}$$

Equation (5.30) tells us that the norm of every small perturbation of P_{st} increases as a function of time if P_{st} corresponds to a free energy maximum with $\delta^2 F < 0$, which means that P_{st} is unstable. In sum, we realize that Lyapunov's direct method and linear stability analysis yield consistent results irrespective of the form of internal energy functionals U. From both methods it follows that stationary probability densities are asymptotically stable if they correspond to free energy minima with positive definite second variations. In contrast, if there is a perturbation of a stationary probability distribution that involves a decrease of the free energy (5.1) then the stationary distribution is (usually) an unstable one.

5.1.3 Self-Consistency Equation Analysis

In Sect. (2.3) we have shown that self-consistency equations can be used to determine the stationary probability densities of nonlinear Fokker–Planck equations. Self-consistency equations can also be used to examine the stability of the stationary probability densities. Let us consider the one-dimensional self-consistency equation

$$m = R(m) \ . \tag{5.31}$$

We assume that solutions of (5.31) determine the stationary order parameter values $\langle A \rangle_{st}$ that, in turn, determine the stationary distributions $P_{st}(x; \langle A \rangle_{st})$. In this context, frequently the claim is made that the derivative of $R(m)$ at $m = \langle A \rangle_{st}$ determines the stability of a stationary distribution. In particular, the hypothesis is made that the following implications hold:

$$\left. \frac{dR}{dm} \right|_{\langle A \rangle_{st}} < 1 \Rightarrow P_{st} \text{ is asymptotically stable,}$$

$$\left. \frac{dR}{dm} \right|_{\langle A \rangle_{st}} > 1 \Rightarrow P_{st} \text{ is unstable.} \tag{5.32}$$

Equation (5.32) states that if the slope of $R(m)$ at an intersection point of $R(m)$ with the diagonal is larger than unity, we deal with an unstable stationary probability density. If the slope of $R(m)$ at $m = \langle A \rangle$ is smaller than unity then we deal with an asymptotically stable stationary distribution, see Fig. 5.2. Note that there is no rigorous proof for the statements "P_{st} is asymptotically stable" and "P_{st} is unstable" in (5.32). Equation (5.32) provides us with a helpful hint that has to be verified or falsified by alternative considerations (e.g., by solving numerically the equations under consideration). Note also that we do not make any claims about stationary distributions related to $dR(\langle A \rangle_{st})/dm = 1$.

Where does (5.32) come from? We may say that the proposition (5.32) is inspired by the theory of iterative maps [409]. Let $R(z)$ denote a function that defines the mapping

$$x_n = R(x_{n-1}) \ . \tag{5.33}$$

Equation (5.33) defines for every initial value x_0 a sequence of values $\{x_0, x_1, \ldots\}$ and is referred to as an iterative map. The stationary points of the map (5.33) satisfy

$$x_{st} = R(x_{st}) \ . \tag{5.34}$$

Figure 5.3 shows an example of a function $R(z)$ and the stationary points of the corresponding map. For the iterative map (5.34) one can prove that the stability of a stationary point x_{st} is determined by the derivative of R at the point x_{st}. A stationary point x_{st} is asymptotically stable if $|dR(x_{st})/dx| < 1$ holds and unstable if $|dR(x_{st})/dx| > 1$ holds. Consequently, the stability criteria of the iterative map resemble those given by (5.32). Comparing the stability analysis by means of self-consistency equations and the stability analysis for iterative maps, we realize that there are two differences. First, as we have mentioned earlier, the validity of (5.32) has not yet been proven for the general case. In contrast, for iterative maps it can be proven rigorously

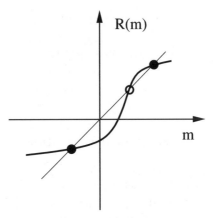

Fig. 5.2. Left- and right-hand sides of a self-consistency equation. Intersection points describe stationary distributions. Self-consistency equation analysis makes the hypothesis that intersection points marked by full (empty) dots describe asymptotically stable (unstable) stationary distributions

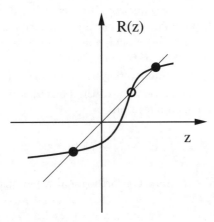

Fig. 5.3. Left- and right-hand sides of the self-consistency equation of an iterative map. Intersection points describe stationary points. Intersection points marked by full (empty) dots describe asymptotically stable (unstable) stationary points

that self-consistency equation analysis yields the correct results. Second, stationary points x_{st} of iterative maps are unstable for $dR(x_{\text{st}})/dx < -1$. That is, the amount of the slope is a crucial issue here. In contrast, as we will show in Sect. 5.1.5, stationary probability densities of nonlinear Fokker–Planck equations can be stable if the inequality $dR(\langle A \rangle_{\text{st}})/dm < -1$ holds. It seems that the crucial issue for the stability of stationary probability densities of nonlinear Fokker–Planck equations is whether or not the slope is larger than unity.

5.1.4 Shiino's Decomposition of Perturbations

Shiino proposed to decompose perturbations of stationary probability densities into particular functions such that, on the one hand, all kinds of perturbations can be represented and, on the other hand, the decomposition can be used to determine the character of extrema of free energy measures [522, 523]. Let us illustrate Shiino's decomposition for a perturbation $\epsilon(x)$. First, we write $\epsilon(x)$ as

$$\epsilon(x) = \sqrt{P_{\text{st}}(x)}\, \epsilon'(x)\,, \tag{5.35}$$

where P_{st} describes a critical point (extrema) of a free energy measure. Let us proceed with three remarks concerning (5.35). First, we assume that $P_{\text{st}} > 0$ for $x \in \Omega$. Therefore, (5.35) describes a one-to-one mapping between the functions $\epsilon(x)$ and $\epsilon'(x)$. Consequently, if we consider $\epsilon'(x)$ rather than $\epsilon(x)$, then we still take all possible perturbations into account. Second, if we examine different stationary or critical points of free energy measures then we are dealing with different stationary distributions P_{st}, say, P_{st}^1, P_{st}^2, and so on. Consequently, we obtain different mappings: $\epsilon(x) = \sqrt{P_{\text{st}}^1(x)}\,\epsilon'(x)$, $\epsilon(x) = \sqrt{P_{\text{st}}^2(x)}\,\epsilon'(x)$, and so on. Third, the normalization condition for $\epsilon(x)$ reads $\int_\Omega \epsilon(x)\, dx = 0$ and implies that

$$\int_\Omega \sqrt{P_{\text{st}}(x)}\, \epsilon'(x)\, dx = 0\,. \tag{5.36}$$

Next, we decompose $\epsilon'(x)$ into two orthogonal functions: $\chi_0(x)$ and $\chi_\perp(x)$. That is, we put

$$\epsilon'(x) = \chi_0(x) + \chi_\perp(x) \tag{5.37}$$

with

$$\int_\Omega \chi_0(x)\chi_\perp(x)\, dx = 0\,. \tag{5.38}$$

The key idea is to relate $\chi_0(x)$ to the order parameter functions of the problem at hand, while $\chi_\perp(x)$ is used to guarantee that we still consider all possible perturbations. Let us assume that we have an order parameter $\langle A \rangle$. Then, we put

$$\chi_0(x) = \beta[A(x) - \langle A \rangle]\sqrt{P_{\text{st}}(x)}\,, \tag{5.39}$$

where β describes an amplitude. As we will show in a moment, this choice of $\chi_0(x)$ is a very useful one. From (5.39) it follows that $\epsilon(x)$ can be written as

$$\epsilon(x) = \beta[A(x) - \langle A \rangle]P_{\text{st}}(x) + \chi_\perp(x)\sqrt{P_{\text{st}}(x)} \ . \tag{5.40}$$

The normalization condition $\int_\Omega \epsilon(x)\,dx = 0$ implies that the integral relation

$$\int_\Omega \chi_\perp(x)\sqrt{P_{\text{st}}(x)}\,dx = 0 \tag{5.41}$$

holds. Equations (5.38), (5.39), and (5.41), in turn, imply that $\chi_\perp(x)$ satisfies

$$\int_\Omega A(x)\chi_\perp(x)\sqrt{P_{\text{st}}(x)}\,dx = 0 \ . \tag{5.42}$$

Now, let us demonstrate the power of this kind of expansion. In order to discuss the character of free energy extrema we will often be concerned with two kinds of integrals:

$$\langle A \rangle_\epsilon = \int_\Omega A(x)\epsilon(x)\,dx \ , \tag{5.43}$$

$$\delta^2 S_{\text{B}}[P_{\text{st}}] = \int_\Omega \frac{\epsilon^2(x)}{P_{\text{st}}(x)}\,dx \ , \tag{5.44}$$

where $S_{\text{B}} = {}^{\text{B}}S$ denotes the Boltzmann entropy. Using (5.40) and (5.42), we find

$$\langle A \rangle_\epsilon = \beta K_A(X) \ . \tag{5.45}$$

In (5.45) the variable $K_A(X)$ denotes the generalized variance defined by

$$K_A(X) = \langle A(X)^2 \rangle - \langle A(X) \rangle^2 = \langle [A - \langle A \rangle]^2 \rangle \ . \tag{5.46}$$

Note that by definition we have $K_A(X) \geq 0$. Moreover, for $A(x) = x$ the variance $K_A(X)$ reduces to the conventional one: $K(X) = \langle X^2 \rangle - \langle X \rangle^2$. Using (5.40-5.42), we find

$$\delta^2 S_{\text{B}}[P_{\text{st}}] = \beta^2 K_A(X) + \int_\Omega \chi_\perp^2(x)\,dx \ . \tag{5.47}$$

Let us summarize. If we need to determine the character of stationary points of F by evaluating $\delta^2 F$, then we will often deal with expressions like $\langle A \rangle_\epsilon$ and $\delta^2 S_{\text{B}}[P_{\text{st}}]$. If we use Shiino's decomposition, these expressions can conveniently be expressed in terms of (5.45) and (5.47). The amplitude β measures the strength of a perturbation $\epsilon \propto [A(x) - \langle A \rangle]P_{\text{st}}(x)$. There are two extreme cases. First, we may put $\chi_\perp = 0$ and $\beta \neq 0$. Then, we consider perturbations

$$\epsilon(x) = \beta[A(x) - \langle A \rangle]P_{\text{st}}(x) \ . \tag{5.48}$$

Second, we may put $\chi_\perp \neq 0$ and $\beta = 0$ and consider perturbations

$$\epsilon(x) = \chi_\perp(x)\sqrt{P_{\mathrm{st}}(x)}, \qquad (5.49)$$

where $\chi_\perp(x)$ denotes an arbitrary function satisfying the constraints (5.41) and (5.42).

Shiino's decomposition can be generalized to the case that we deal with several order parameters $A_1, \ldots, A_d$ [199, 526, 527]. Then, we need to introduce several amplitudes β_i, see also below. In addition, the decomposition can be generalized for free energy measures involving entropies different from the Boltzmann entropy [525, 526, 527].

5.1.5 Generic Cases

K-Model: $F \propto K$

A free energy Fokker–Planck equation that has frequently been studied is given by

$$\frac{\partial}{\partial t} P(x,t;u) = \frac{\partial}{\partial x}\left[\frac{\mathrm{d}V(x)}{\mathrm{d}x} + \kappa\left(x - \langle X \rangle_P\right)\right] P + Q \frac{\partial^2}{\partial x^2} P, \qquad (5.50)$$

with $\kappa \geq 0$. We can verify that (5.50) can be written as

$$\frac{\partial}{\partial t} P(x,t;u) = \frac{\partial}{\partial x} P \frac{\partial}{\partial x} \frac{\delta F}{\delta P}, \qquad (5.51)$$

with F given by

$$F[P] = \int_\Omega V(x)P(x)\,\mathrm{d}x + \frac{\kappa}{2} K(X) - Q\,^{\mathrm{B}}S, \qquad (5.52)$$

where K denotes the variance $K = \langle X^2 \rangle - \langle X \rangle^2$ and $^{\mathrm{B}}S$ is the Boltzmann entropy $^{\mathrm{B}}S = -\int_\Omega P \ln P\,\mathrm{d}x$. We will refer to (5.50) as the K-model. Note that the K-model is usually discussed in the context of natural boundary conditions. That is, we have $x \in \Omega = \mathbb{R}$. Note also that (5.51) is the univariate version of (5.3). Finally, note that for the K-model a fluctuation–dissipation theorem can be found, see Sect. 3.11. From $U_{\mathrm{NL}} = \kappa K/2 > 0$ it follows that the nonlinear energy functional U_{NL} is bounded from below, which implies that F is bounded from below for all potentials V for which the Boltzmann distribution of V exists (see type E in Table 4.2). Consequently, the H-theorem for free energy Fokker–Planck equations applies and we conclude that transient solutions of the K-model converge to stationary ones in the long time limit. The free energy (5.52) may alternatively be expressed as

$$F[P] = \int_\Omega V(x)P(x)\,\mathrm{d}x + \frac{\kappa}{4} \int_\Omega \int_\Omega [x-y]^2 P(x)P(y)\,\mathrm{d}x\,\mathrm{d}y - Q\,^{\mathrm{B}}S \qquad (5.53)$$

because of $2K(X) = \int_\Omega [x-y]^2 P(x)P(y)\,dx\,dy$. Re-arranging the terms, we obtain

$$F[P] = \int_\Omega [V(x)+\frac{\kappa}{2}x^2]P(x)\,dx - \frac{\kappa}{2}\int_\Omega\int_\Omega xy P(x)P(y)\,dx\,dy - Q^{\mathrm{B}}S \;. \quad (5.54)$$

In this case, comparing (5.1) and (5.54), we find

$$U_0(x) = V(x) + \frac{\kappa}{2}x^2 \;, \quad U_{\mathrm{NL}} = -\frac{\kappa}{2}\langle X\rangle^2 \;. \quad (5.55)$$

As a result, (5.5) reads

$$P_{\mathrm{st}}(x) = \frac{1}{Z}\exp\left\{-\frac{V(x)+\kappa x^2/2 - \kappa x\langle X\rangle_{\mathrm{st}}}{Q}\right\} \;, \quad (5.56)$$

which can be solved by means of the corresponding self-consistency equation for the order parameter $m = \langle X\rangle_{\mathrm{st}}$. In line with (5.9), we define $P(x,m)$ and $R(m)$ by

$$P(x,m) = \frac{1}{Z(m)}\exp\left\{-\frac{V(x)+\kappa x^2/2 - \kappa x m}{Q}\right\} \;,$$

$$R(m) = \int_\Omega x P(x,m)\,dx \;, \quad (5.57)$$

where $Z(m)$ is given by $Z(m) = \int_\Omega \exp\{-[V(x)+\kappa x^2/2 - \kappa x m]/Q\}\,dx$. Then, $\langle X\rangle_{\mathrm{st}}$ is given by the solutions of $m = R(m)$. For the K-model the second variation of F reads

$$\delta^2 F[P_{\mathrm{st}}](\epsilon) = -\kappa\int_\Omega x\epsilon(x)\,dx\int_\Omega y\epsilon(y)\,dy + Q\int_\Omega \frac{\epsilon^2(x)}{P_{\mathrm{st}}(x)}\,dx$$

$$= -\kappa\left[\int_\Omega x\epsilon(x)\,dx\right]^2 + Q\int_\Omega \frac{\epsilon^2(x)}{P_{\mathrm{st}}(x)}\,dx \;, \quad (5.58)$$

and can be evaluated using Shiino's decomposition

$$\epsilon(x) = \beta\left[x - \langle X\rangle_{\mathrm{st}}\right]P_{\mathrm{st}}(x) + \chi_\perp(x)\sqrt{P_{\mathrm{st}}(x)} \;, \quad (5.59)$$

where $\chi_\perp$ satisfies the orthogonality relations $\int_\Omega \chi_\perp(x)\sqrt{P_{\mathrm{st}}(x)}\,dx = 0$ and $\int_\Omega x\chi_\perp(x)\sqrt{P_{\mathrm{st}}(x)}\,dx = 0$ (see Sect. 5.1.4). Then, one obtains

$$\delta^2 F[P_{\mathrm{st}}](\epsilon) = \beta^2 K_{\mathrm{st}}(X)\left[Q - \kappa K_{\mathrm{st}}(X)\right] + Q\int_\Omega [\chi_\perp]^2\,dx \;, \quad (5.60)$$

where $K_{\mathrm{st}}(X)$ denotes the variance of a stationary distribution of the form (5.56). In view of (5.60) we introduce the stability coefficient

$$\tilde{\lambda} = Q - \kappa K_{\mathrm{st}}(X) \;. \quad (5.61)$$

It can be seen now that for a distribution P_{st} with $\tilde{\lambda} = Q - \kappa K_{\text{st}}(X) > 0$ we have $\delta^2 F > 0$ for all $\epsilon \neq 0$. Consequently, from the stability analysis carried out in Sect. 5.1, we conclude that in this case P_{st} corresponds to a stable stationary probability density. In contrast, for every distribution P_{st} that yields $\tilde{\lambda} < 0$ there exists an ϵ^* (namely, $\epsilon^* = \beta(x - \langle X \rangle_{\text{st}}) P_{\text{st}}$ with $\chi_\perp = 0$) such that $\delta^2 F[P_{\text{st}}(\epsilon^*)] < 0$ and we deal with an unstable stationary distribution. Critical parameter values of κ and Q define a bifurcation line, which can be computed from $\tilde{\lambda} = 0 \Rightarrow Q = \kappa K_{\text{st}}(X; \kappa, Q)$.

In the preceding, we have exploited linear stability analysis and the stability analysis by means of Lyapunov's direct method. Now, let us turn to the stability analysis by means of self-consistency equations. From (5.57) it follows that the relation

$$\frac{dR(m)}{dm} = \frac{\kappa}{Q} K(X)_{P(x,m)} \geq 0 \qquad (5.62)$$

holds, where the averaging is carried out with respect to $P(x, m)$. That is, $R(m)$ is a monotonically increasing function. In particular, we have

$$\left.\frac{dR}{dm}\right|_{\langle X \rangle_{\text{st}}} = \frac{\kappa}{Q} K_{\text{st}}(X) . \qquad (5.63)$$

In view of the hypothesis (5.32), we are inclined to say that stationary distributions with $\kappa K_{\text{st}}/Q > 1$ (< 1) are unstable (asymptotically stable). In fact, this is the result derived by means of Lyapunov's direct method and linear stability analysis. Therefore, we have proven that for the K-model the proposition (5.32) made by self-consistency equation analysis is correct.

Compensated $B(A)$-Model: $F \propto B(\langle A \rangle)$

Next, we consider systems with internal energy functionals

$$U_{\text{NL}}[P] = \langle B_0(X) \rangle + B(\langle A \rangle) \geq B_{\min} \qquad (5.64)$$

and free energy measures described by

$$F[P] = \int_\Omega V(x) P(x) \, dx + \int_\Omega B_0(x) P(x) \, dx + B(\langle A \rangle) - Q^{\text{B}} S[P] , \qquad (5.65)$$

where $A(x)$ denotes an arbitrary function. Here, we consider arbitrary phase spaces Ω (e.g., $\Omega = \mathbb{R}$, or $\Omega = [a, b]$ with $b > a$). For F given by (5.65) the univariate free energy Fokker–Planck equation (5.51) reads

$$\frac{\partial}{\partial t} P(x, t; u) = \frac{\partial}{\partial x}\left[\frac{dV(x)}{dx} + \frac{dB_0(x)}{dx} + \frac{dA(x)}{dx} \left.\frac{dB(z)}{dz}\right|_{z=\langle A \rangle}\right] P + Q \frac{\partial^2}{\partial x^2} P . \qquad (5.66)$$

5.1 Stability Analysis

From (5.64) we read off that the nonlinearity $B(\langle A \rangle)$ is balanced by a potential $B_0(x)$ such that $\int_\Omega B_0 P \mathrm{d}x + B(\langle A \rangle)$ is bounded from below. Consequently, we do not require that $B(z)$ itself is bounded from below. If $V(x)$ admits for a Boltzmann distribution $W \propto \exp\{-V/Q\}$, then F is a type-F free energy of Table 4.2, which implies that F is bounded from below. In this case, on account of the H-theorem for free energy Fokker–Planck equations (see Sect. 4.3), we conclude that the limiting case $\lim_{t \to \infty} \partial P/\partial t = 0$ holds. Since we have $\delta U_{\mathrm{NL}}[P]/\delta P = B_0(x) + A(x)\mathrm{d}B(\langle A \rangle)/\mathrm{d}z$, (5.5) becomes

$$P_{\mathrm{st}}(x) = \frac{1}{Z} \exp\left\{ -\frac{V(x) + B_0 + A(x)\mathrm{d}B(\langle A \rangle_{\mathrm{st}})/\mathrm{d}z}{Q} \right\} . \tag{5.67}$$

Let us introduce the functions

$$P(x, m) = \frac{1}{Z(m)} \exp\left\{ -\frac{V(x) + B_0(x) + A(x)\mathrm{d}B(m)/\mathrm{d}m}{Q} \right\} ,$$

$$R(m) = \int_\Omega A(x) P(x, m) \, \mathrm{d}x , \tag{5.68}$$

where $Z(m)$ is the normalization constant of $P(x, m)$. Then, $\langle A \rangle_{\mathrm{st}}$ is given by the solutions of the self-consistency equation

$$m = R(m) . \tag{5.69}$$

Let us determine the stability of the distributions (5.67) by means of linear stability analysis and Lyapunov's direct method. Since the second variational derivative of U_{NL} reads $\delta^2 U_{\mathrm{NL}}[P]/\delta P(x) \delta P(y) = A(x)A(y)\mathrm{d}^2 B(\langle A \rangle)/\mathrm{d}z^2$, (5.20) can be found as

$$\delta^2 F[P_{\mathrm{st}}](\epsilon) = \left.\frac{\mathrm{d}^2 B(z)}{\mathrm{d}z^2}\right|_{\langle A \rangle_{\mathrm{st}}} \left[\int_\Omega A(x) \epsilon(x) \, \mathrm{d}x\right]^2 + Q \int_\Omega \frac{\epsilon^2(x)}{P_{\mathrm{st}}(x)} \, \mathrm{d}x . \tag{5.70}$$

In order to evaluate (5.70), we use perturbations given by

$$\epsilon(x) = \beta \left[A(x) - \langle A(X) \rangle_{\mathrm{st}}\right] P_{\mathrm{st}}(x) + \chi_\perp(x) \sqrt{P_{\mathrm{st}}(x)} , \tag{5.71}$$

where the function $\chi_\perp(x)$ satisfies the constraints $\int_\Omega \chi_\perp(x) \sqrt{P_{\mathrm{st}}(x)} \, \mathrm{d}x = 0$ and $\int_\Omega A(x) \chi_\perp(x) \sqrt{P_{\mathrm{st}}(x)} \, \mathrm{d}x = 0$ (see Sect. 5.1.4). Substituting (5.71) into (5.70), one can determine the sign of $\delta^2 F$ from

$$\delta^2 F[P_{\mathrm{st}}](\epsilon) = \beta^2 K_{A,\mathrm{st}}(X) \left[Q + K_{A,\mathrm{st}}(X) \left.\frac{\mathrm{d}^2 B(z)}{\mathrm{d}z^2}\right|_{\langle A \rangle_{\mathrm{st}}}\right] + Q \int_\Omega [\chi_\perp]^2 \, \mathrm{d}x , \tag{5.72}$$

with $K_A(X)$ defined by (5.46). For the sake of convenience, we introduce the stability coefficient

$$\tilde{\lambda} = Q + K_{A,\text{st}}(X) \left.\frac{\mathrm{d}^2 B(z)}{\mathrm{d}z^2}\right|_{\langle A \rangle_{\text{st}}} . \tag{5.73}$$

For $\tilde{\lambda} > 0$ we obtain $\delta^2 F > 0$ indicating that the stationary distribution being studied is asymptotically stable. For $\tilde{\lambda} < 0$ there is a perturbation that yields $\delta^2 F < 0$, which tells us that we are dealing with an unstable stationary distribution. The critical parameter value of Q can be computed from $\tilde{\lambda} = 0 \Rightarrow Q + \mathrm{d}^2 B(m)/\mathrm{d}m^2 K_{A,\text{st}}(X) = 0$ with $m = \langle A \rangle_{\text{st}}$.

Now, let us determine the stability of distributions satisfying (5.67) by means of self-consistency equation analysis. To this end, we define the parameter

$$\tilde{\lambda}' = Q\left(1 - \left.\frac{\mathrm{d}R}{\mathrm{d}m}\right|_{\langle A \rangle_{\text{st}}}\right), \tag{5.74}$$

and compute the derivative of $R(m)$ from (5.68):

$$\left.\frac{\mathrm{d}R(m)}{\mathrm{d}m}\right|_{\langle A \rangle_{\text{st}}} = -\frac{1}{Q} K_{A,\text{st}}(X) \left.\frac{\mathrm{d}^2 B(z)}{\mathrm{d}z^2}\right|_{\langle A \rangle_{\text{st}}} . \tag{5.75}$$

Based on the hypothesis (5.32) made by self-consistency equation analysis, we may conclude that for $\tilde{\lambda}' = Q + K_{A,\text{st}} \mathrm{d}^2 B/\mathrm{d}z^2 > 0$ (< 0) the stationary distributions (5.67) are asymptotically stable (unstable). In fact, we have $\tilde{\lambda}' = \tilde{\lambda}$, which means that self-consistency equation analysis, Lyapunov's direct method, and linear stability analysis yield the same result. In other words, for the $B(A)$-model we have found rigorous proof that the proposition (5.32) made by self-consistency equation analysis is correct. Let us briefly address two special cases of the $B(A)$-model.

K_A-Model: $F \propto K_A = \kappa \langle [A - \langle A \rangle]^2 \rangle$

Here, we have

$$\frac{\partial}{\partial t} P(x, t; u) = \frac{\partial}{\partial x}\left[\frac{\mathrm{d}V(x)}{\mathrm{d}x} + \kappa\left(A(x) - \frac{\mathrm{d}A(x)}{\mathrm{d}x}\langle A(X) \rangle_P\right)\right] P + Q \frac{\partial^2}{\partial x^2} P , \tag{5.76}$$

and

$$P_{\text{st}}(x) = \frac{1}{Z} \exp\left\{-\frac{V(x) + \kappa A^2(x)/2 - \kappa A(x) \langle A \rangle_{\text{st}}}{Q}\right\} . \tag{5.77}$$

Transient solutions converge to stationary solutions of the form (5.77). Using

$$P(x, m) = \frac{1}{Z(m)} \exp\left\{-\frac{V(x) + \kappa A(x)^2/2 - \kappa A(x) m}{Q}\right\} ,$$

$$R(m) = \int_\Omega A(x) P(x, m)\, \mathrm{d}x , \tag{5.78}$$

the order parameter $m = \langle A \rangle_{\text{st}}$ can be computed from $m = R(m)$. Note that we have

$$\frac{dR(m)}{dm} = \frac{\kappa}{Q} K_A(X)|_{P(x,m)} \geq 0 \qquad (5.79)$$

and

$$\left.\frac{dR}{dm}\right|_{\langle A \rangle_{\text{st}}} = \frac{\kappa}{Q} K_{A,\text{st}}(X) . \qquad (5.80)$$

The stability coefficient $\tilde{\lambda}$ reads

$$\tilde{\lambda} = Q - \kappa K_{A,\text{st}}(X) \qquad (5.81)$$

and can be obtained from

$$\tilde{\lambda} = Q\left(1 - \left.\frac{dR}{dm}\right|_{\langle A \rangle_{\text{st}}}\right). \qquad (5.82)$$

For $\tilde{\lambda} > 0$ ($\tilde{\lambda} < 0$) we have $\delta^2 F[P_{\text{st}}] > 0$ ($\delta^2 F[P_{\text{st}}] < 0$) and stationary distributions are asymptotically stable (unstable). Moreover, the proposition (5.32) made by self-consistency equation analysis is correct. The free energy measure of the K_A-model is reported in Table 5.1.

Bounded $B(A)$-Model: $F \propto B(\langle A \rangle) \geq B_{\min}$

We consider the free energy Fokker–Planck equation

$$\frac{\partial}{\partial t} P(x,t;u) = \frac{\partial}{\partial x}\left[\frac{dV(x)}{dx} + \frac{dA(x)}{dx}\left.\frac{dB(z)}{dz}\right|_{z=\langle A\rangle}\right] P + Q\frac{\partial^2}{\partial x^2} P , \qquad (5.83)$$

for $B(z) \geq B_{\min}$. Transient solutions converge to stationary ones for $t \to \infty$. Stationary distributions are given by

$$P_{\text{st}}(\mathbf{x}) = \frac{1}{Z} \exp\left\{-\frac{V(x) + A(x)dB(\langle A\rangle_{\text{st}})/dz}{Q}\right\} . \qquad (5.84)$$

The order parameter $\langle A \rangle_{\text{st}}$ corresponds to solutions of $m = R(m)$ with

$$P(x,m) = \frac{1}{Z(m)} \exp\left\{-\frac{V(x) + A(x)dB(m)/dm}{Q}\right\} ,$$

$$R(m) = \int_\Omega A(x) P(x,m)\, dx . \qquad (5.85)$$

The stability coefficient reads

$$\tilde{\lambda} = Q + K_{A,\text{st}}(X) \left.\frac{d^2 B(z)}{dz^2}\right|_{\langle A\rangle_{\text{st}}} \qquad (5.86)$$

and can alternatively expressed by

$$\tilde{\lambda} = Q\left(1 - \left.\frac{dR}{dm}\right|_{\langle A \rangle_{st}}\right). \tag{5.87}$$

For $\tilde{\lambda} > 0$ ($\tilde{\lambda} < 0$) we have $\delta^2 F[P_{st}] > 0$ ($\delta^2 F[P_{st}] < 0$) and stationary distributions are asymptotically stable (unstable). Moreover, the proposition (5.32) made by self-consistency equation analysis is correct. For the free energy measure of the bounded $B(A)$-model, see Table 5.1.

Summary

In Table 5.1 we have summarized the four models for which we have proven above the consistence between self-consistency equation analysis, on the one hand, and Lyapunov's direct method and linear stability analysis, on the other. In doing so, for these four models we have given rigorous proof that the proposition (5.32) made by self-consistency equation analysis is correct. For further examples, the reader is also referred to [526, 527].

Table 5.1. Generic free energy cases [187] ($S = {}^B S$=Boltzmann entropy)

	K-model	K_A-model	Bounded $B(A)$-model	Compensated $B(A)$-model
U_{NL}	$\frac{\kappa K}{2}$	$\frac{\kappa K_A}{2}$	$B \geq B_{min}$	$\langle B_0 \rangle + B \geq B_{min}$
F	$\langle V \rangle + \frac{\kappa K}{2} - QS$	$\langle V \rangle + \frac{\kappa K_A}{2} - QS$	$\langle V \rangle + B - QS$	$\langle V + B_0 \rangle + B - QS$
$\tilde{\lambda}$	$Q - \kappa K_{st}$	$Q - \kappa K_{A,st}$	$Q + K_{A,st}\frac{d^2 B}{dm^2}$	$Q + K_{A,st}\frac{d^2 B}{dm^2}$

5.1.6 Higher-Dimensional Nonlinearities: $d > 1$

Compensated Case: $U_{NL} = \sum_{i=1}^{d} \kappa_i \langle [A_i - \langle A_i \rangle]^2 \rangle / 2$

In the case of higher nonlinearities one may consider stochastic processes described by the free energy Fokker–Planck equation [526, 527]

$$\frac{\partial}{\partial t} P(x, t; u) = \frac{\partial}{\partial x}\left[\frac{dV(x)}{dx} + \sum_{i=1}^{d} \kappa_i \left(A_i(x) - \frac{dA_i}{dx}\langle A_i(X) \rangle_P\right)\right] P + Q\frac{\partial^2}{\partial x^2} P, \tag{5.88}$$

which involves the free energy

$$F[P] = \int_\Omega V(x)P(x)\,\mathrm{d}x + \sum_{i=1}^d \frac{\kappa_i}{2} K_{A_i}(X) - Q^{\mathrm{B}} S[P] \;, \tag{5.89}$$

and the internal energy functional

$$U_{\mathrm{NL}}[P] = \frac{1}{2} \sum_{i=1}^d \kappa_i \left\langle [A_i - \langle A_i \rangle]^2 \right\rangle = \frac{1}{2} \sum_{i=1}^d \kappa_i K_{A_i}(X) \;. \tag{5.90}$$

For $\kappa_i \geq 0$ and natural boundary conditions and for arbitrary κ_i but periodic boundary conditions, we deal with free energy functionals of the type F (or E) in Table 4.2 and, consequently, we have $F \geq F_{\min}$ and $\partial P/\partial t = 0$ for $t \to \infty$. If we cast (5.89) into the form

$$F[P] = \int_\Omega \underbrace{\left[V(x) + \sum_{i=1}^d \kappa_i \frac{x^2}{2} \right]}_{U_0(x)} P(x)\,\mathrm{d}x - \sum_{i=1}^d \frac{\kappa_i}{2} \langle A_i \rangle^2 - Q^{\mathrm{B}} S[P] \;, \tag{5.91}$$

we obtain from (5.8) the implicit equation

$$P_{\mathrm{st}}(x) = \frac{1}{Z} \exp\left\{ -\frac{V(x) + \sum_{i=1}^d \left[\kappa_i A_i^2(x)/2 - \kappa A_i(x) \langle A_i \rangle_{\mathrm{st}} \right]}{Q} \right\} \;, \tag{5.92}$$

which describes stationary probability densities of (5.88). In order to determine the order parameter values $\langle A_i \rangle_{\mathrm{st}}$, we define the functions

$$R_i(\mathbf{m}) = \int_\Omega A_i(X) P(x, \mathbf{m})\,\mathrm{d}x \;, \tag{5.93}$$

for $i = 1, \ldots, d$ involving the vector $\mathbf{m} = (m_1, \ldots, m_d)$ in combination with

$$P(x, \mathbf{m}) = \frac{1}{Z(\mathbf{m})} \exp\left\{ -\frac{V(x) + \sum_{i=1}^d \left[\kappa_i A_i^2(x)/2 - \kappa A_i(x) m_i \right]}{Q} \right\} \;,$$

$$Z(\mathbf{m}) = \int_\Omega \exp\left\{ -\frac{V(x) + \sum_{i=1}^d \left[\kappa_i A_i^2(x)/2 - \kappa A_i(x) m_i \right]}{Q} \right\} \mathrm{d}x \;. \tag{5.94}$$

Then, solutions of the self-consistency equation

$$\mathbf{m} = \mathbf{R}(\mathbf{m}) \;, \tag{5.95}$$

with $\mathbf{R} = (R_1, \ldots, R_d)$ correspond to order parameters $\langle A_i \rangle_{\mathrm{st}} = m_i$ satisfying (5.92). The second variation of (5.91) reads

$$\delta^2 F[P_{\mathrm{st}}](\epsilon) = -\sum_{i=1}^d \kappa_i \left[\int_\Omega A_i(x) \epsilon(x)\,\mathrm{d}x \right]^2 + Q \int_\Omega \frac{\epsilon^2(x)}{P_{\mathrm{st}}(x)}\,\mathrm{d}x \;. \tag{5.96}$$

Now, let us put $\epsilon(x) = \sqrt{P_{st}(x)}\epsilon'(x)$ and

$$\epsilon'(x) = \sum_{i=1}^{d} a_i[A_i - \langle A_i\rangle_{st}]\sqrt{P_{st}(x)} + \chi_\perp(x) , \qquad (5.97)$$

where $\chi_\perp(x)$ is orthogonal to the functions $\chi_i(x)$ defined by $\chi_i(x) = [A_i(x) - \langle A_i\rangle_{st}]\sqrt{P_{st}(x)}$. Then, $\chi_\perp(x)$ satisfies

$$\int_\Omega [A_i(x) - \langle A_i\rangle]\sqrt{P_{st}(x)}\chi_\perp(x)\,dx = 0 , \qquad (5.98)$$

and, on account of $\int_\Omega \epsilon(x)\,dx = 0$, we have

$$\int_\Omega \chi_\perp(x)\sqrt{P_{st}(x)}\,dx = 0 , \qquad (5.99)$$

which leads to

$$\int_\Omega A_i(x)\chi_\perp(x)\sqrt{P_{st}(x)}\,dx = 0 . \qquad (5.100)$$

In sum, without any loss of generality we consider perturbations $\epsilon(x)$ described by

$$\epsilon(x) = \sum_{i=1}^{d} a_i \left[A_i(x) - \langle A_i(X)\rangle_{st}\right] P_{st}(x) + \chi_\perp(x)\sqrt{P_{st}(x)} . \qquad (5.101)$$

Using (5.96) and exploiting the relations (5.98-5.100), we can determine $\int_\Omega \epsilon^2/P_{st}\,dx$ and $\int_\Omega A_i(x)\epsilon(x)\,dx$ and obtain

$$\int_\Omega \frac{\epsilon^2}{P_{st}}\,dx = \int_\Omega \left[\sum_{i=1}^{d} a_i \left(A_i(x) - \langle A_i\rangle_{st}\right)\right]^2 P_{st}(x)\,dx + \int_\Omega [\chi_\perp]^2\,dx$$

$$= \sum_{i,k=1}^{d} a_i a_k c_{ik} + \int_\Omega [\chi_\perp]^2\,dx \qquad (5.102)$$

and

$$\int_\Omega A_i(x)\,\epsilon(x)\,dx = \sum_{k=1}^{d} a_k c_{ik} , \qquad (5.103)$$

where the stationary cross-correlations c_{ik} are defined by

$$c_{ik} = c_{ki} = \langle A_i A_k\rangle_{st} - \langle A_i\rangle_{st}\langle A_k\rangle_{st} . \qquad (5.104)$$

Note that for $A(x) = x$ the coefficients c_{ik} reduce to the cross-correlation coefficients $\langle (X_i - \langle X_i\rangle_{st})(X_k - \langle X_k\rangle_{st})\rangle_{st}$. Substituting these results into (5.96) gives us

$$\delta^2 F[P_{\text{st}}](\epsilon) = \sum_{i,k=1}^{d} a_i a_k \left[Q c_{ik} - \sum_{l=1}^{d} \kappa_l c_{il} c_{kl} \right] + Q \int_{\Omega} [\chi_\perp]^2 \, dx \ . \quad (5.105)$$

By means of the stability matrix

$$\tilde{\lambda}_{ik} = Q c_{ik} - \sum_{l=1}^{d} \kappa_l c_{il} c_{kl} \quad (5.106)$$

extrema of F can be classified and the stability of stationary distributions can be determined. If $\tilde{\lambda}_{ik}$ is positive definite then we have $\delta F[P_{\text{st}}] > 0$ and P_{st} describes a minimum of F, which implies that P_{st} corresponds to an asymptotically stable stationary probability density. If there exists at least one vector $\mathbf{y} \in \mathbb{R}^d$ with $\sum_{i,k=1}^{d} y_i y_K \tilde{\lambda}_{ik} < 0$, then there exists at least one perturbation ϵ which yields $\delta F[P_{\text{st}}](\epsilon) < 0$. In this case, P_{st} describes a saddle-point or a maximum of F and corresponds to an unstable stationary distribution.

Bounded Case: $U_{\text{NL}}[P] = -\sum_{i=1}^{d} \kappa_i \langle A_i \rangle^2$

In this case, we deal with the evolution equation

$$\frac{\partial}{\partial t} P(x,t;u) = \frac{\partial}{\partial x} \left[\frac{dU_0(x)}{dx} - \sum_{i=1}^{d} \kappa_i \frac{dA_i(x)}{dx} \langle A_i(X) \rangle_P \right] P + Q \frac{\partial^2}{\partial x^2} P \quad (5.107)$$

related to the free energy measure

$$F[P] = \int_{\Omega} U_0(x) P(x) \, dx - \sum_{i=1}^{d} \frac{\kappa_i}{2} \langle A_i \rangle^2 - Q^{\,\text{B}} S[P] \ . \quad (5.108)$$

For bounded potentials $U_{\text{NL}}[P] = -\sum_{i=1}^{d} \kappa_i \langle A_i \rangle^2 \geq U_{\text{NL,min}}$ we deal with free energies of type F in Table 4.2 (or E with $U_{\text{MF}}(x,y) = -\kappa_i A_i(x) A_i(y)/2$). Consequently, we conclude that the relations $F \geq F_{\min}$ and $\lim_{t \to \infty} \partial P/\partial t = 0$ hold. For example, for $\kappa_i \geq 0$ we have the inequality $-\sum_{i=1}^{d} \kappa_i \langle A_i \rangle^2 \geq 0$, which means that $U_{\text{NL}}[P]$ is bounded from below. For $A_i \in C^\infty(\Omega)$ and systems defined on finite phase spaces Ω (e.g. systems subjected to periodic boundary conditions) the functions A_i assume minimum values on Ω and, consequently, the expression $-\sum_{i=1}^{d} \kappa_i \langle A_i \rangle^2$ is bounded from below for arbitrary κ_i. Equation (5.107) can be treated just like the previous case when we replace $V(x) + \sum_{i=1}^{d} \kappa_i A_i^2(x)/2$ by $U_0(x)$. Thus, we obtain

$$P_{\text{st}}(x) = \frac{1}{Z} \exp \left\{ -\frac{U_0(x) - \sum_{i=1}^{d} \kappa_i A_i(x) \langle A_i \rangle_{\text{st}}}{Q} \right\} , \quad (5.109)$$

and

$$R_i(\mathbf{m}) = \int_\Omega A_i(x) P(x, \mathbf{m})\, dx , \qquad (5.110)$$

for $\mathbf{R} = (R_1, \ldots, R_d)$ and $\mathbf{m} = (m_1, \ldots, m_d)$ with

$$P(x, \mathbf{m}) = \frac{1}{Z(\mathbf{m})} \exp\left\{ -\frac{U_0(x) - \sum_{i=1}^d \kappa_i A_i(x) m_i}{Q} \right\},$$

$$Z(\mathbf{m}) = \int_\Omega \exp\left\{ -\frac{U_0(x) - \sum_{i=1}^d \kappa_i A_i(x) m_i}{Q} \right\} dx . \qquad (5.111)$$

The stationary order parameter $\langle \mathbf{A} \rangle_{\mathrm{st}}$ in (5.109) corresponds to the solutions of the self-consistency equation $\mathbf{m} = \mathbf{R}(\mathbf{m})$. The second variation $\delta^2 F$ of F given by (5.108) is described by (5.96). Therefore, the stability of the stationary distributions defined by (5.109) can be determined as in the previous case. Accordingly, if the matrix $\tilde{\lambda}_{ik}$ defined by (5.106) is positive definite for a distribution P_{st}, then P_{st} is asymptotically stable. If there exists a $\mathbf{y}$ with $\sum_{i,k=1}^d y_i y_K \tilde{\lambda}_{ik} < 0$, then P_{st} is unstable.

Special Case: Diagonal Cross-correlation Matrix

Of particular interest are stochastic processes for which the cross-correlation matrix given by the coefficients c_{ik} assumes diagonal form:

$$c_{ik} = \delta_{ik} c_{ii} , \quad c_{ii} = K_{A_i,\mathrm{st}}(X) \geq 0 . \qquad (5.112)$$

In this case, (5.105) and (5.106) reduce to

$$\delta^2 F[P_{\mathrm{st}}](\epsilon) = \sum_{i=1}^d [a_i]^2 K_{A_i,\mathrm{st}}(X) [Q - \kappa_i K_{A_i,\mathrm{st}}(X)] + Q \int_\Omega [\chi_\perp]^2\, dx \quad (5.113)$$

and

$$\tilde{\lambda}_{ik} = \delta_{ik} K_{A_i,\mathrm{st}}(X) \tilde{\lambda}_i , \quad \tilde{\lambda}_i = Q - \kappa_i K_{A_i,\mathrm{st}}(X) . \qquad (5.114)$$

For systems of this kind we conclude that if $\tilde{\lambda}_i > 0$ holds for all i, then we have $\delta F[P_{\mathrm{st}}] > 0$ and P_{st} is asymptotically stable. In contrast, if there is at least one index i^* with $\tilde{\lambda}_{i^*} < 0$, then there exists a perturbation $\epsilon(x)$ (namely, $\epsilon(x) = a_{i^*}(A_{i^*}(x) - \langle A_{i^*}\rangle_{\mathrm{st}}) P_{\mathrm{st}}$ with $a_i = 0$ for $i \neq i^*$ and $\chi_\perp = 0$) with $\delta F[P_{\mathrm{st}}](\epsilon) < 0$ and, consequently, P_{st} is unstable.

5.1.7 Multiplicative Noise

We consider now systems subjected to multiplicative noise [298, 498]. Our objective is to analyze multiplicative noise systems by means of the methods developed in the previous section. To this end, we consider the univariate free energy Fokker–Planck equation

$$\frac{\partial}{\partial t}P(x,t;u) = \frac{\partial}{\partial x}M(x)P\frac{\partial}{\partial x}\frac{\delta F}{\delta P} , \qquad (5.115)$$

which involves a state-dependent mobility coefficient $M > 0$. We assume that stationary distributions of (5.115) are defined by $\delta F/\delta P = \mu$.

Lyapunov's Direct Method

Since (5.115) corresponds to a special case of (4.49) the relations $dF/dt \leq 0$ and $dF/dt = 0 \Leftrightarrow \partial P/\partial t = 0$ hold. Using Lyapunov's direct method, we conclude that stationary distributions P_{st} are asymptotically stable if the inequality $\forall \epsilon \neq 0 : \delta^2 F[P_{\text{st}}](\epsilon) > 0$ holds. If there is at least one perturbation ϵ that yields $\delta^2 F[P_{\text{st}}](\epsilon) < 0$, then we deal with an unstable stationary distribution.

Linear Stability Analysis

From (5.115) it follows that perturbations of stationary distributions satisfy the evolution equation

$$\frac{\partial}{\partial t}\epsilon(x,t) = \frac{\partial}{\partial x}M(x)P_{\text{st}}\frac{\partial}{\partial x}\int_\Omega \frac{\delta^2 F}{\delta P(x)\delta P(y)}\epsilon(y,t)\,dy . \qquad (5.116)$$

Just as for systems with $M = 1$, we can show that if P_{st} is related to a free energy minimum with $\delta^2 F[P_{\text{st}}](\epsilon) > 0$ for $\epsilon \neq 0$, then $\|\epsilon\| = \sqrt{\delta^2 F[P_{\text{st}}](\epsilon)}$ decreases as a function of time and P_{st} is asymptotically stable. For free energy maximum distributions with $\delta^2 F[P_{\text{st}}](\epsilon) < 0$ for $\epsilon \neq 0$ one finds that the norm $\|\epsilon\| = \sqrt{-\delta^2 F[P_{\text{st}}](\epsilon)}$ increases as a function of time which means that distributions of this kind are unstable. From this discussion we see that Lyapunov's direct method and linear stability analysis yield consistent results. Let us interpret next the state-dependent mobility coefficient $M(x)$ in terms of a multiplicative noise source.

Multiplicative Noise Sources

It is clear that due to the state-dependency of M we deal with stochastic processes with multiplicative noise. However, M also occurs in the drift term. For example, for the free energy measure $F = \langle V(X)\rangle + \kappa K_A(X)/2 - Q^{\text{B}}S$ (see Table 5.1, K_A-model), (5.115) becomes

$$\frac{\partial}{\partial t}P(x,t;u) =$$
$$\frac{\partial}{\partial x}M(x)\left[\frac{dV(x)}{dx} + \kappa\left(A(x) - \frac{dA(x)}{dx}\langle A(X)\rangle_P\right)\right]P + Q\frac{\partial}{\partial x}M(x)\frac{\partial}{\partial x}P.$$
$$(5.117)$$

Using the transformation $(M, A, V) \to (D, A', V')$ given by

$$\frac{dV'}{dx} = M\frac{dV}{dx} - Q\frac{dM}{dx} + \kappa M A,$$
$$D(x) = QM(x),$$
$$A'(x) = \sqrt{\frac{\kappa}{Q}} A(x), \qquad (5.118)$$

we can transform (5.117) into

$$\frac{\partial}{\partial t} P(x, t; u) = \frac{\partial}{\partial x}\left[\underbrace{\frac{dV'(x)}{dx} - D(x)\frac{dA'(x)}{dx}\langle A'(X)\rangle_P}_{A_{\text{eff}}(x,P)} P + \frac{\partial^2}{\partial x^2} D(x) P\right].$$

(5.119)

Now, let us assume that we start off with a multiplicative noise system described by means of the nonlinear Fokker–Planck equation (5.119). Then, in a first step, we transform (5.119) into (5.117) by means of the backwards transformation $(D', A', V') \to (M, A, V)$:

$$\frac{dV}{dx} = Q\frac{1}{D}\left[\frac{dV'}{dx} + \frac{dD}{dx}\right] - \sqrt{\kappa Q} A',$$
$$M(x) = \frac{D(x)}{Q},$$
$$A(x) = \sqrt{\frac{Q}{\kappa}} A'(x). \qquad (5.120)$$

Since (5.117) corresponds to the free energy Fokker–Planck equation (5.115), in a second step, the stability of the stationary distributions of (5.119) can be determined using the free energy approach. That is, stationary distributions with stability coefficients $\tilde{\lambda} = Q - \kappa K_{A,\text{st}}(X)$ are asymptotically stable for $\tilde{\lambda} > 0$ and unstable for $\tilde{\lambda} < 0$ (see Table 5.1, K_A-model). In other words, from (5.119) it follows that multiplicative noise systems can be investigated by means of the free energy approach if the effective mean field force $A_{\text{eff}}(x, P)$ can be written as $A_{\text{eff}}(x, P) = D(x)\langle A'\rangle dA'(x)/dx$, that is, if the diffusion coefficient $D(x)$ and $A_{\text{eff}}(x, P)$ satisfy a matching condition.

5.1.8 Norm for Perturbations*

Let $C_\epsilon(\Omega) = \{f(\mathbf{x}) \mid f \in C^\infty(\Omega) \wedge \int_\Omega f(\mathbf{x}) d^M x = 0\}$ denote the function space of the deviations of probability densities. Then the functions $\epsilon \in C_\epsilon$ are the vectors of a linear vector space and satisfy for $a_1, a_2 \in \mathbb{R}$ and $\epsilon_1, \epsilon_2 \in C_\epsilon$ the relation $a_1\epsilon_1 + a_2\epsilon_2 = \epsilon_3 \in C_\epsilon$. Consider a free energy functional F with second variational derivatives that are symmetric at a stationary point P_{st} of F,

$$\frac{\delta^2 F[P_{\text{st}}]}{\delta P_1(\mathbf{x}) \delta P_2(\mathbf{x})} = \chi(\mathbf{x}, \mathbf{y}) \ , \ \chi(\mathbf{x}, \mathbf{y}) = \chi(\mathbf{y}, \mathbf{x}) \ , \tag{5.121}$$

and second variations that are positive definite at P_{st}:

$$\epsilon \neq 0 \Leftrightarrow \delta^2 F[P_{\text{st}}](\epsilon) > 0 \ , \ \epsilon = 0 \Leftrightarrow \delta^2 F[P_{\text{st}}](\epsilon) = 0 \ . \tag{5.122}$$

Now, we introduce the functional

$$(\epsilon_1, \epsilon_2) = \delta^2 F[P_{\text{st}}](\epsilon_1, \epsilon_2) = \int_\Omega \frac{\delta^2 F[P_{\text{st}}]}{\delta P_1(\mathbf{x}) \delta P_2(\mathbf{x})} \epsilon_1(\mathbf{x}) \epsilon_2(\mathbf{x}) \, \mathrm{d}^M x \mathrm{d}^M y \ . \tag{5.123}$$

This expression is a bilinear form satisfying

$$\begin{aligned}(\epsilon_1, \epsilon_2) &= (\epsilon_2, \epsilon_1) \ , \\ (a_1 \epsilon_1 + a_2 \epsilon_2, \epsilon_3) &= a_1 (\epsilon_1, \epsilon_3) + a_2 (\epsilon_2, \epsilon_3)\end{aligned} \tag{5.124}$$

for $a_1, a_2 \in \mathbb{R}$. Then, we introduce the functional $\|\cdot\|$ defined by

$$\|\epsilon\| = \sqrt{\delta^2 F[P_{\text{st}}](\epsilon)} = \sqrt{(\epsilon, \epsilon)} \ . \tag{5.125}$$

Our objective is to show that the relations

$$\begin{aligned}\|\epsilon\| &\geq 0 \ , \\ \|\epsilon\| &= 0 \Leftrightarrow \epsilon = 0 \ , \\ \|\epsilon_1 + \epsilon_2\| &\leq \|\epsilon_1\| + \|\epsilon_2\| \ , \\ \|a\epsilon\| &= |a| \, \|\epsilon\|\end{aligned} \tag{5.126}$$

are satisfied for $a \in \mathbb{R}$, which means that (5.125) is a norm related to a free energy minimum. It is clear from (5.122-5.125) that the first two properties and the last property in (5.126) are satisfied. The triangle inequality can be proven in line with the proof of the triangle inequality for vectors of the Euclidean space. To this end, let us first derive the Cauchy–Schwarz inequality

$$\|\epsilon_1\| \, \|\epsilon_2\| \geq (\epsilon_1, \epsilon_2) \ . \tag{5.127}$$

First, consider $\epsilon_2 = 0$. Then, (5.127) is satisfied. Second, consider $\epsilon_2 \neq 0$. Then, we have $\|\epsilon_2\| > 0$. Next, use

$$\left\| \epsilon_1 \|\epsilon_2\|^2 - (\epsilon_1, \epsilon_2) \epsilon_2 \right\|^2 \geq 0 \ , \tag{5.128}$$

$\|\epsilon\|^2 = (\epsilon, \epsilon)$, and (5.124) to obtain

$$\|\epsilon_1\|^2 \|\epsilon_2\|^4 - (\epsilon_1, \epsilon_2)^2 \|\epsilon_2\|^2 \geq 0 \ . \tag{5.129}$$

Since we have $\|\epsilon_2\| > 0$, we get

$$\|\epsilon_1\|^2 \|\epsilon_2\|^2 \geq (\epsilon_1, \epsilon_2)^2 . \tag{5.130}$$

For $(\epsilon_1, \epsilon_2) \geq 0$ we get from (5.130) the relation

$$\|\epsilon_1\| \|\epsilon_2\| \geq (\epsilon_1, \epsilon_2) . \tag{5.131}$$

For $(\epsilon_1, \epsilon_2) < 0$ we conclude from the inequality $\|\epsilon_1\|^2 \|\epsilon_2\|^2 \geq 0$ that the relation

$$\|\epsilon_1\| \|\epsilon_2\| \geq 0 > (\epsilon_1, \epsilon_2) \tag{5.132}$$

holds. In sum, (5.127) holds for $\epsilon_2 \neq 0$ and $\epsilon_2 = 0$. Consider now

$$\|\epsilon_1 + \epsilon_2\|^2 = (\epsilon_1 + \epsilon_2, \epsilon_1 + \epsilon_2) . \tag{5.133}$$

We use (5.124) to find that

$$\|\epsilon_1 + \epsilon_2\|^2 = \|\epsilon_1\|^2 + 2(\epsilon_1, \epsilon_2) + \|\epsilon_2\|^2 , \tag{5.134}$$

which implies that we have

$$\|\epsilon_1 + \epsilon_2\|^2 - \|\epsilon_1\|^2 - \|\epsilon_2\|^2 = 2(\epsilon_1, \epsilon_2) . \tag{5.135}$$

Since the Cauchy–Schwarz inequality (5.127) holds, we get

$$\|\epsilon_1 + \epsilon_2\|^2 - \|\epsilon_1\|^2 - \|\epsilon_2\|^2 \leq 2 \|\epsilon_1\| \|\epsilon_2\| \tag{5.136}$$

and

$$\|\epsilon_1 + \epsilon_2\|^2 \leq \|\epsilon_1\|^2 + 2 \|\epsilon_1\| \|\epsilon_2\| + \|\epsilon_2\|^2 = (\|\epsilon_1\| + \|\epsilon_2\|)^2 . \tag{5.137}$$

Since the inequalities $\|\epsilon_1 + \epsilon_2\| \geq 0$ and $\|\epsilon_1\| + \|\epsilon_2\| \geq 0$ hold, we can derive from (5.137) the triangle inequality

$$\|\epsilon_1 + \epsilon_2\| \leq \|\epsilon_1\| + \|\epsilon_2\| \tag{5.138}$$

listed in (5.126). Finally note that by analogy one can show that (5.29) describes a norm related to a free energy maximum.

5.2 Natural Boundary Conditions

We will study now examples of free energy Fokker–Planck equations that are described by

$$\frac{\partial}{\partial t} P(x, t; u) = -\frac{\partial}{\partial x} \left(h(x) + b(\langle X \rangle) \right) P + Q \frac{\partial^2}{\partial x^2} P \tag{5.139}$$

and have solutions $P(x, t; u) = \langle \delta(x - X(t)) \rangle$ subjected to natural boundary conditions with $X(t) \in \Omega = \mathbb{R}$. We will distinguish between models that involve linear and nonlinear functions $h(x)$ and $b(z)$, see Table 5.2.

Table 5.2. Models discussed in Sect. 5.2

Model	h	b	Type
Shimizu–Yamada model	linear	linear	compensated, (5.50)
Dynamical Takatsuji model	linear	nonlinear	compensated, (5.66)
Desai–Zwanzig model	nonlinear	linear	compensated, (5.66)
Bounded $B(M_1)$-model	nonlinear	nonlinear	bounded, (5.83)

5.2.1 Shimizu–Yamada Model – Stationary Solutions

The Fokker–Planck equation of the Shimizu–Yamada model reads

$$\frac{\partial}{\partial t}P(x,t;u) = \frac{\partial}{\partial x}\left[\gamma x + \kappa(x - \langle X \rangle)\right] P + Q\frac{\partial^2}{\partial x^2}P , \qquad (5.140)$$

for $\kappa \geq 0$ and $t \geq t_0$. Transient solutions of the Shimizu–Yamada model have already been derived in Sect. 3.10. Here, we will focus on the stationary case with $\kappa > 0$. Equation (5.140) can be written as a free energy Fokker–Planck equation (5.3) with

$$F[P] = \frac{\gamma}{2}\langle X^2 \rangle + \frac{\kappa}{2}K(X) - Q^{\mathrm{B}}S[P] . \qquad (5.141)$$

It is clear from (5.140) that we deal with a special case of (5.50). Consequently, the implicit equation

$$P_{\mathrm{st}}(x) = \frac{1}{Z}\exp\left\{-\frac{(\gamma+\kappa)x^2/2 - \kappa x \langle X \rangle_{\mathrm{st}}}{Q}\right\} \qquad (5.142)$$

describes stationary probability densities of (5.140). Equation (5.142) can be transformed into

$$P_{\mathrm{st}}(x) = \sqrt{\frac{\gamma+\kappa}{2\pi Q}}\exp\left\{-\frac{(\gamma+\kappa)}{2Q}\left[x - \frac{\kappa}{\kappa+\gamma}\langle X \rangle_{\mathrm{st}}\right]^2\right\} . \qquad (5.143)$$

Stationary distributions of this kind exist only for $\gamma > -\kappa$. In order to derive the self-consistency equation of the model (5.140), we define

$$P(x,m) = \sqrt{\frac{\gamma+\kappa}{2\pi Q}}\exp\left\{-\frac{(\gamma+\kappa)}{2Q}\left[x - \frac{\kappa}{\kappa+\gamma}m\right]^2\right\} \qquad (5.144)$$

and the integral $R(m) = \int_\Omega x P(x,m)\,\mathrm{d}x$ which reads

$$R(m) = \frac{\kappa}{\kappa+\gamma}m . \qquad (5.145)$$

Since we have $\kappa > 0$ and $\gamma > -\kappa$ the self-consistency equation $m = R(m)$ has a unique solution given by $m = 0$. This implies that there is a unique solution of the implicit equation (5.143) given by $\langle X \rangle_{\text{st}} = 0$ and

$$P_{\text{st}}(x) = \sqrt{\frac{\gamma + \kappa}{2\pi Q}} \exp\left\{-\frac{(\gamma + \kappa)}{2Q} x^2\right\} . \tag{5.146}$$

Using $K_{\text{st}} = Q/(\gamma + \kappa)$, (5.61) becomes

$$\tilde{\lambda} = Q - \kappa K_{\text{st}} = \frac{\gamma Q}{\gamma + \kappa} \tag{5.147}$$

and we conclude that for $\gamma > 0$ the Gaussian distribution P_{st} is asymptotically stable, whereas for $-\kappa < \gamma < 0$ the stationary distribution is unstable. In sum, the Shimizu–Yamada model (5.140) describes stochastic processes that exhibit stable stationary distributions for $\gamma > 0$, whereas these solutions become unstable for $\gamma \in (-\kappa, 0)$ and cease to exist for $\gamma < -\kappa$. The stability of the stationary distribution (5.146) can be illustrated by several means.

Cumulants

First, the evolution of the first and second cumulants M_1 and K can be obtained from (5.140). To this end, we multiply the nonlinear Fokker–Planck equation (5.140) with x and x^2, respectively, integrate the result with respect to x, and integrate by parts. Assuming that the surface terms thus obtained vanish, for the mean value $M_1(t) = \langle X(t) \rangle$ and the variance $K(t)$ we obtain

$$\frac{d}{dt} M_1(t) = -\gamma M_1(t) , \tag{5.148}$$

$$\frac{d}{dt} K(t) = -2(\gamma + \kappa) K(t) + 2Q . \tag{5.149}$$

We see now that for $\gamma > 0$ both cumulants are asymptotically stable and converge to $M_{1,\text{st}} = 0$ and $K_{\text{st}} = Q/(\gamma + \kappa)$ for $t \to \infty$, which is in line with our observation that P_{st} is asymptotically stable for $\gamma > 0$. For $-\kappa < \gamma < 0$ the variance converges to $K_{\text{st}} = Q/(\gamma + \kappa)$ in the long time limit but the stationary mean value $M_{1,\text{st}} = 0$ becomes unstable indicating that P_{st} is unstable. In addition, for $\gamma < -\kappa$ the variance increases exponentially.

Time-Dependent Free Energy

The stability of the stationary distribution (5.146) can also be illustrated by means of the free energy (5.141). For example, let us consider the time-dependent behavior of the free energy (5.141) for an exact time-dependent solution of (5.140). As stated in Sect. 3.10, substituting

$$P(x,t;\delta(x-x_0)) = \frac{1}{\sqrt{2\pi K(t)}} \exp\left\{-\frac{[x-x_0 m(t)]^2}{2K(t)}\right\} \quad (5.150)$$

into (5.140), we obtain for $\gamma \neq -\kappa$

$$m(t) = \exp\{-\gamma(t-t_0)\}, \quad (5.151)$$

$$K(t) = \frac{Q}{\gamma+\kappa}[1 - e^{-2(\gamma+\kappa)(t-t_0)}]. \quad (5.152)$$

For $\gamma > 0$ the time-dependent Gaussian distribution converges to the stationary one in the long time limit. Moreover, if we substitute (5.150-5.152) into the free energy (5.141), we obtain

$$F(t) = \frac{\gamma+\kappa}{2} K(t) - \frac{Q}{2}(1 + \ln[2\pi K(t)]) + \frac{\gamma}{2} x_0^2 m(t)^2. \quad (5.153)$$

Taking the derivative with respect to t, we get

$$\frac{d}{dt} F(t) = -Q(\gamma+\kappa)e^{-4(\gamma+\kappa)t} - \gamma x_0^2 e^{-2\gamma t} \leq 0. \quad (5.154)$$

We realize that for $\gamma > 0$ the free energy monotonically decreases as predicted by the H-theorem of free energy Fokker–Planck equations (see Sect. 4.3). In addition, we see that the limiting case $dF/dt = 0$ for $t \to \infty$ holds.

Free Energy as a Potential

Next, let us examine the extrema of F. If we plot F over the function space of probability densities, then we see that for $\gamma > 0$ there is a minimum at P_{st} given by (5.146). Unfortunately, we cannot plot F as a function of probability densities. However, we can plot F for a suitably chosen family of probability densities that is described by some parameters and plot F as a function of these parameters. For this reason, we consider now the Gaussian distributions

$$P_G(x) = \frac{1}{\sqrt{2\pi K}} \exp\left\{-\frac{[x-M_1]^2}{2K}\right\} \quad (5.155)$$

described by M_1 and K. For these distributions the free energy (5.141) reads

$$F[P_G] = \frac{\gamma+\kappa}{2} K - \frac{Q}{2}(1 + \ln[2\pi K]) + \frac{\gamma}{2} M_1^2. \quad (5.156)$$

Focusing on the M_1-dependency, F can be written as

$$F[P_G] = F(M_1 = 0) + \frac{\gamma}{2}(M_1)^2 \quad (5.157)$$

and represents a parabolic function, see Fig. 5.4. For $\gamma > 0$ (< 0) there is a free energy minimum (maximum) at $M_1 = 0$. From (5.156) we read off that

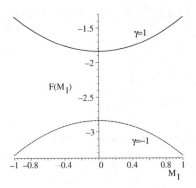

Fig. 5.4. Free energy $F[P_G]$ as a function of M_1 for $Q = \kappa = K = 1$ and $\gamma = 1$ (*solid line*) and $\gamma = -1$ (*dashed line*)

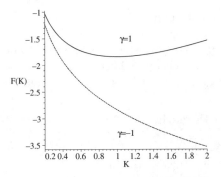

Fig. 5.5. Free energy $F[P_G]$ as a function of the variance K for $Q = \kappa = 1$ and $M_1 = 0$ and $\gamma = -1$ (*solid line*) $\gamma = 1$ (*dashed line*)

the limiting cases $F(K \to 0) \to \infty$ and $F(K \to \infty) \to \infty$ hold for $\gamma > -\kappa$. In addition, we have

$$\frac{\mathrm{d}F[P_G]}{\mathrm{d}K} = \frac{\gamma + \kappa}{2} - \frac{Q}{2K}, \tag{5.158}$$

which implies that there is a minimum at $\mathrm{d}F[P_G]/\mathrm{d}K = 0 \Rightarrow K = Q/(\gamma+\kappa)$ for $\gamma > -\kappa$, whereas for $\gamma < -\kappa$ the free energy $F[P_G]$ is a monotonically decreasing function with respect to K, see Fig. 5.5. We can also plot $F[P_G]$ in the parameter space (K, M_1). For $\gamma > 0$ there is a unique minimum at $M_1 = 0$ and $K = K_{\mathrm{st}}$, see Fig. 5.6.

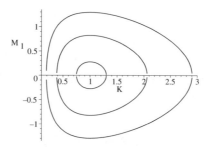

Fig. 5.6. Contour lines of the free energy $F[P_G]$ in the parameter space (K, M_1) for $Q = \kappa = \gamma = 1$. There is a unique minimum at $K = 1$ and $M_1 = 0$ with $F = -\ln(2\pi) = -1.83$. Lines denote free energy values of $F = -1.8$, $F = -1.5$, and $F = -1$ (*from inside to outside*)

Self-Consistency Equation Analysis

Finally, the stability of the stationary distribution may be illustrated by means of self-consistency equation analysis and the proposition (5.32). From (5.145) we obtain

$$\frac{dR(0)}{dm} = \frac{\kappa}{\gamma + \kappa} \ . \tag{5.159}$$

Consequently, for $-\kappa < \gamma < 0$ we have $dR(0)/dm > 1$ indicating that we are dealing with an unstable stationary solution and for $\gamma > 0$ we have

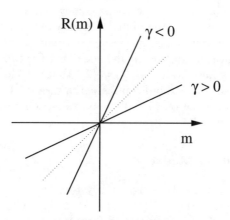

Fig. 5.7. Self-consistency equation analysis based on (5.145) and $m = R(m)$. *Solid lines*: $R(m)$ for $\gamma > 0$ and $\gamma < 0$. *Dashed line*: diagonal of the (m, R)-plane

$dR(0)/dm < 1$ indicating that the stationary distribution (5.144) is asymptotically stable, see Fig. 5.7. These predictions are in line with the results obtained from linear stability analysis and Lyapunov's direct method.

5.2.2 Dynamical Takatsuji Model – Basins of Attraction

Order Parameter Equation of Ising Ferromagnets

The magnetization $M(t)$ of Ising ferromagnets is often described by means of the order parameter equation

$$\frac{d}{dt} M(t) = -a_1 M(t) + a_2 \tanh(a_3 M(t)) \,, \tag{5.160}$$

with $a_1, a_2, a_3 > 0$ [7, 18, 580]. Depending on the parameters a_1, a_2, a_3, (5.160) either exhibits a unique stationary solution $M_{\text{st}} = 0$, which describes the paramagnetic phase of Ising ferromagnets, or two stable stationary solutions with $M_{\text{st}} \neq 0$, which describe ferromagnetic phases that are related to spin-up and spin-down configurations. In the latter case, the stationary solution $M_{\text{st}} = 0$ exists as well but is an unstable one.

Order Parameter Equation for Group Behavior

In the context of human group behavior, Takatsuji [561] has studied a mean field model that shows the same bifurcation diagram as the order parameter equation (5.160). According to the model by Takatsuji, a vanishing order parameter M reflects a human population in which individual behavior dominates. That is, there is no significant collective behavior. In contrast, if $M \neq 0$ the population exhibits a collective behavior, which arises due to the mean field interactions between the individuals. In view of this interpretation of the order parameter equation (5.160), at issue is to define a stochastic process by means of a dynamical mean field model such that the random variable $X(t)$ of the process has a mean value $M_1(t) = \langle X \rangle$ that satisfies (5.160). In short, we will introduce next in a top-down fashion a mean field model that yields the order parameter equation (5.160). A bottom-up approach to this model can be found in Sect. 5.5.5.

Dynamical Takatsuji Model

Let us describe the state of a single subsystem of a many-body system by means of the random variable $X(t) \in \Omega = \mathbb{R}$. In line with mean field theory, we assume that the single subsystem free energy reads

$$F[P] = \frac{\gamma + c}{2} \langle X^2 \rangle + B(\langle X \rangle) - Q^{\text{B}} S[P] \,, \tag{5.161}$$

5.2 Natural Boundary Conditions

with

$$B(z) = -\ln\cosh(\sqrt{c}\,z). \tag{5.162}$$

Substituting the free energy (5.161) into (5.3), we obtain the free energy Fokker–Planck equation

$$\frac{\partial}{\partial t}P(x,t;u) = \frac{\partial}{\partial x}\left[(\gamma+c)x - \sqrt{c}\tanh(\sqrt{c}\,\langle X\rangle)\right]P + Q\frac{\partial^2}{\partial x^2}P. \tag{5.163}$$

The nonlinearity $B(z)$ satisfies $B(0) = 0$ and describes a parabola for small z because of $\cosh(\sqrt{c}\,z) \approx 1 + cz^2/2$ and $B(z) \approx -cz^2/2$ for $z \approx 0$. For large $|z|$ the function $B(z)$ behaves like $B(z) = -\sqrt{c}|z|$ and is linear with respect to $|z|$. In addition, one can show that the inequality $\ln\cosh(z') < [z']^2/2$ holds for $z' \neq 0$, which implies that $-B(z) < cz^2/2$ holds for $z \neq 0$ and $c > 0$, see also Fig. 5.8.

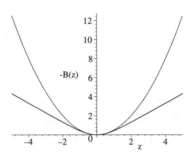

Fig. 5.8. $-B(z)$ (*solid line*) and $cz^2/2$ (*dotted line*) as a function of z for $c=1$.

For $\gamma > -c$ the free energy is bounded from below. To see this we write F as

$$F[P] = \frac{\gamma+c}{4}\langle X^2\rangle + \frac{\gamma+c}{4}K + \frac{\gamma+c}{4}\langle X\rangle^2 + B(\langle X\rangle) - Q^{\mathrm{B}}S[P]. \tag{5.164}$$

Since $B(z) \propto -|z|$ for large $|z|$, the expression $Y = (\gamma+c)z^2/4 + B(z)$ is bounded from below for $\gamma > -c$, which implies that the nonlinear functional $U_{\mathrm{NL}} = (\gamma+c)K/4 + (\gamma+c)\langle X\rangle^2/4 + B(\langle X\rangle)$ is bounded from below, whereas the linear part $U_L[P] = \langle U_0\rangle$ with $U_0 = (\gamma+c)x^2/4$ yields a Boltzmann distribution $W \propto \exp\{-U_0(x)/Q\}$. In sum, the free energy (5.161) corresponds to a measure of type F in Table 4.2. Since F is bounded from below the H-theorem for free energy Fokker–Planck equations applies, which means that we have $\lim_{t\to\infty} \partial P(x,t;u)/\partial t = 0$. Furthermore, the first moment of the solutions $P(x,t;u)$ satisfies

$$\frac{\mathrm{d}}{\mathrm{d}t}M_1 = -(\gamma+c)M_1 + \sqrt{c}\tanh(\sqrt{c}M_1) = -\frac{\mathrm{d}V_M}{\mathrm{d}M_1}, \tag{5.165}$$

where V is defined by

$$V_M(z) = \frac{\gamma}{2}z^2 + \underbrace{\frac{c}{2}z^2 - \ln\cosh(\sqrt{c}z)}_{V_c(z)}. \tag{5.166}$$

Equation (5.165) corresponds to the order parameter equation (5.160) of the Ising model for $a_1 = \gamma + c$, $a_2 = \sqrt{c}$, and $a_3 = \sqrt{c}$. The potential $V_c(z)$ vanishes at $z = 0$ and increases monotonically with $|z|$ both for $z > 0$ and $z < 0$. Consequently, the potential $V_M(z)$ is monostable for $\gamma > 0$ and bistable for $\gamma < 0$ (and $\gamma > -c$, see below), see Fig. 5.9. That is, for $\gamma \in (-c, 0)$ the potential V_M describes a double-well potential and there are multiple stationary solutions of M_1.

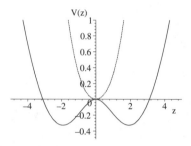

Fig. 5.9. $V_M(z)$ given by (5.166) for $c = 1$ and several values of the parameter γ: $\gamma = 0.5$ (*dotted line*) and $\gamma = -0.5$ (*solid line*)

Equation (5.163) corresponds to the generic free energy Fokker–Planck equation (5.66) for $V(x) = \gamma x^2/2$, $B_0(x) = cx^2/2$, $B(z)$ given by (5.162), and $A(x) = x$. Consequently, stationary solutions satisfy the implicit equation

$$P_{st}(x) = \sqrt{\frac{\gamma + c}{2\pi Q}} \exp\left\{-\frac{(\gamma + c)}{2Q}\left[x - \frac{\sqrt{c}}{c + \gamma}\tanh(\sqrt{c}\langle X\rangle_{st})\right]^2\right\} \tag{5.167}$$

(hint: use (5.67) and determine the quadratic complement). Equation (5.68) now reads

$$P(x, m) = \sqrt{\frac{\gamma + c}{2\pi Q}} \exp\left\{-\frac{(\gamma + c)}{2Q}\left[x - \frac{1}{c + \gamma}\tanh(\sqrt{c}m)\right]^2\right\}, \tag{5.168}$$

which implies that $R(m) = \int_\Omega xP(x, m)\,dx$ is given by

$$R(m) = \frac{\sqrt{c}}{c+\gamma} \tanh(\sqrt{c}\,m) \qquad (5.169)$$

(see also (5.165) for the stationary case). Let us determine now the stationary order parameter values $\langle X \rangle_{\text{st}}$ from the self-consistency equation $\langle X \rangle_{\text{st}} = R(\langle X \rangle_{\text{st}})$. It is clear that $R(0) = 0$ holds and, consequently, a stationary solution of the dynamical Takatsuji model is described by

$$P_{\text{st}}(x) = \sqrt{\frac{\gamma+c}{2\pi Q}} \exp\left\{-\frac{(\gamma+c)x^2}{2Q}\right\}. \qquad (5.170)$$

In view of the antisymmetry property $R(m) = -R(-m)$, we conclude that if $m \neq 0$ is a solution of the self-consistency equation $m = R(m)$ then $-m$ corresponds to a solution as well. Now, let us discuss under which conditions the self-consistency equation has nonvanishing solutions. First, we note that $R(m)$ is a monotonically increasing function with

$$\frac{dR(m)}{dm} = \frac{c}{c+\gamma}\left[1 - \tanh^2(\sqrt{c}\,m)\right] \geq 0 \qquad (5.171)$$

because we have $1 - [\tanh(z)]^2 \geq 0$. It is clear that if $dR/dm < 1$ holds for all m, then $m = R(m)$ is only solved by $m = 0$. The reason for this is that if $R(m)$ with $R(0) = 0$ is smaller than m for $m > 0$ and larger than m for $m < 0$, then $m = R(m)$ is only solved by $m = 0$. Since $0 \leq 1 - \tanh^2(z') \leq 1$ for $z' \in \mathbb{R}$, for $\gamma > 0$ we have $\forall m : dR/dm < 1$ and the self-consistency equation has a unique solution given by $m = 0$. In contrast, for $-c < \gamma < 0$ we have $dR(0)/dm = c/(c+\gamma) > 1$ and we find that the self-consistency equation exhibits solutions with $m \neq 0$, see Fig. 5.10.

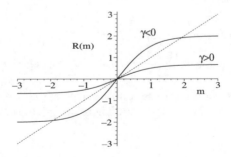

Fig. 5.10. Illustration of the self-consistency equation $m = R(m)$ with $R(m)$ given by (5.169). *Solid lines:* $R(m)$ for $c = 1$ and $\gamma = -0.5$ and $\gamma > 0.5$. *Dashed line:* diagonal of the (m, R)-plane; reprinted from [182], © 2004, with permission from Elsevier

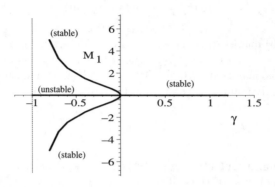

Fig. 5.11. Bifurcation diagram for the order parameter $M_1 = \langle X \rangle_{\rm st}$ of the Takatsuji model (5.163); reprinted from [182], © 2004, with permission from Elsevier

Since we are dealing with a special case of the compensated $B(\langle A \rangle)$-model, the stability of stationary distributions can be determined by means of the slopes of $R(m)$ at intersection points $m = \langle X \rangle_{\rm st}$. For $\gamma > 0$ we have $\mathrm{d}R(0)/\mathrm{d}m = c/(c+\gamma)$, which implies that the stationary probability density (5.170) is asymptotically stable (because of $\mathrm{d}R(0)/\mathrm{d}m < 1$), whereas for $\gamma < 0$ the distribution (5.170) becomes unstable (because of $\mathrm{d}R(0)/\mathrm{d}m > 1$ for $\gamma < 0$). Moreover, from Fig. 5.10 we read off that for $-c < \gamma < 0$ we have $\mathrm{d}R(m)/\mathrm{d}m < 1$ at solutions $m \neq 0$ of $m = R(m)$. Consequently, the stationary probability densities given by (5.167) with $\langle X \rangle_{\rm st} \neq 0$ are asymptotically stable if they exist. The mean values $M_{1,\rm st} = m$ computed from $m = R(m)$ for several values of γ are shown in Fig. 5.11. For $\gamma \downarrow -c$ the stationary mean values $M_{1,\rm st} \neq 0$ behave like $M_{1,\rm st} \to \pm\infty$. For $\gamma \leq -c$ stationary solutions cease to exist.

Autocorrelation Function $\langle X(t)X(t') \rangle$

Our aim now is to discuss the evolution of the autocorrelation function $C(t, t') = \langle X(t)X(t') \rangle$ defined for $t \geq t' \geq t_0$. From (5.165) it follows that for every initial distribution $u(x)$ the first moment M_1 corresponds to a continuous function of time: $M_1 = M_1(t; u)$. Therefore, (5.163) can equivalently be expressed as

$$\frac{\partial}{\partial t}P(x,t;u) = \frac{\partial}{\partial x}\left[(\gamma+c)x - \sqrt{c}\tanh(\sqrt{c}M_1(t;u))\right]P + Q\frac{\partial^2}{\partial x^2}P \quad (5.172)$$

involving the drift coefficient $D_1'(x,t;u) = -(\gamma+c)x + \sqrt{c}\tanh(\sqrt{c}M_1(t;u))$. Since D_1' can be regarded as the first Kramers–Moyal coefficient of a Markov diffusion process, (5.163) is a strongly nonlinear Fokker–Planck equation and the transition probability density of the Markov diffusion process satisfies

5.2 Natural Boundary Conditions

$$\frac{\partial}{\partial t} P(x,t|x',t';u) = \frac{\partial}{\partial x}\left[(\gamma+c)x - \sqrt{c}\tanh(\sqrt{c}\langle X\rangle_{P(x,t;u)})\right] P(x,t|x',t';u)$$
$$+ Q\frac{\partial^2}{\partial x^2} P(x,t|x',t';u) \,. \tag{5.173}$$

Multiplying (5.173) with $P(x',t';u)$, we obtain

$$\frac{\partial}{\partial t} P(x,t;x',t';u) = \frac{\partial}{\partial x}\left[(\gamma+c)x - \sqrt{c}\tanh(\sqrt{c}\langle X\rangle_{P(x,t;u)})\right] P(x,t;x',t';u)$$
$$+ Q\frac{\partial^2}{\partial x^2} P(x,t;x',t';u) \,. \tag{5.174}$$

Multiplying this evolution equation with x and x' and integrating with respect to x and x', we obtain (with the help of partial integration) the evolution equation

$$\frac{\partial}{\partial t} C(t,t') = -(\gamma+c)C(t,t') + \sqrt{c}M_1(t')\tanh(\sqrt{c}M_1(t))\,, \tag{5.175}$$

which has to be solved under the initial condition $C(t',t') = M_2(t') = \langle X^2(t')\rangle$. Thus, we get

$$C(t,t') = M_2(t')\exp\{-(\gamma+c)(t-t')\}$$
$$+ \sqrt{c}M_1(t')\int_{t'}^{t}\exp\{-(\gamma+c)(t-z)\}\tanh(\sqrt{c}M_1(z))\,\mathrm{d}z\,. \tag{5.176}$$

We need to determine now the unknown second moment M_2. From (5.163) it follows that

$$\frac{\mathrm{d}}{\mathrm{d}t} M_2(t) = -2(\gamma+c)M_2(t) + 2Q + 2\sqrt{c}M_1(t)\tanh(\sqrt{c}M_1(t))\,. \tag{5.177}$$

Note that in the stationary case this equation reduces to $M_{2,\mathrm{st}} = Q/(\gamma+c) - M_{1,\mathrm{st}}^2$, which again gives us the variance $K_\mathrm{st} = Q/(\gamma+c)$ of the stationary distribution (5.167). In sum, solving the evolution equations for $M_1(t)$ and $M_2(t)$ given by (5.165) and (5.177) and substituting the result into (5.176), we can obtain for all pairs (t,t') the autocorrelation function $C(t,t')$, see Fig. 5.12.

Now let us show that the autocorrelation function $C(t,t')$ depends only on the time difference $t-t'$ in the stationary case (i.e., it becomes stationary): $\lim_{t\to\infty, t'\to\infty} C(t,t') = C_\mathrm{st}(t-t')$. To this end, let us first note that if $z \in [t',t]$, $t \geq t'$, and $t'\to\infty$ holds then we conclude that $z\to\infty$. Consequently, the integral in (5.176) becomes $\sqrt{c}M_{1,\mathrm{st}}\tanh(\sqrt{c}M_{1,\mathrm{st}})\int_{t'}^{t}\exp\{-(\gamma+c)(t-z)\}\,\mathrm{d}z$. If we use $\sqrt{c}\tanh(\sqrt{c}M_{1,\mathrm{st}}) = (\gamma+c)M_{1,\mathrm{st}}$ (see 5.165)), we obtain

$$\lim_{t'\to\infty} C(t,t') = M_{1,\mathrm{st}}^2 + K_\mathrm{st}\exp\{-(\gamma+c)(t-t')\} = C_\mathrm{st}(t-t')\,. \tag{5.178}$$

That is, $C(t,t')$ indeed becomes stationary in the stationary case. Furthermore, $C_\mathrm{st}(z)$ satisfies the special cases $C_\mathrm{st}(0) = M_{2,\mathrm{st}}$ and $\lim_{z\to\infty} C_\mathrm{st}(z) = M_{1,\mathrm{st}}^2$, see also Fig. 5.13.

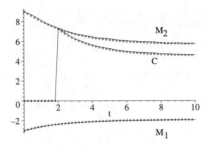

Fig. 5.12. $M_1(t)$, $M_2(t)$, and $C(t,t') = \langle X(t)X(t')\rangle$ as functions of t for a stochastic processes with $u(x) = \delta(x - x_0)$. Solid lines represent results obtained from the Fokker–Planck approach, see (5.165), (5.176), and (5.177). Diamonds represent results obtained from a simulation of the Langevin equation (5.179). using the averaging method for self-consistent Langevin equations. The function $\langle X(t)X(t')\rangle$ is given for $t \geq t'$. For $t < t'$ we have put $\langle X(t)X(t')\rangle = 0$. Parameters: $Q = 1$, $x_0 = -3$, $\gamma = -0.5$, $c = 1$, and $t' = 2$, $t_0 = 0$ ($L = 10^5$, $\Delta t = 0.01$, w_n via Box–Muller); reprinted from [182], © 2004, with permission from Elsevier

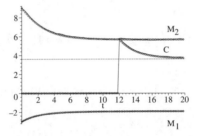

Fig. 5.13. As in Fig. 5.12 but for $t' = 12$. As shown by the first and second moments in this case we are in the stationary regime. The thin horizontal line describes $M_{1,\text{st}}^2$; reprinted from [182], © 2004, with permission from Elsevier

Basins of Attraction

In case of the dynamical Takatsuji model the basins of attraction for stationary solutions can be determined. The reason for this is that for the unknown order parameter $\langle X \rangle_{\text{st}}$ occurring in the stationary distributions of the dynamical Takatsuji model, we have with (5.165) a closed evolution equation for $M_1(t) = \langle X \rangle$ at hand that determines $\langle X \rangle_{\text{st}}$ for every initial distribution $u(x)$ (see also Sect. 2.4). As mentioned earlier the potential $V_M(z)$ given by (5.166) is bounded from below for $\gamma > -c$. It is monostable for $\gamma \geq 0$ and bistable for $-c < \gamma < 0$, see Fig. 5.9. For $-c < \gamma < 0$ we conclude from (5.165) that probability densities $P(x,t;u)$ with $M_1(0) > 0$ converge to the asymptotically stable, asymmetric stationary distribution (5.167) with $M_{1,\text{st}} > 0$, whereas $P(x,t;u)$ with $M_1(0) < 0$ converge to (5.167) with $M_{1,\text{st}} < 0$, see Fig. 5.14.

Probability densities $P(x,t;u)$ with a vanishing first moment converge to the unstable symmetric solution (5.170). Consequently, in the function space of probability densities the distributions $u(x)$ with $\int_\Omega xu(x)\,\mathrm{d}x = 0$ describe some kind of separatrix. We would like to point out that not only do all distribution functions $P(x,t;u)$ with $\int_\Omega xu(x)\,\mathrm{d}x \neq 0$ converge to the stationary distribution (5.167) with $M_{1,\mathrm{st}} > 0$ or $M_{1,\mathrm{st}} < 0$, but also all stochastic processes described by (5.163) and (5.173) and the Markov hierarchy (3.18) converge to only two kinds of stationary stochastic processes exhibiting either $M_{1,\mathrm{st}} > 0$ or $M_{1,\mathrm{st}} < 0$.

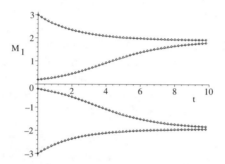

Fig. 5.14. $M_1(t)$ as a function of t as obtained from a simulation of (5.179) (diamonds) and from solving numerically (5.165) (*solid lines*) for delta distributions $u(x)$ and different $M_1(0)$ ($t_0 = 0$). For distributions with $M_1(0) > 0$ (< 0) the stochastic processes converge to the positive (negative) solution of $(\gamma + c)M_{1,\mathrm{st}} = \sqrt{c}\tanh(\sqrt{c}M_{1,\mathrm{st}})$ given for our parameter values by $M_{1,\mathrm{st}} \approx 2$ (-2). Initial values: $M_1(0) = 3$, $M_1(0) = 0.2$, $M_1(0) = -0.2$, and $M_1(0) = -3$ (*from top to bottom*). Parameters: $Q = 1$, $c = 1$, $\gamma = -0.5$ ($L = 10^5$, $\Delta t = 0.01$, w_n via Box–Muller); reprinted from [182], © 2004, with permission from Elsevier

Langevin Equation

The Langevin equation that corresponds to the strongly nonlinear Fokker–Planck equation given by (5.163) and (5.173) reads

$$\frac{\mathrm{d}}{\mathrm{d}t}X(t) = -(\gamma + c)X + \sqrt{c}\tanh(\sqrt{c}\langle X\rangle) + \sqrt{Q}\Gamma(t)\ . \qquad (5.179)$$

We can now compute the autocorrelation function $C(t,t')$ from (5.165), (5.176), and (5.177) obtained from the Fokker–Planck description or alternatively from the Langevin equation (5.179). As shown in Figs. 5.12 and 5.13 both methods yield the same results. In particular, Fig. 5.13 illustrates the convergence of the stationary autocorrelation function $C_{\mathrm{st}}(\Delta t)$ to the limit value $M_{1,\mathrm{st}}^2$ in the limit $\Delta t \to \infty$. In order to illustrate the basins of attraction, we solved the Langevin equation (5.179) for several initial values

5.2.3 Desai–Zwanzig Model

The K-model (5.50) has been studied in detail for $V(x) = -ax/2 + bx^4/4$ by Kometani and Shimizu [352] and Desai and Zwanzig [135]. This model is explicitly given by

$$\frac{\partial}{\partial t}P(x,t;u) = -\frac{\partial}{\partial x}\left[ax - bx^3 - \kappa(x - \langle X \rangle)\right]P + Q\frac{\partial^2}{\partial x^2}P, \quad (5.180)$$

for $\kappa \geq 0$, $a \geq 0$, $b \geq 0$, and $t \geq t_0$ and is often referred to as the Desai–Zwanzig model. For $\kappa = 0$ we deal with a linear Fokker–Planck equation. For $b = 0$ the model (5.180) reduces to the Shimizu–Yamada model discussed in Sect. 5.2.1. In what follows, we consider the case $\kappa > 0$ and $b > 0$.

First of all, we note that by means of the variable transformations $t' = at$ and $x' = x\sqrt{b/a}$ and the operator relations

$$\frac{\partial}{\partial t} = a\frac{\partial}{\partial t'}, \quad \frac{\partial}{\partial x} = \sqrt{\frac{b}{a}}\frac{\partial}{\partial x'}, \quad \frac{\partial^2}{\partial x^2} = \frac{b}{a}\frac{\partial^2}{\partial x'^2}, \quad (5.181)$$

we can write (5.180) as

$$\frac{\partial}{\partial t'}P'(x',t') = -\frac{\partial}{\partial x'}\left[x' - x'^3 - \kappa'(x' - \langle X' \rangle)\right]P' + Q'\frac{\partial^2}{\partial x'^2}P', \quad (5.182)$$

with $\kappa' = \kappa/a$ and $Q' = Qb/a^2$ for the probability density

$$P'(x',t') = \sqrt{\frac{a}{b}}P\left(\sqrt{\frac{a}{b}}x', \frac{t'}{a}; u\right). \quad (5.183)$$

Consequently, we will primarily examine solutions of (5.180) for $a = b = 1$. Equation (5.180) can be written as the free energy Fokker–Planck equation (5.3) with F given by [128, 523]

$$F[P] = \int_\Omega V(x)\,dx + \frac{\kappa}{2}K(X) - Q^B S[P],$$
$$V(x) = -\frac{a}{2}x^2 + \frac{b}{4}x^4. \quad (5.184)$$

The free energy corresponds to a measure of type E in Table 4.2, which implies that we have $F \geq F_{\min}$ and $\lim_{t \to \infty} \partial P/\partial t = 0$. The free energy F can equivalently be expressed as

$$F[P] = \int_\Omega U_0(x)\,dx - \frac{\kappa}{2}\langle X \rangle^2 - Q^B S[P],$$
$$U_0(x) = -\frac{a-\kappa}{2}x^2 + \frac{b}{4}x^4. \quad (5.185)$$

The effective potential U_0 corresponds to a double-well potential for $a > \kappa$ and a potential with a single minimum for $a < \kappa$, see Fig. 5.15.

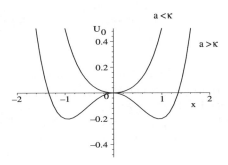

Fig. 5.15. Potential $U_0(x)$ defined by (5.185) for $a = b = 1$, $Q = 0.1$ and two values of κ: $\kappa = 0.1$ and $\kappa = 1.5$

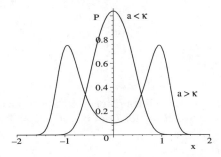

Fig. 5.16. Symmetric stationary solutions (5.188) of the Desai–Zwanzig model for different values of κ. Parameters correspond to those used in Fig. 5.15

Substituting $V(x) = -ax^2/2 + bx^4/4$ into (5.56) and (5.57), we obtain

$$P_{\rm st}(x) = \frac{1}{Z} \exp\left\{ -\frac{[bx^4/4 - (a-\kappa)x^2/2 - \kappa x \langle X \rangle_{\rm st}]}{Q} \right\} \quad (5.186)$$

and

$$P(x,m) = \frac{1}{Z(m)} \exp\left\{ -\frac{[bx^4/4 - (a-\kappa)x^2/2 - \kappa x m]}{Q} \right\},$$

$$Z(m) = \int_\Omega \exp\left\{ -\frac{[bx^4/4 - (a-\kappa)x^2/2 - \kappa x m]}{Q} \right\} dx, \quad (5.187)$$

as well as the self-consistency equation $m = R(m)$ with $R(m) = \int_\Omega x P(x,m)\,dx$. From (5.187) it is clear that $Z(0) > 0$ and $R(0) = 0$, which implies that $m = 0$ is a solution of $m = R(m)$ and

$$P_{\text{st}}(x) = \frac{1}{Z} \exp\left\{-\frac{[bx^4/4 - (a-\kappa)x^2/2]}{Q}\right\} \tag{5.188}$$

describes a stationary probability density of (5.180). We refer to (5.180) as the symmetric stationary solution of the Desai–Zwanzig model. There are two types of symmetric solutions. As shown in Fig. 5.16, for $\kappa < a$ we obtain an unimodal distribution, whereas for $\kappa > a$ we have a bimodal distribution. Stationary mean values $\langle X \rangle_{\text{st}} \neq 0$ satisfying (5.186) can be computed from $m = R(m)$.

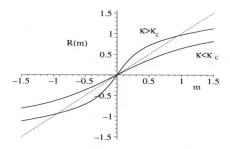

Fig. 5.17. Illustration of the self-consistency equation $m = R(m)$ of the Desai–Zwanzig model. Solid lines: $R(m)$ for $Q = 0.1$, $a = b = 1$ and two values of κ; $\kappa = 0.1 < \kappa_c$ and $\kappa = 1 > \kappa_c$

Since $R(m)$ is an antisymmetric function (i.e, we have $R(m) = -R(-m)$), every solution $m \neq 0$ of $m = R(m)$ gives us another solution of $m = R(m)$, which is $-m$. The stationary probability densities (5.186) with $\langle X \rangle_{\text{st}} = m \neq 0$ describe asymmetric functions (i.e., we have $P_{\text{st}}(x) \neq P_{\text{st}}(-x)$). The function $R(m)$ increases monotonically (see (5.62)) and is shown in Fig. 5.17 for two typical cases. From Fig. 5.17 we read off that

- if there are solutions of $m = R(m)$ with $m \neq 0$, then the slope of $R(m)$ at $m = 0$ is larger than unity;
- if there are solutions of $m = R(m)$ with $m \neq 0$, then the slope of $R(m)$ at these solutions is smaller than unity.

Consequently, $\tilde{\lambda} = Q - \kappa K_{\text{st}}(X) = Q(1 - dR(0)/dm)$ is positive if there is only one solution of the self-consistency equation and negative if there are multiple solutions. This implies that the symmetric stationary solution (5.188) corresponds to an asymptotically stable stationary solution if it is

5.2 Natural Boundary Conditions   151

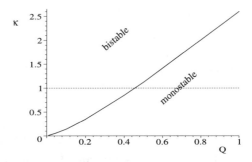

Fig. 5.18. *Solid line*: bifurcation line in the parameter space described by κ and Q for $a = b = 1$. The bifurcation line separates a monostable and a bistable region. *Dashed line*: for $\kappa > 1$ ($\kappa < 1$) the symmetric solution (5.188) is unimodal (bimodal), see Fig. 5.16

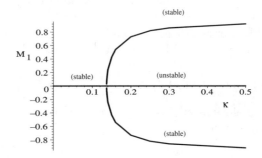

Fig. 5.19. Bifurcation diagram: $M_1 = \langle X \rangle_{\text{st}}$ as a function of κ for $a = b = 1$ and $Q = 0.1$

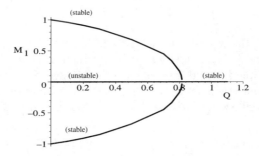

Fig. 5.20. Bifurcation diagram: $M_1 = \langle X \rangle_{\text{st}}$ as a function of Q for $a = b = 1$ and $\kappa = 2.0$

the only solution of the implicit equation (5.186). In contrast, the symmetric solution corresponds to an unstable solution if there are asymmetric solutions satisfying (5.186). In the latter case, these additional asymmetric solutions are stable distribution functions.

The critical parameter values at which the symmetric solution (5.188) becomes unstable and multiple solutions occur can be computed from $\tilde{\lambda} = 0 \Rightarrow \mathrm{d}R(0)/\mathrm{d}m = 1 \Rightarrow Q = \kappa K(X)_{P_{\mathrm{st},m=0}}$, which leads to

$$Q = \kappa \left\langle X^2 \right\rangle_{P_{\mathrm{st},m=0}}, \qquad (5.189)$$

where $P_{\mathrm{st},m=0}$ is defined by (5.188). The function $\kappa(Q)$ thus obtained is shown in Fig. 5.18. Below the bifurcation line the system is monostable and exhibits a unique, symmetric, and asymptotically stable stationary probability density. Above the bifurcation line the system is bistable and exhibits two asymmetric and asymptotically stable stationary probability densities. Note also that for large Q it can be shown that the bifurcation line is described by $\kappa = 3Q$ [135]. Finally, the self-consistency equation $m = R(m)$ can be solved numerically and bifurcation diagrams can be obtained. The order parameter $\langle X \rangle_{\mathrm{st}}$ as a function of κ and Q is shown in Figs. 5.19 and 5.20.

5.2.4 Bounded $B(M_1)$-Model

We examine now solutions of (5.83) for $A(x) = x$ and $b(z) = -\mathrm{d}B/\mathrm{d}z$ [164]. That is, we study solutions of

$$\frac{\partial}{\partial t} P(x,t;u) = \frac{\partial}{\partial x}\left[\frac{\mathrm{d}U_0(x)}{\mathrm{d}x} - b(\langle X \rangle)\right] P + Q \frac{\partial^2}{\partial x^2} P \qquad (5.190)$$

that describe the evolution of systems with free energy measures given by

$$F[P] = \int_\Omega U_0(x) P(x)\,\mathrm{d}x + B(\langle X \rangle) - Q\,{}^B\!S[P]\,. \qquad (5.191)$$

We require that the potentials $U_0 \in C^\infty(\Omega)$ and $B \in C^\infty(\mathbb{R})$ are bounded from below: $U_0(z) \ge U_{\min}$ and $B(z) \ge B_{\min}$. Then, F is bounded from below (see type F in Table 4.2) and transient solutions converge to stationary ones in the long time limit (H-theorem, see Sect. 4.3). In order to describe stationary solutions of (5.190), we substitute $A(x) = x$ into (5.84), which yields

$$P_{\mathrm{st}}(x) = \frac{1}{Z} \exp\left\{-\frac{U_0(x) - x b(\langle X \rangle_{\mathrm{st}})}{Q}\right\}\,. \qquad (5.192)$$

Likewise, (5.85) becomes

$$P(x,m) = \frac{1}{Z(m)} \exp\left\{-\frac{U_0(x) - x b(m)}{Q}\right\},$$

$$Z(m) = \int_\Omega \exp\left\{-\frac{U_0(x) - x b(m)}{Q}\right\} \mathrm{d}x\,. \qquad (5.193)$$

Then, $R(m)$ is defined with the help of (5.193) by $R(m) = \int_\Omega x P(x,m)\,\mathrm{d}x$ and the mean values $\langle X \rangle_{\mathrm{st}}$ of (5.192) can be determined by means of the self-consistency equation $m = R(m)$. Since we deal with a special case of the bounded $B(A)$-model, the stability of the stationary distributions (5.192) is determined by the sign of $\tilde\lambda$ given by $\tilde\lambda = Q + K_{\mathrm{st}}(X)\mathrm{d}^2 B(\langle X \rangle_{\mathrm{st}})/\mathrm{d}m^2$ (see (5.73) for $A(x)=x$). Since we have

$$\frac{\mathrm{d}R(m)}{\mathrm{d}m} = -\frac{1}{Q} K(X)_{P(x,m)} \frac{\mathrm{d}^2 B(m)}{\mathrm{d}m^2} = \frac{1}{Q} K(X)_{P(x,m)} \frac{\mathrm{d}b(m)}{\mathrm{d}m}, \quad (5.194)$$

the coefficient $\tilde\lambda$ can equivalently be expressed as $\tilde\lambda = Q(1 - \mathrm{d}R(m)/\mathrm{d}m)$ with $m = \langle X \rangle_{\mathrm{st}}$ (equivalence between self-consistency equation analysis and Lyapunov's direct method).

Existence of stationary solutions

Let us show now that there is at least one stationary solution of (5.190). For infinitely differentiable functions B and U_0 it is reasonable to assume that $R(m)$ is a continuous function with respect to m. Then, our objective is to show that $R(m)$ is bounded from above for $m > 0$ and bounded from below for $m < 0$, which implies that there is at least one intersection point of $R(m)$ and the diagonal of the (m, R)-plane (i.e., there is at least one m solving $m = R(m)$).

Proof*

In order to prove the aforementioned kind of boundedness, we assume that the Boltzmann distribution

$$W(x) = \frac{1}{Z_B} \exp\left\{-\frac{U_0(x)}{Q}\right\} \quad (5.195)$$

exists (where Z_B is a normalization constant $Z_B = \int_\Omega \exp\{-U_0/Q\}\,\mathrm{d}x$) and that $B(z)$ is globally attractive in the sense that it exhibits the asymptotic behavior $B(z \to \pm\infty) \to \infty$. Next, we realize that the mean of the distribution

$$P_b(x) = \frac{\exp\{-[U_0(x)-xb]/Q\}}{\int_\Omega \exp\{-[U_0(x)-xb]/Q\}\,\mathrm{d}x} \quad (5.196)$$

is smaller than or equal to the mean of $W(x)$ for parameter values $b \leq 0$ and larger than or equal to the mean of $W(x)$ for parameter values $b \geq 0$:

$$\int_\Omega x P_{b \leq 0}(x)\,\mathrm{d}x \leq \langle X \rangle_W \leq \int_\Omega x P_{b \geq 0}(x)\,\mathrm{d}x. \quad (5.197)$$

The reason for this is that for $b = 0$ we have $\int_\Omega P_b(x)\,\mathrm{d}x = \langle X \rangle_W$ and for $b \in \mathbb{R}$ the derivative

$$\frac{\mathrm{d}}{\mathrm{d}b} \int_\Omega x P_b(x)\, \mathrm{d}x = \frac{1}{Q} K(X)_{P_b} \geq 0 \qquad (5.198)$$

is positive, which implies that $\int_\Omega x P_b(x)\, \mathrm{d}x$ is a monotonically increasing function with respect to b. Next, we note that from the asymptotic behavior $B(z \to \pm\infty) \to \infty$ of $B(z)$ it follows that for $z \to \infty$ we have $b(z) < 0$ and $P(x,m)$ corresponds to a distribution P_b with $b(m) \leq 0$: $P(x,m) = P_{b(m)\leq 0}(x)$. Likewise, in the limit $z \to -\infty$ we have $b(z) > 0$ and $P(x,m) = P_{b(m)\geq 0}(x)$. Consequently, the limiting cases $\lim_{m\to\infty} R(m) = \int_\Omega x P_{b(m)\leq 0}(x)\, \mathrm{d}x \leq \langle X \rangle_W$ and $\lim_{m\to-\infty} R(m) = \int_\Omega x P_{b(m)\geq 0}(x)\, \mathrm{d}x \geq \langle X \rangle_W$ hold. That is, $R(m)$ is bounded from above for $m \to \infty$ and from below for $m \to -\infty$. Since $R(m)$ is assumed to be a continuous function, we conclude that $R(m)$ is bounded from above for $m > 0$ and from below for $m < 0$. That is, we have proven the claim made above.

Example

Let us illustrate this issue by a more explicit example. We consider the potential
$$U_0(x) = \frac{\gamma}{2} x^2 \qquad (5.199)$$

with $\gamma > 0$ for which (5.192) can be written as

$$P_{\mathrm{st}}(x) = \sqrt{\frac{\gamma}{2\pi Q}} \exp\left\{ -\frac{\gamma[x - \gamma^{-1}b(\langle X \rangle_{\mathrm{st}})]^2}{2Q} \right\}. \qquad (5.200)$$

In addition, we consider a double-well potential

$$B(z) = -\frac{z^2}{2} + \frac{z^4}{4} \qquad (5.201)$$

leading to $b(z) = z - z^3$. Since the variance of the stationary distributions (5.200) is given by $K_{\mathrm{st}}(X) = Q/\gamma$, the function

$$R(m) = \sqrt{\frac{\gamma}{2\pi Q}} \int_\Omega x \exp\left\{ -\frac{\gamma}{2Q}(x - \gamma^{-1}b(m))^2 \right\} \mathrm{d}x \qquad (5.202)$$

reads $R(m) = b(m)/\gamma = (m - m^3)/\gamma$ and the self-consistency equation $m = R(m)$ is solved by $m = 0$ for $\gamma > 0$ and $m = \pm\sqrt{1-\gamma}$ for $\gamma \in (0,1)$, which means that the critical value of γ is $\gamma_c = 1$. Furthermore, we have $\mathrm{d}R(m)/\mathrm{d}m = (1 - 3m^2)/\gamma$. From this it follows that

$$\left.\frac{\mathrm{d}R(m)}{\mathrm{d}m}\right|_{\langle X \rangle_{\mathrm{st}}=0} > 1\,(<1) \quad \text{for} \quad \gamma \in (0,1)\,(\gamma > 1),$$

$$\left.\frac{\mathrm{d}R(m)}{\mathrm{d}m}\right|_{\langle X \rangle_{\mathrm{st}}=\pm\sqrt{1-\gamma}} > 1 \quad \text{for} \quad \gamma \in (0,1). \qquad (5.203)$$

As a result, we obtain the self-consistency equations illustrated in Fig. 5.21 and the bifurcation diagram depicted in Fig. 5.22. For $\gamma \to 0$ the variance of the distribution (5.200) increases like $K_{st} \to \infty$, which is in line with the fact that for $\gamma < 0$ Gaussian distributions of the form (5.200) no longer exist. Finally, note that in this example we have $\langle X \rangle_W = 0$ and, consequently, the limiting cases $R(m) < 0$ for $m \to \infty$ and $R(m) > 0$ for $m \to -\infty$ hold, see also Fig. 5.21.

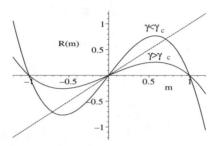

Fig. 5.21. Illustration of the self-consistency equation $m = R(m)$ with $R(m) = (m - m^3)/\gamma$. Solid lines: $R(m)$ for two values of γ: $\gamma = 0.5 < \gamma_c$ and $\gamma = 1.5 > \gamma_c$ with $\gamma_c = 1$

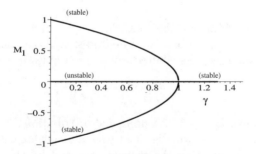

Fig. 5.22. Bifurcation diagram of the order parameter M_1 for the $B(M_1)$-model (5.190) with U_0 and B given by (5.199) and (5.201)

5.3 Periodic Boundary Conditions

In this section we will discuss stochastic processes described by periodic random variables defined on $\Omega = [0, 2\pi]$. We will examine the Kuramoto–

Shinomoto–Sakaguchi (KSS) model and the mean field Haken–Kelso–Bunz (HKB) model. Before we discuss these models, we will introduce two helpful statistical measures: the cluster amplitude and the cluster phase.

5.3.1 Cluster Amplitude and Cluster Phase

The cluster amplitude r and the cluster phase θ of a 2π-periodic probability density P are defined by [365]

$$re^{i\theta} = \langle e^{iX} \rangle = \int_\Omega \exp\{ix\} P(x) \, \mathrm{d}x \; . \tag{5.204}$$

Equation (5.204) can alternatively be written as

$$r\cos\theta = \langle \cos(X) \rangle \; , \quad r\sin\theta = \langle \sin(X) \rangle \; . \tag{5.205}$$

and

$$r = \sqrt{\langle \cos(X) \rangle^2 + \langle \sin(X) \rangle^2} = |\langle \exp\{iX\} \rangle| \geq 0 \; ,$$
$$\theta = \arctan\left(\frac{\langle \sin(X) \rangle}{\langle \cos(X) \rangle}\right) \; . \tag{5.206}$$

Note that $r = |\langle \exp\{iX\} \rangle|$ is not equivalent to $\langle |\exp\{iX\}| \rangle$, which is equal to unity. The cluster amplitude r satisfies

$$P(x) = \delta(x - x_0) \Rightarrow r = 1 \; ,$$
$$P(x) \in C^0(\Omega) \Rightarrow r < 1 \; . \tag{5.207}$$

That is, for continuous distributions $P(x)$ the cluster amplitude is smaller than unity. If $P(x)$ corresponds to a delta distribution, then r equals unity. Moreover, for uniformly distributed random variables in the interval $[0, 2\pi]$ we obtain

$$P(x) = \frac{1}{2\pi} \Rightarrow r = 0 \; . \tag{5.208}$$

In Fig. 5.23 realizations X^k of a unimodal distribution $P(x)$ are plotted on the unit circle. As shown in Fig. 5.23, roughly speaking, the cluster phase describes in which direction the realizations point on the average, whereas the cluster amplitude describes the degree of uniformity of the distribution and tells us how pronounced the peak of the distribution is.

Similar to (5.204), one can define generalized cluster amplitudes r_n and cluster phases θ_n [116, 117, 118, 119, 568]:

$$r_n \exp\{in\theta_n\} = \int_0^{2\pi} \exp\{inx\} P(x) \, \mathrm{d}x \; . \tag{5.209}$$

The amplitudes r_n satisfy $0 \leq r_n \leq 1$ with $r_n^2 = \langle \cos(nX) \rangle^2 + \langle \sin(nX) \rangle^2$. We have $r_n = 0$ for all n if and only if P corresponds to the uniform distribution.

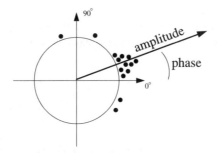

Fig. 5.23. Interpretation of the cluster amplitude and cluster phase. Dots represent realizations X^k of a 2π-periodic random variable X

5.3.2 Kuramoto–Shinomoto–Sakaguchi (KSS) Model – Cluster Amplitude Dynamics

The nonlinear Fokker–Planck equation

$$\frac{\partial}{\partial t}P(x,t;u) = \frac{\partial}{\partial x}\left[\int_\Omega \frac{\mathrm{d}U_{\mathrm{MF}}(x-x')}{\mathrm{d}x}P(x',t;u)\,\mathrm{d}x'\,P\right] + Q\frac{\partial^2}{\partial x^2}P \quad (5.210)$$

has been introduced in works by Kuramoto, Shinomoto and Sakaguchi [364, 365, 366, 506, 531, 530] and since then has extensively been studied (see Sect. 7.5). Solutions of (5.210) are subjected to periodic boundary conditions. We assume that the mean field potential satisfies the relations $U_{\mathrm{MF}}(0) = 0$ and $U_{\mathrm{MF}}(z) = U_{\mathrm{MF}}(z+2\pi)$. We will refer to (5.210) as the KSS model.

Irrespective of the explicit form of $U_{\mathrm{MF}}(z)$, the uniform probability density is a stationary solution of (5.210):

$$P_{\mathrm{st}}(x) = \frac{1}{2\pi} \,. \quad (5.211)$$

This can be shown by substituting (5.211) into (5.210) and taking into account that $\int_\Omega U_{\mathrm{MF}}(x-x')\,\mathrm{d}x' = 0$, which is due to the fact that $U_{\mathrm{MF}}(z)$ can be written as (7.149). In general, stationary probability densities of (5.210) satisfy

$$J = -\int_\Omega \frac{\mathrm{d}U_{\mathrm{MF}}(x-x')}{\mathrm{d}x}P_{\mathrm{st}}(x')\,\mathrm{d}x'\,P_{\mathrm{st}}(x) - Q\frac{\partial}{\partial x}P_{\mathrm{st}}(x)\,, \quad (5.212)$$

where J = constant and $P_{\mathrm{st}}(x) = P_{\mathrm{st}}(x+2\pi)$, see (2.23). Equation (5.212) is solved by $J = 0$ and

$$P_{\mathrm{st}}(x) = \frac{1}{Z}e^{-\frac{\int_\Omega U_{\mathrm{MF}}(x-x')P_{\mathrm{st}}(x')\,\mathrm{d}x}{Q}}\,. \quad (5.213)$$

Free Energy Case

We assume now that U_{MF} is a symmetric function: $U_{\text{MF}}(z) = U_{\text{MF}}(-z)$. The mean field potential can then be expanded into a Fourier series given by

$$U_{\text{MF}}(z) = -\sum_{n=1}^{\infty} c_n \cos(nz) \tag{5.214}$$

and

$$\cos(n(x - x')) = \cos(nx)\cos(nx') + \sin(nx)\sin(nx') . \tag{5.215}$$

We have chosen a minus sign in (5.214) because in this case for $c_n > 0$ the potential $U_{\text{MF}}(z)$ has a minimum at $z = 0$. We will refer to the evolution equation (5.210) with the potential (5.214) as the free energy case of the KSS model (or the free energy KSS model). For U_{MF} given by (5.214) the nonlinear Fokker–Planck equation (5.210) can be written as a free energy Fokker–Planck equation (5.3) with

$$F[P] = \frac{1}{2} \int_\Omega U_{\text{MF}}(x - x') P(x) P(x') \, dx \, dx' - Q^{\text{B}} S[P] . \tag{5.216}$$

Alternatively, F can be written as

$$F[P] = -\frac{1}{2} \sum_{n=1}^{\infty} c_n \left[\langle \cos(nX) \rangle^2 + \langle \sin(nX) \rangle^2 \right] - Q^{\text{B}} S[P] . \tag{5.217}$$

Stationary solutions are given by

$$P_{\text{st}}(x) = \frac{1}{Z} e^{\frac{\sum_{n=1}^{\infty} [c_n \cos(nx) \langle \cos(nX) \rangle_{\text{st}} + c_n \sin(nx) \langle \sin(nX) \rangle_{\text{st}}]}{Q}} \tag{5.218}$$

and by the uniform distribution (5.211). Note that the uniform distribution $P_{\text{st}}(x) = 1/(2\pi)$ can also be regarded as a solution of (5.218) because we have $\langle \cos(nX) \rangle = \langle \sin(nX) \rangle = 0$ for all n if the averaging is carried out with respect to the uniform distribution. Since the free energy (5.216) corresponds to a measure of type E in Table 4.2, we have $F \geq F_{\text{min}}$ and the H-theorem discussed in Sect. 4.3 applies. That is, the limiting case

$$\lim_{t \to \infty} \frac{\partial}{\partial t} P(x, t; u) = 0 \tag{5.219}$$

holds.

Stability Analysis

We may carry out a stability analysis of the uniform distribution as discussed in Sect. 5.1, which is based on the second variational of F. From (5.217) it is clear that the free energy KSS model is a special case of the free energy Fokker–Planck equation (5.107). Consequently, we could simply apply the results derived in Sect. 5.1.6. In what follows, however, we will reiterate the steps of the procedure outlined in Sect. 5.1.6 in order to gain more insight into the problem.

To begin with, the second variation of F reads

$$\delta^2 F[P_{\text{st}}](\epsilon) = Q \int_\Omega \frac{[\epsilon(x)]^2}{P_{\text{st}}(x)} \, dx + \int_\Omega \int_\Omega U_{\text{MF}}(x - x') \epsilon(x) \epsilon(x') \, dx \, dx' \ . \quad (5.220)$$

Using (5.214), (5.220) gives us

$$\delta^2 F[P_{\text{st}}](\epsilon) = -\sum_{n=1}^\infty c_n \left(\left[\int_\Omega \cos(nx)\epsilon(x) \, dx \right]^2 + \left[\int_\Omega \sin(nx)\epsilon(x) \, dx \right]^2 \right)$$

$$+ Q \int_\Omega \frac{[\epsilon(x)]^2}{P_{\text{st}}(x)} \, dx \ . \quad (5.221)$$

From (5.221) we see that the set of linearly independent functions $A_i(x)$ is given by

$$A_n(x) = \begin{cases} \cos(nx/2) & \text{for} \quad n = \text{even} \\ \sin((n+1)x/2) & \text{for} \quad n = \text{odd} \end{cases}, \quad (5.222)$$

with $n = 1, 2, \ldots$. From (5.101) it is clear that we need to consider perturbations $\epsilon(x)$ of the form

$$\epsilon(x) = \sum_{i=1}^r a_{2n_i} \left[\cos(n_i x) - \langle \cos(n_i X) \rangle_{\text{st}} \right] P_{\text{st}}(x)$$

$$+ \sum_{i=1}^r a_{2n_i - 1} \left[\sin(n_i x) - \langle \sin(n_i X) \rangle_{\text{st}} \right] P_{\text{st}}(x) + \chi_\perp(x) \sqrt{P_{\text{st}}(x)} \ .$$

$$(5.223)$$

In (5.223) the indices n_i correspond to nonvanishing Fourier coefficients c_{n_i}. That is, we denote by I_r the set $I_r = \{n_1, \ldots, n_r\}$ of r indices related to the $d = 2r$ expectation values $\langle \cos(n_i X) \rangle$ and $\langle \sin(n_i X) \rangle$ that are assumed to occur in the free energy (5.217) and in the drift coefficient of the nonlinear Fokker–Planck equations (5.210). For the sake of convenience, however, we will use in what follows $\chi_\perp = 0$ and

$$\epsilon(x) = \sum_{n=1}^\infty a_{2n} \left[\cos(nx) - \langle \cos(nX) \rangle_{\text{st}} \right] P_{\text{st}}(x)$$

$$+ \sum_{n=1}^\infty a_{2n-1} \left[\sin(nx) - \langle \sin(nX) \rangle_{\text{st}} \right] P_{\text{st}}(x) \ . \quad (5.224)$$

In the case of the uniform distribution $P_{st} = 1/(2\pi)$ we obtain $\langle A_i \rangle = 0$ and $\langle A_i A_{i'} \rangle_{st} = \delta_{ii'} \langle [A_i]^2 \rangle_{st}$. Consequently, the cross-correlation matrix (5.104) assumes diagonal form and we are concerned with the special case reported in Sect. 5.1.6 that leads to (5.112-5.114). In addition, (5.224) reduces to

$$\epsilon(x) = \frac{1}{2\pi} \sum_{n=1}^{\infty} [a_{2n} \cos(nx) + a_{2n-1} \sin(nx)] , \qquad (5.225)$$

which corresponds to an ordinary Fourier expansion of the perturbation ϵ. In view of these considerations, it does not come as a surprise that if we substitute (5.225) into (5.221) we obtain [172]

$$\delta^2 F \left[P_{st} = \frac{1}{2\pi} \right] (\epsilon) = \sum_{n=1}^{\infty} a_{2n}^2 \langle \cos^2(nX) \rangle_{st} [Q - c_n \langle \cos^2(nX) \rangle]$$

$$+ \sum_{n=1}^{\infty} a_{2n-1}^2 \langle \sin^2(nX) \rangle_{st} [Q - c_n \langle \sin^2(nX) \rangle] \qquad (5.226)$$

(see (5.113)), which can be transformed into

$$\delta^2 F \left[P_{st} = \frac{1}{2\pi} \right] (\epsilon) = \frac{1}{2} \sum_{n=1}^{\infty} (a_{2n}^2 + a_{2n-1}^2) \left[Q - \frac{c_n}{2} \right] \qquad (5.227)$$

because of $[2\pi]^{-1} \int_\Omega \cos^2(nz)\,dz = [2\pi]^{-1} \int_\Omega \sin^2(nz)\,dz = 1/2$. It can be seen from (5.227) that if $\tilde{\lambda} = Q - c_n/2 > 0$ holds for all n, then F has a minimum at $P_{st} = 1/(2\pi)$ and, consequently, the uniform probability density is asymptotically stable. If there is an index n^* such that $\tilde{\lambda} = Q - c_{n^*}/2 < 0$, then F decreases for perturbations $\epsilon(x) = [a_{2n^*} \cos(n^* x) + a_{2n^* +1} \sin(n^* x)]/(2\pi)$ with $a_{2n^*} \neq 0$ or $a_{2n^* +1} \neq 0$. This means that F has a maximum or saddle-point at $P_{st} = 1/(2\pi)$ and the uniform probability density is unstable. Finally, we would like to point out that using the generalized cluster amplitudes r_n and cluster phases θ_n (5.209), the free energy (5.217) can be written as

$$F[P] = -\frac{1}{2} \sum_{n=1}^{\infty} c_n r_n^2 - Q^B S[P] . \qquad (5.228)$$

Therefore, the KSS model is closely related to dynamics of cluster amplitudes.

Free Energy KSS Model with Sine Coupling-force

For the mean field potential $U_{MF}(x - x') = -\kappa \cos(x - x')$ with $\kappa \geq 0$ we find that (5.210) becomes

$$\frac{\partial}{\partial t} P(x, t; u) = \kappa \frac{\partial}{\partial x} \left[\int_\Omega \sin(x - x') P(x', t; u)\,dx' P \right] + Q \frac{\partial^2}{\partial x^2} P \qquad (5.229)$$

5.3 Periodic Boundary Conditions

and can be regarded as a free energy Fokker–Planck equation with

$$F[P] = -\frac{\kappa}{2} \int_\Omega \int_\Omega \cos(x - x') P(x) P(x') \, dx \, dx' - Q^B S[P] \;, \qquad (5.230)$$

$F \geq F_{\min}$, and transient solutions that satisfy $\lim_{t \to \infty} \partial P(x,t)/\partial t = 0$. From (5.213) it follows that stationary probability densities of (5.229) are described by

$$P_{\text{st}}(x) = \frac{1}{Z} e^{\frac{\kappa \int_\Omega \cos(x - x') P_{\text{st}}(x') dx'}{Q}} \;. \qquad (5.231)$$

By means of $r \exp(i\theta) = \int_\Omega \exp(ix) P(x) \, dx$, (5.229) can be written as

$$\frac{\partial}{\partial t} P(x, t; u) = \kappa \frac{\partial}{\partial x} r(t) \sin(x - \theta(t)) P + Q \frac{\partial^2}{\partial x^2} P \;, \qquad (5.232)$$

the free energy (5.230) is found as

$$F[P] = -\frac{\kappa}{2} r^2 - Q^B S[P] \;, \qquad (5.233)$$

and (5.231) can be cast into the form

$$P_{\text{st}}(x) = \frac{1}{Z(r_{\text{st}})} e^{\frac{\kappa r_{\text{st}} \cos(x - \theta_{\text{st}})}{Q}} \;, \qquad (5.234)$$

where Z is independent of θ_{st}:

$$Z(r_{\text{st}}) = \int_\Omega e^{\frac{\kappa r_{\text{st}} \cos(x)}{Q}} \, dx \;. \qquad (5.235)$$

The stationary cluster amplitude r_{st} corresponds to solutions m of the self-consistency equation

$$m \exp(i\theta_{\text{st}}) = \int_\Omega \exp(ix) P(x, m) \, dx \;, \qquad (5.236)$$

with

$$P(x, m) = \frac{1}{Z} e^{\frac{\kappa m \cos(x - \theta_{\text{st}})}{Q}} \qquad (5.237)$$

and Z given by the normalization constant $Z(m) = \int_\Omega \exp\{\kappa m \cos(x)/Q\} dx$. Using the variable transformation $x' = x - \theta_{\text{st}}$, from the self-consistency equation (5.236) we obtain

$$m = R(m) \;, \qquad (5.238)$$

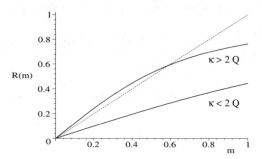

Fig. 5.24. Illustration of the self-consistency equation given by (5.238) and (5.239) of the KSS model (5.229). *Solid lines*: $R(m)$ for $\kappa/Q = 2.5$ and $\kappa/Q = 1$

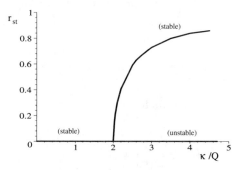

Fig. 5.25. Bifurcation diagram for the order parameter r_{st} of the KSS model (5.229): r_{st} versus κ/Q as computed from (5.238)

with

$$R(m) = \int_\Omega [\cos(x') + i\sin(x')] P(x' + \theta_{\text{st}}, m) \, dx'$$

$$= \frac{1}{Z(m)} \int_\Omega [\cos(x') + i\sin(x')] e^{\frac{\kappa m \cos(x')}{Q}} \, dx'$$

$$= \frac{1}{Z(m)} \int_\Omega \cos(x') e^{\frac{\kappa m \cos(x')}{Q}} \, dx' \,. \qquad (5.239)$$

Since the relations $Z(0) > 0$ and $R(0) = 0$ hold, the self-consistency equation (5.238) is satisfied for $m = 0$, which tells us once more that $P_{\text{st}}(x) = 1/(2\pi)$ describes a stationary probability density of (5.229). Solutions of (5.238) with $m > 0$ can be obtained by solving the self-consistency equation numerically, see Figs. 5.24 and 5.25 (bifurcation diagram). Solutions $m > 0$ emerge when the inequality $dR(0)/dm > 1$ holds. Since we have $dR(0)/dm = \kappa \int \cos^2(x) \, dx/(2\pi Q) = \kappa/(2Q)$, the critical parameter values

5.3 Periodic Boundary Conditions

for κ and Q that yield $\mathrm{d}R(0)/\mathrm{d}m = 1$ are given by $\kappa = 2Q$. As far as stationary probability densities with $r_{\mathrm{st}} = m > 0$ are concerned, we realize that r_{st} does not depend on θ_{st}. Moreover, if θ_{st} describes a stationary probability density then also $\theta_{\mathrm{st}} + \Delta x$. The reason for this is that (5.229) is invariant under transformations $x \to x + \Delta x$.

Stability Analysis

Having determined the stationary distributions of the form (5.234), let us now determine their stability. From (5.227), $c_1 = \kappa$, and $c_{n>1} = 0$, it follows that the stability of the uniform distribution is determined by the sign of

$$\tilde{\lambda} = Q - \kappa/2 . \tag{5.240}$$

Therefore, for $2Q > \kappa$ we deal with an asymptotically stable distribution, whereas for $2Q < \kappa$ the uniform distribution is unstable. In order to examine the stability of stationary distributions with $r_{\mathrm{st}} > 0$, we use perturbations given by

$$\epsilon(x) = a_1 \left[\cos(x) - \langle\cos(X)\rangle_{\mathrm{st}}\right] P_{\mathrm{st}}(x)$$
$$+ a_2 \left[\sin(x) - \langle\sin(X)\rangle_{\mathrm{st}}\right] P_{\mathrm{st}}(x) + \chi_\perp \sqrt{P_{\mathrm{st}}(x)} , \tag{5.241}$$

see (5.101). Substituting (5.241) into (5.221) for $c_n = \kappa$ and $c_{n>1} = 0$, we obtain

$$\delta^2 F[P_{\mathrm{st}}](\epsilon) = a_1^2 K_{\cos,\mathrm{st}}(X)[Q - \kappa K_{\cos,\mathrm{st}}(X)]$$
$$+ a_2^2 K_{\sin,\mathrm{st}}(X)[Q - \kappa K_{\sin,\mathrm{st}}(X)] + \int_\Omega [\chi_\perp(x)]^2 \, \mathrm{d}x , \tag{5.242}$$

where $K_{\cos}$ and $K_{\sin}$, respectively, denote generalized variances of the form (5.46) with $A(x) = \cos(x)$ and $A(x) = \sin(x)$. Due to the aforementioned invariance of (5.232) under transformations $x \to x + \Delta x$, all distribution (5.234) with a particular r_{st} but different θ_{st} have the same stability properties. For this reason, in what follows we study the case $\theta_{\mathrm{st}} = 0$, that is, distributions that are given by $P_{\mathrm{st}} = Z^{-1} \exp\{\kappa r_{\mathrm{st}} \cos(x)/Q\}$ and satisfy $P_{\mathrm{st}}(x) = P_{\mathrm{st}}(-x)$. In this case, (5.241) and (5.242) read

$$\epsilon(x) = a_1 \left[\cos(x) - r_{\mathrm{st}}\right] P_{\mathrm{st}}(x) + a_2 \sin(x) P_{\mathrm{st}}(x) + \chi_\perp \sqrt{P_{\mathrm{st}}(x)} \tag{5.243}$$

and

$$\delta^2 F[P_{\mathrm{st}}](\epsilon) = a_1^2 K_{\cos,\mathrm{st}}(X)[Q - \kappa K_{\cos,\mathrm{st}}(X)]$$
$$+ a_2^2 \langle\sin^2(X)\rangle_{\mathrm{st}} [Q - \kappa \langle\sin^2(X)\rangle_{\mathrm{st}}] + \int_\Omega [\chi_\perp(x)]^2 \, \mathrm{d}x \tag{5.244}$$

and we obtain the stability coefficients

$$\tilde{\lambda}_1 = Q - \kappa K_{\cos,\text{st}}(X) \,, \tag{5.245}$$

$$\tilde{\lambda}_2 = Q - \kappa \left\langle \sin^2(X) \right\rangle_{\text{st}} . \tag{5.246}$$

Note that $\tilde{\lambda}_1$ and $\tilde{\lambda}_2$ reduce to (5.240) for $P_{\text{st}} = 1/(2\pi)$. Let us evaluate the expression $\tilde{\lambda}_2 = Q - \kappa \left\langle \sin^2(X) \right\rangle_{\text{st}}$ for $P_{\text{st}} = Z^{-1} \exp\{\kappa r_{\text{st}} \cos(x)/Q\}$. First, the second derivative of the exponential function $\exp\{\kappa r_{\text{st}} \cos(x)/Q\}$ reads

$$\frac{d^2}{dx^2} e^{\kappa r_{\text{st}} \cos(x)/Q} = -\frac{\kappa r_{\text{st}} \cos(x)}{Q} e^{\kappa r_{\text{st}} \cos(x)/Q}$$
$$+ \left(\frac{\kappa r_{\text{st}}}{Q}\right)^2 \sin^2(x) e^{\kappa r_{\text{st}} \cos(x)/Q} . \tag{5.247}$$

Integrating this equation with respect to x and dividing it by Z yields

$$Q = \kappa \left\langle \sin^2(X) \right\rangle_{\text{st}} . \tag{5.248}$$

Consequently, for $P_{\text{st}}(x) = Z^{-1} \exp\{\kappa r_{\text{st}} \cos(x)/Q\}$ we have

$$\tilde{\lambda}_2 = 0 \tag{5.249}$$

and the second variation of F reads

$$\delta^2 F[P_{\text{st}}](a_1, a_2, \chi_\perp) = a_1^2 K_{\cos,\text{st}}(X)[Q - \kappa K_{\cos,\text{st}}(X)] + \int_\Omega [\chi_\perp(x)]^2 \, dx . \tag{5.250}$$

Using the stability coefficient (5.245), we draw the following conclusions. For $\tilde{\lambda}_1 > 0$ the free energy F increases for all small perturbations $\epsilon \neq 0$ with $a_2 = 0$ because if $a_2 = 0$ we have $a_1 \neq 0$ or $\chi_\perp \neq 0$. Perturbations ϵ with $a_1 = 0$, $\chi_\perp = 0$, and $a_2 \neq 0$ given by

$$\epsilon(x; a_2) = a_2 \sin(x) P_{\text{st}}(x) \tag{5.251}$$

yield $\delta^2 F = 0$ and do not lead to a change F up to the order ϵ^2.

Let us illustrate this issue from another perspective. First, we note that a shifted distribution $P_{\text{st}}(x + \Delta x)$ can be written as a Taylor expansion like $P_{\text{st}}(x - \Delta x) = \exp\{-\Delta x \, d/dx\} P_{\text{st}}(x)$. For $P_{\text{st}}(x) = Z^{-1} \exp\{\kappa r_{\text{st}} \cos(x)/Q\}$ we get $P_{\text{st}}(x - \Delta x) = P_{\text{st}}(x) + \kappa r_{\text{st}} \Delta x Q^{-1} \sin(x) P_{\text{st}}(x) + w(x, \Delta x)$, where $w \propto O(\Delta x^2)$. Let us put $a_2 = \kappa r_{\text{st}} \Delta x Q^{-1}$. Then, we have $P_{\text{st}}(x - Qa_2/[\kappa r_{\text{st}}]) = P_{\text{st}}(x) + a_2 \sin(x) P_{\text{st}}(x) + w(x, Qa_2/[\kappa r_{\text{st}}])$, where $w \propto O(a_2^2)$. Since a_2 is assumed to be a small parameter, we are allowed to add on the right-hand side of (5.251) an expression of order a_2^2 and of higher order without changing the results previously derived by means of linear stability analysis. That is, we add now the function $w(x, Qa_2/[\kappa r_{\text{st}}])$ on the right-hand side of (5.251) and consider the perturbation $\epsilon(x; a_2) = a_2 \sin(x) P_{\text{st}}(x) + w(x, Qa_2/[\kappa r_{\text{st}}])$:

$$P_{\text{st}}(x) + \epsilon(x; a_2) = P_{\text{st}}(x - Qa_2/[\kappa r_{\text{st}}]) = \frac{1}{Z(r_{\text{st}})} e^{\kappa r_{\text{st}} \cos(x - a_2')/Q} , \tag{5.252}$$

5.3 Periodic Boundary Conditions

with $a_2' = a_2 Q/(\kappa r_{st})$ and $P_{st}(x) = Z^{-1} \exp\{\kappa r_{st} \cos(x)/Q\}$. That is, the perturbation simply yields a translation of the distribution $P_{st}(x)$ and a shift of the cluster phase θ_{st} from $\theta_{st} = 0$ to $\theta_{st} = a_2'$, whereas the cluster amplitude r_{st} is not affected by the perturbation.

In sum, we conclude that the perturbation (5.251) does not change F up to the order ϵ^2 because F is invariant under translations $x \to x + \Delta x$ and the perturbation (5.251) basically induces such a translation. As a result, the stationary distribution (5.234) with $\theta_{st} = 0$ is stable for $\tilde{\lambda}_1 > 0$ but not asymptotically stable. We would like to point out that this result carries over to the case of stationary distributions (5.234) with $\theta_{st} \neq 0$ (as explained above). In other words, we may regard the distributions (5.234) as a one-parametric family of distributions with a continuous parameter $\theta_{st} \in [0, 2\pi)$. For $\tilde{\lambda}_1 > 0$ this one-parametric family of distributions describes a "valley" of the free energy measure (5.233) and therefore every member of this family corresponds to a stable but not asymptotically stable stationary distribution (see (5.13) and Sect. 5.1).

For $\tilde{\lambda}_1 > 0$ all perturbations of the distributions $P_{st}(x)$ given by (5.234) decrease as a function of time expect those related to a change of the cluster phase θ_{st}. In contrast, for $\tilde{\lambda}_1 < 0$ perturbations with $a_1 \neq 0$ increase as a function of time and the stationary distributions $P_{st}(x)$ are unstable. Using

$$\frac{dR}{dm} = \frac{\kappa}{Q} K_{\cos,st}(X) , \qquad (5.253)$$

we can write $\tilde{\lambda}_1$ as

$$\tilde{\lambda}_1 = Q \left(1 - \frac{dR}{dm}\bigg|_{m=r_{st}}\right) , \qquad (5.254)$$

which implies that stationary probability densities with $dR/dm > 1$ at $m = r_{st}$ correspond to unstable distribution, whereas distributions with $dR/dm < 1$ at $m = r_{st}$ correspond to stable ones. Then, from Fig. 5.24 we read off that if the self-consistency equation exhibits multiple solutions we have $dR/dm > 1$ at $m = 0$ and $dR/dm < 1$ at $m > 0$. Since (5.245) and (5.254) reduce to (5.240) for $P_{st} = 1/(2\pi)$, the nonlinear Fokker–Planck equation (5.229) exhibits for $\kappa < 2Q$ an asymptotically stable uniform stationary distribution. For $\kappa > 2Q$ the KSS model (5.229) exhibits an unstable uniform stationary distribution and a family of stable (but not asymptotically stable) nonuniform stationary distribution functions given by (5.234) with $r_{st} > 0$ and arbitrary θ_{st}. These findings are summarized in the bifurcation diagram depicted in Fig. 5.25.

In closing our considerations about the KSS model, let us illustrate that for the sine coupling-force the bifurcation from a uniform to a nonuniform stationary distribution indeed results from the fact that a free energy minimum becomes a free energy saddle point or maximum. To this end, we consider the one-parametric family of probability densities

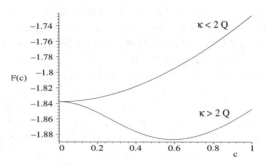

Fig. 5.26. Free energy (5.256) as a function of the parameter c. For $Q = 1$ two cases are shown: $\kappa = 1$ and $\kappa = 2.5$. For $\kappa = 2.5$ there is a minimum at $c_{\min} \approx 0.6$ (which corresponds to the intersection point at $m \approx 0.6$ in Fig. 5.24) [171]

$$P_c(x;\theta) = \frac{1}{Z} \exp\left\{\frac{\kappa c}{Q} \cos(x-\theta)\right\}, \qquad (5.255)$$

with $c \in [0,1]$ and $Z = \int_\Omega \exp\{\kappa c \cos(x)/Q\}\,\mathrm{d}x$ [171]. Substituting (5.255) into (5.233), we obtain

$$F(c) = -\frac{\kappa}{2}\left[\langle\cos(X)\rangle_{P_c(x;0)}\right]^2 + \kappa c \langle\cos(X)\rangle_{P_c(x,0)}$$
$$- Q \ln\left[\int_\Omega \exp\left\{\frac{\kappa c}{Q}\cos(x)\right\}\,\mathrm{d}x\right], \qquad (5.256)$$

where averages are computed from (5.255) with $\theta = 0$. Differentiating (5.256) with respect to c, gives us

$$\frac{\mathrm{d}}{\mathrm{d}c}F(c) = \frac{\kappa^2}{Q} K_{\cos}(X)_{P_c(x;0)}\left[c - \langle\cos(X)\rangle_{P_c(x;0)}\right]. \qquad (5.257)$$

For $c = 0$ we have $\mathrm{d}F/\mathrm{d}c = 0$ because of $c - \langle\cos(X)\rangle = 0$ for $c = 0$, which means that the free energy $F(c)$ is stationary at the uniform distribution. In order the check whether this critical point corresponds to a maximum or a minimum, we differentiate F with respect to c again. Thus, we obtain

$$\frac{\mathrm{d}^2}{\mathrm{d}c^2}F(c) = \frac{\kappa^2}{Q}\left\{K_{\cos}(X)_{P_c(x;0)}\left(1 - \frac{\kappa}{Q}K_{\cos}(X)_{P_c(x;0)}\right)\right.$$
$$\left. + \left(c - \langle\cos(X)\rangle_{P_c(x;0)}\right)\frac{\mathrm{d}}{\mathrm{d}c}K_{\cos}(X)_{P_c(x;0)}\right\}. \qquad (5.258)$$

For the uniform distribution (i.e. for $c = 0$) we obtain $K_{\cos} = 1/2$ and (5.258) reads

$$\left.\frac{\mathrm{d}^2}{\mathrm{d}c^2}F(c)\right|_{c=0} = \frac{\kappa^2}{2Q}\left(1 - \frac{\kappa}{2Q}\right). \qquad (5.259)$$

It is clear from (5.259) that for $\kappa < 2Q$ the free energy $F(c)$ has a minimum at $c = 0$. This is a necessary condition for $F[P]$ to exhibit a minimum at $P_{\rm st} = 1/(2\pi)$ with respect to the whole function space of probability densities. Furthermore, as shown in Fig. 5.26, for $\kappa < 2Q$ the function $F(c)$ increases monotonically. That is, there is no further minimum. For $\kappa > 2Q$ the function $F(c)$ has a maximum at $c = 0$, which implies that $F[P]$ has a maximum or a saddle point at $P = 1/(2\pi)$ with respect to the whole function space of probability densities. As shown in Fig. 5.26, for $\kappa > 2Q$ a new minimum emerges at $c = c_{\min}$. This minimum describes a family of stable stationary probability density given by (5.234) with $r_{\rm st} = c_{\min}$ and arbitrary $\theta_{\rm st}$.

5.3.3 Mean Field Haken–Kelso–Bunz (HKB) Model – Cluster Phase Dynamics

As mentioned earlier, the KSS model with a sine coupling-force (5.229) is invariant under transformations $x \to x + \Delta x$. In this section we break this invariance. To this end, we supplement (5.229) with a drift force $h(x)$ that is not invariant under the transformation $x \to x + \Delta x$. We consider a force $h(x)$ that has attractive points $x_i^* \in \Omega$ (i.e., we have $h(x_i^*) = 0$ and $dh(x_i^*)/dx < 0$). In doing so, we assume that the deterministic evolution equation $dq(t)/dq = h(q)$ of a dynamical system has stable fixed points at $q = x_i^*$. In this case, $h(x)$ "pins" the state variable $q(t)$ of the system to a fixed point x_i^* and is referred to as a pinning force. As a pinning force we choose $h(x) = -\alpha \sin(x) - 2\beta \sin(2x)$, which gives us the nonlinear Fokker–Planck equation

$$\frac{\partial}{\partial t} P(x, t; u)$$
$$= \frac{\partial}{\partial x}\left[\alpha \sin(x) + 2\beta \sin(2x) + \kappa \int_\Omega \sin(x - x') P(x', t; u) \, dx'\right] P + Q \frac{\partial^2}{\partial x^2} P ,$$
(5.260)

with $\kappa \geq 0$ and $\alpha, \beta \geq 0$ [198, 199]. The pinning force $h(x)$ can be regarded as the force $h(x) = -dV_{\rm HKB}/dx$ of the potential

$$V_{\rm HKB}(x) = -\alpha \cos(x) - \beta \cos(2x) . \tag{5.261}$$

The potential $V_{\rm HKB}$ was introduced by Haken, Kelso, and Bunz [259, 264, 335] in biophysics, which is the reason why we will refer to (5.260) as mean field HKB model. We would like to point out that the model given by (5.260) can also be regarded as a generalization of the stochastic HKB model proposed by Schöner et al. [512], which has been extensively studied in the literature, see, for example, [16, 111, 214, 461, 487]. The potential $V_{\rm HKB}$ has extrema at $x = 0$ and $x = \pm\pi$. For $\alpha < 4\beta$ there are two additional extrema at $\cos(x) = -\alpha/(4\beta)$. For $\alpha < 4\beta$ the potential is bistable, exhibits minima at $x = 0$ and $x = \pm\pi$ and has two maxima related to $\cos(x) - \alpha/(4\beta)$, see Fig. 5.27. For $\alpha \uparrow 4\beta$ the maxima merge with the minima at $\pm\pi$. For $\alpha > 4\beta$ the potential is monostable and exhibits a minimum at $x = 0$ and two maxima at $x = \pm\pi$, see Fig. 5.27.

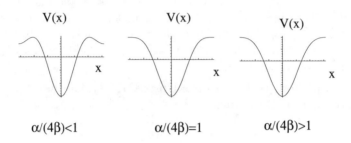

Fig. 5.27. HKB potential for several ratios $\alpha/(4\beta)$ and $x \in [-\pi, \pi]$

We anticipate that the mean field HKB model exhibits two qualitatively different parameter regimes. First, there should be a parameter regime with two stable distributions related to the minima at $x = 0$ and $x = \pi$ (or $x = -\pi$) of the potential V_{HKB} for $\alpha < 4\beta$. Second, there should be a parameter regime with a single stable distribution related to the unique potential minimum of V_{HKB} at $x = 0$ for $\alpha > 4\beta$. Consequently, we anticipate that the model describes a phase transition between a bistable and a monostable regime. This behavior should arise from the interplay between the potential V_{HKB} and the mean field coupling term $h_{\mathrm{MF}}(x) = -\kappa \langle \sin(x - x') \rangle$. The idea is that the HKB potential makes the system potentially bistable, whereas the mean field coupling term transforms this potential bistability into an actual bistability by forcing the subsystems k (represented by the realizations X^k) of a many-body system to assume a common phase.

Equation (5.260) is the free energy Fokker–Planck equation (5.3) for

$$F[P] = \int_\Omega V_{\mathrm{HKB}}(x) P(x) \, \mathrm{d}x - \frac{\kappa}{2} \int_\Omega \cos(x - x') P(x) P(x') \, \mathrm{d}x \, \mathrm{d}x' - Q^{\mathrm{B}} S[P] \,. \tag{5.262}$$

The free energy is of type E in Table 4.2, which implies that the inequality $F \geq F_{\min}$ and the limiting case $\lim_{t\to\infty} \partial P(x,t;u)/\partial t = 0$ holds (H-theorem, see Sect. 4.3). From (5.8) and $\delta U/\delta P = V_{\mathrm{HKB}}(x) - \kappa \int_\Omega \cos(x - x') P(x') \, \mathrm{d}x'$ it follows that

$$P_{\mathrm{st}}(x) = \frac{1}{Z} e^{-\frac{V_{\mathrm{HKB}}(x) - \kappa \int_\Omega \cos(x - x') P_{\mathrm{st}}(x') \, \mathrm{d}x'}{Q}} \,. \tag{5.263}$$

Furthermore, (5.260) can be written as a special case of (5.107):

$$\frac{\partial}{\partial t} P(x, t; u) =$$
$$\frac{\partial}{\partial x} \left[\alpha \sin(x) + 2\beta \sin(2x) + \kappa \left(\sin(x) \langle \cos(X) \rangle - \cos(x) \langle \sin(X) \rangle \right) \right] P$$
$$+ Q \frac{\partial^2}{\partial x^2} P \qquad (5.264)$$

with

$$F[P] = \int_\Omega V_{\text{HKB}}(x) P(x) \, dx - \frac{\kappa}{2} \left[\langle \cos(X) \rangle^2 + \langle \sin(X) \rangle^2 \right] - Q^{\text{B}} S[P] \,. \quad (5.265)$$

Then, stationary solutions can be obtained from (5.109) and satisfy

$$P_{\text{st}}(x) = \frac{1}{Z} e^{-\frac{V_{\text{HKB}}(x) - \kappa \left(\cos(x) \langle \cos(X) \rangle_{\text{st}} + \sin(x) \langle \sin(X) \rangle_{\text{st}} \right)}{Q}} \,. \quad (5.266)$$

In order to determine $\langle \cos(X) \rangle_{\text{st}}$ and $\langle \sin(X) \rangle_{\text{st}}$, we define

$$P(x, \mathbf{m}) = \frac{1}{Z(\mathbf{m})} e^{-\frac{V_{\text{HKB}}(x) - \kappa \left(\cos(x) m_1 + \sin(x) m_2 \right)}{Q}},$$

$$Z(\mathbf{m}) = \int_\Omega e^{-\frac{V_{\text{HKB}}(x) - \kappa \left(\cos(x) m_1 + \sin(x) m_2 \right)}{Q}} \, dx \,, \quad (5.267)$$

for $\mathbf{m} = (m_1, m_2)$ and use $\mathbf{R} = (R_1, R_2)$ given by

$$R_1(\mathbf{m}) = \int_\Omega \cos(x) P(x, \mathbf{m}) \, dx \,, \quad R_2(\mathbf{m}) = \int_\Omega \sin(x) P(x, \mathbf{m}) \, dx \quad (5.268)$$

to write the self-consistency equation

$$\mathbf{m} = \mathbf{R}(\mathbf{m}) \,. \quad (5.269)$$

Then, solutions of (5.269) yield $\langle \cos(X) \rangle_{\text{st}} = m_1$ and $\langle \sin(X) \rangle_{\text{st}} = m_2$.

Stability Analysis

It is clear from (5.265) that the second variation of F does not depend on the linear term $\int_\Omega V_{\text{HKB}}(x) P(x) \, dx$ and therefore is equivalent to $\delta^2 F$ of the free energy equation (5.230) of the KSS model with sine coupling-force. Using (5.221) for $c_1 = \kappa$ and $c_{n>1} = 0$, the second variation reads explicitly

$$\delta^2 F[P_{\text{st}}](\epsilon) = -\kappa \left(\left[\int_\Omega \cos(y) \epsilon(x) \, dy \right]^2 + \left[\int_\Omega \sin(y) \epsilon(x) \, dy \right]^2 \right)$$
$$+ Q \int_\Omega \frac{[\epsilon(x)]^2}{P_{\text{st}}(x)} \, dy \,. \quad (5.270)$$

Substituting perturbations given by (5.241) into (5.270), we obtain (5.242).

Symmetric Stationary Solutions

From the symmetry of the potential $V_{\text{HKB}}(x)$ given by $V(_{\text{HKB}}(x) = V_{\text{HKB}}(-x)$ it follows that $R_2(m_1, 0) = 0$ and, consequently, we can study solutions of (5.269) of the form

$$m_2 = 0 \ , \ m_1 = R_1(m_1, 0) \ . \tag{5.271}$$

In this case, we are dealing with symmetric stationary solutions that satisfy

$$P_{\text{st}}(x) = = \frac{1}{Z} e^{-\frac{V_{\text{HKB}}(x) - \kappa \cos(x) \langle \cos(X) \rangle_{\text{st}}}{Q}} . \tag{5.272}$$

That is, the symmetry of the potential V_{HKB} determines the symmetry of a class of stationary probability densities of the mean field HKB model (5.260). Note that if we are dealing with a pinning force $h(x)$ that satisfies $h(x^* + \Delta x) = -h(x^* - \Delta x)$ and is related to a symmetric potential around $x^* \neq 0$, then the mean field HKB model would exhibit stationary probability densities satisfying $P_{\text{st}}(x^* + \Delta x) = P_{\text{st}}(x^* - \Delta x)$. In order to determine the stability of the distributions (5.272), we proceed as in the case of the KSS model with sine coupling-force. For symmetric solutions (5.272) the perturbation (5.241) reads

$$\epsilon(x) = a_1 \left[\cos(x) - r_{\text{st}} \right] P_{\text{st}}(x) + a_2 \sin(x) P_{\text{st}}(x) + \chi_\perp \sqrt{P_{\text{st}}(x)} \ , \tag{5.273}$$

with $r_{\text{st}} = \int_\Omega \cos(x) P_{\text{st}}(x) \, dx$. Likewise, (5.242) becomes

$$\delta^2 F[P_{\text{st}}](\epsilon) = a_1^2 K_{\cos,\text{st}}(X)[Q - \kappa K_{\cos,\text{st}}(X)]$$
$$+ a_2^2 \langle \sin^2(X) \rangle_{\text{st}} [Q - \kappa \langle \sin^2(X) \rangle_{\text{st}}] + \int_\Omega [\chi_\perp(x)]^2 \, dx \ . \tag{5.274}$$

Using

$$\tilde{\lambda}_1 = Q - \kappa K_{\cos,\text{st}}(X) \ , \tag{5.275}$$

$$\tilde{\lambda}_2 = Q - \kappa \langle \sin^2(X) \rangle_{\text{st}} \ , \tag{5.276}$$

the stability of symmetric stationary probability densities can be conveniently determined. If $\tilde{\lambda}_1 > 0$ and $\tilde{\lambda}_2 > 0$, then stationary solutions (5.272) are asymptotically stable. If $\tilde{\lambda}_1 < 0$ or $\tilde{\lambda}_2 < 0$, then distributions of the form (5.272) are unstable.

What is the nature of the stability conditions related to $\tilde{\lambda}_1$ and $\tilde{\lambda}_2$? In what follows, we will relate $\tilde{\lambda}_1$ and $\tilde{\lambda}_2$ to bifurcations of the cluster amplitude and the cluster phase of symmetric stationary probability densities (5.272) in particular and of stationary probability densities of the mean field HKB model (5.260) in general.

Cluster Amplitude and Cluster Phase Description

Using (5.204), we transform (5.260), (5.265), and (5.266) into

$$\frac{\partial}{\partial t}P(x,t;u) = \frac{\partial}{\partial x}\left[\alpha \sin(x) + 2\beta \sin(2x) + \kappa r(t)\sin(x-\theta(t))\right]P + Q\frac{\partial^2}{\partial x^2}P, \tag{5.277}$$

$$F[P] = \int_\Omega V_{\mathrm{HKB}}(x)P(x)\,\mathrm{d}x - \frac{\kappa}{2}r^2[P] - Q^{\mathrm{B}}S[P], \tag{5.278}$$

and

$$P_{\mathrm{st}}(x) = \frac{1}{Z}e^{-\frac{V_{\mathrm{HKB}}(x) - \kappa r_{\mathrm{st}}\cos(x-\theta_{\mathrm{st}})}{Q}}. \tag{5.279}$$

It is helpful to rescale these equations by means of

$$\tilde{t} = Qt,\ \tilde{\kappa} = \frac{\kappa}{Q},\ \tilde{\alpha} = \frac{\alpha}{Q},\ \tilde{\beta} = \frac{\beta}{Q},\ \tilde{P}(x,\tilde{t};u) = P(x,t;u). \tag{5.280}$$

Moreover, let us introduce the force

$$\tilde{h}(x) = -\tilde{\alpha}\sin(x) - 2\tilde{\beta}\sin(2x) \tag{5.281}$$

and the potential

$$\tilde{V}_{\mathrm{HKB}}(x) = \frac{V_{\mathrm{HKB}}(x)}{Q} = -\tilde{\alpha}\cos(x) - \tilde{\beta}\cos(2x) \tag{5.282}$$

of the deterministic HKB model

$$\frac{\mathrm{d}}{\mathrm{d}\tilde{t}}q(\tilde{t}) = \tilde{h}(q) \tag{5.283}$$

with respect to the rescaled variables. Then, we have

$$\frac{\partial}{\partial \tilde{t}}\tilde{P}(x,\tilde{t};u) = \frac{\partial}{\partial x}\left[\tilde{\alpha}\sin(x) + 2\tilde{\beta}\sin(2x) + \tilde{\kappa}r(\tilde{t})\sin(x-\theta(\tilde{t}))\right]\tilde{P} + \frac{\partial^2}{\partial x^2}\tilde{P} \tag{5.284}$$

for $r(\tilde{t})\exp\{i\theta(\tilde{t})\} = \int_\Omega \exp\{ix\}\tilde{P}(x,\tilde{t},u)\,\mathrm{d}x$ and

$$\tilde{P}_{\mathrm{st}}(x) = \frac{1}{Z}e^{-\tilde{V}_{\mathrm{HKB}}(x) + \tilde{\kappa}r_{\mathrm{st}}\cos(x-\theta_{\mathrm{st}})}. \tag{5.285}$$

Saddle Node Bifurcation of r'_{st}

For the sake of convenience, we introduce now a modified cluster phase $\theta' \in [0,\pi)$ and a signed cluster amplitude $r' \in [-1,1]$ defined by $r'(\tilde{t})\exp\{i\theta'(\tilde{t})\} = \int_\Omega \exp\{ix\}\tilde{P}(x,\tilde{t};u)\,\mathrm{d}x$. Note that the variables θ' and r', on the one hand,

and the variables θ and r, on the other hand, are only different with respect to their domains of definition. For the stationary case this implies:

$$\begin{aligned}
& r_{\text{st}} \geq 0, \\
& \theta'_{\text{st}} = 0, \\
& \theta_{\text{st}} = 0 \Rightarrow r'_{\text{st}} = r_{\text{st}} \geq 0, \\
& \theta_{\text{st}} = \pi \Rightarrow r'_{\text{st}} = -r_{\text{st}} \leq 0, \\
& r'_{\text{st}} > 0 \Rightarrow \theta_{\text{st}} = 0, \\
& r'_{\text{st}} < 0 \Rightarrow \theta_{\text{st}} = \pi.
\end{aligned} \quad (5.286)$$

In addition, (5.285) reads

$$\tilde{P}_{\text{st}}(x) = \frac{1}{Z} e^{-\tilde{V}_{\text{HKB}}(x) + \kappa r'_{\text{st}} \cos(x - \theta'_{\text{st}})}, \quad (5.287)$$

with $\theta'_{\text{st}} \in [0, \pi)$. In order to examine symmetric stationary solutions of the form (5.272), we impose a symmetry constraint on (transient and stationary) solutions of (5.284):

$$\tilde{P}(x, \tilde{t}; u) = \tilde{P}(-x, \tilde{t}; u). \quad (5.288)$$

In this case, from $r(\tilde{t}) \exp\{i\theta(\tilde{t})\} = \int_{\Omega} \exp\{ix\} \tilde{P}(x, \tilde{t}, u)\, dx$ it follows that θ can assume only two values, namely, $\theta(\tilde{t}) = 0$ and $\theta(\tilde{t}) = \pi$, and that θ' and, in particular, θ'_{st} are fixed by $\theta' = \theta'_{\text{st}} = 0$. The objective is to determine stationary solutions of (5.284) and their stability under the constraint (5.288). From (5.287) and $\theta'_{\text{st}} = 0$ we can derive the self-consistency equation

$$m = R(m; \tilde{\kappa}, \tilde{\alpha}, \tilde{\beta}), \quad (5.289)$$

for $m \in [-1, 1]$ with

$$R(m; \tilde{\kappa}, \tilde{\alpha}, \tilde{\beta}) = \int_{\Omega} \cos(x) \tilde{P}(x, m)\, dx,$$

$$\tilde{P}(x, m) = \frac{1}{Z(m)} e^{-\tilde{V}_{\text{HKB}}(x) - \kappa m \cos(x)}. \quad (5.290)$$

Then, the cluster amplitude r'_{st} occurring in (5.287) corresponds to solutions of (5.289). One can verify that R satisfies the following properties:

- $R(x; \ldots)$ is a monotonously increasing but bounded function in $x \in [-1, 1]$:

$$|R(x; \cdot)| \leq 1 \quad \text{and} \quad \frac{d}{dm} R(m; \cdot) = \tilde{\kappa} K_{\cos} \geq 0, \quad (5.291)$$

with $K_{\cos}$ defined by (5.46). In particular, for $\tilde{P}(x, m) = \delta(x)$ and $\tilde{P}(x, m) = \delta(x \pm \pi)$ we have $|R| = 1$, whereas for $\tilde{P}(x, \cdot) \in C^0(\Omega)$ we have $|R| < 1$.

- We have

$$\frac{\partial}{\partial \tilde{\kappa}} R(m; \tilde{\kappa}, \tilde{\alpha}, \tilde{\beta}) = m\tilde{\kappa} K_{\cos} . \quad (5.292)$$

- R is a monotonically increasing function with respect to $\tilde{\alpha}$:

$$\frac{\partial}{\partial \tilde{\alpha}} R(m; \tilde{\kappa}, \tilde{\alpha}, \tilde{\beta}) = K_{\cos} \geq 0 . \quad (5.293)$$

Having discussed some general properties of R, let us consider now the solution $m = r'_{\text{st}} = 0$ of the self-consistency equation (5.289). Since for $\tilde{\alpha} = 0$ we have

$$\int_\Omega \cos(x) \, \tilde{V}_{\text{HKB}}(x) \, dx = 0 , \quad (5.294)$$

from (5.290) and (5.282) we conclude that

$$R(m; \tilde{\kappa}, \tilde{\alpha} = 0, \tilde{\beta}) = 0 . \quad (5.295)$$

Consequently, $r'_{\text{st}} = 0$ solves the self-consistency equation (5.289) for $\tilde{\alpha} = 0$ and the corresponding stationary probability density can be calculated from (5.287). From (5.291), (5.293) and (5.295) we further conclude that for $\tilde{\alpha} > 0$ there exists at least one positive solution of (5.289). In order to see this recall that, as shown above, for $\tilde{\alpha} = 0$ we have $R(m = 0; \cdot) = 0$. For increasing $\tilde{\alpha}$, (5.293) results in $R(m = 0; ., \tilde{\alpha} > 0, .) > 0$. Since R is a monotonously increasing function with $|R| < 1$, there must be at least one intersection point between the graph of $m \to R(m; ., \tilde{\alpha} > 0, .)$ and the diagonal of the (m, R)-plane within the range $m \in [0, 1]$ that corresponds to a positive solution r'_{st}. According to (5.286), this implies the existence of a stationary probability density (5.285) with $\theta_{\text{st}} = 0$. For $\tilde{\alpha}$ smaller than a critical value $\tilde{\alpha}_c$ we obtain two negative solutions of (5.289), which merge for $\tilde{\alpha} = \tilde{\alpha}_c$ and entirely disappear for $\tilde{\alpha} > \tilde{\alpha}_c$. Figure 5.28c shows $R(m; \tilde{\kappa}, \tilde{\alpha}, \tilde{\beta})$ for three values of $\tilde{\alpha}$ while $\tilde{\kappa}$ and the ratio $4\tilde{\beta}/\tilde{\alpha}$ assume fixed values. The intersection points of the solid lines and the diagonal represent the solutions of (5.289). The figure illustrates that for $\tilde{\alpha} > 0$ an intersection point with $m > 0$ always exists.

Now that the existence of stationary probability densities has been demonstrated, we need to determine their stability. Analogously to (5.252), we can show that the perturbation (5.273) of a stationary probability density (5.290) for $a_2 \neq 0$, $a_1 = 0$, and $\chi_\perp = 0$ results in a shift of the cluster phase θ'_{st} from $\theta'_{\text{st}} = 0$ to $\theta'_{\text{st}} \neq 0$. Since we have frozen the cluster phase θ'_{st} by the constraint (5.288) such perturbations are forbidden and the stability is exclusively determined by the coefficient λ_1 given in (5.275). From (5.275), (5.280), and (5.291) it follows that

$$\tilde{\lambda}_1 = Q \left(1 - \frac{dR}{dm} \bigg|_{m=r_{\text{st}}} \right) . \quad (5.296)$$

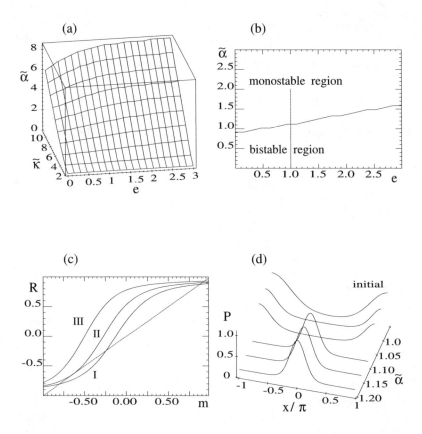

Fig. 5.28. *Panel* (**a**): three-dimensional plot of the critical parameter $\tilde{\alpha}$ versus $\tilde{\kappa}$ and $e = 4\tilde{\alpha}/\tilde{\beta}$ with $\tilde{\alpha} > 0$ as calculated from (5.298). The surface represents the bifurcation points of a saddle node bifurcation in the parameter space $(\tilde{\kappa}, \tilde{\alpha}, e)$ and divides the parameter space into a monostable and a bistable region. *Panel* (**b**): a cut through the surface of Fig. 5.28a at $\tilde{\kappa} = 4$. *Panel* (**c**): illustration of the self-consistency equation (5.289). *Solid lines*: $R(m)$ defined by (5.290) for three different values of $\tilde{\alpha}$ (0.5, 1.1 and 2.0) and $\tilde{\kappa} = 4, e = 1$. *Dashed line*: diagonal of the (R, m) plane. *Panel* (**d**): simulation of the mean field model (5.284) under the constraint (5.288); reprinted from [198], © 2000, with permission from Elsevier

Consequently, from Fig. 5.28c we read off that the stationary probability density that corresponds to the intersection point with $m = r'_{\rm st} > 0$ represents an asymptotically stable solution. Likewise, the two negative solutions of (5.289) represent an asymptotically stable and an unstable solution. The vanishing of this pair of solutions reveals a saddle node bifurcation.

Having obtained the self-consistency equation (5.289), we can numerically compute the number of positive and negative solutions for every particular triple $(\tilde{\kappa}, \tilde{\alpha}, \tilde{\beta})$. For the sake of convenience, we introduce the parameter $e =$

$4\tilde{\beta}/\tilde{\alpha} \geq 0$ (which implies $\tilde{\beta} = \tilde{\alpha}e/4$ for $\tilde{\alpha} \geq 0, e \geq 0$) and use the HKB potential (5.282) described by

$$\tilde{V}_{\text{HKB}}(x) = -\tilde{\alpha}\left(\cos(x) + \frac{e}{4}\cos(2x)\right) . \qquad (5.297)$$

In the parameter space spanned by the parameters $\tilde{\kappa}, \tilde{\alpha}, e$ a surface can be identified that divides the parameter space into two regions: a region which allows for two negative solutions ($r'_{\text{st}} < 0$) and a positive solution ($r'_{\text{st}} > 0$) and another region which only allows for a single positive solution. The surface in Fig. 5.28a represents those points ($\tilde{\kappa}, \tilde{\alpha}, e$) at which both negative solutions of (5.289) merge into a single solution. As a result, the points on the surface satisfy the relations:

$$r'_{\text{st}} = R(r'_{\text{st}};\cdot), \quad \frac{\mathrm{d}}{\mathrm{d}m} R(m;\cdot)|_{m=r'_{\text{st}}} = 1, \quad r'_{\text{st}} < 0 . \qquad (5.298)$$

Panel b) shows a detail of Fig. 5.28a for $\tilde{\kappa} = 4$. The three solid lines in Fig. 5.28c have been calculated for parameters $\tilde{\kappa}, \tilde{\alpha}, e$ chosen along the dashed line shown in panel b). Curve I in Fig. 5.28c corresponds to $e = 1, \tilde{\alpha} = 0.5$ (bistable region); curve II to $e = 1, \tilde{\alpha} = 1.1$ (phase transition point); and curve III to $e = 1, \tilde{\alpha} = 2.0$ (monostable region).

Panel d) shows the results of a numerical simulation of the mean field HKB model (5.284) under the constraint (5.288). This result has been obtained by using the Fourier expansion technique described in Sect. 2.8.2. That is, the probability density $\tilde{P}$ was expanded into a Fourier series with complex time-dependent coefficients. The expansion was truncated after the L first modes. We used $L = 30$. Substituting this approximation into (5.284), we obtained a system of coupled first order differential equations. This system of differential equations was integrated by means of an Euler forward scheme with a single time step of $\Delta t = 0.1/L^2$. The advantage of such a simulation scheme with respect to a simulation by means of Langevin equations is that the symmetry constraint (5.288) can easily be implemented. We simply need to neglect the imaginary parts of the time-dependent coefficients. A Gaussian probability density served as initial distribution (see the label "initial" in panel d). We set $\tilde{\kappa} = 4$ and $e = 1$ and increased $\tilde{\alpha}$ along the dashed line shown in panel b) from $\tilde{\alpha} = 1$ to $\tilde{\alpha} = 1.2$. The parameter $\tilde{\alpha}$ was increased in steps of 0.05 after each computation of 10^6 single time steps. The probability densities shown in Fig. 5.28d can be regarded as the stationary probability densities of (5.284) under the constraint (5.288) for the selected values of $\tilde{\alpha}$. In agreement with the results presented in Figs. 5.28c and 5.28b, the bifurcation occurred at a critical value $\tilde{\alpha}_c$ of about $\tilde{\alpha}_c \approx 1.1$.

Pitchfork Bifurcation of θ_{st}

Now let us dispense with the symmetry constraint (5.288). In this case, the stability of the symmetric stationary distributions given by (5.290) is also determined by the sign of $\tilde{\lambda}_2$. In what follows, we prove that $\tilde{\lambda}_2 > 0$ holds for distributions (5.290) with $r'_{st} > -\tilde{\alpha}/\tilde{\kappa}$. By analogy to (5.248), one finds for $r'_{st} > -\tilde{\alpha}/\tilde{\kappa}$, $\tilde{\alpha} > 0$, and $\tilde{\beta} = 0$ that

$$\langle \sin^2(X) \rangle_{st} = \frac{r'_{st}}{\tilde{\alpha} + \tilde{\kappa} r'_{st}} < \frac{1}{\tilde{\kappa}} . \tag{5.299}$$

If $\tilde{\beta}$ is increased then the distributions (5.290) become smaller at positions x for which $\sin^2(x)$ is large (e.g., at $x = \pi/2$ and $x = 3\pi/2$) and become larger at positions for which $\sin^2(x)$ is small (e.g., at $x = 0$ and $x = \pm\pi$). Consequently, $\langle \sin^2(X) \rangle_{st}$ for $\tilde{\beta} > 0$ is smaller than $\langle \sin^2(X) \rangle_{st}$ for $\tilde{\beta} = 0$ and we conclude that

$$\langle \sin^2(X) \rangle_{st} < \frac{1}{\tilde{\kappa}} , \tag{5.300}$$

for all symmetric stationary distributions with $\tilde{\beta} \geq 0$, $\tilde{\alpha} > 0$, and $r'_{st} > -\tilde{\alpha}/\tilde{\kappa}$. Finally, for $\tilde{\beta} \to \infty$ all distributions satisfying (5.290) reduce to a delta distribution with peaks at $x = 0$ and $x = \pm\pi$. Therefore, $\langle \sin^2(X) \rangle_{st} = 0$ for $\tilde{\beta} \to \infty$.

From (5.276), (5.280), and (5.300), it is clear that $\tilde{\lambda}_2 > 0$ for $r'_{st} > -\tilde{\alpha}/\tilde{\kappa}$. Consequently, the symmetric distribution (5.285) with $\theta_{st} = 0$ (i.e., the distribution (5.287) with $\theta'_{st} = 0$ and $r'_{st} > 0$) is asymptotically stable. In contrast, distributions (5.285) with $\theta_{st} = \pi$ (i.e., distributions (5.287) with $\theta'_{st} = 0$ and $r'_{st} < 0$) and $\tilde{\lambda}_1 > 0$ can yield both positive and negative values for $\tilde{\lambda}_2$. Therefore, they may correspond to asymptotically stable or unstable distributions if we dispense with the symmetry condition (5.288). We will examine this issue next.

Stationary Solutions (5.285) with $\theta_{st} = \pm\pi$

We study now the stability of stationary solutions of the form (5.285) with $\theta_{st} = \pi$ (or likewise $\theta_{st} = -\pi$). We will illustrate that in a particular range of the parameter space, stationary probability densities with $\theta_{st} = \pi$ that have been considered in the preceding as asymptotically stable solutions become unstable if we omit the symmetry constraint (5.288) and take perturbations of the cluster phase θ into account. This change of stability will be interpreted in terms of a pitchfork bifurcation and investigated in two distinct ways: linear stability analysis and the branching of stationary solutions in the function space of probability densities.

Linear Stability Analysis

Since perturbations of asymptotically stable distributions vanish, deviations from such distributions vanish in the long time limit. In particular, expectation values describing theses deviations vanish for $t \to \infty$. Therefore, a

5.3 Periodic Boundary Conditions

necessary condition for a stationary distribution to correspond to an asymptotically stable one is that expectation values of perturbations vanish for $t \to \infty$. In contrast, if there is at least one expectation value that describes a perturbation and increases with time, we have a strong indication that the norm of the perturbation increases and the distribution is an unstable one, see (5.16). In order to examine the stability of distributions (5.285) with $\theta_{\text{st}} = \pi$, we introduce the mean deviation $\langle \Delta x \rangle (\tilde{t})$ given by

$$\langle \Delta x \rangle (\tilde{t}) = \int_0^{2\pi} (x - \pi) \tilde{P}(x, \tilde{t}) \, dx = M_1(\tilde{t}) - \pi , \qquad (5.301)$$

which can be regarded as a shifted first moment $M_1(\tilde{t})$ as indicated. At the instance $\tilde{t}_0$ we consider a perturbation of the probability density $\tilde{P}_{\text{st}}(x; \theta_{\text{st}} = \pi)$ defined by (5.285), which yields

$$u(x) = \tilde{P}_{\text{st}}(x - \epsilon(\tilde{t}_0); \theta_{\text{st}} = \pi) , \qquad (5.302)$$

where $\epsilon(t_0)$ is a small number. We can verify that (5.302) exhibits the cluster phase

$$\theta = \pi + \epsilon(\tilde{t}_0) . \qquad (5.303)$$

From (5.301) and (5.302) we can determine $\langle \Delta x \rangle$ at time $\tilde{t}_0$ and obtain by means of partial integration

$$\langle \Delta x \rangle (\tilde{t}_0) = \epsilon \left[1 - 2\pi \tilde{P}_{\text{st}}(0; \theta_{\text{st}} = \pi) \right] + O(\epsilon^2) . \qquad (5.304)$$

Using (5.285), (5.301), and (5.302), we can determine the evolution of $\langle \Delta x \rangle (\tilde{t})$ at $\tilde{t} = \tilde{t}_0$:

$$\frac{d}{d\tilde{t}} \langle \Delta x \rangle (\tilde{t}) \bigg|_{\tilde{t}=\tilde{t}_0} = \epsilon \int_0^{2\pi} \frac{d\tilde{h}(x)}{dx} \tilde{P}_{\text{st}}(x; \cdot) \, dx + O(\epsilon^2) , \qquad (5.305)$$

with $\tilde{h}$ given by (5.281). Neglecting higher order terms of ϵ and combining (5.304) and (5.305) results in

$$\frac{d}{d\tilde{t}} \langle \Delta x \rangle (\tilde{t}) \bigg|_{\tilde{t}=\tilde{t}_0} = \lambda_{\theta_{\text{st}}=\pi} \langle \Delta x \rangle (\tilde{t}_0) , \qquad (5.306)$$

where $\lambda_{\theta_{\text{st}}=\pi}$ is defined by

$$\lambda_{\theta_{\text{st}}=\pi} = \frac{1}{1 - 2\pi \tilde{P}_{\text{st}}(0; \theta_{\text{st}} = \pi)} \left\langle \frac{d\tilde{h}(x)}{dx} \right\rangle = \frac{\int_\Omega \tilde{P}_{\text{st}}(x; \theta_{\text{st}} = \pi) \frac{d\tilde{h}(x)}{dx} \, dx}{1 - 2\pi \tilde{P}_{\text{st}}(0; \theta_{\text{st}} = \pi)} . \qquad (5.307)$$

Equation (5.306) can be further evaluated for the important special case of unimodal distributions (5.285) that decrease monotonously on either side of the maximum located at $x = \pi$ and satisfy

$$\frac{\mathrm{d}}{\mathrm{d}x}\tilde{P}_{\mathrm{st}}(x) = \begin{cases} 0 & \text{for } x = 0 \vee x = \pi \\ c > 0 & \text{for } x \in (0, \pi) \end{cases}. \tag{5.308}$$

Note that the symmetry of the HKB potential implies that $\mathrm{d}\tilde{P}_{\mathrm{st}}(x)/\mathrm{d}x = -\mathrm{d}\tilde{P}_{\mathrm{st}}(x-\pi)/\mathrm{d}x < 0$ holds for $x \in (\pi, 2\pi)$. The condition (5.308) is satisfied only for $\alpha_{\mathrm{eff}} = \tilde{\alpha} - r_{\mathrm{st}}\tilde{\kappa} < 0$ and $|\alpha_{\mathrm{eff}}| > 4\tilde{\beta}$ (because in this case the corresponding effective potential $V_{\mathrm{eff}}(x) = \tilde{V}_{\mathrm{HKB}}(x) - \tilde{\kappa} r_{\mathrm{st}} \cos(x)$ occurring in (5.287) has a unique minimum at $x = \pi$). Since (5.308) holds, the shifted initial probability density $u(x)$ exhibits a unique maximum. The evolution of the mean deviation $\langle \Delta x \rangle$ at the instance $\tilde{t}_0$ may indicate whether or not the position of the maximum of $u(x)$ relaxes back towards π at time $\tilde{t}_0$. From (5.308) and the normalization of $\tilde{P}(x, \theta_{\mathrm{st}} = \pi)$ it follows that the inequality $\tilde{P}(0, \cdot) < 1/2\pi$ holds. Consequently, the denominator in (5.307) is positive. With (5.307) at hand, we realize that a stationary probability density that is regarded as an asymptotically stable solution under the constraint (5.288) can also become unstable. For a Lyapunov exponent $\lambda_{\theta_{\mathrm{st}}=\pi} > 0$ we deal with an unstable stationary solution, whereas $\lambda_{\theta_{\mathrm{st}}=\pi} < 0$ means that $\tilde{P}_{\mathrm{st}}(x, \theta_{\mathrm{st}} = \pi)$ satisfies a necessary condition for an asymptotically stable stationary solution.

We would like to point out that in the limit of vanishing noise, all unimodal stationary probability densities of the from $\tilde{P}_{\mathrm{st}}(x, \theta_{\mathrm{st}} = \pi)$ reduce to a delta distribution with a peak at $x = \pi$ (i.e., we have $\tilde{P}_{\mathrm{st}}(x; \theta_{\mathrm{st}} = \pi) = \delta(x - \pi)$ and $\tilde{P}_{\mathrm{st}}(0; \theta_{\mathrm{st}} = \pi) = 0$). Then, the averaging in (5.307) can be replaced by the evaluation of the expression within the brackets at $x = \pi$ and we obtain $\lambda_{\theta_{\mathrm{st}}=\pi} = \mathrm{d}\tilde{h}(\pi)/\mathrm{d}x$, which means that we obtain the Lyapunov exponent of the deterministic HKB model (5.283) with respect to the stationary point $q_{\mathrm{st}} = \pi$.

Branching of Stationary Solutions

We now draw attention to the special case in which a change of stability is mediated by a pitchfork bifurcation. We first note that in view of the symmetry and periodicity of the HKB potential (5.282) we conclude that if the pair $(r_{\mathrm{st}}, \theta_{\mathrm{st}} = \pi + \Delta)$ describes a stationary probability density (5.285), then (5.285) for $(r_{\mathrm{st}}, \theta_{\mathrm{st}} = \pi - \Delta)$ describes a stationary distribution as well. Let us rewrite the self-consistency equation $r_{\mathrm{st}} \exp\{i\theta_{\mathrm{st}}\} = \int_\Omega \exp\{ix\}\tilde{P}_{\mathrm{st}}(x)\,\mathrm{d}x$ in terms of the functions f and f':

$$f(r_{\mathrm{st}}, \theta_{\mathrm{st}}, \tilde{\alpha}, \tilde{\beta}, \tilde{\kappa}) = \int \sin(x - \theta_{\mathrm{st}})\tilde{P}_{\mathrm{st}}(x)\,\mathrm{d}x = 0 , \tag{5.309}$$

$$f'(r_{\mathrm{st}}, \theta_{\mathrm{st}}; \tilde{\alpha}, \tilde{\beta}, \tilde{\kappa}) = r_{\mathrm{st}} - \int \cos(x - \theta_{\mathrm{st}})\tilde{P}_{\mathrm{st}}(x)\,\mathrm{d}x = 0 . \tag{5.310}$$

For critical parameters $\tilde{\alpha}_c$, $\tilde{\beta}_c$, and $\tilde{\kappa}_c$ that describe a pitchfork bifurcation point, we consider the normal form $f(r_{\mathrm{st}}, \theta_{\mathrm{st}}; \tilde{\alpha}_c, \tilde{\beta}_c, \tilde{\kappa}_c) \propto (\theta_{\mathrm{st}} - \pi)^3$ that yields

$$\frac{\partial}{\partial \theta_{\text{st}}} f(r_{\text{st}}, \theta_{\text{st}}; \tilde{\alpha}_c, \tilde{\beta}_c, \tilde{\kappa}_c) = 0 \ . \tag{5.311}$$

From (5.285), (5.309), and (5.311) we obtain

$$1 - \frac{2}{\tilde{\kappa}_c} = \langle \cos(2X) \rangle_{\text{st},c} \ , \tag{5.312}$$

which can equivalently be expressed as

$$\langle \sin^2(X) \rangle_{\text{st},c} = \frac{1}{\tilde{\kappa}_c} \ , \tag{5.313}$$

where averaging is carried out with respect to the stationary probability density (5.285) at the bifurcation point described by $\tilde{\alpha}_c$, $\tilde{\beta}_c$, $\tilde{\kappa}_c$, and $\theta_{\text{st}} = \pi$. Note that we have derived this result for all kinds of symmetric distributions (5.285) with $\theta_{\text{st}} = \pi$ that do not necessarily satisfy (5.308).

From (5.280) and (5.313) it follows that a pitchfork bifurcation point of a symmetric stationary distribution with $\theta_{\text{st}} = \pi$ is described by the condition $\kappa \langle \sin^2(x) \rangle_{\text{st}} = Q$. For this condition $\tilde{\lambda}_2$ vanishes, see (5.276). On account of the result obtained from the linear stability analysis of $\langle \Delta x \rangle$, the analysis of (5.309), and the observation mentioned earlier that perturbations related to $\tilde{\lambda}_2$ shift the cluster phase θ_{st} of stationary distributions, we have strong support for the hypothesis that the stability coefficients $\tilde{\lambda}_2$ corresponds to the control parameter of a pitchfork bifurcation.

In order to demonstrate that we are indeed dealing with a pitchfork bifurcation, we solved numerically (5.309) and (5.310) for a particular value of $\tilde{\kappa}$ ($\tilde{\kappa}=4$) and thus obtained the cluster phases θ_{st} and cluster amplitudes r_{st}. Figure 5.29c) depicts the solutions θ_{st} of (5.309) and (5.310). Note that Fig. 5.29c) does not show the solutions that corresponds to the unstable stationary probability density with $dR(m)/dm$ and $m = r'_{\text{st}} < 1$. There are two side branches with $\theta_{\text{st}} = \pi \pm \Delta$ that merge at e_c with the solution $\theta_{\text{st}} = \pi$. The critical value e_c of about $e_c \approx 1.6$ (panel b) coincides with the critical value derived by the linear stability analysis (see below).

Branching of Solutions and Convolution Integrals

Emergence and vanishing of the side branches become transparent for the HKB potential (5.282) when interpreting (5.309) as the convolution of an odd function $g(x)$ and an even function $u(x)$. Substituting (5.282), (5.285), and $z = x - \theta_{\text{st}}$ into (5.309) we obtain:

$$0 = f = \int_\Omega g(z) u(z - \theta_{\text{st}}) \, dz \ , \tag{5.314}$$

with

$$g(x) = \sin(x) e^{\tilde{\kappa} r_{\text{st}} \cos(x)}, \quad u(x) = e^{\tilde{\alpha} \cos(x) + \tilde{\beta} \cos(2x)} \ . \tag{5.315}$$

180   5 Free Energy Fokker–Planck Equations with Boltzmann Statistics

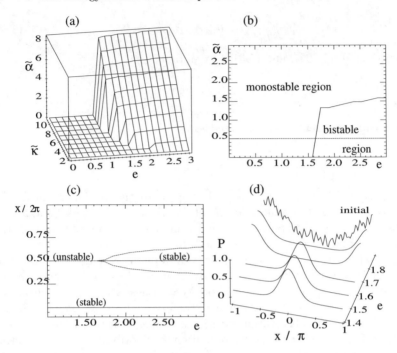

Fig. 5.29. *Panel* (**a**): the "roof-like" surface describes a saddle node bifurcation just as in Fig. 5.28a; "wall-like" surface represents the critical points of pitchfork bifurcations. *Panel* (**b**): the cut through the surface of Fig. 5.29a at $\tilde{\kappa} = 4$. *Panel* (**c**): stationary cluster phases θ_{st} calculated from (5.309) and (5.310) for several e with $\tilde{\kappa} = 4, \tilde{\alpha} = 0.5$. *Panel* (**d**): simulation of the mean field HKB model (5.284); reprinted from [198], © 2000, with permission from Elsevier

Due to the symmetry properties of $g(x)$ and $u(x)$, the integral relation (5.314) is solved in any case by $\theta_{st} = 0$ and $\theta_{st} = \pi$. It is clear from Fig. 5.27 that for $4\tilde{\beta} \leq \tilde{\alpha}$ the function $u(x)$ exhibits a unique maximum at the origin and decreases monotonously on both sides of that maximum. Since $g(z)$ is a antisymmetric function, the convolution integral vanish only if $u(z - \theta_{st})$ is a symmetric function. That is, for $4\tilde{\beta}/\tilde{\alpha} \leq 1$ the integral (5.314) only admits the solutions $\theta_{st} = 0$ and $\theta_{st} = \pi$. In order to study the parameter range $4\tilde{\beta}/\tilde{\alpha} \in (1, \infty)$, we first consider the case $\tilde{\alpha} = 0$. In this case the function $u(z)$ reduces to $u'(z) = \exp[\tilde{\beta}\cos(2z)]$ and, consequently, (5.314) is also solved by $\theta_{st} = \pm \pi/2$. Figure 5.30 shows $g(z)$ and $z \to u'(z - \pi/2)$ for some typical parameter values[1]. If we increase $\tilde{\alpha}$, the degeneration of the two maxima at $\pm\pi/2$ of $z \to u(z - \pi/2)$ disappears. To satisfy (5.314), the larger maximum of the non-degenerated function has to be shifted into the shallower region of $g(z)$ by shifting the whole function. Hence, θ_{st} differs from $\pm\pi/2$. Without any loss of generality, let us assume that the maximum at $\pi/2$ is higher than the maximum at $-\pi/2$ (which is the case shown in Fig. 5.30). In that case, we have to shift the absolute maximum from $\pi/2$ towards π, and, consequently, the whole function shifts towards the right-hand side. Figure 5.30 shows an example of a function $u(z - \theta_{st})$, where $\theta_{st} > \pi/2$ has been chosen such that the integral over the product $g(z)u(z - \theta_{st})$ vanishes (the integral relation (5.314) is satisfied). Therefore, we conclude that in the range $4\tilde{\beta}/\tilde{\alpha} \in (1, \infty)$ the additional solutions $\theta_{st} = \pi \pm \Delta$ (given that they exist) are restricted by $\Delta \in [0, \pi/2]$. In sum, for $4\tilde{\beta}/\tilde{\alpha} \in (1, \infty)$ the deviation Δ characterizing the side branches decreases from $\Delta = \pi/2$ to $\Delta = 0$ when $e = 4\tilde{\beta}/\tilde{\alpha}$ decreases from ∞ to $e_c \geq 1$. The additional solutions with $\Delta \neq 0$ are considered as the unstable side branches of a pitchfork bifurcation (see Fig. 5.29c).

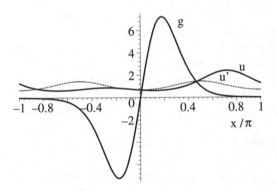

Fig. 5.30. Interpretation of (5.309) as equating a convolution integral to zero. The solid line labeled g represents $g(z)$ with $\tilde{\kappa} = 4$ and $r_{st} = 0.77$. The lines labeled u' and u represents $z \to u' = u(z - \theta_{st})$ for $\tilde{\beta} = 0.375$, $\tilde{\alpha} = 0$, and $\theta_{st} = \pi/2$ and $z \to u(z - \theta_{st})$ for $\tilde{\beta} = 0.375$, $\tilde{\alpha} = 0.5$, and $\theta_{st} = 2.26$. The integrals over the products gu and gu' vanish. The shifts of the maxima of u and u' with respect to the origin represent possible stationary cluster phases that satisfy the necessary condition (5.309)

---

[1] For all parameters $g(x)$ looks qualitatively like the function shown in Fig. 5.30 because for $\tilde{\kappa} > 0$ it follows that $g(x + \pi/2) \leq g(x)$ for $x \in [0, \pi/2]$ and $|g(x - \pi/2)| \leq |g(x)|$ for $x \in [-\pi/2, 0]$.

Saddle Node and Pitchfork Bifurcation

Let us exploit the results obtained so far in order to generalize the bifurcation diagram shown in Fig. 5.28a) to the case in which the symmetry constraint (5.288) does not hold. Substituting (5.285) for $\theta_{st} = \pi$ into (5.307), we obtain

$$\lambda_{\theta_{st}=\pi} = \frac{\tilde{\alpha}|r_{st}| - 4\tilde{\beta}\langle\cos(2x)\rangle_{\tilde{P}_{st}(x;\theta_{st}=\pi)}}{1 - 2\pi\tilde{P}_{st}(0;\theta_{st}=\pi)} . \qquad (5.316)$$

For vanishing noise, we have $r_{st} = 1$ and, following the preceding reasoning, we obtain $\lambda_{\theta_{st}=\pi} = \tilde{\alpha} - 4\tilde{\beta}$, which is the Lyapunov exponent derived from the linear stability analysis of the stationary point $q_{st} = \pi$ of the deterministic HKB model (5.283). Equation (5.316) was evaluated numerically using the potential (5.297) rather than (5.282). Figure 5.29a shows two surfaces dividing the parameter space $(\tilde{\kappa}, \tilde{\alpha}, e)$ into two regions. The "roof-like" surface was adopted from Fig. 5.28a and marks the critical points at which the stationary solutions with $\theta_{st} = \pi$ cease to exist (saddle node bifurcation points). The "wall-like" surface separates the region in which the stationary probability density with $\theta_{st} = \pi$ and $\tilde{\lambda}_1 > 0$ yields, according to (5.316), $\lambda_{\theta_{st}=\pi} < 0$ (to the right of the wall-like surface) and the region in which the stationary probability density yields $\lambda_{\theta_{st}=\pi} > 0$ (to the left of the wall-like surface). From our previous analysis we conclude that the "wall-like" surface in Fig. 5.29a describes points related to a pitchfork bifurcation. In sum, within the cuboid subspace that is shaped by the two surfaces an asymptotically stable stationary probability density with $\theta_{st} \equiv \pi$ exists. Outside this subspace, the stationary probability densities with $\theta_{st} = \pi$ are either unstable or entirely absent. In contrast, the stationary probability density with $\theta_{st} = 0$ describes an asymptotically stable solution in the whole parameter space $(\tilde{\kappa}, \tilde{\alpha}, e)$ for $\tilde{\alpha} > 0$, $\tilde{\kappa} \geq 0$, $e \geq 0$ (or equivalently in the space $(\tilde{\kappa}, \tilde{\alpha}, \tilde{\beta})$ for $\tilde{\alpha} > 0$, $\tilde{\kappa} \geq 0$, $\tilde{\beta} \geq 0$).

Finally, Fig. 5.29d shows the result of a simulation of the mean field HKB model (5.284) with $\tilde{\beta} = \tilde{\alpha}e/4$, where the parameter e was varied along the dashed line depicted in Fig. 5.29b. That is, we use $\tilde{\kappa} = 4$ and $\tilde{\alpha} = 0.5$. A simulation scheme was used as described above in the context of the saddle node bifurcation of the cluster amplitude. However, this time the imaginary parts of the time-dependent coefficients were included. The noisy Gaussian probability density shown in panel d) served as the initial distribution. The parameter e was decreased in steps of 0.1 from 1.8 to 1.4 after each computation of 10^6 single time steps. In agreement with the results obtained from the linear stability analysis shown in panel b) and in accordance with the branching of solutions in the bifurcation diagram shown in panel c), the phase transition occurred at a critical value of about $e_c = 1.6$.

5.4 Characteristics of Bifurcations

5.4.1 Stability and Disorder

There is a close relationship between the disorder and the stability of stationary distributions of nonlinear Fokker–Planck equations when disorder is measured in terms of the variance and similar statistical measures. Accordingly, states with a large degree of order are described by distributions with small variances, whereas states with a large degree of disorder are described by distributions with larger variances. Since the free energy of the Desai–Zwanzig model reads

$$F = U_\mathrm{L} + \frac{\kappa}{2} K - Q^\mathrm{B} S \qquad (5.317)$$

(see (5.184)), we find that the smaller K, the smaller the free energy F. That is, the free energy is low for states with a small degree of disorder. In the case of the KSS model and the mean field HKB model, the disorder can be measured in terms of the cluster amplitude r. For unimodal distributions large values of r correspond to states of high order (low disorder), whereas small values of r are related to states of low order (large disorder), see Sect. 5.3.1. Since we have $r \in [0,1]$, we may say that for unimodal distributions small values of $1 - r$ correspond to states of high order, whereas large values of $1 - r$ describe states of low order. The free energies of the KSS model and the mean field HKB model read

$$F = -\frac{\kappa}{2} r^2 - Q^\mathrm{B} S \,,$$
$$F = U_\mathrm{L} - \frac{\kappa}{2} r^2 - Q^\mathrm{B} S \qquad (5.318)$$

(see (5.233) and (5.278)). We find that the larger r (i.e., the smaller $1-r$), the smaller F. Roughly speaking, this implies that the free energies are smaller for ordered states than for disordered states – just as in the case of the Desai–Zwanzig model. Since we study systems that converge to states of minimal free energy, a naive but useful notion is that stable states are related to states with minimal disorder.

In fact, as shown in Fig. 5.31, in the multistable regime states with a relatively high degree of order are stable, whereas states with a relatively high degree of disorder are unstable ones. The diagram for the Desai–Zwanzig model shown in Fig. 5.31 can be deduced from (5.63). Equation (5.63) states that for fixed control parameters κ and Q the expression $\mathrm{d}R/\mathrm{d}m$ is proportional to the variance K of the stationary distributions. Since we have $\mathrm{d}R/\mathrm{d}m > 1$ for unstable distributions and $\mathrm{d}R/\mathrm{d}m < 1$ for stable distributions, it follows that in the bistable regime the variance of stable solutions is smaller than the variance of unstable solutions. Note that for all K_A-models we can obtain a similar result: in the multistable regime stable stationary solutions have smaller K_A-values than unstable stationary solutions. The reason for

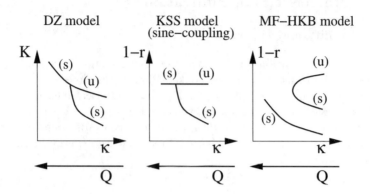

Fig. 5.31. Stability and degree of disorder of the stationary distributions of three mean field models: the Desai–Zwanzig model, the KSS model with a sine coupling-force, and the mean field HKB model; (u) and (s) denote unstable and stable distributions

this is that for the K_A-model (5.76) the relation (5.80) holds. The diagram for the KSS model corresponds to Fig. 5.25. The diagram for the mean field HKB model shows only the symmetric stationary solutions and can be read off from panel c) in Fig. 5.28. There we see that at the saddle node bifurcation point (graph II) the signed cluster amplitude r'_u of the unstable stationary solution is equivalent to the cluster amplitude r'_s of the stable stationary solution with $r'_s < 0$, whereas in the multistable regime (graph I) r'_u lies closer to zero than r'_s, which implies that the original cluster amplitudes $r_u = -r'_u$ and $r_s = -r'_s$ satisfy the relation $r_s > r_u$.

In sum, for the aforementioned models there are two stability criteria. First, stable distributions correspond to free energy minima, whereas unstable distributions correspond to free energy maxima or saddle points. Second, in the multistable regime unstable distributions correspond to states of large disorder, whereas stable distributions correspond to states of small disorder, when disorder is measured in terms of the variance of distributions or similar statistical quantities.

5.4.2 Emergence of Collective Behavior

The bifurcation from a stationary distribution $P_{st}^{(1)}$ to a stationary distribution $P_{st}^{(2)}$ can often be interpreted as the emergence of a collective behavior in a many-body system or the change between different kinds of collective behavior. To this end, it is assumed that the solutions of nonlinear Fokker–Planck equations describe the distributions of subsystems of many-body systems (μ-space description, see Sect. 1.1.2). For example, let us consider the Desai–Zwanzig model in the parameter range in which it exhibits unimodal

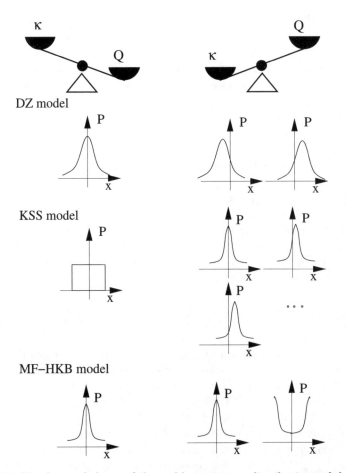

Fig. 5.32. Number and shape of the stable stationary distributions of three mean field models (Desai–Zwanzig model, KSS model, and mean field HKB model) in two different parameter regimes: $\kappa \ll Q$ (left column) and $\kappa \gg Q$ (right column)

stationary distributions only. Then, the symmetric stationary distribution $P_{st}^{(1)}$ of the Desai–Zwanzig model is given by a bell-shaped distribution with vanishing mean value that describes a many-body system with subsystems without any preference or collective behavior. The asymmetric stationary distributions $P_{st}^{(2)}$ describe subsystems that assume on the average positive (or negative) state values and in this sense they describe subsystems that exhibit a common tendency or a collective behavior, see Fig. 5.32. Likewise, the uniform distribution $P_{st}^{(1)}$ of the KSS model with sine coupling-force describes a many-body system that does not exhibit any collective behavior. In contrast, unimodal distributions $P_{st}^{(1)}$ of the KSS model describe subsystems that are centered around the cluster phase θ_{st} (see also Sect. 5.3.1) and in this sense show a collective behavior. Finally, let us consider the HKB model in the parameter range where it exhibits unimodal stationary distributions. In this case, for $\tilde{\kappa} = \kappa/Q < \tilde{\kappa}_c$ and fixed values of $\tilde{\alpha}$ and $\tilde{\beta}$, the mean field HKB model exhibits a unique stationary distribution $P_{st}^{(1)}$ with a single peak at $x = 0$. However, if $\tilde{\kappa}$ exceeds a critical value $\tilde{\kappa}_c$ there is another asymptotically stable, unimodal distribution that exhibits a peak at $x = \pi$. Consequently, if the model is prepared in a parameter regime with $\tilde{\kappa} > \tilde{\kappa}_c$ and subsequently parameters are changed such that the model becomes monostable (i.e., $\tilde{\kappa} \to \tilde{\kappa} < \tilde{\kappa}_c$), then we observe a phase transition between two different kinds of collective behavior.

5.4.3 Multistability and Symmetry

Multistabilities often arise due to symmetries of many-body systems. Let us illustrate this relationship by means of a few examples. A many-body system related to the Desai–Zwanzig model discussed in Sect. 5.2.3 may be defined by means of the multivariate Langevin

$$\frac{d}{dt}X_i(t) = aX_i - bX_i^3 - \frac{\kappa}{N_0}\sum_{k=1}^{N_0}(X_i - X_k) + \sqrt{Q}\,\Gamma_i(t)\,, \qquad (5.319)$$

with $i = 1, \ldots, N_0$. In fact, in the limit $N_0 \to \infty$ we may replace the subsystem ensemble $\{X_1, \ldots, X_{N_0}\}$ in the interaction term by means of a statistical ensemble $\{X^1, \ldots, X^N\}$ of a single subsystem and thus obtain the Desai–Zwanzig model (see Sect. 3.8). The many-body system defined by (5.319) is invariant under the inversion defined by $X_i \to -X_i$. On account of this inversion symmetry, we conclude that if the many-body system exhibits a collective behavior in terms of a phase with a nonvanishing order parameter $q = N^{-1}\sum_{k=1}^{N_0} X_k$, then this collective behavior comes in two types: one type with $q > 0$ and another one with $q < 0$. The situation is reminiscent of the behavior of an Ising ferromagnet, where in the ferromagnetic phase the spins preferably point up or point down (see Sec. 5.5.1). The inversion symmetry of the many-body system described by (5.319) carries over to the Desai–Zwanzig model (5.180). The nonlinear Fokker–Planck equation (5.180) is invariant under inversion $x \to -x$. Therefore, as shown in Fig. 5.32 in the strong coupling regime (i.e. for $\kappa \gg Q$) there is not only one asymmetric stationary distribution but there are two of them.

A many-body system related to the KSS model with sine coupling-force (see Sect. 5.3.2) is given by

$$\frac{d}{dt}X_i(t) = -\frac{\kappa}{N_0}\sum_{k=1}^{N_0}\sin(X_i - X_k) + \sqrt{Q}\,\Gamma_i(t)\,, \qquad (5.320)$$

with $i = 1, \ldots, N_0$. The many-body system is invariant under translations defined by $X_i \to X_i + \Delta x$. Consequently, if there is a collective behavior with a nonvanishing cluster amplitude $r = |N_0^{-1} \sum_{k=1}^{N_0} \exp\{iX_k\}|$ then there are infinite many types of this collective behavior. The situation here is reminiscent of a planar rotator ferromagnet, where the magnetization in the ferromagnetic phase can point in any direction of the XY-plane (see Sect. 5.5.1). The translation symmetry of the many-body system given by (5.320) can also be found in the KSS model (5.229). Therefore, as pointed out in Sect. 5.3.2 and as shown in Fig. 5.32 in the strong coupling regime (i.e. for $\kappa \gg Q$) there is a continuous family of stationary distributions.

The mean field HKB model (5.260) for vanishing α-term is invariant under the operation $x \to x + \pi$. Therefore, as shown in Fig. 5.32 in the strong coupling regime (i.e. for $\kappa \gg Q$) there are two stationary distributions. The α-term breaks the symmetry property of the mean field HKB model for $\alpha = 0$, which is the reason why for $\alpha > 0$ and weak coupling (i.e. for $\kappa \ll Q$) only one distribution survives (see Fig. 5.32 again).

5.4.4 Continuous and Discontinuous Phase Transitions

We refer to phase transitions of the solutions of mean field models as continuous if the order parameters of the models change continuously at phase transition points. In other words, if we regard the stationary order parameter values M_{st} of a mean field model as a function of a control parameter α, $M_{st} = M_{st}(\alpha)$, and if M_{st} is a continuous function of α at a phase transition point α_c, then the phase transition is a continuous one. If there is a discontinuous change (i.e., if $M_{st}(\alpha)$ "jumps" from one value to another) then we are dealing with a discontinuous phase transition. For example, the Desai–Zwanzig model exhibits a continuous phase transition. As shown in the bifurcation diagrams 5.19 and 5.20 the order parameter $\langle X \rangle_{st}$ changes continuously at the phase transition point. In contrast, the mean field HKB model exhibits a discontinuous phase transition. As shown in the bifurcation diagram 5.29c and in Figs. 5.28d and 5.29d the cluster phase θ_{st} changes discontinuously from π to zero. Some models and the type of phase transition they exhibit are listed in Table 5.3. The liquid crystal model will be discussed in Sect. 5.5.3.

Table 5.3. Some mean field models and the type of bifurcations that they describe

Continuous phase transitions	dynamical Takatsuji model
	Desai–Zwanzig model
	bounded $B(M_1)$ model with double-well potential $B(z)$
	KSS model with sine coupling-force
Discontinuous phase transitions	mean field HKB model
	liquid crystal model

5.5 Applications

5.5.1 Ferromagnetism

Ferromagnetic materials exhibit a magnetization at temperatures below a critical temperature T_c. For temperatures larger than T_c this magnetization vanishes. That is, there is a phase transition between two phases: a ferromagnetic phase (finite magnetization) and a paramagnetic phase (no magnetization). The reason for this behavior is that ferromagnetic materials are composed of a large number of interacting elementary magnets given by atomic spins. The exchange energy between pairs of spins is lower for parallel spin orientations than for anti-parallel spin orientations [11, 89, 229, 490, 546, 579]. Consequently, spin-spin interactions tend to establish a highly ordered state with all spins in parallel. The spins exchange heat energy with their environment, which leads to a thermal motion of the spins. The thermal motion of the spins tends to establish random spin orientations. In sum, if thermal spin fluctuations dominate, the overall magnetization vanishes and ferromagnetic materials are in the paramagnetic phase, whereas if spin-spin interactions dominate, spins align themselves into a state with parallel orientations and, as a result, there is a non-vanishing magnetization and ferromagnets are in their ferromagnetic phases, see Fig. 5.33.

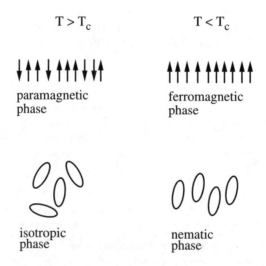

Fig. 5.33. Ordered and disordered states of ferromagnetic materials (upper panels) and liquid crystals (lower panels)

Spins can be described by three-dimensional vectors $\mathbf{s} = (s_x, s_y, s_z)$ subjected to the constraint $|\mathbf{s}| = S$ (note that S does not denote an entropy measure here). The components s_x, s_y, s_z play different roles for different kinds of ferromagnetic materials. In some materials only the z-components of spins are involved in spin-spin interactions. Materials of this kind are referred to as Ising ferromagnets. For Ising ferromagnets the magnetization M_z points along the z-axis and is given by $M_z = \langle s_z \rangle$, where the average is carried out with respect to the ensemble of spins. For $T > T_c$ we have $M_z = \langle s_z \rangle = 0$. For $T < T_c$ the spins point on the average either in the direction of the z-axis ($M_z = \langle s_z \rangle > 0$) or in the opposite direction ($M_z = \langle s_z \rangle < 0$). That is, for $T > T_c$ the system is monostable, whereas for $T < T_c$ it is bistable. In other ferromagnetic materials spins are predominantly affected by spin-spin interactions within a plane, the so-called "easy plane" or XY-plane. Materials of this kind exhibit spin-spin coupling forces involving the components s_x and s_y. We will refer to materials with easy planes as XY-ferromagnets.

Soft-spin Ising Ferromagnets and the Desai–Zwanzig Model

Let us consider an ensemble of N_0 spins and describe the components s_z of the spins by means of the state variables X_k for $k = 1, \ldots, N_0$. Furthermore, let us regard the fluctuation strength Q as a temperature measure T and consider the first moment $M_1 = N_0^{-1} \sum_k X_k$ as a measure for the magnetization M_z. Then, if we neglect correlations between subsystems and identify the state variables X_k with the statistically-independent realization X^k of a random variable $X \in \Omega = \mathbb{R}$, we arrive at mean field models for the one-dimensional Ising ferromagnet.

In Sect. 5.2.2 we have already elucidated the relationship between the dynamical Takatsuji model and the Ising ferromagnet. Now, let us focus on the Desai–Zwanzig model. If we regard the Desai–Zwanzig model as a model for a one-dimensional Ising ferromagnet, then the bifurcation diagram depicted in Fig. 5.20 may be regarded as the bifurcation diagram of an Ising ferromagnet. Moreover, since the Desai–Zwanzig model is a nonlinear Fokker–Planck equation and the nonlinearities of nonlinear Fokker–Planck equations are assumed to reflect subsystem-subsystem interactions, the Desai–Zwanzig model indeed accounts for the mechanism that leads to the ferromagnetic phase, namely, the interactions between spins. As shown in Fig. 5.32, the ferromagnetic phases exist when spin-spin interactions dominate fluctuations, whereas the paramagnetic phase exists when fluctuations dominate the spin-spin interactions (which is in line with Fig. 5.33 when we put $Q = T$). Finally, we would like to point out that we are dealing with a soft-spin model [143] because in contrast to the spin variables that are subjected to the constraint $|\mathbf{s}| = S$, the random variables X_k are not subjected to such a constraint. However, on account of the potential $V(x) = -ax^2/2 + bx^4/4$ the variables X_k are attracted to the positions $x_{1,2} = \pm\sqrt{2a/b}$ that describe the potential minima. That is, the Desai–Zwanzig model involves a weak (or soft) version of the constraint to which spin variables are subjected (whence the name soft-spin model).

Planar Rotator Ferromagnets and the KSS Model

There are two kinds of models for XY-ferromagnets: planar rotator models and XY-models. In the case of planar rotator models, the three-dimensional spin vectors are confined to the XY-plane and therefore, can be described in terms of a single angular variable $X \in \Omega = [0, 2\pi]$. More precisely, we have $s_z = 0$ and $\mathbf{s} = (s_x, s_y) = (\cos X, \sin X)$ [307, 318, 331, 345, 407]. Note that this representation implies that $|\mathbf{s}| = S = 1$. In contrast, the XY-model describes ferromagnet materials with spins that can have a nonvanishing z-component [234, 417, 548]. The KSS model with sine coupling-force and the mean field HKB model can be regarded as planar rotator models for ferromagnets with infinite-range spin-spin interactions [172, 199, 603]. In the case of infinite-range interactions [546], there is an all-to-all coupling between the spins. Consequently, the interaction energy H_{ss} is given by $H_{ss} = -0.5\kappa N_0^{-1} \sum_{i,k=1}^{N_0} \mathbf{s}_i \cdot \mathbf{s}_k = -0.5\kappa N_0^{-1} \sum_{i,k=1}^{N_0} \cos(X_i - X_k)$ with $\mathbf{s}_n = (\cos X_n, \sin X_n)$. Here, $\kappa > 0$ is a proportional factor used to compute the energy E_i of a single spin i in the mean field $\mathbf{S} = N_0^{-1} \sum_{k=1}^{N_0} \mathbf{s}_k$ like $E_i = -\kappa \mathbf{s}_i \cdot \mathbf{S}$. In the sum $\sum_{i,k=1}^{N_0} \mathbf{s}_i \cdot \mathbf{s}_k$ the exchange energy of each spin-pair is counted twice. Therefore, we have introduced in H_{ss} the factor 0.5. Taking external magnetic fields and other symmetry breaking (crystalline) fields into account, we obtain additional Hamiltonians such as $\sum_{i=1}^{N_0} \cos(X_i)$ (magnetic fields), $\sum_{i=1}^{N_0} \cos(2X_i)$ (uniaxial anisotropies), and $\sum_{i=1}^{N_0} \cos(nX_i)$ (n-fold anisotropies) [229, 316, 406, 611]. For example, a constant magnetic field given by $\mathbf{f} = (1, 0)$ gives rise to the interaction energies $\mathbf{f} \cdot \mathbf{s}_i = \cos X_i$. Furthermore, if an anisotropy of an XY-ferromagnet can be described by means of a 2×2-matrix A with matrix elements $A_{1,1} = -A_{2,2} = 1$ and $A_{1,2} = A_{2,1} = 0$, then the potential energy of a spin $\mathbf{s}_i = (\cos X_i, \sin X_i)$ with respect to this anisotropy is given by $\mathbf{s}_i \cdot A \cdot \mathbf{s}_i = \cos(2X_i)$. In sum, the total Hamiltonian of a planar rotator model with infinite-range interactions and symmetry breaking fields reads $H_{\text{tot}} = -\kappa N_0^{-1} \sum_{i,k}^{N_0} \cos(X_i - X_k) - \sum_n c_n \sum_{i=1}^{N_0} \cos(nX_i)$, where the parameters c_n denote coupling factors. Taking fluctuating forces into account, we may describe the stochastic behavior of the coupled planar rotators by means of the multivariate Langevin equation

$$\frac{d}{dt}X_i(t) = -\frac{\partial}{\partial X_i} H_{\text{tot}} + \sqrt{Q}\,\Gamma_i(t) \ . \tag{5.321}$$

If at time $t = t_0$ the random variables X_i are statistically-independent and described by the same distribution $u(x)$, then in the limit $N_0 \to \infty$ the random variables X_k correspond to the realizations X^k of a random variable X and (5.321) becomes

$$\frac{d}{dt}X(t) = -\sum_n c_n \sin(nX) - \kappa \langle \sin(X - X') \rangle + \sqrt{Q}\,\Gamma(t) \ , \tag{5.322}$$

see Sect. 3.8. The stochastic process given by $X(t)$ can then equivalently be described in terms of the free energy Fokker–Planck equation

$$\frac{\partial}{\partial t}P(x,t;u) = \frac{\partial}{\partial x}\left[\sum_n c_n \sin(nx) + \kappa \int_\Omega \sin(x-x')P(x',t;u)\,\mathrm{d}x'\right]P + Q\frac{\partial^2}{\partial x^2}P \tag{5.323}$$

and the evolution equation

$$\frac{\partial}{\partial t}P(x,t|x',t';u) =$$

$$\frac{\partial}{\partial x}\left[\sum_n c_n \sin(nx) + \kappa \int_\Omega \sin(x-x')P(x',t;u)\,\mathrm{d}x'\right]P(x,t|x',t';u)$$

$$+Q\frac{\partial^2}{\partial x^2}P(x,t|x',t';u) \ . \tag{5.324}$$

Equation (5.323) reduces to the KSS model with sine coupling-force for $c_{n \geq 0} = 0$ and the mean field HKB model for $c_1 = \alpha$, $c_2 = \beta$, and $c_{n \geq 3} = 0$. According to this approach to ferromagnetic materials, the nonuniform stationary solutions of the KSS model describe ferromagnetic phases of planar rotator ferromagnets, whereas the uniform solution describes paramagnetic phases. In order to take the impact of the symmetry breaking field into account, we would need to supplement the KSS model with pinning forces. That is, we would need to consider models such as the mean field HKB model.

5.5.2 Synchronization

There are numerous examples of many-body systems with subsystems that exhibit synchronized oscillatory behavior. In particular, there are various biological and social systems showing behavioral synchronization and there are many examples of synchronized neural activity. Let us mention a few of them.

Applause

The applause after a concert is a typical example for mass-behavior of humans. Applause usually exhibits the following two states pattern. There is an initial period in which people clap with their preferred frequencies and, thus, produce an unsynchronized applause. Subsequently, the audience finds a common clapping frequency. During this second period, not only are individuals entrained by the common clapping frequency, they also perform their clapping movements with roughly the same phase angles [441]. In view of these characteristics, the unsynchronized and synchronized applause can be distinguished using a mean harmonic correlation function introduced in [441], which reflects a collective property of applause. More precisely, the correlations between the signals of the members of an audience and a harmonic function with fixed frequency and phase are computed and averaged. Frequency and phase of the harmonic function are adjusted in order to obtain a maximum value. Unsynchronized applause produces low values irrespective of the parameter values for frequency and phase. In contrast, synchronized applause yields large function values. By comparing the clapping behavior of isolated individuals with the clapping behavior of individual members of audiences, one can show that the clapping behavior changes due to the acoustic interactions between individuals [441]. In short, in this example acoustic coupling forces result in the emergence of a collective synchronized behavior.

Tree Crickets and Fireflies

Snowy tree crickets chirp in a rhythmic pattern with frequencies between 50 and 200 chirps per minute depending on the temperature of their environments [616]. With their tympanic organs (receptors for acoustic stimuli) removed, they chirp with individual phase lags. Consequently, populations of these crickets produce de-synchronized chirping patterns [616]. In contrast, populations of intact snowy tree crickets are known to chirp in unison. That is, these crickets usually synchronize their chirping behavior [616]. To this end, a cricket shortens the interval between two consecutive chirps if its chirping call follows the calls of other crickets such that the subsequent call is closer to the calls of the other crickets. Likewise, a cricket lengthens a chirp interval if its own call is ahead of the calls of other crickets [616, 633]. In view of these observations, it has been suggested that cricket populations feature a collective behavior, namely, they form choruses producing synchronized calling songs [557, 633]. The formation of choruses can be regarded as a process of self-organization that is established by means of acoustic couplings between the individuals. In the absence of these couplings there is no chorusing. In contrast, if the coupling forces dominate, then synchronized behavior can emerge. The emergence of the collective behavior can be described by means of the non-uniformity of the phase distribution of cricket chirps. The degree of non-uniformity can easily be quantified in terms of the cluster amplitude r, see Sect. 5.3.1. Unsynchronized chirping corresponds to a uniform phase distribution and a vanishing degree of non-uniformity (i.e., $r = 0$), whereas chorusing and synchronized chirping involves a nonuniform phase distribution (i.e., $r > 0$).

It is known that some fireflies flash in isolation with rhythmic patterns at frequencies in the range from 1 to 2 Hz. When a population of these fireflies is put together, for example, on a tree, then the fireflies flash with a common frequency and almost vanishing phase differences, that is, the population exhibits a communal synchronized flashing [81]. It has been observed that the fireflies are entrained by the flashes of their fellow-fireflies. In response to flash-like stimuli, a firefly shortens or lengthens the intervals between its consecutive flashes such that its own flashing pattern approaches the stimulus pattern [82, 277, 633]. Therefore, it is generally assumed that the synchronous flashing of firefly populations results from self-organization and the visual couplings between the population members. The establishment of a communal synchronous rhythm can quantitatively be assessed in terms of the non-uniformity or the cluster amplitude r of firefly flashes. Just as in the previous examples, fireflies that do not synchronize their flashings produce a uniform phase distribution with a vanishing degree of non-uniformity (i.e., we have $r = 0$), whereas fireflies that synchronize their flashings produce a unimodal phase distribution with a finite degree of non-uniformity (i.e., we have $r > 0$).

Synchronization of Brain Activity

An important concept in neuroscience is that the neocortex of humans and animals is organized as a map. Particular areas correspond to functional abilities (e.g., memory) and sensory and motoric organs (see e.g. [578]). We may regard these areas as "black boxes", neural dipoles, or elementary neural units and describe their functioning by means of zero-dimensional models (just as we would describe the functioning of an Ohmic resistance). According to a new paradigm in neuroscience, however, neural computation involves the cooperative activity of many spatially distributed neurons. In particular, it is believed that perception and action involve spatially extended systems composed of many neurons that are bound to task-related entities by synchronizing their activity. Using cross-correlation analysis, in animal studies with cats, synchronized neural activity has been found in functional columns of the visual cortex [239, 538], among several unilateral areas of the visual cortex [150], and between left- and right-hemispheric visual areas [151]. Although synchronization typically occurs during the presentation of an object, it is assumed that the visual stimulus does not act as a driving force in order to establish a population of synchronized neurons. On the basis of theoretical reasoning and experimental evidence, it has been suggested that synchronization emerges due to mutual couplings between so-called neural oscillators. Accordingly, the synchronization of particular populations of neurons corresponds to particular perceptions [122, 539, 614]. Moreover, if in animal experiments the corticocortical connections between cortical hemispheres are removed, then synchronization between the hemispheres vanishes despite of the presence of a visual stimulus [151].

Synchronized oscillatory neural activity has also been found during the performance of manual tasks. There are significant cross-correlations between neural populations of different unilateral cortical regions when monkeys perform perceptual-motor tasks [75]. Such correlations are almost entirely absent before and after task performance. In similar animal studies, it has been found that arm movements involve both uni- and bilateral synchronized brain activity [439]. Synchronized brain activity has also been found in man. For example, it has been shown that visual flicker signals can induce synchronized oscillatory neural activity over the whole cortex [537]. In addition, it has been shown that synchronization between different cortical areas occurs during tremor [567, 566], the production of isometric forces [245], and when humans listen to music [56].

The conflict between zero-dimensional and spatially distributed neural units is a conflict between localized and nonlocalized task-related neural activity. Probably the truth is to be found somewhere in between these extremes. In order to illustrate this, we may compare the situation with the theories of electromagnetism and gravity. Both theories deal with localized, zero-dimensional elements (charged particles and particles with mass, respectively) as well as nonlocalized, spatially extended elements (electromagnetic fields and gravitational fields, respectively). There are problems that can be treated solely by reference to the localized elements. Other problems only involve the field theoretical aspects of electromagnetism and gravity. Similarly, a comprehensive account of neural task-related activity may combine zero-dimensional and spatially distributed task-related units. Nevertheless, in special cases it may be sufficient to model neural systems as either zero-dimensional or spatially extended systems. In the latter case, mean field models such as the KSS model and extensions of it are tailored to describe the synchronization of neural activity produced by populations of neurons and neuronal units.

Synchronization, Planar Rotators and Phase Oscillators

The KSS model and modifications of it have frequently been used to describe the emergence of synchronized states of neural activity. The reason for this is that it often seems sufficient to describe ensembles of oscillatory constituents by means of coupled planar rotators in order to obtain a good understanding of the observed phenomenon. In the context of neural activity, one often refers to planar rotators as phase oscillators. The state $Y_i(t)$ of a phase oscillator is given by

$$Y_i(t) = A\sin(\omega t + X_i(t)) . \tag{5.325}$$

Here, A and ω are amplitude and frequency of the oscillator and assume fixed values. We realize that the temporal evolution of a phase oscillator is completely described by the phase X_i. Let us label the ensemble members by $i = 1, 2, \ldots$. Then, it is clear from (5.325) that in this simple case all phase oscillators have the same amplitude and eigenfrequency. Moreover, the oscillator ensemble is completely de-synchronized if the phases X_i are uniformly distributed. If the distribution of X_i differs from the uniform distribution, then there is a partial synchronization of the oscillator ensemble. If we neglect statistical correlations between the oscillators, we can describe the states X_i by means of the realizations X^i of the random variable X defined by the KSS model. Then, the synchronized states emerge when the uniform stationary distribution of the KSS model becomes unstable and gives rise to a stable stationary non-uniform distribution.

Bibliographic Note

Our understanding of how synchronization can emerge in coupled oscillator systems of the mean field type was in particular advanced by studies of Winfree [632, 633] and Kuramoto [365]. Synchronization in the context of nonlinear Fokker–Planck equations has been extensively studied in the literature. The reader may consult the review articles and monographs [477, 552, 557, 568]. For further information on this topic see also Sect. 7.5.

5.5.3 Isotropic-nematic Phase Transitions and Maier–Saupe Model

Liquid crystals combine properties of solids and liquids. They can exhibit both some kind of microscopic order (just as in solids) and fluidity (just as in liquids) [129, 131]. Liquid crystals can exhibit an isotropic and a nematic phase. In the isotropic phase the orientations of the molecules are randomly distributed, whereas in the nematic phase all molecules point in roughly the same direction, see Fig. 5.33. The isotropic phase can be regarded as a state of orientational disorder, whereas the nematic phase exhibits orientational order. The degree of alignment of the molecules is a statistical measure that describes how the orientations of the molecules are distributed and can be regarded as an order parameter q. If the molecules' orientations are uniformly distributed (isotropic phase), then there is no alignment and, consequently, we put the degree of alignment equal to zero (i.e., we have $q = 0$). If all molecules point in exactly the same direction, the degree of alignment becomes unity (i.e., we have $q = 1$). The order-disorder phase transition from the isotropic phase to the nematic phase involves the emergence of a collective (or cooperative) behavior: the alignment of molecules. That is, the isotropic-nematic phase transition can be described in terms of q by means of a change from $q = 0$ to $q > 0$ [90, 129]. The alignment of molecules is an example of self-organization because it is caused by the interaction forces between the molecules of a liquid crystal. The molecules themselves produce an overall torque force that can be described, for example, in terms of the Maier–Saupe potential (see below). This torque force acts back on the individual molecules and tends to establish a nematic phase. If fluctuating forces dominate, then the overall torque force cannot be established and, consequently, the crystal settles down in the isotropic phase. In contrast, if the interaction forces dominate, the collective torque force emerges and results in an alignment of the molecules and the emergence of a nematic phase.

The orientation of a simple macromolecule can be described by means of a director $\mathbf{n} = (n_x, n_y, n_z) \in \Omega$. The director is normalized to unity $|\mathbf{n}| = 1$, which implies that $n_x, n_y, n_z \in [-1, 1]$ and that there are only two free parameters at our disposal. This becomes obvious if $\mathbf{n}$ is given in spherical polar coordinates: $\mathbf{n} = (\sin \vartheta \sin \varphi, \sin \vartheta \cos \varphi, \cos \vartheta)$ with $\vartheta \in [0, \pi]$ and $\varphi \in [0, 2\pi]$.

Accordingly, the phase space Ω of $\mathbf{n}$ is the surface of a sphere with radius 1. Orientational order can be described by means of the order parameter

$$q = \langle A(\vartheta) \rangle = \frac{1}{2} \langle 3\cos^2(\vartheta) - 1 \rangle_{P(\mathbf{n})} = \frac{1}{2} \oint_\Omega (3\cos^2(\vartheta) - 1) P(\mathbf{n}) \mathrm{d}O \quad (5.326)$$

defined as the mean value of $A(\vartheta) = [3\cos^2(\vartheta) - 1]/2$ [90, 129, 485, 551].

In order to elucidate the role of the order parameter q, let us consider liquid crystals with a rotational symmetry around an axis, say, the z-axis. Then, φ is uniformly distributed: $P(\varphi) = 1/(2\pi)$. In this case the statistical properties of liquid crystals are primarily described by the director distribution $P(n_z)$ with $n_z \in \Omega_z = [-1, 1]$. For macromolecules with rotational symmetry the order parameter q becomes

$$q = \frac{1}{2} \int_{-1}^{1} (3n_z^2 - 1) P(n_z) \mathrm{d}n_z . \quad (5.327)$$

We assume that the director is a non-polar vector, which means that a macromolecule with $\mathbf{n}^*$ has the same properties as a molecule with $-\mathbf{n}^*$. In this case, director distributions $P(n_z)$ are symmetric with respect to the origin: $P(n_z) = P(-n_z)$. In the isotropic phase the director components n_z are uniformly distributed in $[-1, 1]$, that is, we have $P(n_z) = 1/2$. Consequently, for the isotropic phase we obtain $q = 0$. Note that the distribution $P(\vartheta)$ of the isotropic phase is not uniform because the transformation $n_z = \cos(\vartheta)$ yields $P(n_z) = 1/2 \Rightarrow P(\vartheta) = \sin(\vartheta)/2$ (for the transformation of random variables see e.g. [498]). If $P(n_z)$ differs from the uniform distribution, we have orientational order to a certain extent and refer to the phase as nematic. We choose our coordinate system in such a way that in the nematic phase the directors have the tendency to be parallel to the z-axis. Then, in the nematic phase we have $q > 0$. As a result, the emergence of the nematic phase from the isotropic phase goes along with a bifurcation of q from $q = 0$ to $q > 0$ – as mentioned previously. If there is a complete alignment of the molecules, we have only molecules with $n_z = 1$ and $n_z = -1$. The corresponding probability distribution is $P(n_z) = [\delta(n_z - 1) + \delta(n_z + 1)]/2$ and the order parameter assumes its maximum value: $q = 1$.

Maier–Saupe Mean Field Theory

We consider here liquid crystals to which the Maier–Saupe theory applies [90, 129, 399, 400, 485, 551]. The macromolecules are assumed to interact with each other by means of electronic dipole forces. The strength of the dipole-dipole interactions depends on the orientations of the molecules. The interaction energy between one macromolecule and all other macromolecules depends, roughly speaking, on the orientation of the macromolecule with respect to the mean orientation of all other molecules and is given by $E_i = -\kappa A(\vartheta_i) \langle A(\vartheta) \rangle$. Consequently, the total interaction energy reads $U = -\kappa q^2/2$. Likewise, the free energy is given by

$$F[P(\mathbf{n})] = -\frac{\kappa}{2}q^2 - T\,^{\mathrm{or}}S[P(\mathbf{n})]\,, \tag{5.328}$$

where $^{\mathrm{or}}S[P(\mathbf{n})]$ denotes the orientational entropy defined by [485]

$$^{\mathrm{or}}S[P] = -\oint_{\Omega} P[\mathbf{n}(\varphi,\theta)] \ln P[\mathbf{n}(\varphi,\theta)]\,\mathrm{d}O$$
$$= -\int_0^{2\pi} \mathrm{d}\varphi \int_0^{\pi} \mathrm{d}\theta \sin(\theta) P[\mathbf{n}(\varphi,\theta)] \ln P[\mathbf{n}(\varphi,\theta)]\,. \tag{5.329}$$

Assuming now rotational symmetry of the macromolecules and using $X = n_z$, we obtain

$$F[P] = -\frac{\kappa}{2}q^2 - T\,^{\mathrm{B}}S[P]$$
$$= -\frac{\kappa}{2}\langle A\rangle^2 - T\,^{\mathrm{B}}S[P] \tag{5.330}$$

with $q = \langle A \rangle$, $^{\mathrm{B}}S[P] = -\int_{\Omega_z} P \ln P\,\mathrm{d}x$, and

$$A(x) = \frac{3x^2 - 1}{2}\,. \tag{5.331}$$

The free energy (5.330) belongs to the class of free energies given by (5.317) and (5.318). The free energy Fokker–Planck equation (5.3) can be found as

$$\frac{\partial}{\partial t}P(x,t;u) = -\frac{9}{2}\kappa\frac{\partial}{\partial x}x\left[\langle X^2\rangle - \frac{1}{3}\right]P + T\frac{\partial^2}{\partial x^2}P\,. \tag{5.332}$$

Note that $X = 1$ ($\vartheta = 0$) is equivalent to $X = -1$ ($\vartheta = \pi$) because (as mentioned earlier) a director $\mathbf{n}^*$ is equivalent to a director $-\mathbf{n}^*$ that points in the opposite direction. Consequently, we study solutions of (5.332) that are defined on $x \in [-1,1]$ and satisfy both periodic boundary conditions and the additional constraint that $P(x)$ is a symmetric function. In view of these considerations, one may say that the whole physics of the liquid crystal model is given by a random walk defined on $[0,1]$ with reflective boundaries at $x = 0$ and $x = 1$. However, in what follows, for the sake of convenience, we will continue to use X defined on $[-1,1]$. For appropriate initial distributions $u(x)$ the second moment $\langle X^2 \rangle$ will correspond to a continuous nonsingular function with respect to t. For such initial distributions the evolution equation (5.332) is a strongly nonlinear Fokker–Planck equation and Markov diffusion processes can be defined by means of the solutions of (5.332) and the evolution equation

$$\frac{\partial}{\partial t}P(x,t|x',t';u) =$$
$$-\frac{9}{2}\kappa\frac{\partial}{\partial x}x\left[\langle X^2\rangle_{P(x,t;u)} - \frac{1}{3}\right]P(x,t|x',t';u) + T\frac{\partial^2}{\partial x^2}P(x,t|x',t';u)$$
$$\tag{5.333}$$

or, alternatively, in terms of the Langevin equation

$$\frac{d}{dt}X(t) = \frac{9}{2}\kappa X\left[\langle X^2\rangle - \frac{1}{3}\right] + \sqrt{T}\Gamma(t) . \qquad (5.334)$$

Equation (5.332) can be regarded as a special case of the generic model (5.107) for $U_0 = 0$, $d = 1$, $\kappa_1 = \kappa$, and $T = Q$. Consequently, stationary solutions of (5.332) can be obtained from (5.109) and are given by

$$P_{st}(x) = \frac{1}{Z}\exp\left\{\kappa\frac{A(x)\langle A\rangle_{st}}{T}\right\}$$

$$= \frac{1}{Z'}\exp\left\{\frac{3\kappa x^2\langle (3X^2-1)/2\rangle_{st}}{2T}\right\} . \qquad (5.335)$$

The order parameter $q_{st} = \langle A\rangle_{st}$ corresponds to solutions of the self-consistency equation $m = R(m)$ defined by

$$P(x, m) = \frac{1}{Z'(m)}\exp\left\{\frac{3\kappa x^2 m}{2T}\right\} , \qquad (5.336)$$

$$Z'(m) = \int_{-1}^{1}\exp\left\{\frac{3\kappa x^2 m}{2T}\right\}dx \qquad (5.337)$$

and

$$R(m) = \int_{-1}^{1}\frac{3x^2-1}{2}P(x, m)\,dx , \qquad (5.338)$$

see also (5.110) and (5.111).

Stationary Solutions and Stability Analysis

Since we are dealing with the special case $d = 1$, the stability matrix given by (5.104) and (5.106) of the generic model (5.107) reduces to the stability coefficient of the model (5.83), which is $\tilde\lambda = Q + K_{A,st}d^2B/dm^2$, see Table 5.1 (third column). In our case we have $B(z) = -\kappa z^2/2$ and $Q = T$, which leads to $\tilde\lambda = T - \kappa K_{A,st}$. Consequently, the stability of the stationary distributions can be read off from the sign of $\tilde\lambda$ with

$$K_{A,st} = \frac{1}{4}\left[\langle(3X^2-1)^2\rangle_{st} - \langle(3X^2-1)\rangle_{st}^2\right] . \qquad (5.339)$$

We realize that $m = 0$ is a solution of the self-consistency equation $R(m) = m$ for all ratios κ/T. From $m = 0$ it follows that $P_{st} = 1/2$ and $K_{A,st} = 1/5$, which, in turn, gives us the stability coefficient $\tilde\lambda = T - \kappa/5$. Consequently, the isotropic phase is asymptotically stable for $T > \kappa/5$. Conversely, the isotropic phase is unstable for $T < \kappa/5$. In order to study the emergence

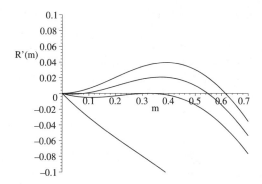

Fig. 5.34. Illustration of the self-consistency equation $R'(m) = R(m) - m = 0$ (see (5.340)). Zeros of $R'(m)$ correspond to stationary distributions; $\mathrm{d}R'(m)/\mathrm{d}m < 0$ (> 0) at $R'(m) = 0$ describe asymptotically stable (unstable) stationary distributions. *From bottom to top*: $T/\kappa = 0.3$, $T/\kappa \approx 0.223$ (saddle node bifurcation point; emergence of the nematic phase), $T/\kappa = 0.21$, and $T/\kappa = 1/5$ (the isotropic phase becomes unstable)

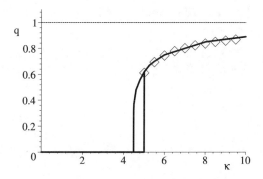

Fig. 5.35. Bifurcation diagram of the order parameter q of the Maier–Saupe model obtained from the self-consistency equation $R'(q) = R(q) - q = 0$ (*solid line*). Only stable stationary solutions are shown. The upper branch of the bifurcation diagram was also determined by solving numerically the Langevin equation (5.334) using the averaging method for self-consistent Langevin equations and periodic boundary conditions for $X \in [-1, 1]$ (*diamonds*). Parameters: $T = 1$ ($L = 50000$, $\Delta t = 10^{-4}$, w_n via Box-Muller, initial distribution $u(x) = [\delta(x - 0.8) + \delta(x + 0.8)]/2$, κ was gradually increased in steps of 0.5 starting with $\kappa = 5.0$)

of the nematic phase, we evaluate numerically the self-consistency equation $m = R(m)$ in the form of

$$R'(m) = R(m) - m = 0 \ . \tag{5.340}$$

It is clear from our discussion of the bounded potential model (5.83) that solutions $R'(m) = 0$ with $dR'/dm < 0$ $(dR'/dm < 0)$ correspond to asymptotically stable (unstable) distributions. From Fig. 5.34 we read off that there are three qualitatively different parameter regimes. For $T/\kappa > C$ with $C \approx 0.223$ the system is monostable and exhibits the isotropic phase only. For $1/5 < T/\kappa < C$ the system is bistable and exhibits a stable isotropic phase, an unstable nematic phase and a stable nematic phase. For $T/\kappa < 1/5$ the system is monostable and exhibits a stable nematic phase and an unstable isotropic phase. Finally, Fig. 5.35 shows the bifurcation diagram of the order parameter q. As indicated in Fig. 5.35 for $\kappa/T \in [1/C, 5]$ with $1/C \approx 4.488$ both the isotropic phase with $q = 0$ and the nematic phase with $q > 0$ are stable solutions of the mean field model. Note that the loop in Fig. 5.35 describes a hysteresis loop (see also [139]).

Let us illustrate the stability of stationary solutions by means of the free energy measure F given by (5.330) (see also [90] and [139, Sect. 10.3]). To this end, we consider a one-parametric family of probability densities

$$P_c(x) = \frac{1}{Z} \exp\left\{\kappa \frac{A(x)c}{T}\right\} \tag{5.341}$$

with $Z = \int_\Omega \exp\{\kappa A(x)c/T\}\,dx$ and substitute $P_c(x)$ into (5.330). Thus, we obtain $F(c) = F[P_c]$ with

$$F(c) = -\frac{\kappa}{2}\left[\langle A \rangle_{P_c(x)}\right]^2 + \kappa c \langle A \rangle_{P_c(x)} - T\ln\left[\int_\Omega \exp\left\{\frac{\kappa c}{T}A(x)\right\}\,dx\right]. \tag{5.342}$$

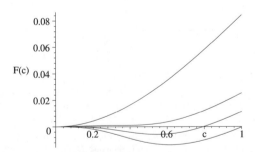

Fig. 5.36. Free energy $F(c)$ computed from (5.342). We have plotted the four cases shown in Fig. 5.34. *From top to bottom:* $T/\kappa = 0.3$ (isotropic phase; unique minimum at $c = 0$), $T/\kappa \approx 0.223$ (saddle node bifurcation point; emergence of the nematic phase), $T/\kappa = 0.21$ (bistable regime, two minima at $c = 0$ and $c > 0$), and $T/\kappa = 1/5 (= 0.2)$ (the isotropic phase becomes unstable, unique minima at $c > 0$). Note that the free energies are shifted such that we have $F(c = 0) = 0$

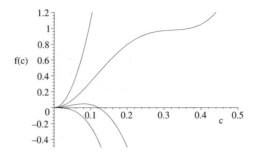

Fig. 5.37. A detail of Fig. 5.36 in order to illustrate the bistable regime; $f(c)$ corresponds to $10^3 \times F(c)$. We see that at the saddle node bifurcation point (*second graph from above*) a second minimum occurs. For $T/\kappa = 0.21$ (*third graph from above*) there is a minimum at $c = 0$ in addition to the minimum at $c > 0$ shown in Fig. 5.36. For $T/\kappa = 1/5$ (*bottom graph*) the minimum at $F(c)$ vanishes and becomes a maximum

Note that (5.341) and (5.342) are the counterparts to (5.255) and (5.256) discussed in the context of the KSS model. Figures 5.36 and 5.37 illustrate that with respect to the family of distributions P_c the free energy behaves as predicted by the analysis of the self-consistency equation. For $T/\kappa > C$ the free energy $F(c)$ has a unique minimum at $c = 0$ indicating that the system is monostable and exhibits the isotropic phase only. For $1/5 < T/\kappa < C$ the measure $F(c)$ has a minimum at $c = 0$ and a minimum at $c > 0$ related to an isotropic phase and a nematic phase. For $T/\kappa < 1/5$ the free energy $F(c)$ has a unique minimum at $c > 0$, which indicates that the system is monostable and exhibits only one stable phase, namely, the nematic phase.

Rotational Diffusion Models

Above, we have first derived a free energy measure for the director component n_z and, subsequently, we have written a free energy Fokker–Planck equation. More sophisticated models describing liquid crystals in terms of nonlinear Fokker–Planck equations can be obtained by approaching the problem in the opposite way. According to studies by Doi and Edwards [139] and Hess [289], the starting point is a multivariate nonlinear Fokker–Planck equation with a diffusion term that accounts for rotational diffusion. Subsequently, one may reduce the evolution equation to a univariate nonlinear Fokker–Planck equation by exploiting the assumed cylindrical symmetry of the director field [155, 157, 158, 304, 303, 378, 458, 618]. In line with this second approach, we may modify (5.332) and supplement it with a state-dependent mobility coefficient $M(x) > 0$ (see e.g. [156]).

5.5.4 Muscular Contraction

In line with a proposal by Shimizu and Yamada [528, 529], we will consider next a hydrodynamic muscle model and derive a phenomenological equation for muscular contraction from this model. Figure 5.38 depicts a muscle attached at the bones of two limbs. Before contraction, the muscle has a particular rest length. During contraction the muscle shortens and the relative position of the limbs change. As shown in Fig. 5.38, muscles consist of thin

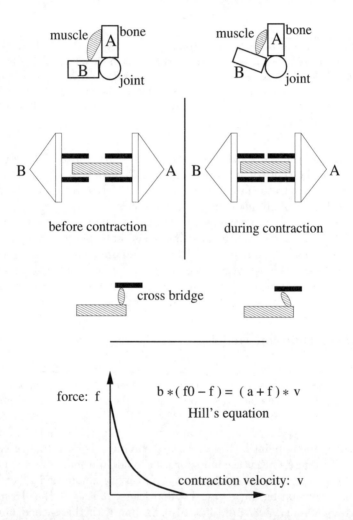

Fig. 5.38. Simple model of a muscle (*upper panel*) and Hill's equation (*lower panel*) describing the force-velocity relationship during muscular contraction

and thick filaments [595]. According to the sliding-filament theory, during contraction the thin filaments are pulled closer together, which leads to a shortening of the muscle. The pulling force is established by cross-bridges. Cross-bridges have an active and a passive state. In the active state, cross-bridges connect thin and thick filaments and pull the thin filaments closer together by exerting power strokes. In the passive state, they do not connect the filaments but prepare themselves for another power stroke.

As a result of this basic mechanism, muscles shorten with particular contraction velocities. The contraction velocity v depends primarily on the force f that is produced during the contraction. Hill's equation (see Fig. 5.38) is a phenomenological description of the force-velocity relationship. There are two points of the force-velocity curve that have a simple interpretation. First, if a muscle produces its maximum force, it does not shorten at all (i.e., we are dealing with an isometric contraction). Consequently, for $f = f_{\max}$ we have $v = 0$. Second, if muscles are unloaded, that is, if they do not need to produce a force, then they can contract as fast as possible. In this case, we have $f = 0$ and $v = v_{\max}$. In between these two extremes the velocity v monotonically decreases as a function of f. In order to derive Hill's equation from the sliding-filament theory, Shimizu and Yamada [528, 529] exploited an analogy between sliding filaments and fluids, see Fig. 5.39. Following Shimizu and Yamada, we consider N_0 particles that are dispersed in a fluid. Let v and u_i denote the velocity of the fluid and the velocities of the particles. If the velocities of the particles differ significantly from the velocity of the fluid, then there is an interaction between the fluid and the particles (hydrodynamic interaction [458]).

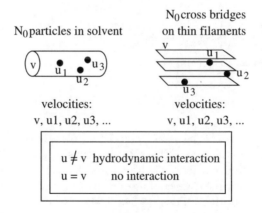

Fig. 5.39. Scheme of a hydrodynamic muscle model

In a muscle the thin filaments may play the role of the fluid and the cross-bridges may be regarded as the counterparts of the particles. Just as in the case of hydrodynamic interaction, we assume that there is an interaction between the cross-bridges and the thin filaments. It is reasonable to assume that active cross-bridges (i.e., those attached to the thin filaments) are accelerated if their velocities fall behind the filament velocity (which is the contraction velocity) and are de-accelerated if they move faster than the filaments move and the muscle shortens. Likewise, we can assume that the contraction velocity v increases (decreases) if the cross-bridges move on the average faster (slower) than v. Finally, according to the hydrodynamic muscle model, the percentage $r \in [0,1]$ of active muscles depends on the contraction velocity v. These basic relations are summarized in Table 5.4.

Table 5.4. Basic relations of the hydrodynamic muscle model

Variable	Change	Parameters
$u_i(t)$	$u_i \Uparrow$ if $u_i < v$ $u_i \Downarrow$ if $u_i > v$	u_0
$v(t)$	$v \Uparrow$ if $v <$ mean of u_i $v \Downarrow$ if $v >$ mean of u_i	f
$r(t)$	$r \Uparrow$ with k^+	k^+
$r(t)$	$r \Downarrow$ with $k^-(v) = c_1 + c_2 v$	$k^-(v)$

In detail, the evolution equations for u_i, v and r are given by

$$\frac{d}{dt}u_i(t) = \chi_i(t)\left[\lambda(u_0 - u_i) - \gamma u_i - \kappa(u_i - v) + \sqrt{Q}\Gamma_i(t)\right], \quad (5.343)$$

$$M\frac{d}{dt}v(t) = \sum_{i=1}^{N_0} \chi_i(t)[u_i - v] - f, \quad (5.344)$$

$$\frac{d}{dt}r(t) = k^+(1-r) - (b_1 + b_2 v)r = k^+ - k^-(v)r \quad (5.345)$$

with $\lambda, \gamma, \kappa, k^+, b_1, b_2, M > 0$ and $i = 1, \ldots, N_0$. Here, we put $\chi_k = 1$ if the cross-bridge k is active. Otherwise, we have $\chi_k = 0$. Since we are not interested in the information, which cross-bridge can be found in an active state at a particular time t, we confine ourselves to studying only the behavior of the active cross-bridges. Then, (5.343-5.345) become

$$\frac{d}{dt}u_i(t) = -\lambda(u_i - u_0) - \gamma u_i - \kappa(u_i - v) + \sqrt{Q}\Gamma_i(t), \quad (5.346)$$

$$M\frac{d}{dt}v(t) = r\sum_{i=1}^{N_0}(u_i - v) - f, \quad (5.347)$$

$$\frac{d}{dt} r(t) = k^+(1-r) - (b_1 + b_2 v) r \ . \tag{5.348}$$

Let us first evaluate the deterministic case: $Q = 0$. In this case, the stationary states are given by $u_{i,\text{st}} = (\lambda u_0 + \kappa v_{\text{st}})/(\lambda + \gamma + \kappa)$, $r_{\text{st}} \sum_i (u_{i,\text{st}} - v_{\text{st}}) = f$, and $r_{\text{st}} = k^+/(k^+ + b_1 + b_2 v_{\text{st}})$, which gives us

$$b[f_0 - f] = (a + \underbrace{f)v_{\text{st}}}_{k^-(v)} \ , \tag{5.349}$$

with $b = (k^+ + b_1)/b_2$, $f_0 = k^+ N_0 \lambda u_0 / [(k^+ + b_1)(\lambda + \gamma + \kappa)]$ and $a = k^+ N_0 (\lambda + \gamma) / [b_2 (\lambda + \gamma + \kappa)]$. Consequently, from the hydrodynamic muscle model we can derive Hill's empirical equation. As indicated in (5.349) the mixed term $f v_{\text{st}}$ results from the assumption that the decrease of the number of active cross-bridges depends on the contraction velocity.

In order to study time-dependent solutions of (5.346-5.348), we put again $Q = 0$ and assume that $v(t)$ and $r(t)$ evolve much faster than $u_i(t)$. Using adiabatic elimination [254, 255], we obtain $r(t)[\sum_i (u_i(t) - N_0 v(t)] = f' N_0$ and $r(t) = k^+/[k^+ + b_1 + b_2 v(t)]$, where $f' = f/N_0$ denotes the contraction force per available cross-bridge. Consequently, we can eliminate $r(t)$, express $v(t)$ in terms of $v(t) = [k^+ N_0^{-1} \sum_i u_i(t) - f'(k^+ + b_1)]/[f' b_2 + k^+]$, and transform (5.346) into

$$\frac{d}{dt} u_i(t) = -\lambda_{\text{eff}} (u_i - u_{i,\text{st}}) - \kappa_{\text{eff}} \left(u_i - \frac{1}{N_0} \sum_{k=1}^{N_0} u_k \right) \ , \tag{5.350}$$

with $\lambda_{\text{eff}} = \lambda + \gamma + \kappa [1 - k^+/(k^+ + b_2 f')] > 0$ and $\kappa_{\text{eff}} = \kappa \, k^+/(k^+ + b_2 f')$. Taking the fluctuating force $\sqrt{Q} \Gamma_i$ into account, for $X_i = u_i - u_{i,\text{st}}$ we obtain the multivariate Langevin equation

$$\frac{d}{dt} X_i(t) = -\lambda_{\text{eff}} X_i - \kappa_{\text{eff}} \left(X_i - \frac{1}{N_0} \sum_{k=1}^{N_0} X_k \right) + \sqrt{Q} \Gamma_i \tag{5.351}$$

that in turn reduces to

$$\frac{d}{dt} X(t) = -\lambda_{\text{eff}} X - \kappa_{\text{eff}} (X - \langle X \rangle) + \sqrt{Q} \Gamma \ , \tag{5.352}$$

if we use the limit $N_0 \to \infty$, assume that X_k are statistically-independent random variables at an initial time t_0, and assume that X_k are distributed the same way at time t_0 (see Sect. 3.8). That is, the hydrodynamic muscle model can be evaluated by means of the Shimizu–Yamada model described in Sect. 5.2.1. Since we have $\lambda_{\text{eff}} > 0$, we obtain in any case a stationary Gaussian distribution for the cross-bridge velocities u_i.

In closing these considerations, we would like to point out that in order to derive (5.352) we have applied adiabatic elimination to the deterministic evolution equation. The method of adiabatic elimination, however, can also be applied to stochastic evolution equations [197, 254, 511].

5.5.5 Network Models for Group Behavior

We briefly address here two network models that can describe the emergence of group behavior in populations of humans and animals and that are used in Sects. 5.2.2 and 7.3.2 to illustrate the basins of attraction of stationary distributions and the emergence of reentrant phase transitions.

Network Interpretation of the Dynamical Takatsuji Model

We discuss first the network model shown in Fig. 5.40 related to the dynamical Takatsuji model discussed in Sect. 5.2.2. Let $X_i \in \Omega = \mathbb{R}$ describe the behavior of the ith member of a population with N_0 members. Let us assume that $X_i = 0$ describes the stationary behavior of an isolated subject of the population and that the relaxation to this stationary behavior in the presence of a fluctuating force can be described by $dX_i/dt = -a_1 X_i + \sqrt{Q}\Gamma_i(t)$ with $a_1 > 0$. Let us further assume that the member i receives input from all other fellow-members at particular interfaces depicted in Fig. 5.40 as circles. In the context of the dynamical Takatsuji model, we consider a lin-

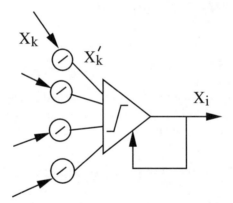

Fig. 5.40. Interpretation of the dynamical Takatsuji model in terms of a network model for interacting population members

ear conversion of the inputs X_k at the interfaces such that we obtain the converted inputs $X'_k = a_3 X_k$ with $a_3 > 0$. The converted inputs X'_k are averaged and the result is coupled to the internal dynamics using a nonlinear conversion function $S(z)$. Therefore, the stochastic differential equation $dX_i/dt = -a_1 X_i + \sqrt{Q}\Gamma_i(t)$ becomes

$$\frac{d}{dt}X_i(t) = -a_1 X_i + S\left(\frac{1}{N_0}\sum_{k=1}^{N_0} X'_k\right) + \sqrt{Q}\Gamma_i(t) , \qquad (5.353)$$

with $i = 1, \ldots, N_0$. A typical nonlinear conversion function $S(z)$ that involves a saturation domain (i.e., we have $S(z \to \pm\infty) = \pm S_{\text{sat}}$) is given by $S(z) = a_2 \tanh(z)$ with $a_2 > 0$. Using this choice together with $X'_k = a_3 X_k$, we obtain the network model

$$\frac{\mathrm{d}}{\mathrm{d}t} X_i(t) = -a_1 X_i + a_2 \tanh\left(\frac{a_3}{N_0} \sum_{k=1}^{N_0} X_k\right) + \sqrt{Q}\Gamma_i(t) \ . \tag{5.354}$$

As show in Sect. 3.8, in the limiting case $N_0 \to \infty$ we may equate the ensemble of members $\{X_1, \ldots, X_{N_0}\}$ with the statistical ensemble $\{X^1, \ldots, X^N\}$ of an arbitrary member and study the many-body system described by (5.354) in terms of the self-consistent Langevin equation

$$\frac{\mathrm{d}}{\mathrm{d}t} X(t) = -a_1 X(t) + a_2 \tanh[a_3 \langle X(t) \rangle] + \sqrt{Q}\Gamma(t) \ . \tag{5.355}$$

This Langevin equation corresponds to the dynamical Takatsuji model (5.163) for $a_1 = \gamma + c$ and $a_2 = a_3 = \sqrt{c}$ (see also (5.179)). As discussed in Sect. 5.2.2 the dynamical Takatsuji model can describe a phase transition between a phase of indecisive members (e.g. undecided voters) with $q = N_0^{-1} \sum_{k=1}^{N_0} X_k = 0$ and a phase of determined members (e.g. decided voters) with $q \neq 0$. The latter phase occurs with two types: $q > 0$ and $q < 0$.

Network Model for Reentrant Phase Transitions of Group Behavior

The previous network model exhibits a linear conversion at the interfaces between population members and a nonlinear conversion of the overall input. Let us discuss now the opposite situation as shown in Fig. 5.41. According to

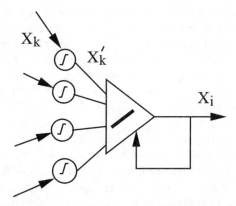

Fig. 5.41. A network model for reentrant phase transitions that occur in populations with interacting members

Fig. 5.41, we assume that the behavior of an isolated subject is determined by the stochastic evolution equation $\mathrm{d}X_i/\mathrm{d}t = f(X_i, \Gamma_i)$, where $\Gamma_i(t)$ denotes a Langevin forces. We assume that the dynamics involves multiplicative noise: $f(X, \Gamma) = h(x) + g(x; Q)\Gamma$, where Q denotes the overall amplitude of the noise. Such multiplicative noise typically reflects the impact of a fluctuating pumping force (e.g., a supply with food or information that involves some kind of erraticness) [298]. We assume a nonlinear conversion of the inputs from other population members such that the converted inputs X'_k are given by $X'_k = S(X_k)$ (see circles in Fig. 5.41). We further assume that the converted inputs are averaged and that the result couples linearly to the internal dynamics. Thus, we obtain

$$\frac{\mathrm{d}}{\mathrm{d}t}X_i(t) = h(X_i) + \frac{\kappa}{N_0}\sum_{k=1}^{N_0} X'_k + g(X_i; Q)\Gamma_i(t) , \qquad (5.356)$$

with $i = 1, \ldots, N_0$. Apart from the tanh-function, we may use the arctan-function as a nonlinear conversion function that involves a saturation domain. For $S(z) = \arctan(\sqrt{b}z)/\sqrt{b}$ we obtain

$$\frac{\mathrm{d}}{\mathrm{d}t}X_i(t) = h(X_i) + \frac{\kappa}{\sqrt{b}N_0}\sum_{k=1}^{N_0} \arctan(\sqrt{b}X_k) + g(X_i; Q)\Gamma_i(t) . \qquad (5.357)$$

Using mean field theory, in the limit $N_0 \to \infty$ we may study the network model defined by (5.357) in terms of a single subject described by $X(t)$, where the behavior $X(t)$ is determined by the Langevin equation

$$\frac{\mathrm{d}}{\mathrm{d}t}X(t) = h(X) + \frac{\kappa}{\sqrt{b}}\left\langle \arctan(\sqrt{b}X)\right\rangle + g(X; Q)\Gamma(t) \qquad (5.358)$$

and the corresponding nonlinear Fokker-Planck equation. In Sect. 7.3.2 we will study the properties of this network model in detail. In order to analyze the network model conveniently, we will make however two minor modifications. First, we will replace $h(x)$ by $h(x) - \kappa S(x)$. Second, we will use a slightly modified arctan-function that exhibits however the same basic properties as the arctan-function. Both modifications are not considered as key issues of the network model. The replacement $h(x) \to h(x) - \kappa S(x)$ might be interpreted as a feedback of the output X_i to the input interfaces such that in (5.356) we have $\kappa N_0^{-1}\sum_{k=1}^{N_0}[X'_k - X'_i]$ instead of $\kappa N_0^{-1}\sum_{k=1}^{N_0} X'_k$. Taking these two modifications into account, as we will see in Sect. 7.3.2 the network model (5.357) describes two subsequent phase transitions. The first transition is a phase transition from a phase of indecisive members with $q = N_0^{-1}\sum_{k=1}^{N_0} X_k = 0$ to a phase of determined members with $q \neq 0$, when the noise amplitude Q becomes larger than a critical value Q_1. That is, the noise results in a process of collective decision making or in the emergence of a collective behavior. The second transition is a phase transition from the phase of determined members with $q \neq 0$ back to the phase of indecisive members with $q = 0$ and occurs when Q becomes larger than a critical value Q_2 (with $Q_2 > Q_1$). That is, too much noise destroys the previously obtained collective decision or behavior.

5.5.6 Multistable Perception-Action Systems

Perception-action systems of humans and animals that exhibit multiple stable coordination patterns under the same circumstances are said to be multistable. This kind of multistability has been found in animal locomotion [299, 427], rhythmic single limb movements (see below), and rhythmic multilimb movements [15, 16, 51, 264, 334, 412, 461, 476, 487]. Multistability has also been observed in discrete movements such as hitting a ball with a table tennis bat [544] and in coordinated oscillatory movements that involve more than one limb and more than one oscillation frequency (e.g., polyrhythmic drumming movements) [110, 133, 265, 472, 473, 474, 475, 549]. Multistable perception-action systems can be described by dynamical systems involving potential functions with multiple minima. Each minimum of such a potential function (or potential) corresponds to a stable coordination pattern. This is illustrated in Fig. 5.42 for a bistable system. Accordingly, the double-well potential $V(x)$ can be regarded as an energy measure or a measure of effort, where x represents a suitably chosen state variable. Stable perception-action systems occupy states of minimal effort or energy, that is, they occupy a minimum of the potential V [49, 259, 335, 594]. If a perception-action system is initially located outside a potential minimum, then it performs an overdamped energy-decreasing motion and finally converges to a minimum. Consequently, in the deterministic case, we find a bistable perception-action system in either one of the two minima depicted in Fig. 5.42. In order to take fluctuations into account, we may follow the conventional approach and regard the multistable perception-action system as a zero-dimensional unit. In this case, we describe the system in terms of a linear Fokker–Planck equation with additive noise involving the potential $V(x)$ [108, 111, 197, 214, 461, 512]. The stationary state of the perception-action system is then described by a Boltzmann distribution of $V(x)$, see Fig. 5.42. This approach predicts a unique solution and, therefore, fails to account for the phenomenon of multistability. Therefore, we may regard multistable perception-action systems as spatially extended many-body systems composed of interacting components (e.g., neural oscillators) [189]. In this case, we can describe multistable perception-action systems by means of nonlinear Fokker–Planck equations involving potentials $V(x)$ and stochastic feedback in terms of mean field forces arising from the interactions among the components. Such models may exhibit for the same control parameters multiple stationary probability densities as depicted in Fig. 5.42 and as demonstrated in detail, for example, for the Desai–Zwanzig model. Consequently, spatially extended models exhibiting stochastic feedback can describe multistable perception-action system in the presence of noise.

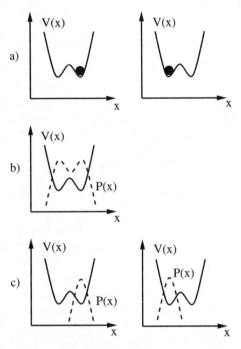

Fig. 5.42. Examples of dynamical models involving a double-well potential $V(x)$: a) deterministic model, b) stochastic zero-dimensional model, c) stochastic spatially extended model. The black balls describe the stationary states of the model a); $P(x)$ corresponds to the stationary probability densities of the models b) and c) [189]

Paced Rhythmic Single Limb Movements

Let us briefly discuss an example of a multistable action-perception system. In a seminal work, Kelso and his colleagues have proposed to study paced rhythmic finger movements from the perspective of dynamical systems theory [335, 337]. This multistable perception-action system is of particular importance for human movement sciences because, on the one hand, it shows a variety of interesting phenomena and, on the other hand, it is easily accessible for experimental observation and manipulation. In several experiments subjects were asked to move single limbs such as an index finger or a forearm in a rhythmic fashion along with the beat of a metronome or an oscillating visual pacing signal [51, 109, 110, 337, 335, 338, 259, 631]. They were requested to move on the beat (on-beat condition) or off the beat (off-beat condition).

In this context, off the beat means that they tapped between two consecutive beats of the metronome or moved into the opposite direction of the visual pacing signal. During the experiments the frequency of the beats was gradually increased. At low pacing frequencies both off-beat and on-beat movements could be stably performed, whereas at high pacing frequencies only the on-beat condition could be performed in a stable fashion. If the subjects were asked to move off-beat, then often involuntary switches occurred from the required off-beat pattern to the on-beat pattern when the pacing frequency exceeded critical (subject-dependent) values. This involuntary transition from the off-beat to the on-beat coordination pattern has been interpreted as a nonequilibrium phase transition [49, 215, 216, 262, 312, 313, 314, 315, 335]. From this observation, it has been concluded that paced rhythmic limb movements involve a perception-action system that is bistable in a particular parameter regime and monostable in another.

In order to account for the multistability of paced limb movements and to model the transitions between different patterns of paced limb movements, nonlinear Fokker–Planck equations have been used in a series of studies [171, 173, 198, 199, 201]. In these studies it was assumed that the primary part of the perception-action systems involved in the performance of paced limb movements can be modeled in terms of phase oscillator populations, see Sect. 5.5.2. These phase oscillator ensembles, in turn, have been described by means of the mean field HKB model discussed in Sect. 5.3.3. In the context of the mean field HKB model, the potential force $h(x) = -\mathrm{d}V_{\mathrm{HKB}}(x)/\mathrm{d}x$ describes a driving force that results from the pacing signal and acts on each oscillator phase X_k, whereas the sine coupling-force describes the couplings between the neural oscillators. The phase transitions described by the mean field HKB model and illustrated in Figs. 5.28d and 5.29d are assumed to reflect the phase transitions observed in the experiments.

6 Entropy Fokker–Planck Equations

In Sect. 2.6 free energy Fokker–Planck equations with linear energy functionals U have been referred to as entropy Fokker–Planck equations. In line with (4.49), we define entropy Fokker–Planck equations by

$$\frac{\partial}{\partial t} P(x,t;u) = \frac{\partial}{\partial x} M(P) P \frac{\partial}{\partial x} \frac{\delta F}{\delta P} \quad (6.1)$$

in the univariate case and

$$\frac{\partial}{\partial t} P(\mathbf{x},t;u) = -\nabla \cdot \{\mathbf{I} P\} + \nabla \cdot \mathcal{M}(P) P \cdot \nabla \frac{\delta F}{\delta P} \quad (6.2)$$

in the multivariate case, where F is given by

$$F[P] = U_{\mathrm{L}}[P] - QS[P] \quad (6.3)$$

and U_{L} describes a linear functional. That is, U_{L} is given by $U_{\mathrm{L}}[P] = \int_\Omega U_0(x) P(x)\, dx$ in the univariate case and $U_{\mathrm{L}}[P] = \int_\Omega U_0(\mathbf{x}) P(\mathbf{x})\, d^M x$ in the multivariate case. Using entropy and information measures of the form (4.5), entropy Fokker–Planck equations can be written as

$$\frac{\partial}{\partial t} P(x,t;u) = \frac{\partial}{\partial x} M(P) P \left[\frac{dU_0(x)}{dx} - Q \frac{\partial}{\partial x} \frac{\delta S}{\delta P} \right]$$

$$= \frac{\partial}{\partial x} M(P) \left[\frac{dU_0(x)}{dx} P + Q \left. \frac{dB}{dz} \right|_{z_0} \frac{\partial}{\partial x} \hat{L}_s(P) \right] \quad (6.4)$$

and

$$\frac{\partial}{\partial t} P(\mathbf{x},t;u) = -\nabla \cdot \{[\mathbf{I} - \mathcal{M} \cdot \nabla U_0(\mathbf{x})] P\} + Q \nabla \cdot \mathcal{M} P \cdot \nabla \frac{\delta S}{\delta P}$$

$$= -\nabla \cdot \{[\mathbf{I} - \mathcal{M} \cdot \nabla U_0(\mathbf{x})] P\} + Q \left. \frac{dB}{dz} \right|_{z_0} \nabla \cdot \mathcal{M} \cdot \nabla \hat{L}_s(P) , \quad (6.5)$$

with z_0 given by $z_0 = \int_\Omega s(P)\, d^M x$ and $\hat{L}_s$ defined by (4.6) [190, 191].

Canonical-Dissipative Systems

Nonlinear Fokker–Planck equations of canonical-dissipative systems of type B defined by (4.101) can be regarded as entropy Fokker–Planck equations because they involve linear internal energy functionals (see (4.99)).

Stationary Solutions

We confine ourselves to considering entropy Fokker–Planck equations for which the drift and diffusion coefficients and the boundary conditions are defined such that stationary solutions can be derived from the free energy principle (4.17). Consequently, in the univariate case, stationary solutions can be obtained by substituting the functional $U = U_L[P] = \int_\Omega U_0(x) P(x)\, dx$ into (4.20), which leads to

$$P_{\text{st}}(x) = \left[\frac{ds}{dz}\right]^{-1} \left(\frac{U_0(x) - \mu}{Q\frac{dB(z_{0,\text{st}})}{dz}}\right), \tag{6.6}$$

with $z_{0,\text{st}} = \int_\Omega s(P_{\text{st}})\, dx$. In the multivariate case, from (4.20) stationary solutions of the entropy Fokker–Planck equation (6.5) can be found as

$$P_{\text{st}}(\mathbf{x}) = \left[\frac{ds}{dz}\right]^{-1} \left(\frac{U_0(\mathbf{x}) - \mu}{Q\frac{dB(z_{0,\text{st}})}{dz}}\right), \tag{6.7}$$

with $z_{0,\text{st}} = \int_\Omega s(P_{\text{st}})\, d^M x$.

Approach to Stationary Solutions

If S is concave, then the free energy measure (6.3) is bounded from below (see case A in Table 4.2). Consequently, the H-theorem of free energy Fokker–Planck equations (see Sect. 4.3) can be used to prove that transient solutions of entropy Fokker–Planck equations converge to stationary ones in the long time limit:

$$\lim_{t \to \infty} \frac{\partial}{\partial t} P = 0. \tag{6.8}$$

Stability of Stationary Solutions

As stated in Sect. 4.1, stationary distributions derived from the free energy principle (4.17) correspond to critical points of the free energy F. That is, the first variation of F vanishes with respect to the space of probability densities. In symbols: we have $\delta F[P_{\text{st}}](\epsilon) = 0$ for perturbations ϵ that yield probability

densities again. Due to the linearity of U_L the second variation of F is given by
$$\delta^2 F = -Q \delta^2 S \ . \tag{6.9}$$
In general, for concave entropy and information measures and linear internal energy functionals, stationary distributions defined by the free energy principle correspond to free energy minima, see (4.133). In particular, for entropy and information measures that can be cast into the form $S = B(\int_\Omega s(P) \, d^M x)$ and satisfy the conditions listed in (6.81), it follows that the inequalities $\delta^2 S[P] < 0$ in general and $\delta^2 S[P_\text{st}] < 0$ in particular hold (see Sect. 6.4.1), which implies that we have $\delta^2 F[P] > 0$ in general and $\delta^2 F[P_\text{st}] > 0$ in particular. That is, we are dealing again with stationary distributions that correspond to free energy minima. If we replace (5.20) by means of (6.9), then linear stability analysis and stability analysis by means of Lyapunov's direct method can be carried out as discussed in Sect. 5.1 and one finds that the stationary solutions (6.6) are asymptotically stable because they correspond to free energy minima.

$\delta^2 S$ as a Lyapunov Functional

For entropy and information measures with $\delta^2 S[P_\text{st}] < 0$ the norm (5.27) reads
$$\|\epsilon\| = \sqrt{\delta^2 F[P_\text{st}](\epsilon)} = \sqrt{-Q \, \delta^2 S[P_\text{st}](\epsilon)} \ . \tag{6.10}$$
As argued above, a linear stability analysis similar to the one discussed in Sect. 5.1 can be carried out and thus the inequality $d \|\epsilon\| / dt < 0$ for $\epsilon \neq 0$ and the limiting case $\|\epsilon\| = 0$ for $t \to 0$ can be obtained (see (5.28)). From the inequality $d \|\epsilon\| / dt < 0$ it follows that perturbations $\epsilon(\mathbf{x}, t)$ of stationary solutions of entropy Fokker–Planck equations satisfy
$$\frac{d}{dt} \delta^2 S[P_\text{st}](\epsilon) > 0 \ . \tag{6.11}$$
In particular, for S given by (6.81) we have $\delta^2 S < 0$ for $\epsilon \neq 0$ and $\delta^2 S = 0 \Leftrightarrow \epsilon = 0$ (see (6.82) and (6.83)), which gives us $\delta^2 S \to 0$ in the limit $t \to \infty$. In this case $\delta^2 S$ corresponds to a Lyapunov functional. Note that this interpretation of $\delta^2 S$ has also been given in nonequilibrium thermodynamics for the special case in which S corresponds to the Boltzmann entropy [228, 353, 449].

6.1 Existence and Uniqueness of Stationary Solutions*

As was demonstrated in Sect. 4.7.1, stationary solutions of free energy Fokker–Planck equations can be expressed by means of Boltzmann distributions and distortion functionals. In what follows, we will use distortion functionals in order to show that stationary solutions of entropy Fokker–Planck equations exist and are unique [192]. To this end, we consider measures S given by

$$S[P] = \int_\Omega s(P) \, d^M x \, ,$$

$$\lim_{z \downarrow 0} \frac{ds}{dz} \to \infty \, , \qquad (6.12)$$

and $d^2 s/ds^2 < 0$. The latter condition implies the concavity of the functional S (see Sect. 6.4.1). Substituting (6.12) and $U_{\text{NL}} = 0$ into (4.112), we obtain the relations

$$G(z) = \exp\left\{-\frac{ds}{dz} - 1\right\}, \quad \lim_{z \downarrow 0} G(z) = 0, \quad \frac{dG}{dz} > 0 \, . \qquad (6.13)$$

From these relations, we see that the mapping $G(z)$ is invertible. We distinguish between two cases: $G(z)$ is bounded from above and $G(z)$ is not bounded from above. For the sake of convenience, we formally define the maximum value of G by

$$G_{\max} = \begin{cases} G_c : \lim_{z \to \infty} G(z) = G_c < \infty \\ \infty : \lim_{z \to \infty} G(z) \to \infty \end{cases} . \qquad (6.14)$$

Consequently, if Ω^{-1} denotes the domain of definition of G^{-1}, we have $\Omega^{-1} = (0, G_{\max})$. From (4.116) it follows that

$$P_{\text{st}}(\mathbf{x}) = G^{-1}\left(\frac{1}{Z} W(\mathbf{x})\right) \qquad (6.15)$$

holds. That is, P_{st} can be computed from the Boltzmann distribution $W(\mathbf{x}) = \exp\{-U_0(\mathbf{x})/Q\}/Z_B$ with $Z_B = \int_\Omega \exp\{-U_0(\mathbf{x})/Q\} \, d^M x$ by means of a mapping that distorts the shape of W. Note that in (6.15) the constant Z is confined to the interval $Z > W_{\max}/G_{\max}$, where $W_{\max}$ denotes the maximum of $W(\mathbf{x})$. The next objective is to determine Z. From (6.13) and (6.14) it follows that

$$G^{-1}(z \downarrow 0) = 0, \quad \frac{dG^{-1}}{dz} > 0, \quad \lim_{z \to G_{\max}} G^{-1}(z) \to \infty \, . \qquad (6.16)$$

We define the normalization integral

$$n(Z) = \int_\Omega G^{-1}\left(\frac{1}{Z} W(\mathbf{x})\right) d^M x \qquad (6.17)$$

on $Z \in (W_{\max}/G_{\max}, \infty)$. Due to the relations given in (6.16), the integral $n(z)$ satisfies

6.1 Existence and Uniqueness of Stationary Solutions*

$$\lim_{Z \to \infty} n(Z) = 0 \,, \tag{6.18}$$

$$\frac{\mathrm{d}n(Z)}{\mathrm{d}Z} = -\frac{1}{Z^2} \underbrace{\int_\Omega W(\mathbf{x}) \left.\frac{\mathrm{d}G^{-1}}{\mathrm{d}z}\right|_{z=Z^{-1}W(\mathbf{x})} \mathrm{d}^M x}_{>0} < 0 \,. \tag{6.19}$$

Consequently, $n(Z)$ is a strictly monotonically decreasing function that vanishes in the limit $Z \to \infty$. The question arises whether or not there is a Z such that $n(Z)$ is larger than or equal to unity. To answer this question, we treat the cases $G_{\max} = \infty$ and $G_{\max} = G_c$ separately.

For $G_{\max} = \infty$ we consider a subset $\omega \subset \Omega$ for which $W(\mathbf{x})$ is larger than or equal to a constant C (i.e., $\forall \mathbf{x} \in \omega : W(\mathbf{x}) \geq C$). Then, one obtains the inequality

$$n(Z) \geq \int_\omega G^{-1}\left(\frac{C}{Z}\right) \mathrm{d}^M x = G^{-1}\left(\frac{C}{Z}\right) \underbrace{\int_\omega \mathrm{d}^M x}_{>0} \,. \tag{6.20}$$

Since $G_{\max} = \infty$ the integral $n(Z)$ is defined on $Z \in (0, \infty)$ and we can consider the limiting case $Z \to 0$. It is clear that in this case we obtain

$$\lim_{Z \downarrow 0} n_S(Z) \geq \int_\omega \mathrm{d}^M x \lim_{Z \downarrow 0} G^{-1}\left(\frac{C}{Z}\right) \to \infty \,. \tag{6.21}$$

In sum, for $G_{\max} = \infty$ the integral $n(Z)$ is not bounded from above.

In the case $G_{\max} = G_c$ we exploit the fact that $n(Z)$ is a monotonically decreasing function. Therefore, the maximum value of $n(Z)$ is given for the smallest possible value of Z, which is $Z = W_{\max}/G_{\max}$:

$$\max_{Z \in (W_{\max}/G_c, \infty)} \{n(Z)\} = n\left(Z \downarrow \frac{W_{\max}}{G_c}\right) \,. \tag{6.22}$$

Equations (6.21) and (6.22) can be summarized as follows:

$$n_{\max} = \begin{cases} \lim_{Z \downarrow W_{\max}/G_c} \int_\Omega G^{-1}\left(\frac{1}{Z}W(\mathbf{x})\right) \mathrm{d}^M x & : \lim_{z \to \infty} G(z) = G_c < \infty \\ \infty & : \lim_{z \to \infty} G(z) \to \infty \end{cases} \,. \tag{6.23}$$

If this maximum value equals one or is larger than one, then there is a unique $Z^* \in (W_{\max}/G_{\max}, \infty)$ that yields $n(Z^*) = 1$ and we can write (6.15) as

$$P_{\mathrm{st}}(\mathbf{x}) = G^{-1}\left(\frac{1}{n^{-1}(1)} W(\mathbf{x})\right) \,. \tag{6.24}$$

In sum, for distortion functionals G without an upper boundary the existence of $Z^* = n^{-1}(1)$ is guaranteed. In contrast, if G is bounded from above by G_c, we need to make an additional effort to verify that P_{st} can be written in the form of (6.24): we need to check if there is at least one $Z' > W_{\max}/G_{\max}$ that gives us a value $n(Z') \geq 1$ (which means that we do not necessarily need to determine $n_{\max}$).

6.2 Entropy Increase and Anomalous Diffusion

For $U_L = 0$ the entropy Fokker–Planck equation (6.1) becomes

$$\frac{\partial}{\partial t}P(x,t;u) = -Q\frac{\partial}{\partial x}M(P)P\frac{\partial}{\partial x}\frac{\delta S}{\delta P} , \qquad (6.25)$$

which can be expressed as

$$\frac{\partial}{\partial t}P(x,t;u) = Q\left.\frac{\mathrm{d}B}{\mathrm{d}z}\right|_{z_0}\frac{\partial}{\partial x}M(P)\frac{\partial}{\partial x}\hat{L}_s(P) , \qquad (6.26)$$

for $S[P]$ given by (4.5). In the linear case (i.e., for $S = {}^{\mathrm{B}}S = -\int P\ln P\,\mathrm{d}x$ and $M(P) = 1$) the diffusion term is linear with respect to P and we obtain the Fokker–Planck equation of univariate Wiener processes,

$$\frac{\partial}{\partial t}P(x,t;u) = Q\frac{\partial^2}{\partial x^2}P , \qquad (6.27)$$

see Sect. 3.9.4. Let us study two fundamental properties of the solutions of (6.25).

6.2.1 Entropy Increase

It appeals to our intuition that diffusion processes result in an increase of the disorder of the systems in which they take place. The increase of disorder may be measured in terms of an increase of the entropy of the systems or an increase of the information that is necessary to describe the systems. Indeed, the Boltzmann entropy ${}^{\mathrm{B}}S$ increases for solutions of the linear diffusion equation (6.27). From (6.27) we obtain by means of partial integration

$$\frac{\mathrm{d}}{\mathrm{d}t}{}^{\mathrm{B}}S[P] = Q\int_\Omega \frac{1}{P}\left[\frac{\partial}{\partial x}P\right]^2 \mathrm{d}x \geq 0 . \qquad (6.28)$$

In order to derive (6.28) we have assumed that the surface term that arises due to the integration by parts vanishes. From (6.28) it is clear that ${}^{\mathrm{B}}S$ is a monotonically increasing function with respect to t. For the more general case given by (6.25) a similar result holds [163]. In order to verify this, we differentiate S with respect to t, eliminate $\partial P/\partial t$ by means of (6.25), integrate by parts and assume that the surface term thus obtained vanishes. Then, we obtain

$$\frac{\mathrm{d}}{\mathrm{d}t}S[P] = Q\int_\Omega P\left[\frac{\partial}{\partial x}\frac{\delta S}{\delta P}\right]^2 \mathrm{d}x \geq 0 . \qquad (6.29)$$

That is, in line with linear nonequilibrium thermodynamics, the entropy production $\mathrm{d}_i S = \mathrm{d}S$ is semi-positive (see also Sect. 4.5.1).

6.2.2 Anomalous Diffusion

We consider now stochastic processes subjected to natural boundary conditions and put $\Omega = \mathbb{R}$. As we have shown in Sect. 3.9.4, the variance K of solutions of the linear diffusion equation (6.27) increases linearly with t: $K(t) \propto t$. Diffusion that exhibits such a linear relationship is referred to as normal diffusion. If the graph $K(t)$ differs from a linear function, we deal with anomalous diffusion (see Sect. 1.4.5). In recognition of this classification, the question arises as to what kind of diffusion processes (normal or anomalous diffusion) are described by nonlinear diffusion equations of the form (6.25). Our aim now is to answer this question without knowing the explicit form of the measure S that occurs in (6.25). We will show that the crucial issue is not the explicit form of S but the extensivity or nonextensivity of S.

Composability, Extensivity, and Nonextensivity

Following [170], we will examine this relationship for entropy and information measures that satisfy

$$S[PP'] = S[P] + S[P'] + G(S[P], S[P']) . \tag{6.30}$$

Here, P and P' are distributions defined on Ω and Ω' and the product PP' is regarded as a distribution P'' defined on $\Omega \times \Omega'$. Equation (6.30) says that if we combine two statistically-independent systems described by P and P' and the joint distribution $P'' = PP'$, then the entropy and information of the combined system can be computed from the entropy and information measures of the single systems. We say that measures S that satisfy (6.30) are composable (see also Sect. 6.4.1).

The function $G(u,v)$ denotes an arbitrary function of the subsystem entropies $S[P]$ and $S[P']$. Note that the commutativity of the multiplication (i.e., $PP' = P'P$) implies the symmetry of G (i.e., we have $G(u,v) = G(v,u)$). If G depends on u (or v), then S corresponds to a nonextensive measure, see Sect. 6.4.1. In contrast, for $G = \text{const.}$, we can introduce the shifted functional $S' = S + G$, which satisfies the relation of extensive entropies [261, 625]:

$$S'[PP'] = S'[P] + S'[P'] . \tag{6.31}$$

Therefore, for $G = \text{constant}$ and in particular for $G = 0$ we consider S as an extensive measure.

Evolution Equation for K

From (6.26) we can derive the evolution equation of the first and second moments:

$$\frac{d}{dt}\langle X\rangle = 0,\qquad(6.32)$$

$$\frac{d}{dt}\langle X^2\rangle = 2Q\left.\frac{dB}{dz}\right|_{z_0}\int_\Omega \hat{L}_s\,dx.\qquad(6.33)$$

Substituting the definition (4.6) of $\hat{L}_s$ into (6.33), for $K(t) = \langle X^2\rangle - \langle X\rangle^2$ we then find

$$\frac{d}{dt}K(t) = 2Q\underbrace{\left.\frac{dB}{dz}\right|_{z_0}\int_\Omega \left[s(P(x,t;u)) - P(x,t;u)\left.\frac{ds(z)}{dz}\right|_{z=P(x,t;u)}\right]dx}_{Y[P]}.$$

(6.34)

Consequently, if the functional $Y[P]$ is constant with respect to time we have normal diffusion (i.e., $K(t) \propto t$). Otherwise, we find anomalous diffusion (i.e., $K(t) \not\propto t$). We now derive Y from the assumed nonextensivity/extensivity property (6.30). To this end, let $P(x)$ be defined on the real line and $P'(y)$ be defined on the interval $[0,a]$ with $a > 0$. Furthermore, let $P'(y)$ denote the uniform distribution on $[0,a]$, that is, $P'(y) = 1/a$. Then, from (6.30) and $S = B[\int_\Omega s(P)\,dx]$ it follows that

$$B\left[a\int_\Omega s\left(\frac{P(x)}{a}\right)dx\right] = S[P] + B\left[as\left(\frac{1}{a}\right)\right] + G\left(S[P], B\left[as\left(\frac{1}{a}\right)\right]\right).$$

(6.35)

Next, we differentiate (6.35) with respect to a and put $a = 1$. Thus, we obtain

$$Y[R_1] = C + \left.\frac{\partial G(S[R_1], v)}{\partial v}\right|_{v=C},\qquad(6.36)$$

with

$$C = \left(s(1) - \left.\frac{ds(z)}{dz}\right|_{z=1}\right)\left.\frac{dB(z)}{dz}\right|_{z=s(1)}.\qquad(6.37)$$

Consequently, the variance $K(t)$ satisfies the evolution equation

$$\frac{d}{dt}K(t) = 2QC + 2Q\left.\frac{\partial G(S[P], v)}{\partial v}\right|_{v=C},\qquad(6.38)$$

with $P(x,t;u)$ determined by (6.26). Since C does not depend on the time variable t, we deal with normal diffusion if the second term on the right-hand side of (6.38) does not depend on time. Otherwise, we find anomalous diffusion. Therefore, at issue is how $S[P]$ and $\partial G(S[P],v)/\partial v|_{v=C}$ evolve as functions of time. From (6.29) it follows that $S[P]$ satisfies the relation

$$\frac{d}{dt}S[P] = Q\left(\left.\frac{dB}{dz}\right|_{z_0}\right)^2\int_\Omega P(x,t;u)\left(\left.\frac{\partial}{\partial x}\frac{ds(z)}{dz}\right|_{z=P(x,t;u)}\right)^2 dx \geq 0.$$

(6.39)

6.2 Entropy Increase and Anomalous Diffusion

We require now that the entropy functional S is sensitive to unlikely events. More precisely, we require the existence of an interval $(0, l]$ of finite length l (i.e., $l > 0$) such that the derivative $\mathrm{d}s(z)/\mathrm{d}z$ maps all elements of $(0, l]$ to nonvanishing function values (i.e., $\exists l : \forall z \in (0, l] : \mathrm{d}s(z)/\mathrm{d}z \neq 0$). We find that for entropies of this kind the expression

$$\left.\frac{\partial}{\partial x} \frac{\mathrm{d}s(z)}{\mathrm{d}z}\right|_{z=P(x)} \tag{6.40}$$

can have at most a finite number of zeros for $x \in I_k$, where the intervals I_k are defined by $I_k = [x_l^k, x_r^k]$ with $x_l^k < x_r^k$, I_k mutually disjoint, and $0 < P(z) \leq l$ for all $z \in I_k$. That is, $\partial\mathrm{d}s/\partial x \mathrm{d}z \neq 0$ for $z = P(x)$ and $x \in I_k$. Since we restrict our considerations to continuous probability densities which decay to zero in a continuous fashion (i.e., we assume $P \in C(\Omega)$), the existence of such intervals is guaranteed. From (6.39) we can then read off that

$$\frac{\mathrm{d}}{\mathrm{d}t} S[P] \geq Q \left(\left.\frac{\mathrm{d}B}{\mathrm{d}z}\right|_{z_0}\right)^2 \sum_k \int_{I_k} P(x, t; u) \left(\left.\frac{\partial}{\partial x} \frac{\mathrm{d}s(z)}{\mathrm{d}z}\right|_{z=P(x;t;u)}\right)^2 \mathrm{d}x > 0 . \tag{6.41}$$

In sum, entropies which are sensitive to unlikely events are strictly monotonically increasing functions for solutions of (6.26). We are now in the position to examine the evolution of $\partial G(S[P], v)/\partial v|_{v=C}$. Differentiating this expression with respect to t yields

$$\left.\frac{\mathrm{d}}{\mathrm{d}t} \frac{\partial G(S[P], v)}{\partial v}\right|_{v=C} = \left.\frac{\partial^2 G(u, v)}{\partial u \partial v}\right|_{v=C, u=S[P]} \frac{\mathrm{d}S[P]}{\mathrm{d}t} . \tag{6.42}$$

Using this result and differentiating (6.38) with respect to t, we obtain

$$\frac{\mathrm{d}^2}{\mathrm{d}t^2} K(t) = 2Q \underbrace{\frac{\mathrm{d}S[P]}{\mathrm{d}t}}_{> 0} \left.\frac{\partial^2 G(u, v)}{\partial u \partial v}\right|_{v=C, u=S[P]} . \tag{6.43}$$

Therefore, we have normal diffusion (i.e., we have $\mathrm{d}^2 K(t)/\mathrm{d}t^2 = 0$) if and only if

$$\left.\frac{\partial^2 G(u, v)}{\partial u \partial v}\right|_{v=C, u=S[P]} = 0 . \tag{6.44}$$

Since the extensivity and nonextensivity, respectively, can be defined by means of the derivative $\partial G(u, v)/\partial u|_{u=S[P]}$ (see above), we can now draw our final conclusions (which are only valid within the framework of nonlinear Fokker–Planck equations and composable entropies, of course).

- Extensive systems show normal diffusion. Nonextensive systems with $\partial G(u, v)/\partial v|_{v=C} = 0$ show normal diffusion. Normal diffusion only occurs in nonextensive systems with $\partial G(u, v)/\partial v|_{v=C} = 0$ and extensive systems.

- Nonextensive systems with $\partial G(u,v)/\partial v|_{v=C} \neq 0$ show anomalous diffusion. Anomalous diffusion only occurs in nonextensive systems satisfying $\partial G(u,v)/\partial v|_{v=C} \neq 0$.

Roughly speaking, nonextensivity implies anomalous diffusion and extensivity implies normal diffusion. But there is an exception to this rule: in nonextensive systems for which the first order partial derivative of $G(u,v)$ with respect to u vanishes for $u = C$ normal diffusion can also occur.

Let us conclude our considerations about the relationship between anomalous diffusion and nonextensivity with two remarks. First, in the absence of an outer function B (i.e., for $B(z) = z$) we usually find $s(1) = 0$. The reason for this is that for $B(z) = z$ many entropy functionals S are derived from entropies $S = \sum_{i=1}^{N} s(p_i)$ for discrete sets of probabilities p_i. In these cases, by convention, one often defines $s(z)$ in such a way that it vanishes in the case of a certain event such that $S = 0$ for $p_i = 1$ and $\forall k \neq i : p_k = 0$. For $B(z) = z$ and $s(1) = 0$ the constant C reads $C = -\,\mathrm{d}s(z)/\,\mathrm{d}z|_{z=1}$. In this special case, the evolution of the variance is described by

$$\frac{\mathrm{d}}{\mathrm{d}t} K(t) = -2Q \left.\frac{\mathrm{d}s(z)}{\mathrm{d}z}\right|_{z=1} + 2Q \left.\frac{\partial G(S[P], v)}{\partial v}\right|_{v=-\,\mathrm{d}s(z)/\mathrm{d}z|_{z=1}}. \quad (6.45)$$

In particular, for the Boltzmann entropy $^{B}S = -\int P \ln P \,\mathrm{d}x$ we obtain $\mathrm{d}s(z)/\,\mathrm{d}z = -1 - \ln z$ and the relation $\mathrm{d}K(t)/\,\mathrm{d}t = 2Q \Rightarrow K(t) = 2Q(t-t_0)$ derived earlier for univariate Wiener processes.

Second, the derivation presented here does not hold for diffusion processes with infinite variances such as Lévy distributions [429, 533]. In this context, several limitations of the approach to anomalous diffusion by means of nonlinear Fokker–Planck equations and the calculation of variances have been uncovered [191, 192]. In order to circumvent such problems, one may follow Tsallis and Bukman who introduced the notion of pseudo anomalous diffusion for Lévy distributions and used time-dependent normalization constants instead of variances to address the issue of anomalous diffusion [591].

6.3 Drift- and Diffusion Forms

Univariate Case

There are two special cases of (6.4) in which either the drift term or the diffusion term become linear with respect to P [190]. For $M(P) = 1$ the nonlinear Fokker–Planck equation (6.4) can be written as

$$\frac{\partial}{\partial t} P(x,t;u) = -\frac{\partial}{\partial x} h(x) P + Q \left.\frac{\mathrm{d}B}{\mathrm{d}z}\right|_{z_0} \frac{\partial^2}{\partial x^2} \hat{L}_s(P), \quad (6.46)$$

where $h(x)$ is defined by $h(x) = -\mathrm{d}U_0(x)/\mathrm{d}x$. We refer to (6.46) as the diffusion form of the entropy Fokker–Planck equation (6.4) because it is the diffusion term that makes (6.46) nonlinear with respect to P. Now, let us chose $M(P)$ such that the diffusion term of (6.4) becomes linear. That is, we require that the equivalence

$$M(P) \left.\frac{\mathrm{d}B}{\mathrm{d}z}\right|_{z_0} \frac{\partial}{\partial x} \hat{L}_s(P) = \frac{\partial}{\partial x} P \qquad (6.47)$$

holds. Solving (6.47) for M gives us

$$M(P) = \left[\left.\frac{\mathrm{d}B}{\mathrm{d}z}\right|_{z_0} \frac{\mathrm{d}\hat{L}_s(P)}{\mathrm{d}P} \right]^{-1}. \qquad (6.48)$$

Since $\mathrm{d}\hat{L}_s(P)/\mathrm{d}P = -P\mathrm{d}^2 s(P)/\mathrm{d}P^2$ holds, we write (6.48) as

$$M(P) = - \left| \left.\frac{\mathrm{d}B}{\mathrm{d}z}\right|_{z_0} P \frac{\mathrm{d}^2 s(P)}{\mathrm{d}P^2} \right|^{-1} \qquad (6.49)$$

with $z_0 = \int_\Omega s(P)\,\mathrm{d}x$. In particular, for concave entropy and information measures of the form (6.81), the mobility coefficient (6.49) is nonsingular and well-defined because the concavity property leads to $\mathrm{d}B(z_0)/\mathrm{d}z\,\mathrm{d}^2 s(P)/\mathrm{d}P^2 < 0$. For mobility coefficients $M(P)$ described by (6.48) and (6.49) the entropy Fokker–Planck equation (6.4) can be expressed as

$$\frac{\partial}{\partial t} P(x,t;u) = -\frac{\partial}{\partial x} h(x) \frac{P}{\mathrm{d}B(\int_\Omega s(P)\,\mathrm{d}x)/\mathrm{d}z\,\mathrm{d}\hat{L}_s(P)/\mathrm{d}P} + Q\frac{\partial^2}{\partial x^2} P \quad (6.50)$$

and

$$\frac{\partial}{\partial t} P(x,t;u) = \frac{\partial}{\partial x} \frac{h(x)}{|\mathrm{d}B(\int_\Omega s(P)\,\mathrm{d}x)/\mathrm{d}z\,\mathrm{d}^2 s(P)/\mathrm{d}P^2|} + Q\frac{\partial^2}{\partial x^2} P. \quad (6.51)$$

We refer to these equations as drift forms of the entropy Fokker–Planck equation (6.4).

Multivariate Case

The multivariate case can be treated in a similar vein [193]. For $\mathcal{M}(\mathbf{x},t,P) = \mathcal{M}(\mathbf{x},t)$ and $\mathbf{I}(\mathbf{x},t,P) = \mathbf{I}(\mathbf{x},t)$ the evolution equation (6.5) reads

$$\frac{\partial}{\partial t} P(\mathbf{x},t;u) = -\nabla \cdot \left\{ [\mathbf{I}(\mathbf{x},t) - \mathcal{M}(\mathbf{x},t) \cdot \nabla U_0(\mathbf{x})]P \right\}$$
$$+ Q \left.\frac{\mathrm{d}B}{\mathrm{d}z}\right|_{z_0} \nabla \cdot \mathcal{M}(\mathbf{x},t) \cdot \nabla \hat{L}_s(P) \qquad (6.52)$$

and corresponds to the diffusion form of the entropy Fokker–Planck equation (6.5). In contrast, if we use for $\mathcal{M}$ an expression similar to (6.48), namely,

$$\mathcal{M}(\mathbf{x},t,P) = \left[\left.\frac{\mathrm{d}B}{\mathrm{d}z}\right|_{\int_\Omega s(P)\,\mathrm{d}^M x} \frac{\mathrm{d}\hat{L}_s(P)}{\mathrm{d}P}\right]^{-1} \mathcal{M}'(\mathbf{x},t)\,, \qquad (6.53)$$

then (6.5) becomes

$$\frac{\partial}{\partial t}P(\mathbf{x},t;u) = -\nabla\cdot\left\{\left[\mathbf{I}(\mathbf{x},t,P) - \frac{\mathcal{M}'(\mathbf{x},t)\cdot\nabla U_0(\mathbf{x})}{\mathrm{d}B(z_0)/\mathrm{d}z\,\mathrm{d}\hat{L}_s(P)/\mathrm{d}P}\right]P\right\}$$
$$+Q\nabla\cdot\mathcal{M}'(\mathbf{x},t)\cdot\nabla P\,, \qquad (6.54)$$

with $z_0 = \int_\Omega s(P)\,\mathrm{d}^M x$. Equation (6.54) is the drift form of (6.5).

6.4 Entropy and Information Measures

In this section, we will consider entropy functionals and information measures

$$S = S[P] \qquad (6.55)$$

that describe mappings from continuous probability densities P to the set of real numbers. We will discuss some elementary properties that are usually satisfied by measures of this kind.

6.4.1 Properties*

Continuous and Differentiable

Let us define a measure S that acts on continuous probability densities P by

$$S[P] = B\left[\int_\Omega s[P(\mathbf{x})]\,\mathrm{d}^M x\right]\,,$$
$$s(z) \in C^0([0,\infty)),\ s(z) \in C^\infty((0,\infty))\,,$$
$$B(z) \in C^\infty(\Omega_{\mathrm{cod}})\,. \qquad (6.56)$$

The function $B(z)$ is defined on the range of values of the functional $\int s[P(\mathbf{x})]\,\mathrm{d}^M x$ (i.e., the co-domain) denoted by Ω_{cod}:

$$\Omega_{\mathrm{cod}} = \{y|\ y = \int_\Omega s[P(\mathbf{x})]\,\mathrm{d}^M x\,,\ P(\mathbf{x}) \in C^0(\Omega),\ P > 0\}\,. \qquad (6.57)$$

The kernel $s(z)$ is continuous on $z \in [0,\infty)$ and infinitely differentiable for $z > 0$. Furthermore, we require that the scale function B is infinitely differentiable on Ω_{cod}. On account of these properties, we assume that the variations δS and $\delta^2 S$ and the functional derivatives $\delta S/\delta P$, and $\delta^2 S/\delta P^2$ exist for positive-definite probability densities (i.e., for P with $P > 0$). The first and second variations of S are defined by means of the Taylor expansion

6.4 Entropy and Information Measures

$$S(P+\epsilon) = S(P) + \delta S[P](\epsilon) + \frac{1}{2}\delta^2 S[P](\epsilon^2) + O(\epsilon^3) \;, \tag{6.58}$$

which gives us

$$\delta S[P](\epsilon) = \left.\frac{\mathrm{d}B(z)}{\mathrm{d}z}\right|_{\int_\Omega s(P)\,\mathrm{d}^M x} \int_\Omega \left.\frac{\mathrm{d}s(z)}{\mathrm{d}z}\right|_{P(\mathbf{x})} \epsilon(\mathbf{x})\,\mathrm{d}^M x \tag{6.59}$$

and

$$\delta^2 S[P](\epsilon^2) = \left.\frac{\mathrm{d}^2 B(z)}{\mathrm{d}z^2}\right|_{\int_\Omega s(P)\,\mathrm{d}^M x} \left(\int_\Omega \left.\frac{\mathrm{d}s(z)}{\mathrm{d}z}\right|_{P(\mathbf{x})} \epsilon(\mathbf{x})\,\mathrm{d}^M x\right)^2$$
$$+ \left.\frac{\mathrm{d}B(z)}{\mathrm{d}z}\right|_{\int_\Omega s(P)\,\mathrm{d}^M x} \int_\Omega \left.\frac{\mathrm{d}^2 s(z)}{\mathrm{d}z^2}\right|_{P(\mathbf{x})} [\epsilon(\mathbf{x})]^2\,\mathrm{d}^M x \;, \tag{6.60}$$

where $\epsilon(\mathbf{x})$ describes a small perturbation with $\int_\Omega \epsilon(\mathbf{x})\,\mathrm{d}^M x = 0$. The first and second functional derivatives are given by

$$\delta S[P](\epsilon) = \int_\Omega \frac{\delta S}{\delta P(\mathbf{x})}\epsilon(\mathbf{x})\,\mathrm{d}^M x \;, \tag{6.61}$$

$$\delta^2 S[P](\epsilon^2) = \int_\Omega \frac{\delta^2 S}{\delta P(\mathbf{x})\delta P(\mathbf{y})}\epsilon(\mathbf{x})\epsilon(\mathbf{y})\,\mathrm{d}^M x\,\mathrm{d}^M y \tag{6.62}$$

(note that in general the second variation $\delta^2 S$ involves two different arguments ϵ_1 and ϵ_2, see e.g. Sect. 5.1.8; in our context, however, we have $\epsilon_1 = \epsilon_2 = \epsilon$, which is the perturbation of P). From (6.58-6.60) it follows that

$$\frac{\delta S}{\delta P(\mathbf{x})} = \left.\frac{\mathrm{d}B(z)}{\mathrm{d}z}\right|_{\int_\Omega s(P)\,\mathrm{d}^M x} \left.\frac{\mathrm{d}s(z)}{\mathrm{d}z}\right|_{P(\mathbf{x})} \tag{6.63}$$

and

$$\frac{\delta^2 S}{\delta P(\mathbf{x})\delta P(\mathbf{y})} = \left.\frac{\mathrm{d}^2 B(z)}{\mathrm{d}z^2}\right|_{\int_\Omega s(P)\,\mathrm{d}^M x} \left.\frac{\mathrm{d}s(z)}{\mathrm{d}z}\right|_{P(\mathbf{x})} \left.\frac{\mathrm{d}s(z)}{\mathrm{d}z}\right|_{P(\mathbf{y})}$$
$$+ \left.\frac{\mathrm{d}B(z)}{\mathrm{d}z}\right|_{\int_\Omega s(P)\,\mathrm{d}^M x} \left.\frac{\mathrm{d}^2 s(z)}{\mathrm{d}z^2}\right|_{P(\mathbf{x})} \delta(\mathbf{x}-\mathbf{y}) \;. \tag{6.64}$$

Expansibility

The measure S is said to be expansible if states of the phase space Ω that are certainly not occupied by a system do not contribute to the entropy or information measured by S. Let $\mathrm{supp}_\Omega\{P\}$ denote the support of P, that is, the subspace of Ω for which P is positive: $\text{supp}_\Omega\{P\} = \{\mathbf{x} \in \Omega | P(\mathbf{x}) > 0\}$. For an expansible measure S we need to consider only the support of P:

$$S[P;\Omega] = S[P;\Omega' = \text{supp}_\Omega\{P\}] \ . \tag{6.65}$$

For example, an expansible measure is given by (6.56) with $s(0) = 0$ and $B(0) = 0$.

Concavity

A measure S is referred to as strictly concave if it satisfies for $\lambda \in [0,1]$ the inequality

$$S[\lambda P + (1-\lambda)P'] \geq \lambda S[P] + (1-\lambda)S[P'] \ , \tag{6.66}$$

where the equal sign only holds in three cases: $P = P'$, $\lambda = 0$, and $\lambda = 1$. Consequently, we have

$$P \neq P' \wedge \lambda \in (0,1): \ S[\lambda P + (1-\lambda)P'] > \lambda S[P] + (1-\lambda)S[P'] \ . \tag{6.67}$$

In order to elucidate the meaning of the concavity inequality (6.66), we define a straight line $S_{\text{line}}(\lambda)$ that connects the points $(P', S[P'])$ and $(P, S[P])$,

$$S_{\text{line}}(\lambda) = S[P'] + \lambda \left(S[P] - S[P']\right) \ , \tag{6.68}$$

such that $S_{\text{line}}(0) = S[P']$ and $S_{\text{line}}(1) = S[P]$, see Fig. 6.1. The concavity inequality (6.66) states that $S_{\text{line}}(\lambda)$ is smaller than S

$$S[P' + \lambda(P - P')] > S_{\text{line}}(\lambda) \ , \tag{6.69}$$

for $\lambda \in (0,1)$ and $P \neq P'$, see also Fig. 6.1.

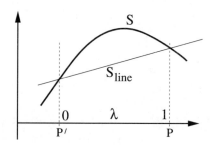

Fig. 6.1. Illustration of the concavity inequality (6.69)

From (6.66) the concavity inequality

$$S[P] \leq S_{\text{lin}}[P,P'] = S[P'] + \int_\Omega \frac{\delta S}{\delta P'(\mathbf{x})} [P(\mathbf{x}) - P'(\mathbf{x})] \, \mathrm{d}^M x \tag{6.70}$$

6.4 Entropy and Information Measures

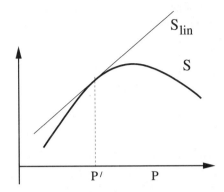

Fig. 6.2. Illustration of the concavity inequality (6.70)

can be derived (see below), where the equal sign holds for $P = P'$ only. $S_{\text{lin}}[P, P']$ denotes the linearization of S at the distribution P'. Note that the concavity inequality (6.70) can be cast into the concise form

$$S[P] \leq S[P'] + \delta S[P'](P - P') \,. \tag{6.71}$$

We realize that if S is strictly concave, then the linearization of S is larger than S except for the "point" P' at which the linearization is carried out, see Fig. 6.2. It can be immediately verified that (6.70) and (6.71) hold for $P = P'$. By means of two different methods, we will show next how to obtain (6.70) for $P \neq P'$:

First Method to Derive (6.70) [168]

Let us write (6.66) as

$$\frac{S[P' + \lambda(P - P')] - S[P']}{\lambda} > S[P] - S[P'] \,, \tag{6.72}$$

with $\lambda \in (0, 1)$. To evaluate (6.72), we use the concept of the directional functional derivative. Note that the directional derivative of an N-dimensional function $f(\mathbf{x})$ in the direction of the vector $\mathbf{h}$ is given by

$$\lim_{\lambda \to 0} \frac{f(\mathbf{x} + \lambda \mathbf{h}) - f(\mathbf{x})}{\lambda} = \mathbf{h} \cdot \nabla f(\mathbf{x}) \,. \tag{6.73}$$

In a similar vein, the directional functional derivative of an operator $A[f] = \int a(f(\mathbf{x})) \, \mathrm{d}^M x$ in the direction of the function $h(\mathbf{x})$ is given by

$$\lim_{\lambda \to 0} \frac{A[f + \lambda h] - A[f]}{\lambda} = \int \frac{\delta A}{\delta f(\mathbf{x})} h(\mathbf{x}) \, \mathrm{d}^M x = \delta A[f](h) \,. \tag{6.74}$$

That is, for the integral operator A the functional derivative in the direction h is given by the first variation $\delta A[f](h)$ of A at f for a perturbation h. Then, we can perform the limiting case

$$\lim_{\lambda \downarrow 0} \frac{S[P' + \lambda(P - P')] - S[P']}{\lambda} = \int \frac{\delta S}{\delta P'(\mathbf{x})} [P(\mathbf{x}) - P'(\mathbf{x})] \, \mathrm{d}^M x \qquad (6.75)$$

(hint: use the identifications $A \to S$, $f \to P'$, and $h \to P - P'$). Substituting (6.72) in the left-hand side of (6.75), we obtain (6.70).

Second Method to Derive (6.70)

Our departure point is the functional

$$R(\lambda; P', P) = \lambda \left(S[P'] - S[P]\right) - S[P'] + S[P' + \lambda(P - P')] , \qquad (6.76)$$

for $\lambda \in [0, 1]$. Note that R satisfies $R(0) = 0$. Since S involves a differentiable scale function B and an integral kernel s that is differentiable for $\lambda \in (0, 1)$ and $P' > 0$, the functional R is differentiable with respect to λ for $\lambda \in (0, 1)$ and $P' > 0$. That is, $\lim_{\lambda \to 0+} \mathrm{d}R(\lambda)/\mathrm{d}\lambda$ exists. Furthermore, we can read off from (6.67) that the inequality $R(\lambda) > 0$ for $\lambda > 0$ and $P \neq P'$ holds. From $R(0) = 0$ and $R(\lambda) > 0$ for $\lambda > 0$ and $P \neq P'$ it follows that $\lim_{\lambda \to 0+} \mathrm{d}R(\lambda)/\mathrm{d}\lambda > 0$ for $P \neq P'$. Differentiation of (6.76) with respect to λ in the limiting case $\lambda \to 0+$ for $P \neq P'$ yields

$$0 < \lim_{\lambda \to 0+} \frac{\mathrm{d}R(\lambda)}{\mathrm{d}\lambda} = S[P'] - S[P] + \lim_{\lambda \to 0+} \frac{S[P' + \lambda(P - P')] - S[P']}{\lambda}$$
$$= S[P'] - S[P] + \delta S[P'](P - P') , \qquad (6.77)$$

which is equivalent to (6.70).

Implications of Concavity

As discussed in Sect. 4.7.2 every concave entropy measure induces a semi-positive definite Kullback measure. Furthermore, as discussed in Sect. 4.2 for concave entropy measures with $\delta^2 S < 0$ stationary distributions always correspond to maximum entropy distributions.

Extensivity, Nonextensivity, and Composability

Let us consider two statistically independent continuous random variables $\mathbf{X} \in \Omega$ and $\mathbf{Y} \in \Omega'$ that are distributed according to the probability densities $P(\mathbf{x}) = \langle \delta(\mathbf{x} - \mathbf{X}) \rangle$ and $P'(\mathbf{y}) = \langle \delta(\mathbf{y} - \mathbf{Y}) \rangle$. A composed system described by the vector $(\mathbf{X}, \mathbf{Y})$ is given by the joint probability density $P(\mathbf{x}, \mathbf{y}) = P(\mathbf{x})P'(\mathbf{y})$. If the entropy or information of the composed system equals the sum of the entropy or information of the single systems like

6.4 Entropy and Information Measures

$$S[PP'; \Omega = \Omega_x \Omega_y] = S[P; \Omega_x] + S[P', \Omega_y] \,, \tag{6.78}$$

then the measure S is said to be extensive [625] (see also Sect. 6.2.2). If we cannot add up the single-system entropy values in order to compute the entropy of the two-variable system, we say, S is nonextensive:

$$S[PP'; \Omega = \Omega_x \Omega_y] \neq S[P; \Omega_x] + S[P', \Omega_y] \,. \tag{6.79}$$

If $S(PP')$ can be computed from the measures $S(P)$ and $S(P')$ of the single systems, then we refer to S as a composable measure [587]. In this case, $S[P(\mathbf{x}, \mathbf{y})]$ can be computed from $S[P]$ and $S[P']$ and a symmetric function $\tilde{G}(u, v) = \tilde{G}(v, u)$ like

$$S[PP'; \Omega = \Omega_x \Omega_y] = \tilde{G}(S[P; \Omega_x], S[P', \Omega_y]) \,. \tag{6.80}$$

On a Fundamental Entropy and Information Measure

Let us consider now an entropy and information measure of the form (6.56) that is described by

$$\begin{aligned}
& S[P] = B\left[\int_\Omega s[P(\mathbf{x})] \, \mathrm{d}^M x \right] \,, \\
& s(z) \in C^0([0, \infty)), \ s(z) \in C^\infty((0, \infty)) \,, \\
& s(0) = 0, \ z \in [0, 1] : s(z) \geq 0 \,, \\
& B(z) \in C^\infty(\Omega_{\text{cod}}), \ z \in \Omega_{\text{cod}} : \frac{\mathrm{d}^2 B(z)}{\mathrm{d} z^2} < 0 \,, \\
& z > 0, \ z' \in \Omega_{\text{cod}} : \\
& \left\{ \frac{\mathrm{d}^2 s(z)}{\mathrm{d} z^2} < 0 \ \wedge \ \frac{\mathrm{d} B(z')}{\mathrm{d} z'} > 0 \right\} \vee \left\{ \frac{\mathrm{d}^2 s(z)}{\mathrm{d} z^2} > 0 \ \wedge \ \frac{\mathrm{d} B(z')}{\mathrm{d} z'} < 0 \right\} \,.
\end{aligned} \tag{6.81}$$

Since S satisfies (6.56), S is continuous and differentiable and we assume that the first and second variations of S and the corresponding functional derivatives exist, see (6.58-6.64). The functional S is expansible because of $s(0) = 0$. In addition, S is related to entropy and information measures for probability distributions of discrete random variables defined by $S = B(\sum_{i=1}^N s(p_i))$ with $\sum_{i=1}^N p_i = 1$ and $p_i \in [0, 1]$ such that the condition $s(p_i) \geq 0$ holds. Note that S satisfies the concavity inequality (6.70) which can be shown by a detailed calculation (see below). Furthermore, from (6.60) and (6.81) it follows that

$$\delta^2 S[P](\epsilon) < 0 \,, \tag{6.82}$$

for $\epsilon \neq 0$ and

$$\delta^2 S[P](\epsilon) = 0 \Leftrightarrow \epsilon = 0 \,. \tag{6.83}$$

Concavity

We prove the concavity inequality for the discrete version of the measure S. For discrete random variables that are defined on N states labeled by $i = 1, \ldots, N$ the measure (6.81) reads

$$S = B\left[\sum_{i=1}^{N} s(p_i)\right],$$

$$s(z) \in C^0([0,1]), \ s(z) \in C^\infty((0,1)), \ s(0) = 0, \ z \in [0,1] : s(z) \geq 0,$$

$$B(z) \in C^\infty(\Omega_{\text{cod}}), \ z \in \Omega_{\text{cod}} : \frac{\mathrm{d}^2 B(z)}{\mathrm{d}z^2} < 0,$$

$$z \in (0,1], \ z' \in \Omega_{\text{cod}} :$$

$$\left\{\frac{\mathrm{d}^2 s(z)}{\mathrm{d}z^2} < 0 \wedge \frac{\mathrm{d}B(z')}{\mathrm{d}z'} > 0\right\} \vee \left\{\frac{\mathrm{d}^2 s(z)}{\mathrm{d}z^2} > 0 \wedge \frac{\mathrm{d}B(z')}{\mathrm{d}z'} < 0\right\}.$$

(6.84)

Here, Ω_{cod} denotes the range of values of the expression $\sum_{i=1}^{N} s(p_i)$ (i.e., the co-domain of $\sum_i \ldots$):

$$\Omega_{\text{cod}} = \{y \mid y = \sum_{i=1}^{N} s(p_i) \text{ for all distributions } \{p_k\}_{k=1}^{N} \}. \quad (6.85)$$

Note also that we have $\sum_{i=1}^{N} p_i = 1$ and $p_i \in [0,1]$. Let us show now that S satisfies the concavity inequality (6.70) in terms of

$$S(p_1, \ldots, p_N) \leq S(p'_1, \ldots, p'_N) + \sum_{i=1}^{N} (p_i - p'_i) \frac{\partial S(p'_1, \ldots, p'_N)}{\partial p'_i}, \quad (6.86)$$

where the equal sign holds for $p_i = p'_i$ only.

Our first objective is to exploit the fact that for a function $f(x)$ with $\mathrm{d}^2 f/\mathrm{d}x^2 < 0 \ (> 0)$, we have the concavity inequality $f(x) < f(x_0) + (x - x_0)\mathrm{d}f(x_0)/\mathrm{d}x$ (convexity inequality $f(x) > f(x_0) + (x - x_0)\mathrm{d}f(x_0)/\mathrm{d}x$) [79]. As far as the entropy kernel $s(z)$ is concerned, we have

$$z \neq z_0, \ z_0 \in (0,1], \ z \in [0,1] : s(z) - s(z_0) - (z - z_0)\frac{\mathrm{d}s(z_0)}{\mathrm{d}z} \gtrless 0. \quad (6.87)$$

For $B(z)$ with $\mathrm{d}^2 B(z)/\mathrm{d}z^2 < 0$ we get

$$z \neq z_0, \ z_0, z \in \mathbb{R} : B(z) < B(z_0) + (z - z_0)\frac{\mathrm{d}B(z_0)}{\mathrm{d}z}. \quad (6.88)$$

If we substitute $z = p_i$ and $z_0 = p'_i$, from (6.87) it follows that

6.4 Entropy and Information Measures

$$\{p_i\} \neq \{p_i'\},\ p_i' > 0: \sum_{i=1}^{N} s(p_i) \gtrless \sum_{i=1}^{N} s(p_i') + \sum_{i=1}^{N}(p_i - p_i')\frac{\mathrm{d}s(p_i')}{\mathrm{d}p_i'}\ . \quad (6.89)$$

For linear scaling functions $B(z)$, that is, for $B(z) = z$, $S(p_1, \ldots, p_N) = \sum_i s(p_i)$, and $\mathrm{d}^2 s(z)/\mathrm{d}z^2 < 0$, the inequality (6.89) becomes the concavity inequality (6.86) because in this case the lower inequality sign holds. Next, let us consider nonlinear scaling functions $B(z)$. Let us first consider the case $\mathrm{d}^2 s(z)/\mathrm{d}z^2 < 0$ and $\mathrm{d}B(z)/\mathrm{d}z > 0$. From $\mathrm{d}^2 s(z)/\mathrm{d}z^2 < 0$ and (6.89) it follows for $z = \sum_i s(p_i)$ and $z' = \sum_{i=1}^{N} s(p_i') + \sum_{i=1}^{N}(p_i - p_i')\mathrm{d}s(p_i')/\mathrm{d}p_i'$ that $z > z'$. From $\mathrm{d}B(z)/\mathrm{d}z > 0$ it then follows that $B(z) < B(z')$, that is, we have

$$\{p_i\} \neq \{p_i'\},\ p_i' > 0: B\left[\sum_{i=1}^{N} s(p_i)\right] < B\left[\sum_{i=1}^{N} s(p_i') + \sum_{i=1}^{N}(p_i - p_i')\frac{\mathrm{d}s(p_i')}{\mathrm{d}p_i'}\right]\ . \quad (6.90)$$

Next, let us first consider the case $\mathrm{d}^2 s(z)/\mathrm{d}z^2 > 0$ and $\mathrm{d}B(z)/\mathrm{d}z < 0$. Now, from $\mathrm{d}^2 s(z)/\mathrm{d}z^2 > 0$ and (6.89) it follows for $z = \sum_i s(p_i)$ and $z' = \sum_{i=1}^{N} s(p_i') + \sum_{i=1}^{N}(p_i - p_i')\mathrm{d}s(p_i')/\mathrm{d}p_i'$ that $z < z'$. From $\mathrm{d}B(z)/\mathrm{d}z > 0$ it then follows that $B(z) < B(z')$, that is, we obtain (6.90) again. The right-hand side of (6.90) can be evaluated by means of the inequality (6.88). We put

$$z = \sum_{i=1}^{N} s(p_i') + \sum_{i=1}^{N}(p_i - p_i')\frac{\mathrm{d}s(p_i')}{\mathrm{d}p_i'}\ ,$$

$$z_0 = \sum_{i=1}^{N} s(p_i')\ ,$$

$$\Rightarrow B\left[\sum_{i=1}^{N} s(p_i') + \sum_{i=1}^{N}(p_i - p_i')\frac{\mathrm{d}s(p_i')}{\mathrm{d}p_i'}\right]$$

$$< B\left[\sum_{i=1}^{N} s(p_i')\right] + \frac{\mathrm{d}B(z_0)}{\mathrm{d}z}\sum_{i=1}^{N}(p_i - p_i')\frac{\mathrm{d}s(p_i')}{\mathrm{d}p_i'}\ . \quad (6.91)$$

This result in combination with (6.90) leads to

$$\{p_i\} \neq \{p_i'\},\ p_i' > 0:$$

$$B\left[\sum_{i=1}^{N} s(p_i)\right] < B\left[\sum_{i=1}^{N} s(p_i')\right] + \left.\frac{\mathrm{d}B(z)}{\mathrm{d}z}\right|_{z_0 = \sum_{i=1}^{N} s(p_i')}\sum_{i=1}^{N}(p_i - p_i')\frac{\mathrm{d}s(p_i')}{\mathrm{d}p_i'}\ , \quad (6.92)$$

which is equivalent to the concavity inequality (6.86).

6.4.2 Examples

In this section we will present some entropy and information measures S that have found applications in various disciplines. Our emphasis will be on the properties of these measures such as extensivity and concavity. We will examine some of these properties by means of probability distributions of discrete random variables. In these cases, probability distributions will be described by means of a set of probabilities p_i with $i = 1, \ldots, N$, $p_i \in [0, 1]$, and $\sum_{i=1}^{N} p_i = 1$.

Boltzmann Entropy

Before we give the definition of the Boltzmann entropy, we highlight a property of the function $f(x) = x \ln x$ which occurs in the definition of the Boltzmann entropy. The function $f(x)$ is not defined at $x = 0$:

$$f(x) = x \ln x = \frac{\ln x}{\frac{1}{x}} \Rightarrow f(0) = \text{"} \left[\frac{-\infty}{\infty} \right] \text{"} . \tag{6.93}$$

However, using the rule of l'Hopital, we see that the limit value $f(x \to 0)$ exists:

$$\lim_{x \to 0} f(x) = \lim_{x \to 0} \frac{\ln x}{\frac{1}{x}} = \lim_{x \to 0} \frac{\frac{1}{x}}{-\frac{1}{x^2}} = -\lim_{x \to 0} x = 0 . \tag{6.94}$$

For this reason, the domain of definition of the function $f(x) = x \ln x$ can be extended from $x \in (0, \infty)$ to $x \in [0, \infty)$ in such a way that we obtain a continuous function for $x \in [0, \infty)$. To this end, the expression $x \ln x$ will be interpreted in what follows as

$$\text{"}[x \ln x]\text{"} = \begin{cases} x \ln x & \text{for } x > 0 \\ 0 & \text{for } x = 0 \end{cases} . \tag{6.95}$$

The Boltzmann entropy is defined by

$$^{B}S[P] = S_B[P] = -\int_\Omega P(\mathbf{x}) \ln P(\mathbf{x}) \, d^M x . \tag{6.96}$$

The entropy kernel reads $s(z) = -z \ln z$ and is a strictly concave function: $d^2 z/dz^2 = -1/z < 0$ for $z > 0$. Furthermore, we have $B(z) = z$, $s(0) = 0$, $s(z) \in C^0([0, \infty))$, and $s(z) \in C^\infty((0, \infty))$. Therefore, the Boltzmann entropy is continuous, differentiable, expansible, and strictly concave. Furthermore, for two statistically-independent random variables $\mathbf{X} \in \Omega_x$ and $\mathbf{Y} \in \Omega_y$ distributed according to P and P' the Boltzmann entropy ^BS satisfies

$$^BS(P P') = -\int_{\Omega_x} \int_{\Omega_y} P(\mathbf{x}) P'(\mathbf{y}) \ln[P(\mathbf{x}) P'(\mathbf{y})] \, d^M x \, d^{M'} y$$

$$= -\int_{\Omega_x} \int_{\Omega_y} P(\mathbf{x}) P'(\mathbf{y}) \ln P(\mathbf{x}) \, d^M x \, d^{M'} y$$

$$-\int_{\Omega_x} \int_{\Omega_y} P(\mathbf{x}) P'(\mathbf{y}) \ln P'(\mathbf{y}) \, d^M x \, d^{M'} y$$

$$= {}^BS[P] + {}^BS[P'] . \tag{6.97}$$

6.4 Entropy and Information Measures

Consequently, the Boltzmann entropy is extensive. Note that as indicated in (6.96) we will use two different symbols for the Boltzmann entropy, namely, $^{\text{B}}S[P]$ and $S_B[P]$. In particular, we will use the notation $S_B[P]$ in the context of the second variation $\delta^2 S_B$.

The Boltzmann entropy can in particular be used to describe the entropy of many-body systems that are in or close to thermodynamic equilibrium and are studied by means of their particle distribution functions (μ-space description) [48, 348, 432, 490, 576, 625]. The functional $S = -\int_{\Omega_\Gamma} P(\mathbf{x}) \ln P(\mathbf{x}) \, \mathrm{d}^{M_\Gamma} x$ can also be applied to describe systems with respect to the Γ-space and is then referred to as the Gibbs entropy [89, 97, 348, 625]. In this case, the vector $\mathbf{x}$ is composed of the vectors $\mathbf{x}_i$ of the particles $i = 1, \ldots, N_0$ like $\mathbf{x} = (\mathbf{x}_1, \ldots, \mathbf{x}_{N_0})$ and the Γ-space Ω_Γ is given by $\Omega_\Gamma = \Omega^{N_0}$, where Ω is the phase space of a single particle. Finally, the functional $S = -\int_{\Omega_\Gamma} P(\mathbf{x}) \ln P(\mathbf{x}) \, \mathrm{d}^{M_\Gamma} x$ can be used to measure the disorder of systems that operate far from thermal equilibrium. In this case S is regarded as the Shannon information measure [261, 309, 310, 348, 509, 518, 625]. For a summary see Table 6.1.

Table 6.1. Applications of the functional $S = -\int P \ln P \, \mathrm{d}\Omega$

	Close to thermal equilibrium	Far from thermal equilibrium
μ-space	Boltzmann	–
Γ-space	Gibbs	Shannon

Entropy for Fermions and Bosons

Many particle systems composed of fermions and bosons can be described by means of the mean occupation number ρ_i of their energy levels i. The mean occupation number ρ_i describes how many particles of a system composed of N_0 particles on the average occupy an energy level i with energy ϵ_i. The phase space is given by $\Omega = \{\epsilon_1, \ldots, \epsilon_N\}$. If the energy levels are not degenerated, the quantum mechanical entropy reads

$$^{\text{FD,BE}}S(\rho_1, \ldots, \rho_N) = -\sum_{i=1}^{N} \rho_i \ln \rho_i \mp \sum_{i=1}^{N} (1 \mp \rho_i) \ln(1 \mp \rho_i) , \qquad (6.98)$$

where the upper sign holds for fermions and the lower sign for bosons [9, 232, 238, 318, 340, 346, 359, 373, 490, 619]. Note that in the case of fermions the maximal mean occupation number ρ_i equals unity, that is, we have the constraint $\rho_i \leq 1$. The reason for this is that in the case of fermions an energy level i cannot be occupied by more than one particle. Note also that the Fermi entropy (6.98) also describes the entropy of the Fermi quasi-particles that occur in the BCS theory of superconductivity [464].

In line with the stochastic description involving discrete energy levels, a classical stochastic description for fermions and bosons can be obtained in the case of continuous energy levels. Let $\epsilon \in \Omega$ describe the energy of a Fermi or Bose particle or subsystem. Let $\rho(\epsilon)$ denote the occupation number density of the energy ϵ. Then, in analogy to (6.98) we define the entropy measure

$$^{\mathrm{FD,BE}}S[\rho] = -\int_\Omega \rho(\epsilon) \ln \rho(\epsilon)\, d\epsilon \mp \int_\Omega [1 \mp \rho(\epsilon)] \ln[1 \mp \rho(\epsilon)]\, d\epsilon . \quad (6.99)$$

In analogy to the constraint $\rho_i \leq 1$, for fermions we require $\rho \leq 1$ [377]. The entropy (6.99) involves a linear entropy scale function $B(z) = z$ and an entropy kernel

$$s(z) = -z \ln z \mp (1 \mp z) \ln(1 \mp z) . \quad (6.100)$$

The derivative of $s(z)$ reads

$$\frac{ds}{dz} = -\ln\left(\frac{z}{1 \mp z}\right) = \ln\left(\frac{1}{z} \mp 1\right) \quad (6.101)$$

and is invertible:

$$\left[\frac{ds}{dz}\right]^{-1}(z) = \frac{1}{\exp\{z\} \pm 1} . \quad (6.102)$$

The kernel satisfies $s(0) = 0$ and

$$\frac{d^2 s}{dz^2} = \frac{-1}{z(1 \mp z)} < 0 . \quad (6.103)$$

For bosons we have $s(z) \in C^0([0, \infty))$ and $s(z) \in C^\infty((0, \infty))$. For fermions we have $0 \leq \rho \leq 1$, $s(z) \in C^0([0, 1])$, and $s(z) \in C^\infty((0, 1))$. Therefore, the entropies $^{\mathrm{FD}}S$ and $^{\mathrm{BE}}S$ are continuous, differentiable, expansible and strictly concave. Let us define now the internal energy functional $U[\rho]$ and the free energy

$$F[\rho] = U[\rho] - T\,^{\mathrm{FD,BE}}S[\rho] , \quad (6.104)$$

where T denotes the temperature of a system. If we assume that stationary distributions ρ_{st} correspond to critical points of F, then from $\delta F/\delta \rho = \mu$ and (4.22) and (6.102) we obtain

$$\rho_{\mathrm{st}}(\epsilon) = \frac{1}{\exp\left\{\dfrac{\delta U[\rho_{\mathrm{st}}]/\delta \rho - \mu}{T}\right\} \pm 1} . \quad (6.105)$$

In particular, for $U[\rho] = \int_\Omega \epsilon \rho(\epsilon)\, d\epsilon$ we obtain the Fermi–Dirac and Bose–Einstein statistics

6.4 Entropy and Information Measures

$$\rho_{\text{st}}(\epsilon) = \frac{1}{\exp\left\{\dfrac{\epsilon - \mu}{T}\right\} \pm 1} . \tag{6.106}$$

Taking a more general point of view, we may consider a Fermi or Bose particle (or subsystem) defined on a continuous M-dimensional phase space Ω. Let $U[\rho]$ denote the internal energy of a particle distribution or subsystem density. More precisely, let $\rho(\mathbf{x})$ describe the occupation number density of the particles or subsystems. Let us assume again that we are dealing with nondegenerated energy states. Then, the quantum mechanical entropy reads

$$^{\text{FD,BE}}S[\rho] = -\int_\Omega \rho(\mathbf{x}) \ln \rho(\mathbf{x}) \, \mathrm{d}^M x \mp \int_\Omega [1 \mp \rho(\mathbf{x})] \ln[1 \mp \rho(\mathbf{x})] \, \mathrm{d}^M x . \tag{6.107}$$

For free energy measures (6.104) involving (6.107) the free energy principle (4.21) gives us distributions

$$\rho_{\text{st}}(\mathbf{x}) = \frac{1}{\exp\left\{\dfrac{\delta U[\rho_{\text{st}}]/\delta\rho - \mu}{T}\right\} \pm 1} \tag{6.108}$$

that are akin to Fermi–Dirac and Bose–Einstein distributions. For example, we may consider $\mathbf{x} = (\mathbf{p}, \mathbf{q})$, where $\mathbf{p} = (p_x, p_y, p_z)$ and $\mathbf{q} = (q_x, q_y, q_z)$ correspond to particle momentum and particle position, and a Hamiltonian energy function $U = \langle H(\mathbf{p}, \mathbf{q}) \rangle$. In this case, (6.108) becomes

$$\rho_{\text{st}}(\mathbf{p}, \mathbf{q}) = \frac{1}{\exp\left\{\dfrac{H(\mathbf{p}, \mathbf{q}) - \mu}{T}\right\} \pm 1} . \tag{6.109}$$

Vorticity Entropy

The two-dimensional flow of fluids can be described by means of the velocity field $\mathbf{v}(\mathbf{x}, t)$. In the turbulent regime, vortices can be observed in the flow field. These vortices can be described in terms of the vorticity function

$$\omega(\mathbf{x}, t) = \mathbf{e}_z \operatorname{rot} \mathbf{v}(\mathbf{x}, t) , \tag{6.110}$$

where $\mathbf{e}_z$ is the unit vector in the z-direction and the fluid flow is confined to the XY-plane. According to a two-state model, we may distinguish only vortex-free positions and positions with vortices of strength ω_0. This implies that in a discrete XY-space we would have either $\omega(\mathbf{x}_i, t) = \omega_0$ or $\omega(\mathbf{x}_i, t) = 0$. In this case, it has been suggested to derive the statistics of the vortices from a vorticity entropy $^{\text{vor}}S$. Let $\rho(\mathbf{x})$ with $\mathbf{x} \in \Omega = \mathbb{R}^2$ describe the density of the vortices in the XY-plane. Then, $^{\text{vor}}S$ is given by [92, 93, 311, 421, 422]

$$^{\text{vor}}S[\rho] = -\int_\Omega \rho(\mathbf{x}) \ln \rho(\mathbf{x}) \, \mathrm{d}^M x - \int_\Omega [1 - \rho(\mathbf{x})] \ln[1 - \rho(\mathbf{x})] \, \mathrm{d}^M x . \tag{6.111}$$

We realize that the vorticity entropy corresponds to the Fermi–Dirac entropy.

Renyi Measure

For a discrete random variable the Renyi information and entropy measure is defined by [48, 495, 625]

$$^{\mathrm{R}}S_r(p_1,\ldots,p_N) = \begin{cases} \dfrac{1}{1-r}\ln\left(\sum_{i=1}^{N}[p_i]^r\right) , & \text{for } r > 0,\ r \neq 1 \\ -\sum_{i=1}^{N} p_i \ln p_i , & \text{for } r = 1 \end{cases}. \tag{6.112}$$

In the continuous case, the Renyi measure reads

$$^{\mathrm{R}}S_r[P] = \begin{cases} \dfrac{1}{1-r}\ln\left(\int_\Omega [P(\mathbf{x})]^r \mathrm{d}^M x\right) , & \text{for } r > 0,\ r \neq 1 \\ -\int_\Omega P(\mathbf{x}) \ln P(\mathbf{x})\, \mathrm{d}^M x , & \text{for } r = 1 \end{cases}. \tag{6.113}$$

The Renyi measure is a one-parametric information measure. The parameter value $r = 1$ requires particular attention. The Renyi measure for $r \neq 1$ can be regarded as a function $f(r)$ involving the function $g(r)$ (with $g = \sum[p_i]^r$ or $g = \int P^r \mathrm{d}^M x$) as shown in (6.114). For $r = 1$ the function $f(r)$ is not defined:

$$\left\{f(r) = \frac{\ln g(r)}{1-r},\ g(1) = 1\right\} \Rightarrow f(1) = ''\begin{bmatrix}0\\0\end{bmatrix}''. \tag{6.114}$$

For this reason, in (6.112) and (6.113) the case $r = 1$ is treated separately. Due to this supplementary definition, the Renyi measure is continuous with respect to the parameter r at $r = 1$. This property can be proven as follows. First, we write $[p_i]^r$ as $p_i \exp\{(r-1)\ln p_i\}$ and expand the exponential function into its Taylor series. Then, we expand the logarithmic function $\ln(1+x)$ into its Taylor series. Thus, we obtain

$$^{\mathrm{R}}S_r = \frac{1}{1-r}\ln\left[1 - \left\{1 - \sum_{i=1}^{N}[p_i]^r\right\}\right] = \frac{1}{1-r}\ln\left[1 - \left\{1 - \sum_{i=1}^{N} p_i e^{(r-1)\ln p_i}\right\}\right]$$

$$= \frac{1}{1-r}\ln\left[1 + (r-1)\sum_{i=1}^{N} p_i \ln p_i + \sum_{i=1}^{N} p_i \sum_{n=2}^{\infty}\frac{(r-1)^n}{n!}[\ln p_i]^n\right]$$

$$= \frac{1}{1-r}\sum_{m=1}^{\infty}(-1)^{m+1}\frac{1}{m}\left[(r-1)\sum_{i=1}^{N} p_i \ln p_i + \sum_{i=1}^{N} p_i \sum_{n=2}^{\infty}\frac{(r-1)^n}{n!}[\ln p_i]^n\right]^m$$

$$= -\sum_{i=1}^{N} p_i \ln p_i + \underbrace{\sum_{i=1}^{N} p_i \sum_{n=2}^{\infty}\frac{(-1)^n(1-r)^{n-1}}{n!}[\ln p_i]^n}_{Y_1(r)}$$

$$+\frac{1}{1-r}\underbrace{\sum_{m=2}^{\infty}(-1)^{m+1}\frac{1}{m}\left[(r-1)\sum_{i=1}^{N}p_i\ln p_i+\sum_{i=1}^{N}p_i\sum_{n=2}^{\infty}\frac{(r-1)^n}{n!}[\ln p_i]^n\right]^m}_{Y_2(r)}.$$
(6.115)

In the limit $r \to 1$ the expressions Y_1 and Y_2 vanish. Consequently, we obtain

$$\lim_{r\to 1} {}^{\mathrm{R}}S_r(p_1,\ldots,p_N) = -\sum_{i=1}^{N} p_i \ln p_i = {}^{\mathrm{B}}S(p_1,\ldots,p_N). \quad (6.116)$$

Equation (6.116) states that in the limit $r \to 1$ the Renyi measure reduces to the Boltzmann entropy. The scaling function $B(z)$ satisfies

$$B(z) = \frac{\ln z}{1-r} \Rightarrow$$
$$\frac{dB}{dz} = \frac{1}{z(1-r)} > 0, \quad \frac{d^2 B}{dz^2} = \frac{-1}{z^2(1-r)} < 0 \quad \text{for } r \in (0,1),\ z > 0.$$
(6.117)

The kernel $s(z)$ satisfies

$$s(z) = z^r \Rightarrow \frac{d^2 s}{dz^2} = r(r-1)z^{r-2} < 0 \quad \text{for } r \in (0,1),\ z > 0. \quad (6.118)$$

Consequently, the Renyi measures is a strictly concave measure in the parameter range $r \in (0,1]$. Furthermore, we have $s(0) = 0$, $s(z) \in C^0([0,\infty))$, $s(z) \in C^\infty((0,\infty))$. The expressions $\sum_i p_i^r$ and $\int_\Omega [P(\mathbf{x})]^r d^M x$ are larger than zero for any distribution $\{p_i\}$ and probability density $P \in C^0(\Omega)$ due to the normalization conditions $\sum_{i=1}^{N} p_i = 1$ and $\int_\Omega P(\mathbf{x}) d^M x = 1$. Therefore, the co-domain Ω_{cod} is given by $\Omega_{\mathrm{cod}} = (0,\infty)$. We have $B(z) \in C^\infty(\Omega_{\mathrm{cod}})$. In view of these properties, we conclude that the Renyi measure is continuous, differentiable and expansible for $r \in (0,1]$. Given two statistically independent random variables X and X' described by the distributions $\{p_i\}_{i=1}^{N}$ and $\{p_i'\}_{i=1}^{N'}$ the Renyi measure of the joint distribution $p_{i,k} = p_i p_k'$ satisfies

$$\begin{aligned}
{}^{\mathrm{R}}S_r(p_1 p_1',\ldots,p_N p_{N'}') &= \frac{1}{1-r}\ln\sum_{i=1}^{N}\sum_{k=1}^{N'}[p_i p_k']^r \\
&= \frac{1}{1-r}\ln\left\{\sum_{i=1}^{N}[p_i]^r \cdot \sum_{k=1}^{N'}[p_k']^r\right\} = \frac{1}{1-r}\ln\sum_{i=1}^{N}[p_i]^r + \frac{1}{1-r}\ln\sum_{k=1}^{N'}[p_k']^r \\
&= {}^{\mathrm{R}}S_r(p_1,\ldots,p_N) + {}^{\mathrm{R}}S_r(p_1',\ldots,p_{N'}').
\end{aligned} \quad (6.119)$$

That is, the Renyi measure is extensive.

Nonextensive Entropy Proposed by Tsallis

A one-parametric entropy measure proposed by Tsallis [582, 587] is given by

$$^{\mathrm{T}}S_q(p_1,\ldots,p_N) = \begin{cases} \dfrac{\sum_{i=1}^{N}[p_i]^q - 1}{1-q} = \dfrac{1 - \sum_{i=1}^{N}[p_i]^q}{q-1} \,, & \text{for } q > 0,\ q \neq 1 \\ ^{\mathrm{B}}S = -\sum_{i=1}^{N} p_i \ln p_i \,, & \text{for } q = 1 \end{cases} \quad (6.120)$$

for the discrete case and

$$^{\mathrm{T}}S_q[P] = \begin{cases} \dfrac{1}{1-q}\left[\displaystyle\int_\Omega [P(\mathbf{x})]^q \mathrm{d}^M x - 1\right] \,, & \text{for } q > 0,\ q \neq 1 \\ ^{\mathrm{B}}S = -\displaystyle\int_\Omega P(\mathbf{x}) \ln P(\mathbf{x})\, \mathrm{d}^M x \,, & \text{for } q = 1 \end{cases} \quad (6.121)$$

for the continuous case. The measure $^{\mathrm{T}}S_q$ for $q \neq 1$ can be written in terms of the functions $f(q)$ and $g(q)$ (with $g = \sum[p_i]^q$ or $g = \int P^q\, \mathrm{d}^M x$) as shown in (6.122). We realize that $f(q)$ is not defined at $q = 1$:

$$\left\{ f(q) = \frac{g(q) - 1}{1 - q}\,,\ g(1) = 1 \right\} \Rightarrow f(1) = ''\begin{bmatrix} 0 \\ 0 \end{bmatrix}''\,. \quad (6.122)$$

For this reason, in the definition of $^{\mathrm{T}}S_q$ the case $q = 1$ is treated separately. The nonextensive entropy is continuous with respect to the parameter q at $q = 1$, which can be seen by writing $^{\mathrm{T}}S_q$ as

$$^{\mathrm{T}}S_q = \frac{\sum_i p_i\left[-1 + e^{-(1-q)\ln p_i}\right]}{1-q} = \sum_{i=1}^{N}\sum_{n=1}^{\infty} p_i \frac{(1-q)^{n-1}}{n!}[-\ln p_i]^n$$

$$= -\sum_{i=1}^{N} p_i \ln p_i + \sum_{n=2}^{\infty}\sum_{i=1}^{N} p_i \frac{(1-q)^{n-1}}{n!}[-\ln p_i]^n \,. \quad (6.123)$$

Consequently, in the limit $q \to 1$ the nonextensive entropy reduces to the Boltzmann entropy:

$$\lim_{q \to 1} {}^{\mathrm{T}}S_q(p_1,\ldots,p_N) = -\sum_{i=1}^{N} p_i \ln p_i = {}^{\mathrm{B}}S(p_1,\ldots,p_N)\,. \quad (6.124)$$

The entropy $^{\mathrm{T}}S_q$ involves a linear scale function $B(z) = z$. The entropy kernel satisfies $s(0) = 0$, $s(z) \in C^0([0,\infty))$, $s(z) \in C^\infty((0,\infty))$, and

$$s(z) = \frac{z^q - z}{1 - q} \Rightarrow \frac{\mathrm{d}^2 s}{\mathrm{d}z^2} = -q z^{q-2} < 0 \quad \text{for } q > 0,\ q \neq 1,\ z > 0\,. \quad (6.125)$$

Consequently, for $q > 0$ the nonextensive entropy is continuous, differentiable, expansible, and strictly concave. Note that the entropy can also be defined for parameter values $q < 0$. In this case, however, the entropy measure is not expansible. Note that the expression $\sum_{i=1}^{N}[p_i]^q$ can be expressed in terms of

$$\sum_{i=1}^{N}[p_i]^q = 1 + (1-q) \, ^{\mathrm{T}}S_q(p_1,\ldots,p_N) \, . \tag{6.126}$$

Then, for two statistically independent random variables involving the distributions $\{p_i\}_{i=1}^{N}$ and $\{p'_i\}_{i=1}^{N'}$ and the joint distribution given by the probabilities $p_{i,k} = p_i p'_k$ we get

$$^{\mathrm{T}}S_q(p_1 p'_1, \ldots, p_N p'_{N'}) = \frac{1}{1-q}\left[\sum_{i=1}^{N}\sum_{k=1}^{N'}[p_i p'_k]^q - 1\right]$$

$$= \frac{1}{1-q}\left[\sum_{i=1}^{N}[p_i]^q \cdot \sum_{k=1}^{N'}[p'_k]^q - 1\right]$$

$$= \frac{1}{1-q}\left[\{1+(1-q)\,^{\mathrm{T}}S_q(p_1,\ldots,p_N)\}\{1+(1-q)\,^{\mathrm{T}}S_q(p'_1,\ldots,p'_{N'})\}-1\right]$$

$$= \,^{\mathrm{T}}S_q(p_1,\ldots,p_N) + \,^{\mathrm{T}}S_q(p'_1,\ldots,p'_{N'})$$
$$+ (1-q)\,^{\mathrm{T}}S_q(p_1,\ldots,p_N)\,^{\mathrm{T}}S_q(p'_1,\ldots,p'_{N'}) \, . \tag{6.127}$$

Equation (6.127) states that the entropy $^{\mathrm{T}}S_q$ is nonextensive and composable. In particular, we have

$$^{\mathrm{T}}S_q(p_1 p'_1, \ldots, p_N p'_{N'}) = \,^{\mathrm{T}}S_q(p_1,\ldots,p_N) + \,^{\mathrm{T}}S_q(p'_1,\ldots,p'_{N'})$$
$$+ G[\,^{\mathrm{T}}S_q(p_1,\ldots,p_N), \,^{\mathrm{T}}S_q(p'_1,\ldots,p'_{N'})] \, , \tag{6.128}$$

with $G(x,y) = (1-q)xy$. Note that in this chapter the nonextensive entropy by Tsallis will be used in the context of the μ-space descriptions of many-body systems. As such, the entropy has been discussed as an entropy measure both with regard to the μ-space and the Γ-space description of systems. Furthermore, the entropy has been discussed as in the context of information theory [8, 124, 598].

Sharma–Mittal Measure

For a discrete phase space $\Omega = \{m_1,\ldots,m_N\}$, the Sharma–Mittal entropy and information measure is defined by [520]

$$^{\mathrm{SM}}S_{q,r}(p_1,\ldots,p_N)$$

$$= \begin{cases} \dfrac{1 - \left[\sum_{i=1}^{N}[p_i]^r\right]^{(q-1)/(r-1)}}{q-1}, & \text{for } q, r > 0,\ q, r \neq 1 \\[2ex] {}^{G}S_q = \dfrac{1 - \exp\left\{(q-1)\sum_{i=1}^{N} p_i \ln p_i\right\}}{q-1}, & \text{for } q > 0,\ q \neq 1,\ r = 1 \\[2ex] {}^{R}S_r = \dfrac{1}{1-r} \ln\left(\sum_{i=1}^{N}[p_i]^r\right), & \text{for } r > 0,\ r \neq 1,\ q = 1 \\[2ex] {}^{B}S = -\sum_{i=1}^{N} p_i \ln p_i, & \text{for } q = 1,\ r = 1 \end{cases}$$

(6.129)

Accordingly, for a continuous phase space Ω, we get

$${}^{SM}S_{q,r}[P]$$
$$= \begin{cases} \dfrac{1 - \left[\int_\Omega [P(\mathbf{x})]^r \, \mathrm{d}^M x\right]^{(q-1)/(r-1)}}{q-1}, & \text{for } q, r > 0,\ q, r \neq 1 \\[2ex] {}^{G}S_q = \dfrac{1 - \exp\left\{(q-1)\sum_{i=1}^{N} \int_\Omega P(\mathbf{x}) \ln P(\mathbf{x}) \mathrm{d}^M x\right\}}{q-1}, & \\ & \text{for } q > 0,\ q \neq 1,\ r = 1 \\[2ex] {}^{R}S_r = \dfrac{1}{1-r} \ln\left(\int_\Omega [P(\mathbf{x})]^r \, \mathrm{d}^M x\right), & \text{for } r > 0,\ r \neq 1,\ q = 1 \\[2ex] {}^{B}S = -\int_\Omega P(\mathbf{x}) \ln P(\mathbf{x}) \, \mathrm{d}^M x, & \text{for } q = 1,\ r = 1 \end{cases}$$

(6.130)

For $q = r$ the Sharma–Mittal measure is equivalent to the entropy ${}^{T}S_q$:

$${}^{SM}S_{q,r=q} = {}^{T}S_q . \tag{6.131}$$

The ${}^{SM}S_{q,r}$ measure includes as special cases the Renyi measure, the nonextensive entropy ${}^{T}S_q$, the Boltzmann entropy and the one-parametric measure ${}^{G}S_q$, see Fig. 6.3. The measure ${}^{G}S_q$ will be referred to as Gaussian entropy.

The two-parametric measure ${}^{SM}S_{q,r}$ is continuous with respect to its parameters q and r at $q = 1$ and $r = 1$. Let us first investigate the case $q \neq 1$ and $r \to 1$. For $q \neq 1$ and $r \neq 1$, the Sharma–Mittal measure can be written as

$${}^{SM}S_{q,r} = \dfrac{1 - \exp\left\{\dfrac{q-1}{r-1} \ln \sum_{i=1}^{N}[p_i]^r\right\}}{q-1} = \dfrac{1 - \exp\{(1-q){}^{R}S_r\}}{q-1} . \tag{6.132}$$

Since the Renyi information measure reduces in the limit $r \to 1$ to the Boltzmann entropy and the measure ${}^{G}S_q$ basically corresponds to an exponential

6.4 Entropy and Information Measures

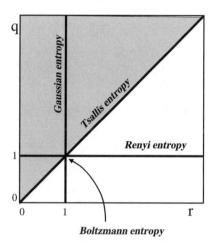

Fig. 6.3. Concavity and limiting cases of the two-parameter entropy $^{SM}S_{r,q}$. In the parameter region colored grey the entropy is strictly concave

function with the Boltzmann entropy as its argument, we realize that the limiting case

$$q \neq 1: \quad \lim_{r \to 1} {}^{SM}S_{q,r} = \frac{1 - \exp\left\{(1-q) \lim_{r \to 1} {}^{R}S_r\right\}}{q-1} = {}^{G}S_q \qquad (6.133)$$

holds. Next, let us consider the limiting case $q = 1$ and $r \to 1$. By definition, for $q = 1$ the Sharma–Mittal measure is equivalent to the Renyi measure. From our considerations on the Renyi measure, it follows that in the limit $q = 1, r \to 1$ the measure $^{SM}S_{q,r}$ is continuous again:

$$^{SM}S_{q=1,r} = {}^{R}S_r \;\Rightarrow\; \lim_{r \to 1} {}^{SM}S_{q=1,r} = {}^{B}S \;. \qquad (6.134)$$

Now, let us consider the limiting case $q \to 1$ and $r \neq 1$. To this end, we write $^{SM}S_{q,r}$ as

$$^{SM}S_{q,r} = \frac{1 - \exp\{(1-q)\,{}^{R}S_r\}}{q-1} = {}^{R}S_r + \sum_{n=2}^{\infty} \frac{(-1)^n (q-1)^{n-1}}{n!} [{}^{R}S_r]^n \;. \qquad (6.135)$$

Then, we have

$$r \neq 1: \quad \lim_{q \to 1} {}^{SM}S_{q,r} = {}^{R}S_r \;. \qquad (6.136)$$

We proceed by considering the case $q \to 1$ and $r \neq 1$. In this case, we cast the measure $^{G}S_q$ into the form

$$^G S_q = \frac{1 - \exp\{(1-q)\,^B S\}}{q-1} = {}^B S + \sum_{n=2}^{\infty} \frac{(-1)^n (q-1)^{n-1}}{n!} [^B S]^n . \quad (6.137)$$

Then, the limiting cases

$$\lim_{q \to 1} {}^G S_q = {}^B S ,$$
$$\lim_{q \to 1} {}^{SM} S_{q,r=1} = \lim_{q \to 1} {}^G S_q = {}^B S \quad (6.138)$$

hold. Finally, we consider the limiting cases $q \to 1, r \to 1$ and $r \to 1, q \to 1$. Then, we have

$$\lim_{r \to 1} \left(\lim_{q \to 1} {}^{SM} S_{q,r} \right) = \lim_{r \to 1} {}^R S_r = {}^B S , \quad (6.139)$$

$$\lim_{q \to 1} \left(\lim_{r \to 1} {}^{SM} S_{q,r} \right) = \lim_{q \to 1} {}^G S_q = {}^B S . \quad (6.140)$$

For $q = 1$ the Sharma–Mittal measure is equivalent to the Renyi measure and, therefore, in this parameter regime ${}^{SM} S_{q,r}$ is continuous, differentiable, and expansible. For $q \neq 1$ and $r \neq 1$ the scaling function B and the kernel s is given in (6.141) (see below). For $q \neq 1$ and $r = 1$ the scaling function B and the kernel s is given in (6.142) (see below). In both cases B and s satisfy $s(0) = 0$, $s(z) \in C^0([0,\infty))$, $s(z) \in C^\infty((0,\infty))$, and $B(z) \in C^\infty(\Omega_{\text{cod}})$ with $\Omega_{\text{cod}} = (0,\infty)$, which implies that for $q \neq 1$ the Sharma–Mittal measure is continuous, differentiable, and expansible, again. As for the concavity of ${}^{SM} S_{q,r}$, for $q = 1$ we are dealing with the Renyi measure. Therefore, ${}^{SM} S_{q,r}$ is strictly concave for $q = 1$ and $r \in (0,1]$. Next, let us explore the case $q \neq 1$. First, we consider the parameter range given by $q > r$ and $r \in (0,1)$. Then, we have

$$q > r, \; r \in (0,1) : \; B(z) = \frac{1 - z^{(q-1)/(r-1)}}{q-1}, \; s(z) = z^s ,$$

$$\frac{d^2 s}{dz^2} = r(r-1) z^{r-2} < 0 ,$$

$$\frac{dB}{dz} = \frac{1}{1-r} z^{(q-r)/(r-1)} > 0 ,$$

$$\frac{d^2 B}{dz^2} = \frac{r-q}{(1-r)^2} z^{(q-2r+1)/(r-1)} < 0 , \quad (6.141)$$

which implies that in this parameter domain ${}^{SM} S_{q,r}$ is a strictly concave measure. Second, we consider

$$q > 1, \; r = 1 : \; B(z) = \frac{1 - \exp\{(1-q)z\}}{q-1}, \; s(z) = -z \ln z ,$$

$$\frac{d^2 s}{dz^2} = -\frac{1}{z} < 0 ,$$

$$\frac{dB}{dz} = \exp\{(1-q)z\} > 0,$$
$$\frac{d^2 B}{dz^2} = (1-q)\exp\{(1-q)z\} < 0. \tag{6.142}$$

Again, we conclude that we are dealing with a strictly concave measure. Finally, we examine the parameter range $q > r > 1$. Then, we get

$$q > r > 1: \quad B(z) = \frac{1 - z^{(q-1)/(r-1)}}{q-1}, \quad s(z) = z^s,$$
$$\frac{d^2 s}{dz^2} = r(r-1)z^{r-2} > 0,$$
$$\frac{dB}{dz} = \frac{1}{1-r} z^{(q-r)/(r-1)} < 0,$$
$$\frac{d^2 B}{dz^2} = \frac{r-q}{(1-r)^2} z^{(q-2r+1)/(r-1)} < 0. \tag{6.143}$$

That is, in this domain $^{SM}S_{q,r}$ corresponds to a strictly concave measure as well. In sum, the Sharma–Mittal measure is a strictly concave measure for $q \geq r > 0$:

$$q \geq r > 0 \;\Rightarrow\; {}^{SM}S_{q,r} \text{ strictly concave}. \tag{6.144}$$

This finding is consistent with the fact that on the diagonal $q = r$ of the parameter space (q, r) we have the strictly concave entropy $^{T}S_q$, see also Fig. 6.3.

In closing our investigation of the Sharma–Mittal measure, let us examine whether or not $^{SM}S_{q,r}$ is an extensive measure. First, we note that for $q \neq 1$ and $r \neq 1$ the relation

$$\left[\sum_{i=1}^{N} [p_i]^r\right]^{(q-1)/(r-1)} = 1 + (1-q)\,{}^{SM}S_{q,r}(p_1,\ldots,p_N) \tag{6.145}$$

holds. Consequently, one obtains

$$^{SM}S_{q,r}(p_1 p'_1,\ldots,p_N p'_{N'}) = \frac{1}{1-q}\left(\left[\sum_{i=1}^{N}\sum_{k=1}^{N'} [p_i p'_k]^r\right]^{(q-1)/(r-1)} - 1\right)$$
$$= \frac{1}{1-q}\left(\left[\sum_{i=1}^{N}[p_i]^q\right]^{(q-1)/(r-1)} \left[\sum_{k=1}^{N'}[p'_k]^q\right]^{(q-1)/(r-1)} - 1\right)$$
$$= \frac{\{1+(1-q)\,{}^{SM}S_{q,r}(p_1,\ldots,p_N)\}\{1+(1-q)\,{}^{SM}S_{q,r}(p'_1,\ldots,p'_{N'})\}-1}{1-q}$$
$$= {}^{SM}S_{q,r}(p_1,\ldots,p_N) + {}^{SM}S_{q,r}(p'_1,\ldots,p'_{N'})$$
$$+ (1-q)\,{}^{SM}S_{q,r}(p_1,\ldots,p_N)\,{}^{SM}S_{q,r}(p'_1,\ldots,p'_{N'}). \tag{6.146}$$

Likewise, for $q \neq 1$ and $r = 1$, we get

$$\exp\{(1-q)\,{}^{\mathrm{B}}S(p_1,\ldots,p_N)\} = 1 + (1-q)\,{}^{\mathrm{G}}S_q(p_1,\ldots,p_N) \tag{6.147}$$

and

$$\begin{aligned}
{}^{\mathrm{G}}S_q(p_1 p'_1,\ldots,p_N p'_{N'}) &= \frac{1}{1-q}\left(\exp\{(1-q)\,{}^{\mathrm{B}}S(p_1 p'_1,\ldots,p_N p'_{N'})\} - 1\right)\\
&= \frac{1}{1-q}\left(\exp\{(1-q)\,[{}^{\mathrm{B}}S(p_1,\ldots,p_N) + {}^{\mathrm{B}}S(p'_1,\ldots,p'_{N'})]\} - 1\right)\\
&= \frac{\exp\{(1-q)\,{}^{\mathrm{B}}S(p_1,\ldots,p_N)\}\exp\{(1-q)\,{}^{\mathrm{B}}S(p'_1,\ldots,p'_{N'})\} - 1}{1-q}\\
&= \frac{\{1 + (1-q)\,{}^{\mathrm{G}}S_q(p_1,\ldots,p_N)\}\{1 + (1-q)\,{}^{\mathrm{G}}S_q(p'_1,\ldots,p'_{N'})\} - 1}{1-q}\\
&= {}^{\mathrm{G}}S_q(p_1,\ldots,p_N) + {}^{\mathrm{G}}S_q(p'_1,\ldots,p'_{N'})\\
&\quad + (1-q)\,{}^{\mathrm{G}}S_q(p_1,\ldots,p_N)\,{}^{\mathrm{G}}S_q(p'_1,\ldots,p'_{N'})\,.
\end{aligned} \tag{6.148}$$

From (6.146) and (6.148), we can read off that for $q \neq 1$ the Sharma–Mittal measure is nonextensive. In addition, ${}^{\mathrm{SM}}S_{q,r}$ is composable and satisfies (6.80) with $G(x,y) = (1-q)xy$. For $q = 1$ the measure ${}^{\mathrm{SM}}S_{q,r}$ reduces to the Renyi measure and, therefore, is extensive. The power of the Sharma–Mittal measure lies in its unifying character. It includes the Renyi measure and the entropy ${}^{\mathrm{T}}S_q$ as special cases. Finally, note that there is also a three-parametric generalization of the Sharma–Mittal measure [136].

Average Information Measure

We consider the measure

$${}^{\mathrm{A}}S_\psi(p_1,\ldots,p_N) = \psi^{-1}\left\{\sum_{i=1}^N p_i\,\psi(-\ln p_i)\right\},$$

$${}^{\mathrm{A}}S_\psi[P] = \psi^{-1}\left\{\int_\Omega P(\mathbf{x})\,\psi[-\ln P(\mathbf{x})]\,\mathrm{d}^M x\right\},$$

$$\psi \in C^\infty(\mathbb{R}),\ \lim_{z\to\infty}[e^{-z}\psi(z)] = 0,\ z \in \mathbb{R}: \frac{\mathrm{d}\psi}{\mathrm{d}z} \gtrless \frac{\mathrm{d}^2\psi}{\mathrm{d}z^2} \gtrless 0\,, \tag{6.149}$$

for discrete and continuous phase spaces. In line with a proposal by Aczel and Daroczy [8, 625, 626], we refer to ${}^{\mathrm{A}}S_\psi$ as average information measure. Analogous to (6.95), the expression $x\,\psi(-\ln x)$ is interpreted as $x\,\psi(-\ln x)$ for $x > 0$ and $x\,\psi(-\ln x) = 0$ for $x = 0$. In particular, the limit $x \downarrow 0$ yields

$$\lim_{x\downarrow 0}[x\,\psi(-\ln x)] = \lim_{z\to\infty}\left[e^{-z}\psi(z)\right] = 0\,. \tag{6.150}$$

The scaling function B and the kernel s are given by

6.4 Entropy and Information Measures

$$s(z) = z\psi(-\ln z), \quad B(z) = \psi^{-1}(z) \,. \tag{6.151}$$

They satisfy $s(0) = 0$, $s(z) \in C^0([0,\infty))$, $s(z) \in C^\infty((0,\infty))$, and

$$\frac{d^2 s}{dz^2} = -\frac{1}{z}\left[\frac{d\psi}{du} - \frac{d^2\psi}{du^2}\right]_{u=-\ln z} \lessgtr 0 \tag{6.152}$$

$$\left.\frac{dB}{dz}\right|_{z=\psi(u)} = \left[\frac{d\psi}{du}\right]^{-1} \gtrless 0 \,, \tag{6.153}$$

$$\frac{d^2 B}{dz^2} < 0 \,. \tag{6.154}$$

Let us derive the inequality (6.154). To this end, we differentiate the identity $\psi^{-1}[\psi(u)] = u$ twice with respect to u and thus obtain

$$\frac{d^2\psi^{-1}(v)}{dv^2}\left[\frac{d\psi(u)}{du}\right]^2 = -\frac{d\psi^{-1}(v)}{dv}\frac{d^2\psi(u)}{du^2}\,, \tag{6.155}$$

with $v = \psi(u)$. Since by definition the first and second derivatives of ψ have the same sign and, consequently, the first derivative of ψ^{-1} and the second derivative of ψ have the same sign, the left-hand side of (6.155) is negative. This implies that the second derivative of ψ^{-1}, which is equivalent to the second derivative of B, is negative.

In view of the inequalities (6.152-6.154) and the aforementioned properties of s and B, we realize that the average entropy $^A S_\psi$ is continuous, differentiable, expansible, and strictly concave. In the special case $\psi(z) = z$, the average information measure reduces to the Boltzmann entropy. For $r \in (0,1)$, $\psi(z) = \exp\{(1-r)z\} \Rightarrow \psi^{-1}(z) = \ln(z)/(1-r)$, the average entropy recovers the Renyi measure and the first and second derivatives of $\psi(z)$ satisfy the inequalities listed in (6.149) (upper signs).

Information Measure of Systems with Negative Stochastic Feedback

In stochastic systems, attractive states may be less affected by noise if the probability is large that the states are occupied. In other words, the more probable it is to find a system in an attractive state, the weaker the fluctuating forces driving the system out of the attractive state. Such a phenomenon has been called negative stochastic feedback. For systems with negative stochastic feedback (NSF) a one-parametric information measure $^{NSF}S_\alpha$ has been proposed [196]. For discrete phase spaces $^{NSF}S_\alpha$ reads

$$^{NSF}S_\alpha(p_1,\ldots,p_N) = -\sum_{i=1}^N p_i \ln\left[\frac{p_i}{1+\alpha p_i}\right] + c' \,, \tag{6.156}$$

with $\alpha \geq 0$. For continuous phase spaces $^{NSF}S_\alpha$ is defined by

$$^{\text{NSF}}S_\alpha[P] = -\int_\Omega P(\mathbf{x}) \ln\left[\frac{P(\mathbf{x})}{1+\alpha P(\mathbf{x})}\right] d^M x + c'$$
$$= -\int_\Omega P(\mathbf{x}) \ln\left[\frac{cP(\mathbf{x})}{1+\alpha P(\mathbf{x})}\right] d^M x , \qquad (6.157)$$

with $\alpha \geq 0$ and $\ln c = c'$. The constant c' determines the minimum value of $^{\text{NSF}}S_\alpha$ in the discrete case. For $c' = \ln(1+\alpha)$ we have $^{\text{NSF}}S_\alpha(p_1, \ldots, p_N) = 0$, if one of the probabilities p_i equals unity and $^{\text{NSF}}S_\alpha(p_1, \ldots, p_N) \geq 0$ for arbitrary distributions $\{p_i\}$. For $\alpha = 0$ the information measure $^{\text{NSF}}S_\alpha$ reduces to the Boltzmann entropy. The scaling function B is linear and the kernel s satisfies $s(0) = 0$, $s(z) \in C^0([0, \infty))$, $s(z) \in C^\infty((0, \infty))$, and

$$s(z) = -z \ln\left[\frac{z}{1+\alpha z}\right] + zc' \Rightarrow \frac{d^2 s}{dz^2} = \frac{-1}{z(1+\alpha z)^2} < 0 . \qquad (6.158)$$

Therefore, $^{\text{NSF}}S_\alpha$ is continuous, differentiable, expansible, and strictly concave. Systems with negative stochastic feedback will be discussed in detail in Sect. 7.4.4.

Further Examples

In Table 6.2 we list further entropy and information measures that have been discussed in the literature. This list should not be considered as a comprehensive survey. Rather it illustrates the variety of measures that can characterize stochastic systems. For the sake of conveniency, we present either the discrete or the continuous version of a measure. Note that the measures # 7, 8, 11, and 12 can be regarded as Sharma–Mittal measures (6.80). The measure # 7 corresponds to the Sharma–Mittal measure when substituting $\tilde{q} = (1-q)(1-r)^{-2} + (r-1)^{-1}$. The measure # 8 can be regarded as a special case of the Sharma–Mittal measure given by $q = 2 - r$. The measure # 11 may be interpreted in terms of a Sharma–Mittal measure $^{\text{SM}}S_{q,r}$ for $q = 1/\tilde{q}$ and $r = \tilde{q}$ which implies $(q-1)/(r-1) = -1/\tilde{q}$. The measure # 12 is a special case of $^{\text{SM}}S_{q,r}$ for $q = 2 - 1/\tilde{q}$ and $r = \tilde{q}$ which implies $(q-1)/(r-1) = 1/\tilde{q}$. Furthermore, note that the Fisher information measure # 13 involves a derivative. Finally, the reader may also consult the review articles [585, 625, 626].

6.5 Examples and Applications

6.5.1 Porous Medium Equation

The porous medium equation describes the diffusion of gas molecules and fluid particles through porous media. Its derivation basically involves three steps [34, 395, 470].

Table 6.2. Further examples of entropy and information measures

#	$S(p_1,\ldots,p_N)$, $S[P]$	Reference
1	$S_{q,n} = \dfrac{n!}{(q-1)^n}\left\{\sum_{i=1}^{N}\sum_{k=0}^{n-1}\dfrac{p_i}{[\ln p_i]^{n-1-k}}\dfrac{(q-1)^k}{k!} - \sum_{i=1}^{N}\dfrac{p_i}{[\ln p_i]^{n-1}}\right\}$	[460]
2	$S_q = \dfrac{(q-1)\,^{\mathrm{T}}S_q + (1-1/q)\,^{\mathrm{T}}S_{1/q}}{q - 1/q}$	[1]
3	$S_\eta = \sum_{i=1}^{N}\left\{\Gamma\left(\dfrac{\eta+1}{\eta}, -\ln p_i\right) - p_i\Gamma\left(\dfrac{\eta+1}{\eta}, 0\right)\right\}$, $\Gamma(\mu,t) = \int_t^\infty y^{\mu-1} e^{-y}\,dy$	[21]
4	$S_q = \dfrac{\sum_{i=1}^{N}[p_i]^{1/q} - [p_i]^q}{q - 1/q}$	[64]
5	$S_{q,q'} = \dfrac{(1-q')\,^{\mathrm{T}}S_{q'} - (1-q)\,^{\mathrm{T}}S_q}{q - q'}$	[64]
6	$S_{\kappa,\alpha} = \dfrac{\alpha^\kappa}{1+\kappa}\,^{\mathrm{T}}S_{1+\kappa} + \dfrac{\alpha^{-\kappa}}{1-\kappa}\,^{\mathrm{T}}S_{1-\kappa}$	[322, 323] [324]
7	$S_{\tilde{q},r} = \dfrac{1}{r-1}\left[\dfrac{\left[\sum_{i=1}^{N}[p_i]^r\right]^{-1-\tilde{q}(1-r)} + \tilde{q}(1-r)}{1+\tilde{q}(1-r)} - 1\right]$	[388]
8	$^{\mathrm{U}}S_r = \dfrac{1}{1-r}\left[1 - \dfrac{1}{\sum_{i=1}^{N}[p_i]^r}\right]$	[375, 376]
9	$S_{q,a,b,c} = \dfrac{1}{(1-q)(a+b+c)}\left[a\,^{\mathrm{R}}S_q + b\,^{\mathrm{T}}S_q + c\,^{\mathrm{U}}S_q\right]$	[375, 376]
10	$S_{r,\beta} = \begin{cases}\dfrac{1}{1-r}\ln\left(\dfrac{\sum_{i=1}^{N}[p_i]^{r-1+\beta}}{\sum_{i=1}^{N}[p_i]^\beta}\right) & \text{for } r \neq 1 \\ \dfrac{\sum_{i=1}^{N}\{[p_i]^\beta \ln p_i\}}{\sum_{i=1}^{N}[p_i]^\beta} & \text{for } r = 1\end{cases}$	[8]
11	$S_{\tilde{q}} = \dfrac{1 - \left[\sum_{i=1}^{N}[p_i]^{\tilde{q}}\right]^{-1/\tilde{q}}}{(1/\tilde{q}) - 1}$	[136, 202] [524]
12	$S_{\tilde{q}} = \dfrac{1 - \left[\sum_{i=1}^{N}[p_i]^{\tilde{q}}\right]^{1/\tilde{q}}}{1 - (1/\tilde{q})}$	[29]
13	$S = \displaystyle\int_{-\infty}^{\infty} \dfrac{1}{P(x)}\left[\dfrac{dP(x)}{dx}\right]^2 dx$	[203]

Continuity Equation

Let $\rho(\mathbf{x}, t)$ denote the density of a gas or fluid flowing through a porous material. We assume that gas particles are neither created nor annihilated and that the fluids under consideration are incompressible. Then, $\rho(\mathbf{x}, t)$ satisfies the continuity equation

$$\omega \frac{\partial}{\partial t} \rho(\mathbf{x}, t) = -\nabla \cdot \{\mathbf{v}(\mathbf{x}, t) \rho(\mathbf{x}, t)\} \ , \tag{6.159}$$

where $\mathbf{v}(\mathbf{x}, t)$ describes the velocity of the flow field and ω denotes the porosity of the media. The product $\rho(\mathbf{x}, t,)\mathbf{v}(\mathbf{x}, t)$ can be regarded as a particle current $\mathbf{J}(\mathbf{x}, t)$, see Sec. 4.5.1.

Darcy's Law

The flow field velocity at a position $\mathbf{x}$ can be related to the pressure p at $\mathbf{x}$ by means of an empirical law, Darcy's law, which states that v is proportional to the pressure gradient and that (gas or fluid) particles flow from regions of high pressure to regions of low pressure. In symbols, Darcy's law reads [34, 395, 470]

$$\mathbf{v}(x, t) = -c \nabla p(\mathbf{x}, t) \ , \tag{6.160}$$

where $c > 0$ is a constant. Note that empirical laws of this form abound in physics. For example, Fick's law states that the flux of particles is proportional to the concentration gradient, Fourier's law states that the flux of the heat energy is proportional to the temperature gradient and, finally, Ohm's law tells us that electric currents are proportional to gradients of electric potentials [395]. In the context of linear nonequilibrium thermodynamics, we have expressed relations of this kind by means of (4.40) and (4.43): the thermodynamic current $\mathbf{J}^{\text{th}}$ is proportional to the gradient of the "thermodynamic potential" $\delta F/\delta P$ or the chemical potential μ.

State Equation of Polytropic Gases

Our third and final ingredient is the state equation of polytropic gases

$$\frac{p}{p'} = \left[\frac{\rho}{\rho'}\right]^\nu \ . \tag{6.161}$$

Equation (6.161) can be used to determine p as a function of ρ provided that we have information about a reference state with p' and ρ'. That is, from (6.161) it follows that $p(\rho) = C\rho^\nu$ holds with $C = p'/[\rho']^\nu$.

Porous Medium Equation

Taking (6.159-6.161) together, we obtain the porous medium equation

$$\frac{\partial}{\partial t}\rho(\mathbf{x},t) = Q\Delta\rho(\mathbf{x},t)^q \;, \qquad (6.162)$$

with $Q = \nu c p'[\rho']^{-\nu}/[\omega(\nu+1)]$ and $q = \nu + 1$.

$^{\mathrm{T}}S_q$-Entropy Fokker–Planck Equation

Let $M_0 = \int_\Omega \rho(\mathbf{x})\,\mathrm{d}^M x$ denote the number of particles or the total mass of the system described by ρ. In order to interpret (6.162) as a Fokker–Planck equation we put $P(\mathbf{x}) = \rho(\mathbf{x})/M_0$, replace QM_0^{q-1} by Q and thus obtain

$$\frac{\partial}{\partial t}P(\mathbf{x},t;u) = Q\Delta P^q \;. \qquad (6.163)$$

Equation (6.163), in turn, can be interpreted as the entropy Fokker–Planck equation (6.5) for the entropy (6.121). In fact, substituting the entropy $^{\mathrm{T}}S_q$ into (6.5) for $\mathbf{I} = 0$, $U_0 = 0$, and $M_{ik} = \delta_{ik}$ gives us (6.163).

Barenblatt–Pattle Solutions

In the univariate case, (6.163) reads

$$\frac{\partial}{\partial t}P(x,t;u) = Q\frac{\partial^2}{\partial x^2}P^q \;. \qquad (6.164)$$

For (6.164) exact time-dependent solutions can be derived [34, 38, 329, 395, 467, 470]. For probability densities that are subjected to natural boundary conditions and correspond to delta distributions at time $t = t_0$, we get

$$P(x,t;\delta(x-x_0)) = D_0 \left[\frac{1}{t-t_0}\right]^{1/(1+q)} \left[\frac{1}{1 + \dfrac{C_0}{2}(1-q)\dfrac{(x-x_0)^2}{[t-t_0]^{2/(1+q)}}}\right]^{1/(1-q)}, \qquad (6.165)$$

for $q \in (0,1)$. For $q = 1$ we deal with Wiener processes described by $P(x,t;\delta(x-x_0)) = P(x,t|x_0,t_0;\delta(x-x_0))$ with $P(x,t|x_0,t_0;\delta(x-x_0))$ defined by (3.114) and $K(t,t_0) = Q(t-t_0)$, $m = 1$, see Sect. 3.9.4. For $q > 1$ we obtain

$$P(x,t;\delta(x-x_0)) = D_0 \left[\frac{1}{t-t_0}\right]^{1/(1+q)} \left[\underbrace{1 - \frac{C_0}{2}(q-1)\frac{(x-x_0)^2}{[t-t_0]^{2/(1+q)}}}_{Y(x,t)>0}\right]^{1/(q-1)}, \qquad (6.166)$$

provided that the expression $Y(x,t)$ is smaller than unity. For $Y(x,t) \geq 1$ we put $P(x,t;\delta(x-x_0)) = 0$. That is, we are dealing with a cut-off distribution. For the sake of convenience, we introduce the operator $\{\cdot\}_+ : g(x,t) \to h(x,t)$ defined by

$$h(x,t) = \{g(x,t)\}_+ = \max[g(x,t), 0] \,. \tag{6.167}$$

That is, $\{\cdot\}_+$ maps real functions onto semi-positive real ones. Then, (6.165) and (6.166) can be summarized as

$$P(x,t;\delta(x-x_0)) =$$
$$D_0 \left[\frac{1}{t-t_0}\right]^{1/(1+q)} \left[\left\{1 - \frac{C_0}{2}(q-1)\frac{(x-x_0)^2}{[t-t_0]^{2/(1+q)}}\right\}_+\right]^{1/(q-1)} . \tag{6.168}$$

The coefficients D_0 and C_0 are given by

$$D_0 = \left[2Qq(1+q)[z_q]^2\right]^{-1/(1+q)} ,$$
$$C_0 = \frac{1}{2}\left[2Qq(1+q)[z_q]^{1-q}\right]^{-2/(1+q)} , \tag{6.169}$$

where z_q is defined in Table 6.3. For $x_0 = 0$ the solution (6.168) is often called the Barenblatt–Pattle solution [38, 467].

Anomalous Diffusion

It has been shown that for $q \neq 1$ the nonlinear diffusion equation (6.164) describes anomalous diffusion processes. To this end, the exact time-dependent solutions (6.168) [591], scaling arguments [99], and Ito–Langevin equations [66] have been used. The observation of anomalous diffusion is in line with the general property of entropy Fokker–Planck equations derived in Sect. 6.2.2, namely, that entropy Fokker–Planck equations with nonextensive entropy measures such as $^T S_q$ describe anomalous diffusion processes. In Sect. 6.5.3 we will show how to derive (6.168) and return to the issue of anomalous diffusion.

Flows in Porous Media

We would like to point out that there are several modifications of the porous medium equation that play a crucial role in the description of gas and fluid flows in porous media, see, for example, [17, 76, 395, 468, 465, 559, 634].

6.5.2 $^T S_q$-Entropy Fokker–Planck Equation by Plastino and Plastino

We may regard stochastic processes described by the porous medium equation as generalized Wiener processes. If so, we may raise the question whether or

6.5 Examples and Applications

not stochastic processes with nonvanishing drift terms, such as Ornstein–Uhlenbeck processes can be generalized in a similar manner. This question can be answered in the affirmative. To this end, we supplement (6.164) with a drift term that is linear with respect to P. Thus, we obtain the nonlinear Fokker–Planck equation

$$\frac{\partial}{\partial t} P(x,t;u) = -\frac{\partial}{\partial x} h(x) P + Q \frac{\partial^2}{\partial x^2} P^q , \qquad (6.170)$$

for $x \in \Omega = \mathbb{R}$ suggested by Plastino and Plastino [484]. Note that in [484] the diffusion term reads $\partial^2 P^{2-\tilde{q}}/\partial x^2$. Equation (6.170) corresponds to the diffusion form of the entropy Fokker–Planck equation (6.4) for the entropy $^{T}S_q$, which can be verified by substituting $^{T}S_q$ given by (6.121) into (6.46).

Stationary Solutions

If we evaluate the entropy kernel (6.125) of $^{T}S_q$, we find $ds/dz = (qz^{q-1} - 1)/(1-q)$ and

$$\left[\frac{ds}{dz}\right]^{-1}(z) = \left[\frac{q}{1+(1-q)z}\right]^{1/(1-q)} . \qquad (6.171)$$

From (6.6) it then follows that the stationary solution of (6.170) is described by

$$P_{\text{st}}(x) = \left[\left\{\frac{1+(1-q)(U_0(x)-\mu)/Q}{q}\right\}_+\right]^{1/(q-1)} ; \qquad (6.172)$$

for $q \neq 1$ and $U_0(x) = -\int^x h(x')\,dx'$. For $q=1$ we are dealing with a linear Fokker–Planck equation and the stationary probability density is given by the Boltzmann distribution of $U_0(x)$. Equation (6.172) can alternatively be expressed as

$$P_{\text{st}}(x) = D_{\text{st}} \left[\left\{1 + \frac{(1-q)}{qQD_{\text{st}}^{q-1}} U_0(x)\right\}_+\right]^{1/(q-1)} , \qquad (6.173)$$

with $D_{\text{st}} = [(1-(1-q)\mu/Q)/q]^{1/(q-1)}$.

Linear Drift Force, Power Law Distributions, and Anomalous Diffusion

For $h(x) = -\gamma x$ and $\gamma > 0$ we have

$$\frac{\partial}{\partial t} P(x,t;u) = \frac{\partial}{\partial x} \gamma x P + Q \frac{\partial^2}{\partial x^2} P^q \qquad (6.174)$$

and the stationary solution is given by

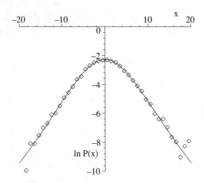

Fig. 6.4. Log-normal plot of the stationary solution P_{st} of (6.174). Diamonds: P_{st} computed from (6.194) with $h(x) = -\gamma x$ using the averaging method for self-consistent Langevin equations. Solid line: exact solution (6.175). Parameters: $\gamma = 0.1$, $Q = 1.0$, $q = 0.8$ ($L = 20000$, $\Delta t = 0.3$, $2[\Delta x]^2 = 0.1$, w_n via Box–Muller)

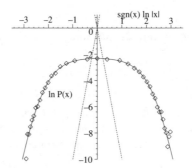

Fig. 6.5. Graphs shown in Fig. 6.4 are depicted here as log-log plots. As predicted by (6.176), we obtain straight lines with slopes $\pm 2/(1-q)$ in the asymptotic limit. For comparison we have also depicted dotted lines with slopes $\pm 2/(1-q)$ [179]

$$P_{st}(x) = D_{st} \left[\left\{ 1 - \frac{\gamma}{2qQ\, D_{st}^{q-1}} (q-1) x^2 \right\}_+ \right]^{1/(q-1)}, \qquad (6.175)$$

for $q \neq 1$. For $q = 1$ we obtain the stationary solution of Ornstein–Uhlenbeck processes: $P_{st}(x) \propto \exp\{-\gamma x^2/Q\}$. For $q \in (0,1)$ we are dealing with solutions that describe power laws in the asymptotic limit:

$$P_{st}(x) \propto |x|^{-\frac{2}{1-q}} \quad \text{for} \quad |x| \to \infty, \qquad (6.176)$$

see Figs. 6.4 and 6.5. In contrast, for $q > 1$ we have cut-off distributions. The coefficient $D_{st} = [\gamma/(2qQz_q^2)]^{1/(1+q)}$ can also be found in Table 6.4 as $^T D_{st}$. A transient solution of (6.174) is given by [191, 484, 591]

$$P(x,t;\delta(x-x_0)) = D(t)\left[\left\{1 - [z_q D(t)]^2 (q-1)(x-m(t))^2\right\}_+\right]^{1/(q-1)}, \tag{6.177}$$

with

$$m(t) = x_0 \exp\{-\gamma(t-t_0)\},$$

$$D(t) = \left[\frac{\gamma}{2qQ}\frac{1}{[z_q]^2}\frac{1}{1-\exp\{-(1+q)\gamma(t-t_0)\}}\right]^{1/(1+q)}. \tag{6.178}$$

The variance $K(t)$ of the solution (6.177) can be obtained. In particular, in the limiting case of a vanishing drift force (i.e., for $\gamma = 0$), we find that

$$K(t) \propto t^{2/(1+q)} \tag{6.179}$$

(see (6.275) below). We have an anomalous diffusion for $q \neq 1$, see Figs. 6.6 and 6.7. A derivation of (6.175-6.179) will be given in Sect. 6.5.3. The transient solution (6.177) reduces in the limit $q \to 1$ to the transient solution of an Ornstein–Uhlenbeck process given by $P(x,t;\delta(x-x_0)) = P(x,t|x_0,t_0;\delta(x-x_0))$, with $P(x,t|x_0,t_0;\delta(x-x_0))$ defined by (3.114), see Sect. 3.9.5. Therefore, (4.37) can be regarded as a generalization of Ornstein–Uhlenbeck processes to systems characterized by the nonextensive entropy measure $^T S_q$.

H-Theorem for $h(x)$

For classical solutions of (6.170) we can apply the H-theorem of free energy Fokker–Planck equations, see Sect. 4.3. Accordingly, (6.170) describes relaxation processes with $P(x,t;u) \to P_{st}$ related to a decrease of the free energy measure $F = U_L - Q{^T S_q}$. Let us demonstrate that the inequality $dF/dt \leq 0$ holds for solutions of (6.170). The evolution of $F[P]$ can conveniently be determined by means of the corresponding generalized Kullback measure $K[P, P_{st}]$ [192]. More precisely, from (4.131) it follows that $dF/dt = QdK/dt$. Therefore, we will examine the time-dependent behavior of $K[P, P_{st}]$ rather than the evolution of $F[P]$. Substituting $S = {^T S_q}$ into (4.126) we obtain

$$^T K[P, P_{st}] = \int_\Omega [P_{st}(x)]^q \, dx + \frac{1}{(1-q)}\int_\Omega P(x)\left\{q[P_{st}(x)]^{q-1} - [P(x)]^{q-1}\right\} dx. \tag{6.180}$$

By means of the generalized logarithm

$$\mathrm{Ln}_q(z) = \frac{z^{q-1}-1}{q-1} \quad \Rightarrow \quad \lim_{q\to 1}\mathrm{Ln}_q(z) = \ln z, \tag{6.181}$$

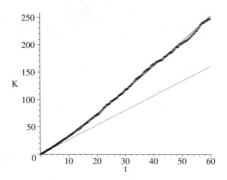

Fig. 6.6. Variance $K(t)$ as a function of t. Diamonds: simulation of (6.195) using the averaging method for self-consistent Langevin equations. Solid line: exact solution $K(t) \propto t^{2/(1+q)}$ given by (6.275). Dotted line: $K(t) \propto t$. Parameters: $Q = 1.0$, $q = 0.8$ ($L = 2000$, $\Delta t = 0.3$, $2[\Delta x]^2 = 0.1$, w_n via Box–Muller)

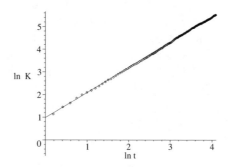

Fig. 6.7. The graph $t \to K(t)$ shown in Fig. 6.6 is depicted here as a log-log plot. As predicted by (6.275), we obtain a straight line

the relation (6.180) can be expressed as

$$^{\mathrm{T}}K[P, P_{\mathrm{st}}] = (1-q) \, ^{\mathrm{T}}S_q[P_{\mathrm{st}}] + \int_\Omega P(x) \left\{ \mathrm{Ln}_q[P(x)] - q\mathrm{Ln}_q[P_{\mathrm{st}}(x)] \right\} \mathrm{d}x \quad (6.182)$$

(note that an alternative generalization of the logarithmic function has been proposed in [592]). From (6.182) we can read off that in the limit $q \to 1$ the functional $^{\mathrm{T}}K$ reduces to $\int_\Omega P(z) \ln \{P(z)/P_{\mathrm{st}}(z)\} \, \mathrm{d}z$, that is, it recovers the Kullback measure based on the Boltzmann entropy. The first derivative of $^{\mathrm{T}}K[P, P_{\mathrm{st}}]$ with respect to t reads

6.5 Examples and Applications 255

$$\frac{\mathrm{d}}{\mathrm{d}t}{}^{\mathrm{T}}K[P, P_{\mathrm{st}}] = \frac{q}{(1-q)} \int_\Omega \{[P_{\mathrm{st}}(x)]^{q-1} - [P(x,t;u)]^{q-1}\} \frac{\partial}{\partial t} P(x,t;u)\,\mathrm{d}x \ . \tag{6.183}$$

Substituting the entropy Fokker–Planck equation (6.170) into (6.183) and integrating by parts gives us

$$\frac{\mathrm{d}}{\mathrm{d}t}{}^{\mathrm{T}}K[P, P_{\mathrm{st}}] = q \int_\Omega P \left\{ [P_{\mathrm{st}}]^{q-2} \frac{\mathrm{d}P_{\mathrm{st}}}{\mathrm{d}x} - [P]^{q-2} \frac{\partial P}{\partial x} \right\} \left\{ -h + \frac{Q}{P} \frac{\partial P^q}{\partial x} \right\} \mathrm{d}x \ , \tag{6.184}$$

when assuming that the surface term arising from the partial integration vanishes. Using $h(x) P_{\mathrm{st}}(x) = Q\,\mathrm{d}[P_{\mathrm{st}}(x)]^q/\mathrm{d}x$, (6.184) can be transformed into

$$\frac{\mathrm{d}}{\mathrm{d}t}{}^{\mathrm{T}}K[P, P_{\mathrm{st}}] = -q^2 Q \int_\Omega P \left[[P_{\mathrm{st}}]^{q-2} \frac{\mathrm{d}P_{\mathrm{st}}}{\mathrm{d}x} - P^{q-2} \frac{\partial P}{\partial x} \right]^2 \mathrm{d}x \le 0 \ , \tag{6.185}$$

implying that the Kullback measure and Lyapunov functional ${}^{\mathrm{T}}K$ is a nonincreasing function. The inequality $\mathrm{d}\,{}^{\mathrm{T}}K/\mathrm{d}t \le 0$, in turn, gives us $\mathrm{d}F[P]/\mathrm{d}t \le 0$ (as argued above).

H-Theorem for $h(x) = -\gamma x$

Let us now focus on the special case (6.174). Substituting the transient solution (6.177) and the corresponding stationary solution (6.175) into (6.180) yields

$${}^{\mathrm{T}}K[P(x,t;\delta(x-x_0)), P_{\mathrm{st}}] = \frac{\gamma}{2Q} m^2(t)$$

$$+ \frac{q}{(3q-1)(1-q)} \left\{ (1+q)[D_{\mathrm{st}}]^{q-1} + \frac{1}{[D(t)]^2} \left((1-q)[D_{\mathrm{st}}]^{q+1} - 2[D(t)]^{q+1} \right) \right\} \tag{6.186}$$

(for details see the Appendix in [192]). Note that D_{st} denotes the stationary value of $D(t)$, that is, we have $D_{\mathrm{st}} = D(t \to \infty) = [\gamma/(2qQz_q^2)]^{1/(1+q)}$ (see Table 6.4). Note also that for solutions (6.177) the integrals involved in (6.180) diverge for $q \in (0, 1/3)$ [192]. Consequently, the generalized Kullback measure (6.180) only applies to solutions (6.177) with $q \in (1/3, 1)$. The functional (6.186) and the evolution of the normalization constant $D(t)$ in (6.178) allow for several conclusions. Since $D(t)$ is a positive and strictly monotonically decreasing function (i.e., we have $D(t) > 0$ and $\mathrm{d}D(t)/\mathrm{d}t < 0$ with $D(t \to t_0) \to \infty$, $D(t \to \infty) = D_{\mathrm{st}}$, and $0 \le t < \infty : D(t) > D_{\mathrm{st}}$, we find

$$\lim_{t \to t_0} {}^{\mathrm{T}}K[P, P_{\mathrm{st}}] = \underbrace{\frac{\gamma}{2Q}[x_0]^2}_{\ge 0} + \underbrace{\frac{q(1+q)}{(3q-1)(1-q)}[D_{\mathrm{st}}]^{q-1}}_{>0} > 0 \ , \tag{6.187}$$

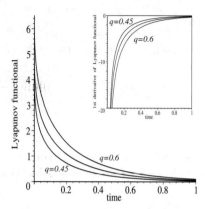

Fig. 6.8. Lyapunov functional $^T K$ given by (6.190) as a function of time for $Q=\gamma=1$, $m_0 = 0$, and different values of q: $q = 0.45$, $q = 0.5$, and $q = 0.6$ (*from bottom to top*). Inlet: First derivative of the Lyapunov functional $^T K$ computed from (6.191): $q = 0.45$, $q = 0.5$, and $q = 0.6$ (*from top to bottom*). Note that the first derivative is singular at $t = 0$, cf. also (6.191); reprinted from [192], © 2001, with permission from Elsevier

$$\lim_{t\to t_0} \frac{d}{dt} {}^T K[P, P_{st}] = \underbrace{\frac{\gamma}{2Q} \frac{dm^2(t)}{dt}}_{\leq 0} + \underbrace{\frac{2q\{[D(t)]^{1+q}-[D_{st}]^{1+q}\}}{(3q-1)[D(t)]^3}}_{>0} \underbrace{\frac{dD(t)}{dt}}_{<0} < 0,$$

(6.188)

$$\lim_{t\to\infty} {}^T K[P, P_{st}] = 0,$$

(6.189)

for $q \in (1/3, 1)$. Substituting the functions $m(t)$ and $D(t)$ given by (6.178) for $t = t_0$ into (6.186) and (6.188), respectively, we obtain the explicit relations

$$^T K[P, P_{st}] = \frac{\gamma}{2Q}[x_0]^2 e^{-2\gamma t}$$
$$+ \frac{g_q}{1-q}\left(1+q+(1-q)\left[1-e^{-(1+q)\gamma t}\right]^{2/(1+q)} - 2\left[1-e^{-(1+q)\gamma t}\right]^{(1-q)/(1+q)}\right)$$

(6.190)

and

$$\frac{d}{dt} {}^T K[P, P_{st}] = -\frac{1}{Q}\gamma^2 [x_0]^2 e^{-2\gamma t}$$
$$- 2\gamma g_q \, e^{-(1+q)\gamma t}\left(\left[\frac{1}{1-e^{-(1+q)\gamma t}}\right]^{2q/(1+q)} - \left[1-e^{-(1+q)\gamma t}\right]^{(1-q)/(1+q)}\right),$$

(6.191)

with

$$g_q = \frac{q}{3q-1}\left[\frac{2\pi q Q}{\gamma(1-q)}\left(\frac{\Gamma\left(\frac{1+q}{2(1-q)}\right)}{\Gamma\left(\frac{1}{1-q}\right)}\right)^2\right]^{(1-q)/(1+q)}, \qquad (6.192)$$

where $\Gamma(z)$ denotes the Gamma-function. Figure 6.8 shows the Lyapunov functional ^TK and the first derivative of ^TK for different values of q, that is, for different degrees of nonextensivity. Figure 6.8 also gives us the qualitative behavior of the free energy $F = U_L - Q\,^TS_q$ because of $F(t) = Q\,^TK(t) + C$, where C is constant (see above).

Langevin Equation

For solutions $P(x, t; u)$ of (6.170), for which the evolution equation

$$\frac{\partial}{\partial t}P(x,t|x',t';u) = -\frac{\partial}{\partial x}h(x)P(x,t|x',t';u)$$
$$+Q\frac{\partial^2}{\partial x^2}P^{q-1}(x,t;u)P(x,t|x',t';u) \qquad (6.193)$$

describes Markov transition probability densities $P(x, t|x', t'; u)$, we can regard (6.170) as a strongly nonlinear Fokker–Planck equation. Then, stochastic processes can be completely described by $P(x, t; u)$ and $P(x, t|x', t'; u)$, on the one hand, and by the Ito–Langevin equation [66]

$$\frac{d}{dt}X(t) = h(X) + [P(x,t;u)]^{(q-1)/2}\bigg|_{x=X(t)}\sqrt{Q}\,\Gamma(t), \qquad (6.194)$$

on the other hand, where $X(t)$ is distributed like $u(x)$ for $t = t_0$. Let us exploit (6.194) in order to compute transient and stationary solutions of (6.170). To this end, we simulate (6.194) by means of the averaging method for self-consistent Langevin equations, see Sect. 3.4.4. Figures 6.4 and 6.5 show the stationary distribution of (6.174) for $q = 0.8$. Diamonds represent results obtained by solving numerically the Ito–Langevin equation (6.194) for $h(x) = -\gamma x$. The solid line is obtained by plotting the exact solution (6.175). Figure 6.4 gives a log-normal plot. In contrast, in Fig. 6.5 results are depicted in a log-log plot. In the log-log plot the asymptotic behavior is described by straight lines with slopes $2/(1-q)$. Figures 6.6 and 6.7 show the increase of the variance $K(t)$ for a diffusion process given by (6.194) and $h = 0$:

$$\frac{d}{dt}X(t) = [P(x,t;u)]^{(q-1)/2}\bigg|_{x=X(t)}\sqrt{Q}\,\Gamma(t). \qquad (6.195)$$

The solid line gives the exact result $K(t) \propto t^{2/(1+q)}$, see (6.179) (see also (6.275) below). The diamonds represent results obtained from a numerical simulation of (6.195). The dotted line in Fig. 6.6 describes a normal diffusion process with $K(t) \propto t$. Figures 6.6 and 6.7 illustrate that we are dealing with an anomalous diffusion process.

Autocorrelation Functions

We can also derive autocorrelation functions $C^{mn}(t,t') = \langle X^m(t) X^n(t') \rangle$ of the Plastino–Plastino model [179]. Multiplying (6.193) with $P(x,t;u)$, we get

$$\frac{\partial}{\partial t} P(x,t;x',t';u) = -\frac{\partial}{\partial x} h(x) P(x,t;x',t';u)$$
$$+ Q \frac{\partial^2}{\partial x^2} P^{q-1}(x,t;u) P(x,t;x',t';u) . \quad (6.196)$$

Substituting $h(x) = -\gamma x$ into (6.196), multiplying with x and x', and integrating with respect to x and x', we find that

$$\frac{\partial}{\partial t} C^{11}(t,t') = -\gamma C^{11}(t,t') . \quad (6.197)$$

Solving this equation for the initial condition $C^{11}(t',t') = \langle X^2(t') \rangle$, we obtain

$$C^{11}(t,t') = \langle X^2(t') \rangle \exp\{-\gamma(t-t')\} . \quad (6.198)$$

By means of (6.198), we can compute the autocorrelation function $C^{11} = \langle X(t) X(t') \rangle$ for every pair (t,t') with $t \geq t'$ provided that the value of the second moment $\langle X^2 \rangle$ at time t' is given. For example, as we will show in the following section, the mean $M_1(t)$ and the variance $K(t)$ of the transient solution (6.177) is given by

$$M_1(t) = x_0 \exp\{-\gamma(t-t_0)\} , \quad (6.199)$$

$$K(t) = \frac{1}{3q-1} \left[\frac{2qQ[z_q]^{(1-q)}}{\gamma} (1 - \exp\{-(1+q)\gamma(t-t_0)\}) \right]^{2/(1+q)} , \quad (6.200)$$

for $q \in (1/3, 1)$. Replacing t by t' in (6.199) and (6.200), we can compute $\langle X^2(t') \rangle = K(t') + M_1^2(t')$. Substituting this result into (6.198), we obtain an analytical expression for $C^{11}(t,t')$. Figure 6.9 shows the exact analytical results and the numerical results for $C^{11}(t,t')$ and $\langle X^2(t) \rangle$.

In the stationary case, for $h(x) = \gamma x$ and $q \in (1/3, 1)$, we may substitute the stationary solution (6.175) into (6.193):

$$\frac{\partial}{\partial t} P(x,t|x',t';P_{\text{st}}) = \gamma \frac{\partial}{\partial x} x P(x,t|x',t';P_{\text{st}})$$
$$+ Q D_{\text{st}}^{q-1} \frac{\partial^2}{\partial x^2} \left[1 + \frac{\gamma}{2qQ\, D_{\text{st}}^{q-1}} (1-q) x^2 \right] P(x,t|x',t';P_{\text{st}}) . \quad (6.201)$$

From (6.201) we read off that in this case the diffusion coefficient $D_2(x,t,P) = Q P^{q-1}$ can indeed be regarded as the second Kramers–Moyal coefficient of a Markov diffusion process. From (6.201) one can derive evolution equations for the stationary autocorrelation functions $C^{mn}(z) = \langle X^m(t+z) X^n(t) \rangle$.

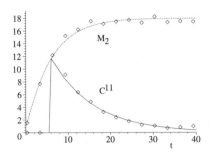

Fig. 6.9. *Solid line*: autocorrelation function $C^{11}(t,t')$ computed from (6.198), (6.199), (6.200). *Dashed line*: second moment $\langle X^2(t) \rangle$ computed from (6.199) and (6.200). *Diamonds*: $C^{11}(t,t')$ and $\langle X^2(t) \rangle$ obtained by solving numerically the Langevin equation (6.194) with $h(x) = -\gamma x$ using the averaging method for self-consistent Langevin equations. For $t < t'$ we put $C^{11} = 0$. Parameters: $x_0 = -1$, $t_0 = 0$, $t' = 6$; other parameters as in Fig. 6.4 except for $L = 3000$ [179]

For example, if we proceed just as in the previously discussed nonstationary case, we find

$$\frac{\mathrm{d}}{\mathrm{d}z} C^{11}(z) = -\gamma C^{11}(z) \,, \tag{6.202}$$

for $C^{11}(z) = \langle X(t+z)X(t) \rangle_{\mathrm{st}}$ and

$$\frac{\mathrm{d}}{\mathrm{d}z} C^{22}(z) = -\frac{\gamma(3q-1)}{q} \left[C^{22}(z) - \langle X^2 \rangle_{\mathrm{st}}^2 \right] \,, \tag{6.203}$$

for $C^{22}(z) = \langle X^2(t+z)X^2(t) \rangle_{\mathrm{st}}$. Solving these equations for the respective initial conditions, we get

$$C^{11}(z) = \langle X^2 \rangle_{\mathrm{st}} \exp\{-\gamma z\} \tag{6.204}$$

and

$$C^{22}(z) = \langle X^2 \rangle_{\mathrm{st}}^2 + \left[\langle X^4 \rangle_{\mathrm{st}} - \langle X^2 \rangle_{\mathrm{st}}^2 \right] \exp\left\{ -\frac{\gamma(3q-1)}{q} z \right\} \,, \tag{6.205}$$

with $\langle X^2 \rangle_{\mathrm{st}} = K_{\mathrm{st}}$. The amplitude of the exponential function in (6.205) is semi-positive because we have $\langle X^4 \rangle_{\mathrm{st}} - \langle X^2 \rangle_{\mathrm{st}}^2 = \langle [X^2 - \langle X^2 \rangle]^2 \rangle_{\mathrm{st}} \geq 0$. Consequently, the autocorrelation function $C^{22}(z)$ decays monotonically from $C^{22}(0) = \langle X^4 \rangle_{\mathrm{st}}$ to $\lim_{z \to \infty} C^{22}(z) = \langle X^2 \rangle_{\mathrm{st}}^2 = K_{\mathrm{st}}^2$. Next, let us compare the analytical results obtained from the Fokker–Planck description with numerical results obtained from simulations of the Ito–Langevin equation (6.194). Figures 6.10 and 6.11 show the correlations functions $C^{11}(z)$ and $C^{22}(z)$ as obtained from our analytical considerations on the nonlinear Fokker–Planck equation (6.174) and as obtained by simulations of the Langevin equation (6.194).

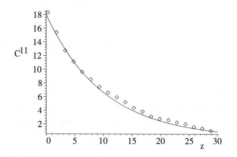

Fig. 6.10. Stationary autocorrelation function C^{11} computed from (6.204) (*solid line*) and from the Langevin equation (6.194) for $h(x) = -\gamma x$ (*diamonds*). Parameters as in Fig. 6.4 except for $L = 3000$ [179]

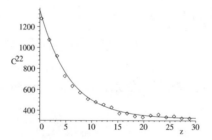

Fig. 6.11. Autocorrelation function C^{22} computed from (6.205) (*solid line*) and from the Langevin equation (6.194) with $h(x) = -\gamma x$ (*diamonds*). Parameters as in Fig. 6.4 except $L = 3000$ [179]

Generalizations

The entropy Fokker–Planck equation (6.174) has been generalized in order to study transitions between potential minima (Kramers escape rate) [387], the multivariate case [469], time-dependent drift coefficients [227], time-dependent coefficients in the multivariate case [99, 100, 401], and diffusion terms that are explicitly state-dependent [68, 401]. The relevance of exact time-dependent solutions of (6.174) has been studied in [405]. Finally, exact time-dependent solutions have been derived for (6.174) supplemented with source terms [140, 481, 497].

6.5.3 Sharma–Mittal Entropy Fokker–Planck Equation

A striking feature of the porous medium equation and the entropy Fokker–Planck equation by Plastino and Plastino is that they admit exact time-dependent solutions. Exact solutions are helpful tools for the illustration of fundamental principles and concepts in physics (for an example, see the decrease of the free energy above). Another entropy Fokker–Planck equation that allows for exact time-dependent solutions is the Fokker–Planck equation related to the Sharma–Mittal entropy [190, 191].

In the univariate case $x \in \Omega$, the Sharma–Mittal entropy $^{SM}S_{q,r}$ (6.130) is defined for $q, r > 0$ and $q, r \neq 1$ by

$$^{SM}S_{q,r} = \frac{1}{q-1}\left[1 - \left(\int_\Omega P^r \, dx\right)^{(q-1)/(r-1)}\right], \quad (6.206)$$

$$^{SM}S_{q=1,r} = \,^{R}S_r = \frac{1}{1-r}\ln\int_\Omega P^r \, dx, \quad (6.207)$$

$$^{SM}S_{q,r=1} = \,^{G}S_q = \frac{1}{q-1}\left[1 - \exp\left\{(q-1)\int_\Omega P\ln P \, dx\right\}\right]. \quad (6.208)$$

As shown in Sect. 6.4.2, the Sharma–Mittal entropy includes as special cases the entropy $^{T}S_q$ and the Boltzmann entropy ^{B}S, which we list here for the sake of completeness:

$$^{SM}S_{q=r,r} = \,^{T}S_q = \frac{1}{1-q}\int_\Omega (P^q - P)\, dx, \quad (6.209)$$

$$^{SM}S_{q=1,r=1} = \,^{B}S = -\int_\Omega P\ln P \, dx. \quad (6.210)$$

Substituting the entropy measures $^{SM}S_{q,r}$, $^{R}S_r$, $^{T}S_q$, and $^{G}S_q$ into (6.46), we obtain the entropy Fokker–Planck equations

$$\frac{\partial}{\partial t}\,^{R}P(x,t;u) = -\frac{\partial}{\partial x}h(x)\,^{R}P + Q\left[\int_\Omega (^{R}P)^r \, dx\right]^{-1}\frac{\partial^2}{\partial x^2}(^{R}P)^r, \quad (6.211)$$

$$\frac{\partial}{\partial t}\,^{T}P(x,t;u) = -\frac{\partial}{\partial x}h(x)\,^{T}P + Q\frac{\partial^2}{\partial x^2}(^{T}P)^q, \quad (6.212)$$

$$\frac{\partial}{\partial t}\,^{G}P(x,t;u) = -\frac{\partial}{\partial x}h(x)\,^{G}P + Q\left[e^{(q-1)\int_\Omega \,^{G}P\ln\,^{G}P\,dx}\right]\frac{\partial^2}{\partial x^2}\,^{G}P,$$

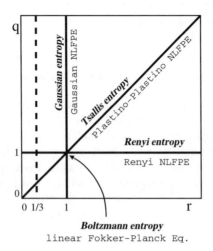

Fig. 6.12. Limiting cases of the two-parametric Sharma–Mittal entropy $^{SM}S_{r,q}$ and limiting cases of the Sharma–Mittal entropy Fokker–Planck equation (6.214) (NLFPE = nonlinear Fokker–Planck equation); reprinted from [191], © 2000, with permission from Elsevier

$$\tag{6.213}$$

$$\frac{\partial}{\partial t}{}^{SM}P(x,t;u) = -\frac{\partial}{\partial x}h(x)\,{}^{SM}P + Q\left[\int_\Omega \left({}^{SM}P\right)^r \mathrm{d}x\right]^{\frac{q-r}{r-1}} \frac{\partial^2}{\partial x^2}\left({}^{SM}P\right)^r , \tag{6.214}$$

for $t \geq t_0$. Figure 6.12 depicts the limiting cases $^{R}S_r$, $^{T}S_q$, and $^{G}S_q$ of the entropy functional $^{SM}S_{q,r}$ and the corresponding limiting cases (6.211-6.213) of the entropy Fokker–Planck equation (6.214) in the parameter-space (r,q). Note that the entropy Fokker–Planck equation (6.212) corresponds to the one discussed in the previous section.

The evolution equations (6.211-6.214) can be used to define generalized Ornstein–Uhlenbeck processes and generalized Wiener processes related to the Sharma–Mittal entropy. To this end, we put $h(x) = -\gamma x$ for $\gamma \geq 0$ and consider the entropy Fokker–Planck equations

$$\frac{\partial}{\partial t}{}^{R}P(x,t;u) = \gamma\frac{\partial}{\partial x}x\,{}^{R}P + Q\left[\int_\Omega \left({}^{R}P\right)^r \mathrm{d}x\right]^{-1} \frac{\partial^2}{\partial x^2}\left({}^{R}P\right)^r , \tag{6.215}$$

$$\frac{\partial}{\partial t}{}^{T}P(x,t;u) = \gamma\frac{\partial}{\partial x}x\,{}^{T}P + Q\frac{\partial^2}{\partial x^2}\left({}^{T}P\right)^q , \tag{6.216}$$

6.5 Examples and Applications

$$\frac{\partial}{\partial t} {}^G P(x,t;u) = \gamma \frac{\partial}{\partial x} x \, {}^G P + Q \left[e^{(q-1) \int_\Omega {}^G P \ln {}^G P \, dx} \right] \frac{\partial^2}{\partial x^2} {}^G P \,, \quad (6.217)$$

$$\frac{\partial}{\partial t} {}^{SM} P(x,t;u) = \gamma \frac{\partial}{\partial x} x \, {}^{SM} P(x,t;u) + Q \left[\int_\Omega ({}^{SM} P)^r \, dx \right]^{\frac{q-r}{r-1}} \frac{\partial^2}{\partial x^2} ({}^{SM} P)^r \,, \quad (6.218)$$

with

$$r > 1/3, \ q > 0, \ \text{and} \ r, q \neq 1 \,. \quad (6.219)$$

The restriction on r will be explained below. We will study probability densities ${}^{(\cdot)} P(x,t;u) = \langle \delta(x - X(t)) \rangle$ subjected to natural boundary conditions that describe a random variable $X(t) \in \Omega = \mathbb{R}$ for $t \geq t_0 = 0$. First, let us discuss the stationary case.

Stationary Solutions

It is convenient to introduce the functionals ${}^{(\cdot)} \mathcal{U}(t) : {}^{(\cdot)} P(x,t;u) \to {}^{(\cdot)} \mathcal{U}[{}^{(\cdot)} P]$, which occur in the diffusion coefficients of (6.215-6.218). These functionals read

$$^R \mathcal{U} = \int_\Omega (^R P)^r \, dx \,, \quad (6.220)$$

$$^T \mathcal{U} = 1 \,, \quad (6.221)$$

$$^G \mathcal{U} = \exp \left\{ (1-q) \int_\Omega {}^G P \ln {}^G P \, dx \right\} \,, \quad (6.222)$$

$$^{SM} \mathcal{U} = \left[\int_\Omega ({}^{SM} P)^r \, dx \right]^{\frac{q-r}{1-r}} \,. \quad (6.223)$$

We assume that ${}^{(\cdot)} \mathcal{U}(t)$ is finite and positive, which implies that we can write the entropy Fokker–Planck equations (6.215-6.218) as

$$\frac{\partial}{\partial t} {}^\varpi P(x,t;u) = \gamma \frac{\partial}{\partial x} x \, {}^\varpi P + \frac{Q}{{}^\varpi \mathcal{U}(t)} \frac{\partial^2}{\partial x^2} ({}^\varpi P)^\nu \,, \quad (6.224)$$

with $(\varpi, \nu) = (R, r)$, $(\varpi, \nu) = (T, q)$, $(\varpi, \nu) = (G, 1)$, and $(\varpi, \nu) = (SM, r)$. In the stationary case, the functionals (6.220-6.223) are positive constants, ${}^{(\cdot)} \mathcal{U}_{\text{st}} > 0$. Consequently, the stationary solutions of (6.224) correspond to the stationary solution (6.173) of (6.170) for $U_0(x) = -\gamma x^2/2$ when replacing q by ν and Q by $Q/{}^{(\cdot)} \mathcal{U}_{\text{st}} > 0$. Thus, we obtain

$$^\varpi P_{\text{st}}(x) = {}^\varpi D_{\text{st}} \left[\left\{ 1 - \frac{b_\nu \, {}^\varpi \mathcal{U}_{\text{st}}}{({}^\varpi D_{\text{st}})^{\nu-1}} (\nu - 1) x^2 \right\}_+ \right]^{\frac{1}{\nu-1}} \,, \quad (6.225)$$

for $(\varpi, \nu) = (R, r)$, $(\varpi, \nu) = (T, q)$, and $(\varpi, \nu) = (SM, r)$. For $(\varpi, \nu) = (G, 1)$ the entropy Fokker–Planck equation (6.224) can be regarded as a linear Fokker–Planck equation when $^G\mathcal{U}(t)$ is considered as a time-dependent coefficient, which means that in the stationary case (i.e, for $^G\mathcal{U}(t) = {}^G\mathcal{U}_{st}$) we are dealing with a Boltzmann distribution

$$^GP_{st}(x) = \underbrace{\sqrt{\frac{\gamma\, {}^G\mathcal{U}_{st}}{2\pi Q}}}_{{}^GD_{st}} \exp\left(-\frac{\gamma\, {}^G\mathcal{U}_{st}}{2Q} x^2\right). \qquad (6.226)$$

The factor b_ν is defined as $b_\nu = \gamma/(2\nu Q)$. We assume that the normalization constants ${}^{(\cdot)}D_{st}$ are finite and positive. Since $b_\nu > 0$, ${}^\varpi\mathcal{U}_{st} > 0$, and $({}^\varpi D_{st})^{\nu-1} > 0$, the probability densities (6.225) correspond to classical solutions for $\nu \in (0, 1)$ (i.e., we have ${}^\varpi P_{st} \in C^\infty$). For $\nu > 1$ the probability densities (6.225) are continuous but not continuously differentiable at $x = \pm\sqrt{({}^\varpi D_{st})^{\nu-1}/[(\nu-1)b_\nu\,{}^\varpi\mathcal{U}_{st}]}$ and describe cut-off distributions.

Returning to our objective to find the stationary solutions of the entropy Fokker–Planck equation (6.224), the normalization constants ${}^{(\cdot)}D_{st}$ along with the functionals ${}^{(\cdot)}\mathcal{U}_{st}$ can be calculated from the conditions (6.220-6.223) and the normalization condition $\int_\Omega {}^{(\cdot)}P_{st}\,dx = 1$. The key step here is to use linear variable transformations of the form $x = cy$ in order to express the integrals $\int_\Omega {}^{(\cdot)}P\,dx = 1$ and $\int_\Omega ({}^{(\cdot)}P)^\nu\,dx$ by means of the constants z_ν and $z_{\nu\nu}$ listed in Table 6.3. The results obtained from this procedure are summarized in Table 6.4. Equations (6.225) and (6.226) in combination with Tables 6.3 and 6.4 provide a complete description of the stationary solutions of the (6.215-6.218) in the parameter range (6.219).

Power Law Distributions

For $\nu \in (1/3, 1)$ and large $|x|$ the stationary distributions (6.225) decay for $|x| \to \infty$ like

$$^\varpi P_{st}(x) \propto |x|^{-\frac{2}{1-\nu}}. \qquad (6.227)$$

This means that we are dealing with power law distributions with an exponent $\delta = -2/(1-\nu)$ that is in the range $(3, \infty)$.

Cut-Off Distributions

For $\nu > 1$ and $\varpi \neq G$ the stationary distributions (6.225) correspond to cut-off distributions. They are positive on the interval $(-x^*, x^*)$, where x^* denotes the argument for which the curled bracket in (6.225) vanishes. That is, we have

$$x^* = \sqrt{\frac{({}^\varpi D_{st})^{\nu-1}}{(\nu-1)b_\nu\,{}^\varpi\mathcal{U}_{st}}}. \qquad (6.228)$$

For $|x| \geq x^*$ we have ${}^\varpi P_{st}(x) = 0$.

6.5 Examples and Applications 265

Table 6.3. Definition of the integrals z_ν and $z_{\nu\nu}$. Representations in terms of Beta- and Gamma-functions and limits for $\nu \to 1$ [191]

$$z_\nu = \frac{1}{\sqrt{|1-\nu|}} \int_{-\infty}^{\infty} \left[\{1 + \mathrm{sgn}(1-\nu)y^2\}_+\right]^{\frac{1}{\nu-1}} dy \,,\ \nu > 0\,,\ \nu \neq 1$$

$$z_{\nu\nu} = \frac{1}{\sqrt{|1-\nu|}} \int_{-\infty}^{\infty} \left[\{1 + \mathrm{sgn}(1-\nu)y^2\}_+\right]^{\frac{\nu}{\nu-1}} dy \,,\ \nu > \frac{1}{3}\,,\ \nu \neq 1$$

$$z_\nu = \begin{cases} \sqrt{\dfrac{1}{1-\nu}}\, B\!\left(\dfrac{1}{2}, \dfrac{1+\nu}{2(1-\nu)}\right) = \sqrt{\dfrac{\pi}{1-\nu}}\, \dfrac{\Gamma\!\left(\dfrac{1+\nu}{2(1-\nu)}\right)}{\Gamma\!\left(\dfrac{1}{1-\nu}\right)} & \text{for } \nu \in (0,1) \\[2ex] \sqrt{\pi} & \text{for } \nu \to 1 \\[2ex] \sqrt{\dfrac{1}{\nu-1}}\, B\!\left(\dfrac{1}{2}, \dfrac{\nu}{\nu-1}\right) = \sqrt{\dfrac{\pi}{\nu-1}}\, \dfrac{\Gamma\!\left(\dfrac{\nu}{\nu-1}\right)}{\Gamma\!\left(\dfrac{3\nu-1}{2(\nu-1)}\right)} & \text{for } \nu > 1 \end{cases}$$

$$z_{\nu\nu} = \begin{cases} \sqrt{\dfrac{1}{1-\nu}}\, B\!\left(\dfrac{1}{2}, \dfrac{3\nu-1}{2(1-\nu)}\right) = \sqrt{\dfrac{\pi}{1-\nu}}\, \dfrac{\Gamma\!\left(\dfrac{3\nu-1}{2(1-\nu)}\right)}{\Gamma\!\left(\dfrac{\nu}{1-\nu}\right)} & \text{for } \nu \in \left(\dfrac{1}{3},1\right) \\[2ex] \sqrt{\pi} & \text{for } \nu \to 1 \\[2ex] \sqrt{\dfrac{1}{\nu-1}}\, B\!\left(\dfrac{1}{2}, \dfrac{2\nu-1}{\nu-1}\right) = \sqrt{\dfrac{\pi}{\nu-1}}\, \dfrac{\Gamma\!\left(\dfrac{2\nu-1}{\nu-1}\right)}{\Gamma\!\left(\dfrac{5\nu-3}{2(\nu-1)}\right)} & \text{for } \nu > 1 \end{cases}$$

$$\frac{z_\nu}{z_{\nu\nu}} = \frac{3\nu-1}{2\nu}\,,\ \nu > \frac{1}{3}$$

Table 6.4. Stationary values for $^\varpi D$, $^\varpi\mathcal{U}$, and the ratio $^\varpi\mathcal{U}/(^\varpi D)^{\nu-1}$ [191]

ϖ	$^\varpi D_{\text{st}}$	$^\varpi\mathcal{U}_{\text{st}}$	$^\varpi\mathcal{U}_{\text{st}}/(^\varpi D_{\text{st}})^{\nu-1}$
R	$\sqrt{b_r \dfrac{z_{rr}}{(z_r)^3}} = \sqrt{\dfrac{2rb_r}{(3r-1)(z_r)^2}}$	$\dfrac{(z_r)^2}{b_r}\left(^R D_{\text{st}}\right)^{r+1}$	$\dfrac{z_{rr}}{z_r} = \dfrac{2r}{3r-1}$
T	$\left(\dfrac{b_q}{(z_q)^2}\right)^{\frac{1}{q+1}}$	1	$\dfrac{(z_q)^2}{b_q}\left(^T D_{\text{st}}\right)^2$
G	$\sqrt{\dfrac{\gamma}{2\pi Q}}\,^G\mathcal{U}_{\text{st}} =$ $\left[\dfrac{\gamma}{2\pi Q}(\sqrt{e})^{(q-1)}\right]^{\frac{1}{1+q}}$	$\left(\dfrac{\gamma}{2\pi e Q}\right)^{\frac{1-q}{1+q}}$	–
SM	$\left[\dfrac{b_r}{(z_r)^2}\left(\dfrac{z_{rr}}{z_r}\right)^{\frac{q-r}{1-r}}\right]^{\frac{1}{1+q}}$ $= \left[\dfrac{b_r}{(z_r)^2}\left(\dfrac{2r}{3r-1}\right)^{\frac{q-r}{1-r}}\right]^{\frac{1}{1+q}}$	$\dfrac{(z_r)^2}{b_r}\left(^{SM} D_{\text{st}}\right)^{r+1}$	$\dfrac{(z_r)^2}{b_r}\left(^{SM} D_{\text{st}}\right)^2$

Border $r = 1/3$

In closing these considerations, we discuss the limiting case $r \downarrow 1/3$. For $\nu = 1/3$ and for large x, the integrand of $z_{\nu\nu}$ is proportional to $1/|x|$, see Table 6.3. Consequently, the integral $z_{\nu\nu}$ diverges. For $0 < \nu < 1/3$ the integrand of $z_{\nu\nu}$ behaves for large x as $1/|x|^m$ with $m \in (0,1)$, which again implies the divergence of the integral. Since $z_{\nu\nu}$ occurs in $^R D_{\text{st}}$ and $^{SM} D_{\text{st}}$, the stationary solutions of the Renyi and Sharma–Mittal entropy Fokker–Planck equations (6.215) and (6.218) are not well-defined for $0 < r \leq 1/3$, respectively. There is no effect on the stationary solutions of Plastino–Plastino Fokker–Planck equation (6.216). However, tracing the entropy Fokker–Planck equation (6.216) back to the entropy functional $^T S_q$ given by (6.209), we realize that the entropy $^T S_q$ of the stationary solution (6.225) with $(\varpi, \nu) = (T, q)$ also diverges for $0 < q \leq 1/3$. For this reason, we require $q > 1/3$, as indicated by (6.219).

Let us briefly discuss the situation in which the dashed line shown in Fig. 6.12 is approached from the right-hand side. In the case of the Renyi

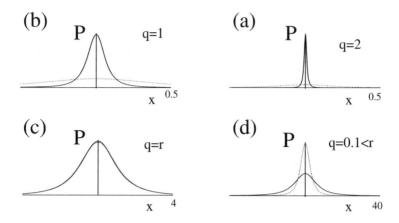

Fig. 6.13. Plots of stationary probability densities typically found close to the border line $r = 1/3$ computed from (6.225) and (6.226). Parameters for all panels are $Q = \gamma = 1.0$, $r = 0.400$ (*dotted lines*), and $r = 0.335$ (*solid lines*). Probability densities within each panel are drawn on the same scale. For each panel the scales of the X- and Y-axes are chosen appropriately. Counterclockwise q decreases like: $q > 1 > r$, $q = 1 > r$, $q = r$, and $q < r$. In detail we have (**a**) $^{SM}P_{st}(x)$ for $q = 2$, (**b**) $^{R}P_{st}(x)$ ($q=1$), (**c**) $^{T}P_{st}(x)$ ($q=r$), and **d**) $^{SM}P_{st}(x)$ for $q = 0.1$; reprinted from [191], © 2000, with permission from Elsevier

entropy Fokker–Planck equation the ratio $^{R}\mathcal{U}_{st}/\left(^{R}D_{st}\right)^{r-1}$, which is a measure of how fast the stationary solution (6.225) falls off at the origin, runs to infinity for $r \downarrow 1/3$ (see Table 6.4), which means that for $r \downarrow 1/3$ the stationary solution (6.225) of The Renyi entropy Fokker–Planck equation (6.215) converges to a delta distribution. For $q > r$, with the same reasoning, we can conclude that the stationary solutions of the Sharma–Mittal entropy Fokker–Planck equation (6.218) converge in the limit $r \downarrow 1/3$ to delta distributions. In contrast, for $q < r$, the normalization constant $^{SM}D_{st}$ vanishes in the limit $r \downarrow 1/3$ and, in turn, the ratio $^{SM}\mathcal{U}_{st}/\left(^{SM}D_{st}\right)^{r-1}$ converges to zero, see Table 6.4. Then, the stationary solution (6.225) converges to the uniform distribution. In addition, the Plastino–Plastino model (6.216) is not at all affected by the limit $r \downarrow 1/3$ and yields a classical solution, that is, an infinitely differentiable function. These results are summarized in (6.229):

$$r \downarrow 1/3 \begin{cases} q > 1 & : {}^{SM}P_{st} \to \delta(x) \\ q = 1 & : {}^{R}P_{st} \to \delta(x) \\ 1 > q > r & : {}^{SM}P_{st} \to \delta(x) \\ q = r & : {}^{T}P_{st} \in C^{\infty} \text{ (classical solution)} \\ q < r & : {}^{SM}P_{st} \to \text{uniform distribution} \end{cases}.$$

(6.229)

Typical examples of stationary solutions close to the border line $r = 1/3$ are shown in Fig. 6.13.

Transient Solutions

Let us first consider (6.224) for $(\varpi, \nu) \neq (G, 1)$. In view of the structure of the stationary solution (6.225), we are looking for solutions of the form

$$\varpi P(x,t;u) = {}^\varpi D(t) \left[\left\{ 1 - \frac{{}^\varpi C(t)}{2}(\nu-1)(x-m(t))^2 \right\}_+ \right]^{\frac{1}{\nu-1}}, \quad (6.230)$$

where ${}^\varpi D(t) > 0$, ${}^\varpi C(t) > 0$, and $m(t)$ are time-dependent functions. Just as in the stationary case, for $\nu > 1$ we regard the probability density (6.230) as a cut-off solution. The normalization of (6.230) determines the relation between ${}^\varpi D$ and ${}^\varpi C$:

$${}^\varpi C(t) = 2 \left[z_\nu \, {}^\varpi D(t) \right]^2 . \quad (6.231)$$

Using (6.231) and inserting the probability density (6.230) into (6.224), we obtain first order differential equations for ${}^\varpi D(t)$ and $m(t)$ described by

$$\frac{d}{dt} {}^\varpi D(t) = \gamma \, {}^\varpi D(t) - \frac{2\nu Q}{{}^\varpi \mathcal{U}(t)} \left[{}^\varpi D(t) \right]^{\nu+2}, \quad (6.232)$$

$$\frac{d}{dt} m(t) = -\gamma \, m(t) . \quad (6.233)$$

Next, we express the functional ${}^\varpi \mathcal{U}(t)$ in terms of ${}^\varpi D(t)$. Therefore, we substitute the probability density (6.230) together with (6.231) into the functionals (6.220), (6.221), and (6.223). We get

$${}^R \mathcal{U}(t) = \frac{z_{rr}}{z_r} \left[{}^R D(t) \right]^{r-1}, \quad (6.234)$$

$${}^T \mathcal{U}(t) = 1 , \quad (6.235)$$

$${}^{SM} \mathcal{U}(t) = \left[{}^{SM} D(t) \right]^{r-q} \left[\frac{z_{rr}}{z_r} \right]^{\frac{q-r}{1-r}} . \quad (6.236)$$

We can now eliminate the functional ${}^\varpi \mathcal{U}(t)$ occurring in (6.232) and the evolution equations for ${}^\varpi D(t)$ become

$$\frac{d}{dt} {}^R D(t) = \gamma \, {}^R D(t) - 2\nu Q \frac{[z_r]^3}{z_{rr}} \left[{}^R D(t) \right]^3 , \quad (6.237)$$

$$\frac{d}{dt} {}^T D(t) = \gamma \, {}^T D(t) - 2\nu Q \left[z_q \right]^2 \left[{}^T D(t) \right]^{q+2} , \quad (6.238)$$

6.5 Examples and Applications

$$\frac{d}{dt}{}^{SM}D(t) = \gamma\,{}^{SM}D(t) - 2\nu Q\,[z_r]^2 \left[\frac{z_r}{z_{rr}}\right]^{\frac{q-r}{1-r}} [{}^{SM}D(t)]^{q+2}. \quad (6.239)$$

For $\gamma > 0$ and ${}^\varpi P(x, 0; u) = u(x) = \delta(x - x_0)$, we can immediately solve the Bernoulli equations (6.237-6.239) as well as (6.233), which yields

$$
\begin{aligned}
{}^R D(t) &= \left[b_r \frac{z_{rr}}{[z_r]^3} \frac{1}{1-\exp\{-2\gamma t\}}\right]^{\frac{1}{2}} \\
&= \left[\frac{2rb_r}{(3r-1)[z_r]^2} \frac{1}{1-\exp\{-2\gamma t\}}\right]^{\frac{1}{2}}, \quad (6.240)
\end{aligned}
$$

$${}^T D(t) = \left[b_q \frac{1}{[z_q]^2}\frac{1}{1-\exp\{-(1+q)\gamma t\}}\right]^{\frac{1}{1+q}}, \quad (6.241)$$

$${}^{SM} D(t) = \left[b_r \frac{1}{K_{r,q}[z_r]^2}\frac{1}{1-\exp\{-(1+q)\gamma t\}}\right]^{\frac{1}{1+q}}, \quad (6.242)$$

and

$$m(t) = x_0\, e^{-\gamma t}, \quad (6.243)$$

where $K_{r,q}$ is given by

$$K_{r,q} = \left[\frac{z_r}{z_{rr}}\right]^{\frac{q-r}{1-r}} = \left[\frac{3r-1}{2r}\right]^{\frac{q-r}{1-r}}$$

$$\Rightarrow K_{1,1} = 1, \quad K_{r,q=1} = \frac{3r-1}{2r}, \quad \lim_{r\to 1} K_{r,q} = \left[\sqrt{e}\right]^{(1-q)}. \quad (6.244)$$

The limiting case $\lim_{r\to 1} K_{r,q}$ listed in (6.244) can be found by rewriting the ratio z_{rr}/z_r as $\int [\{1+(1-r)(z_r)^2 x^2\}_+]^{r/(r-1)}\,dx = \int u_r(x)^r dx$ and calculating the limit $\lim_{r\to 1}[z_r/z_{rr}]^{1/(1-r)} = \lim_{r\to 1}[\int u_r(x)^r dx]^{1/(r-1)}$, which gives us the limiting case $\exp\left[\lim_{r\to 1}\int u_r(x)\ln u_r(x)dx\right] = 1/\sqrt{e}$ along the lines of $\lim_{r\to 1}{}^{SM}S_{q,r} = {}^G S_q$, see Sect. 6.4.2. The probability densities (6.230) along with the relations (6.240-6.244), and (6.231) describe transient probability densities of generalized Ornstein–Uhlenbeck processes for initial delta distributions centered at x_0.

Let us focus on the entropy Fokker–Planck equation (6.217) or alternatively on (6.224) for $(\varpi,\nu) = (G,1)$. In what follows, we will demonstrate that (6.217) admits time-dependent Gaussian distributions (whence the label G). Using a notation similar to (6.230), we are looking for a solution of the form

$$^G P(x,t;u) = {^G}D(t)\exp\left[-\frac{{^G}C(t)}{2}(x-m(t))^2\right]. \tag{6.245}$$

Due to the normalization of the probability density (6.245), we find

$$^G C(t) = 2\left[\sqrt{\pi}\,{^G}D(t)\right]^2 \tag{6.246}$$

(see (3.106)), which is also consistent with the limiting case $\nu \to 1$ of (6.231) (cf. also Table 6.3). Inserting the probability density (6.245) into (6.224), we again obtain $dm/dt = -\gamma m$ and, in addition, the evolution equation

$$\frac{d}{dt}\,{^G}D(t) = \gamma\,{^G}D(t) - \frac{2\pi Q}{^G\mathcal{U}(t)}\left[{^G}D(t)\right]^3. \tag{6.247}$$

Substituting (6.245) into the functional $^G\mathcal{U}$ (see (6.222)) results in

$$^G\mathcal{U}(t) = \left(\sqrt{e}\right)^{q-1}\left[{^G}D(t)\right]^{1-q}. \tag{6.248}$$

Finally, we insert the functional (6.248) into (6.247), which gives us

$$\frac{d}{dt}\,{^G}D(t) = \gamma\,{^G}D(t) - 2\pi Q\left(\sqrt{e}\right)^{1-q}\left[{^G}D(t)\right]^{q+2}. \tag{6.249}$$

Solving (6.233) and (6.249) for an initial delta distribution $^G P(x,0;u) = u(x) = \delta(x-x_0)$ yields (6.243) and

$$^G D(t) = \left[\frac{\gamma}{2\pi Q}\left(\sqrt{e}\right)^{q-1}\frac{1}{1-\exp\{-(1+q)\gamma t\}}\right]^{\frac{1}{1+q}}. \tag{6.250}$$

In the limits $q \to 1$ and $r \to 1$ we obtain the time-dependent solution $P(x,t;\delta(x-x_0))$ of an Ornstein–Uhlenbeck process given by $P(x,t;\delta(x-x_0)) = P(x,t|x_0,t_0;\delta(x-x_0))$ with $P(x,t|x_0,t_0;\delta(x-x_0))$ described by (3.114), see Sect. 3.9.5.

Evolution of Variances

To classify the previously derived solutions into anomalous diffusion processes and normal diffusion processes, we first derive a general expression for the variance $^{\varpi}K = \langle X^2\rangle - \langle X\rangle^2$ of the probability densities in question. The first moment of the probability densities (6.230) and (6.245) equals $m(t)$ and the variance $^G K$ of the Gaussian solution (6.245) is determined by

$$^G K(t) = \frac{1}{^G C(t)}. \tag{6.251}$$

For the probability density (6.230) a similar expression can be found,

6.5 Examples and Applications

Table 6.5. Definition of the integral $z_{\nu\sigma}$. Representations in terms of Beta- and Gamma-functions, and limit for $\nu \to 1$ [191]

$$z_{\nu\sigma} = \frac{1}{\sqrt{|1-\nu|^3}} \int_{-\infty}^{\infty} y^2 \left[\{1 + \text{sgn}(1-\nu)y^2\}_+\right]^{\frac{1}{\nu-1}} dy, \quad \nu > 1/3, \; \nu \neq 1$$

$$z_{\nu\sigma} = \begin{cases} \sqrt{\frac{1}{(1-\nu)^3}} B\left(\frac{3}{2}, \frac{3\nu-1}{2(1-\nu)}\right) = \frac{1}{2}\sqrt{\frac{\pi}{(1-\nu)^3}} \dfrac{\Gamma\left(\frac{3\nu-1}{2(1-\nu)}\right)}{\Gamma\left(\frac{1}{1-\nu}\right)} & \text{for } \nu \in \left(\frac{1}{3}, 1\right) \\[2ex] \frac{1}{2}\sqrt{\pi} & \text{for } \nu \to 1 \\[2ex] \sqrt{\frac{1}{(\nu-1)^3}} B\left(\frac{3}{2}, \frac{\nu}{\nu-1}\right) = \frac{1}{2}\sqrt{\frac{\pi}{(\nu-1)^3}} \dfrac{\Gamma\left(\frac{\nu}{\nu-1}\right)}{\Gamma\left(\frac{5\nu-3}{2(\nu-1)}\right)} & \text{for } \nu > 1 \end{cases}$$

$$^\varpi K(t) = 2\frac{z_{\nu\sigma}}{z_\nu}\frac{1}{^\varpi C(t)} = \frac{2}{3\nu-1}\frac{1}{^\varpi C(t)}, \qquad (6.252)$$

with the integral $z_{\nu\sigma}$ defined in Table 6.5. Notice that for $\nu \in (0,\infty)$ the integral $z_{\nu\sigma}$ converges (diverges) if and only if $z_{\nu\nu}$ converges (diverges), because the integrands of $z_{\nu\nu}$ and $z_{\nu\sigma}$ have the same asymptotic behavior for $|x| \to \infty$ (we can rewrite the exponent $\nu/(\nu-1)$ as $1/(\nu-1)+1$). Furthermore, note that by exploiting the definition of the Gamma-function (i.e., $\Gamma(x+1) = x\Gamma(x)$), we can find simple expressions for the ratios $z_{\nu\sigma}/z_\nu$, $z_{\nu\sigma}/z_{\nu\nu}$, and $z_\nu/z_{\nu\nu}$, see Table 6.6.

Using (6.251) and (6.252) in combination with $^G C = 2\pi(^G D)^2$ and $^\varpi C = 2(z_\nu \, ^\varpi D)^2$ (see (6.246) and (6.231)), we can express the variances $^\varpi K(t)$ in terms of the functions $^\varpi D(t)$. Finally, by means of the time-dependent solutions of $^\varpi D(t)$ given by (6.240-6.242) and (6.250), we can compute the evolution of the respective variances:

$$^R K(t) = \frac{1}{b_r}\frac{z_{r\sigma}}{z_{rr}}(1 - \exp(-2\gamma t)) = \frac{1}{2r\, b_r}(1 - \exp(-2\gamma t)), \qquad (6.253)$$

$$^T K(t) = \frac{z_{q\sigma}}{z_q}\left[\frac{1}{b_q}[z_q]^{(1-q)}(1 - \exp(-[1+q]\gamma t))\right]^{\frac{2}{1+q}}$$

$$= \frac{1}{3q-1}\left[\frac{1}{b_q}[z_q]^{(1-q)}(1 - \exp(-[1+q]\gamma t))\right]^{\frac{2}{1+q}}, \qquad (6.254)$$

Table 6.6. Some relations between Gamma-functions and the ratios $z_{\nu\sigma}/z_\nu$, $z_{\nu\sigma}/z_{\nu\nu}$, and $z_\nu/z_{\nu\nu}$

$$\Gamma\left(\frac{\nu}{1-\nu}\right)\bigg/\Gamma\left(\frac{1}{1-\nu}\right)=\frac{\Gamma(x)}{\Gamma(x+1)}=\frac{1}{x} \qquad x=\frac{\nu}{(1-\nu)} \qquad \nu<1$$

$$\Gamma\left(\frac{3\nu-1}{2(1-\nu)}\right)\bigg/\Gamma\left(\frac{1+\nu}{2(1-\nu)}\right)=\frac{\Gamma(x)}{\Gamma(x+1)}=\frac{1}{x} \qquad x=\frac{3\nu-1}{2(1-\nu)} \qquad \nu \in (1/3, 1)$$

$$\Gamma\left(\frac{\nu}{\nu-1}\right)\bigg/\Gamma\left(\frac{2\nu-1}{\nu-1}\right)=\frac{\Gamma(x)}{\Gamma(x+1)}=\frac{1}{x} \qquad x=\frac{\nu}{(\nu-1)} \qquad \nu>1$$

$$\Gamma\left(\frac{3\nu-1}{2(1-\nu)}\right)\bigg/\Gamma\left(\frac{5\nu-3}{2(\nu-1)}\right)=\frac{\Gamma(x)}{\Gamma(x+1)}=\frac{1}{x} \qquad x=\frac{3\nu-1}{2(\nu-1)} \qquad \nu>1$$

$$\frac{z_{\nu\sigma}}{z_\nu}=\frac{1}{3\nu-1}, \quad \frac{z_{\nu\sigma}}{z_{\nu\nu}}=\frac{1}{2\nu}, \quad \frac{z_\nu}{z_{\nu\nu}}=\frac{3\nu-1}{2\nu} \qquad \nu>1/3$$

$$^G K(t) = \left[\frac{Q\left(\sqrt{2\pi e}\right)^{(1-q)}}{\gamma}(1-\exp(-[1+q]\gamma t))\right]^{\frac{2}{1+q}}, \qquad (6.255)$$

$$^{SM}K(t) = \frac{z_{r\sigma}}{z_r}\left[\frac{K_{r,q}}{b_r}[z_r]^{(1-q)}(1-\exp(-[1+s]\gamma t))\right]^{\frac{2}{1+q}}$$

$$= \frac{1}{3r-1}\left[\frac{K_{r,q}}{b_r}[z_r]^{(1-q)}(1-\exp(-[1+s]\gamma t))\right]^{\frac{2}{1+q}}. \qquad (6.256)$$

Consequently, in the stationary case, we obtain

$$^R K_{st} = \frac{1}{2r\,b_r} = \frac{Q}{\gamma}, \qquad (6.257)$$

$$^T K_{st} = \frac{1}{3q-1}\left[\frac{z_q^{(1-q)}}{b_q}\right]^{\frac{2}{1+q}}, \qquad (6.258)$$

$$^G K_{st} = \left[\frac{Q\left(\sqrt{2\pi e}\right)^{(1-q)}}{\gamma} \right]^{\frac{2}{1+q}}, \qquad (6.259)$$

$$^{SM}K_{st} = \frac{1}{3r-1} \left[\frac{K_{r,q}}{b_r} \, [z_r]^{(1-q)} \right]^{\frac{2}{1+q}}. \qquad (6.260)$$

Diffusion Equations

In the case of free Brownian motion, that is, for $\gamma = 0$, we are dealing with the nonlinear diffusion equations

$$\frac{\partial}{\partial t} {}^R P(x,t;u) = Q \left[\int_\Omega ({}^R P)^r \, dx \right]^{-1} \frac{\partial^2}{\partial x^2} ({}^R P)^r, \qquad (6.261)$$

$$\frac{\partial}{\partial t} {}^T P(x,t;u) = Q \frac{\partial^2}{\partial x^2} ({}^T P)^q, \qquad (6.262)$$

$$\frac{\partial}{\partial t} {}^G P(x,t;u) = Q \, e^{(q-1) \int_\Omega {}^G P \ln {}^G P \, dx} \frac{\partial^2}{\partial x^2} {}^G P, \qquad (6.263)$$

$$\frac{\partial}{\partial t} {}^{SM} P(x,t;u) = Q \left[\int_\Omega ({}^{SM}P)^r \, dx \right]^{\frac{q-r}{r-1}} \frac{\partial^2}{\partial x^2} ({}^{SM}P)^r. \qquad (6.264)$$

These evolution equations are solved again by transient solutions of the form (6.230) and (6.245), because in the preceding discussion we did not exclude the case $\gamma = 0$. In order to determine the evolution of the coefficients $^{\varpi}D(t)$ we can now proceed in two ways: either we perform the limit $\gamma \to 0$ in the time-dependent solutions $^{\varpi}D(t)$ given by (6.240-6.242) and (6.250), or we solve the corresponding evolution equations (6.237-6.239) and (6.247) under the initial condition $u(x) = \delta(x - x_0)$ putting $\gamma = 0$. Both approaches yield the following results:

$$^R D(t) = \left[\frac{1}{4rQ} \frac{z_{rr}}{[z_r]^3} \frac{1}{t} \right]^{\frac{1}{2}} = \left[\frac{1}{2(3r-1)Q \, [z_r]^2} \frac{1}{t} \right]^{\frac{1}{2}}, \qquad (6.265)$$

$$^T D(t) = \left[\frac{1}{2q(1+q)Q} \frac{1}{[z_q]^2} \frac{1}{t} \right]^{\frac{1}{1+q}}, \qquad (6.266)$$

$$^G D(t) = \left[\frac{(\sqrt{e})^{q-1}}{2\pi(1+q)Q} \frac{1}{t} \right]^{\frac{1}{1+q}}, \qquad (6.267)$$

$$^{SM}D(t) = \left[\frac{1}{2r(1+q)Q} \frac{1}{K_{r,q}[z_r]^2} \frac{1}{t}\right]^{\frac{1}{1+q}}, \qquad (6.268)$$

and
$$m(t) = x_0 . \qquad (6.269)$$

The functions $^\varpi C(t)$ can be derived from $^G C = 2\pi (^G D)^2$ and for $\varpi \neq G$ from $^\varpi C = 2(z_\nu\, ^\varpi D)^2$ (see (6.231) and (6.246)) and read

$$^R C(t) = \frac{1}{2rQ} \frac{z_{rr}}{z_r} \frac{1}{t} = \frac{1}{(3r-1)Q} \frac{1}{t}, \qquad (6.270)$$

$$^T C(t) = 2\left[\frac{1}{2q(1+q)Q} \frac{1}{[z_q]^{(1-q)}} \frac{1}{t}\right]^{\frac{2}{1+q}}, \qquad (6.271)$$

$$^G C(t) = 2\pi \left[\frac{(\sqrt{e})^{(q-1)}}{2\pi(1+q)Q} \frac{1}{t}\right]^{\frac{2}{1+q}}, \qquad (6.272)$$

$$^{SM}C(t) = 2\left[\frac{1}{2r(1+q)Q} \frac{1}{K_{r,q}[z_r]^{(1-q)}} \frac{1}{t}\right]^{\frac{2}{1+q}}. \qquad (6.273)$$

The solution (6.230) for $(\varpi, \nu) = (T, q)$ with $^T D(t)$ and $^T C(t)$ given by (6.266) and (6.271) corresponds to the Barenblatt–Pattle solution (6.168). Finally, we compute the variances $^\varpi K$ from (6.251) and (6.252) as

$$^R K(t) = 2Q\, t, \qquad (6.274)$$

$$^T K(t) = \frac{1}{3q-1}\left[2q(1+q)Q\, [z_q]^{(1-q)}\, t\right]^{\frac{2}{1+q}}, \qquad (6.275)$$

$$^G K(t) = \left[(1+q)\left(\sqrt{2\pi e}\right)^{(1-q)} Q\, t\right]^{\frac{2}{1+q}}, \qquad (6.276)$$

$$^{SM}K(t) = \frac{1}{3r-1}\left[2r(1+q)Q\, K_{r,q}[z_r]^{(1-q)}\, t\right]^{\frac{2}{1+q}}. \qquad (6.277)$$

Equations (6.274-6.277) can also be derived directly from (6.253-6.256) by taking the limit $\gamma \to 0$. We find anomalous diffusion processes in three cases

– the entropy Fokker–Planck equations related to the nonextensive entropy functionals $^{T}S_q$, $^{G}S_q$, and $^{SM}S_{q,r}$. For $r, q > 1$ the variances (6.275), (6.276), and (6.277) scale slower than t and, consequently, the corresponding solutions evolve subdiffusively. For $r \in (1/3, 1)$ and $q \in (0, 1)$ the variances scale faster than t, implying that the corresponding solutions evolve superdiffusively. The solution (6.230) of the Renyi Fokker–Planck equation (6.261) related to the extensive Renyi entropy represents a normal diffusion process, see (6.274). Moreover, the variance of the diffusion process defined by the Renyi Fokker–Planck equation (6.261) evolves exactly like the variance of a Wiener process defined by (3.143).

These findings nicely illustrate the general relationship between extensivity and anomalous diffusion addressed in Sect. 6.2.2: entropy Fokker–Planck equations of extensive entropy and information measures describe normal diffusion processes, whereas if we deal with nonextensive entropy and information measures we have anomalous diffusion processes. In this context, it is worth mentioning that (6.274) can also be obtained by substituting the Renyi entropy with $G = 0$ into the relations (6.37) and (6.38) derived in Sect. 6.2.2.

Interpretation of r and q

The meaning of the two parameters of the Sharma–Mittal entropy within the framework of entropy Fokker–Planck equations now becomes apparent. We have shown that anomalous diffusion is directly related to the parameter q, which measures the degree of nonextensivity of the entropy $^{SM}S_{q,r}$. In contrast, the parameter r measures the distortion of the shape of probability densities with respect to the shape of a Gaussian probability density. This distortion can result in cut-off solutions ($r > 1$), can induce the divergence of variances and the divergence of the entropy functional $^{T}S_q$ ($r \downarrow 1/3$, $q < r$), or leads to variances converging to zero ($r \downarrow 1/3$, $q > r$). In the special case of the entropy $^{T}S_q$ and the corresponding Fokker–Planck equation (6.216), r and q collapse into a single parameter. Consequently, the phenomena of shape-distortion and anomalous versus normal diffusion are intertwined.

Figure 6.14 provides a summary of the results obtained so far. The parameter space (r, q) is divided into two regions by the horizontal line $q = 1$ that represents the one-parametric Renyi entropy $^{R}S_r$. For $q > 1$ we have subdiffusion and for $q \in (0, 1)$ superdiffusion. In addition, we can distinguish two further regions. They are separated by the vertical line $r = 1$ representing the one-parametric functional $^{G}S_q$. On the right-hand side, that is, for $r > 1$, there exist stationary cut-off solutions, whereas on the left-hand side, that is, for $r \in (1/3, 1)$, we find power law distributions. Note that power law distributions have in particular been studied in the context of the Renyi entropy [28, 39, 389, 636].

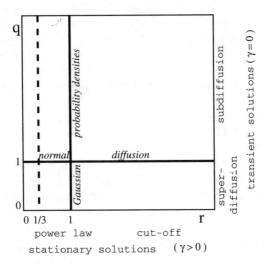

Fig. 6.14. Key properties of the solutions of the entropy Fokker–Planck equation (6.218) as functions of r and q. The parameter q determines dynamic properties of the transient solutions. There are two separated regions: $q>1$ results in subdiffusion; $0<q<1$ in superdiffusion. The parameter r measures the degree of shape-distortion of the stationary solutions with respect to the Gaussian distribution. There are two separated regions: for $r>1$ we have cut-off solutions; for $1/3<r<1$ we have power law distributions; reprinted from [191], © 2000, with permission from Elsevier

Transition Probability Densities of the Gaussian Entropy Fokker–Planck Equation

So far, the focus has been on the evolution of the probability density $P(x,t;u)$. In what follows, we will study the evolution of joint probability densities such as $P(x,t;x',t';u)$ and $P(x,t;x',t';x'',t''u)$ of the entropy Fokker–Planck equation related to the Gaussian entropy (6.208) [181]. To this end, we write (6.217) as

$$\frac{\partial}{\partial t}P(x,t;u) = \frac{\partial}{\partial x}\gamma x P + Q\exp\{(1-q)\,^{\mathrm{B}}S[P]\}\frac{\partial}{\partial x^2}P\ . \qquad (6.278)$$

Note that in this paragraph we will drop in most cases the upper index G. Equation (6.278) involves the diffusion coefficient $D_2(P) = Q\exp\{(1-q)\,^{\mathrm{B}}S[P]\}$. If for a solution $P(x,t;u)$ the coefficient $D'_2(x,t,u) = D_2(P)$ corresponds to the second Kramers–Moyal coefficient of a Markov diffusion process, then this Markov process is described by the solution $P(x,t;u)$ and the transition probability density $P(x,t|x',t';u)$ defined by (see Sect. 3.2)

$$\frac{\partial}{\partial t}P(x,t|x',t';u) =$$

$$\frac{\partial}{\partial x}\gamma x P(x,t|x',t';u) + Q\exp\{(1-q)\,{}^{\mathrm{B}}S[P(x,t;u)]\}\frac{\partial}{\partial x^2}P(x,t|x',t';u)\,, \tag{6.279}$$

Therefore, let us consider the exact transient solution derived above, which reads

$$P(x,t;\delta(x-x_0)) = \sqrt{\frac{1}{2\pi K(t)}}\exp\left\{-\frac{[x-M_1(t)]^2}{2K(t)}\right\} \tag{6.280}$$

and involves the first moment $M_1(t)$,

$$M_1(t) = x_0\exp\{-\gamma(t-t_0)\}\,, \tag{6.281}$$

and the variance $K(t)$,

$$K(t) = \left[\frac{Q}{\gamma}[2\pi e]^{(1-q)/2}\left(1-\exp\{-(1+q)\gamma(t-t_0)\}\right)\right]^{2/(1+q)} \tag{6.282}$$

and converges in the long time limit to

$$P_{\mathrm{st}}(x) = \sqrt{\frac{1}{2\pi K_{\mathrm{st}}}}\exp\left\{-\frac{x^2}{2K_{\mathrm{st}}}\right\}\,, \tag{6.283}$$

with

$$K_{\mathrm{st}} = \left[\frac{Q}{\gamma}[2\pi e]^{(1-q)/2}\right]^{2/(1+q)}. \tag{6.284}$$

From (6.224) it follows that for the transient solution (6.280) the diffusion coefficient reads $D_2'(t,\delta(x-x_0)) = D_2(P) = Q/{}^{\mathrm{G}}\mathcal{U}(t)$. Let us collect (6.246), (6.248), and (6.251):

$${}^{\mathrm{G}}C(t) = 2\left[\sqrt{\pi}\,{}^{\mathrm{G}}D(t)\right]^2,\quad {}^{\mathrm{G}}\mathcal{U}(t) = \left[\frac{{}^{\mathrm{G}}D(t)}{\sqrt{e}}\right]^{1-q},\quad {}^{\mathrm{G}}K(t) = \frac{1}{{}^{\mathrm{G}}C(t)}\,. \tag{6.285}$$

Then, we obtain

$$D_2'(t,\delta(x-x_0)) = D_2(P) = Q/{}^{\mathrm{G}}\mathcal{U}(t) = [2\pi e K(t)]^{(1-q)/2} \tag{6.286}$$

and we see that D_2' can indeed be regarded as the second Kramers–Moyal coefficient of a Markov diffusion process. Consequently, for the transient solution (6.280) the Gaussian entropy Fokker–Planck equation (6.279) reads

$$\frac{\partial}{\partial t}P(x,t|x',t';\delta(x-x_0)) = \frac{\partial}{\partial x}\gamma x P + Q[2\pi e K(t)]^{(1-q)/2}\frac{\partial}{\partial x^2}P \tag{6.287}$$

and defines the transition probability density of a Markov diffusion process. Equation (6.287) is solved by

6 Entropy Fokker–Planck Equations

$$P(x,t|x',t';\delta(x-x_0)) = \sqrt{\frac{1}{2\pi K(t,t')}} \exp\left\{-\frac{[x-M_1(t,t')]^2}{2K(t,t')}\right\}, \quad (6.288)$$

where $M_1(t,t')$ and $K(t,t')$ describe the first moment and variance with respect to x. Substituting (6.288) and $Q[2\pi e]^{(1-q)/2}/\gamma = K_{st}^{(1+q)/2}$ (see (6.284)) into (6.287), we obtain

$$M_1(t,t') = x' \exp\{-\gamma(t-t')\} \quad (6.289)$$

and

$$\frac{\partial}{\partial t} K(t,t') = -2\gamma \left(K(t,t') - K_{st}^{(1+q)/2} K^{(1-q)/2}(t) \right). \quad (6.290)$$

Solving (6.290) for the initial condition $\lim_{t\to t'} K(t,t') = 0$, we obtain

$$K(t,t') = 2\gamma K_{st}^{(1+q)/2} \int_{t'}^{t} \exp\{-2\gamma(t-z)\}[K(z)]^{(1-q)/2} \, dz. \quad (6.291)$$

Equations (6.280), (6.281), (6.282), (6.288), (6.289), and (6.291) provide us with a complete description of the generalized Ornstein–Uhlenbeck process related to the nonextensive entropy measure ${}^{G}S_q$ in terms of the expressions $P(x,t;\delta(x-x_0))$ and $P(x,t|x',t';\delta(x-x_0))$ and $M_1(t), K(t), M_1(t,t'), K(t,t')$.

Let us discuss the approach to the stationary case and some numerical issues. Let us consider the stationary case in terms of the limit $t' \to \infty$ (which implies $t \to \infty$ because of $t \geq t'$). Then, $K(z)$ in (6.291) converges to K_{st} and, consequently, (6.291) becomes

$$\lim_{t'\to\infty} K(t,t') = K_{st}(t-t') = K_{st}[1 - \exp\{-2\gamma(t-t')\}], \quad (6.292)$$

which means that the transition probability density (6.288) becomes stationary:

$$\lim_{t'\to\infty} P(x,t|x',t';\delta(x-x_0)) = P_{st}(x, t-t'|x'). \quad (6.293)$$

In order to discuss the asymptotic behavior in this stationary case, we consider the limiting case given by $t' \to \infty$ and $t - t' \to \infty$. Then, we have $M_1(t,t') = 0$ and $K(t,t') = K_{st}$, which implies that

$$\lim_{t-t'\to\infty, t'\to\infty} P(x,t|x',t';\delta(x-x_0)) = \lim_{t-t'\to\infty} P_{st}(x,t-t'|x') = P_{st}(x). \quad (6.294)$$

The Ito–Langevin equation of (6.279) reads (see Sect. 3.4)

$$\frac{d}{dt} X(t) = -\gamma X(t) + \sqrt{Q \exp\{(1-q) \, {}^{B}S[P]\}} \, \Gamma(t). \quad (6.295)$$

In order to solve (6.295), we will use the averaging method of self-consistent Langevin equations, see Sect. 3.4.4. Let us show in detail how this method

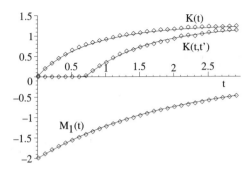

Fig. 6.15. $M_1(t)$, $K(t)$, and $K(t,t')$ for $q = 1.3$. For $t \leq t'$ we have put $K(t,t') = 0$. Parameters: $\gamma = 0.5$, $x_0 = -2$, $t_0 = 0$, $Q = 1$, $t' = 0.7$, $x' = -1$. ($L = 5000$, $\Delta t = 0.01$, $2[\Delta x]^2 = 0.1$, w_n via Box–Muller)

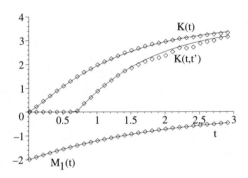

Fig. 6.16. $M_1(t)$, $K(t)$, and $K(t,t')$ for $q = 0.7$ and $L = 20000$; other parameters as in Fig. 6.15

works for (6.295). First, we interpret (6.295) as the Langevin equation of a nonlinear family of Markov diffusion processes, $dX(t)/dt = -\gamma X(t) + \sqrt{D'_2(t)}\,\Gamma(t)$, and use the Euler forward scheme

$$X^l(t_n + \Delta t) = X^l(t_n) - \Delta t\gamma X^l(t_n) + \sqrt{Q\Delta t \exp\{(1-q)\,{}^\mathrm{B}S[P]\}}\,w_n^l \quad (6.296)$$

to compute the realizations X^l of X. Accordingly, the random variable $X(t)$ is evaluated at times $t_n = n\Delta t$ for $n = 0, 1, 2, \ldots$. The variables w_n^l are realizations of Gaussian distributed random variables w_n satisfying $\langle w_n \rangle = 0$ and $\langle w_i w_k \rangle = 2\delta_{ik}$, see Sect. 3.4.4. Next, we use (3.49), that is, $\delta(x -$

$x_0) = \exp\{-(x-x_0)^2/(2[\Delta x]^2)\}/[\sqrt{2\pi}\Delta x]$ for $\Delta x \to 0$, and $P(x,t;u) = \langle \delta(x - X(t)) \rangle$ and write

$$P(x,t;u) = \frac{1}{L\sqrt{2\pi}\Delta x}\sum_{l=1}^{L}\exp\left\{-\frac{1}{2}\left[\frac{x - X^l(t)}{\Delta x}\right]^2\right\} \qquad (6.297)$$

for Δx small and L large. Then, from $^B S[P] = -\langle \ln P \rangle$ it follows that

$$^B S[P] = -\frac{1}{L}\sum_{k=1}^{L}\ln\left[\frac{1}{L\sqrt{2\pi}\Delta x}\sum_{l=1}^{L}\exp\left\{-\frac{1}{2}\left[\frac{X^k(t) - X^l(t)}{\Delta x}\right]^2\right\}\right]. \qquad (6.298)$$

In sum, (6.296) becomes

$$X^l(t_n + \Delta t) = X^l(t_n) - \Delta t \gamma X^l(t_n)$$

$$+ w_n^l \sqrt{Q \Delta t \exp\left\{\frac{(q-1)}{L}\sum_{k=1}^{L}\ln\left[\frac{1}{L\sqrt{2\pi}\Delta x}\sum_{l=1}^{L}\exp\left\{-\frac{[X^k(t_n) - X^l(t_n)]^2}{2[\Delta x]^2}\right\}\right]\right\}}. \qquad (6.299)$$

In Figs. 6.15 and 6.16 we have plotted $M_1(t)$, $K(t)$ and $K(t,t')$ for $q > 1$ and $q < 1$, respectively. We have computed these quantities both from the analytical expressions (6.281), (6.282), and (6.291) (solid lines) and from the Langevin equation (6.295) by solving (6.299) iteratively (diamonds). Note that in order to derive $K(t,t')$ we evaluated only a small subset of all simulated trajectories, namely, the trajectories $X^l(t)$ that assume at time $t = t'$ the value x'. The reason for this is that $K(t,t')$ is the variance of $X(t)$ given that the condition $X(t') = x'$ is satisfied. For further simulations involving other q-parameters see [181].

6.5.4 Fokker–Planck Equations for Fermions and Bosons

In order to account for the the quantum statistics of particles, we will consider now free energy Fokker–Planck equations with free energies that involve quantum mechanical entropies. In doing so, we will obtain nonlinear Fokker–Planck equations that describe statistical properties of many-body systems composed of quantum particles. Moreover, nonlinear Fokker–Planck equations of this kind can describe the evolution of distribution functions of quantum particles.

Drift- and Diffusion Forms

Let $\rho(x)$ denote the mean occupation number density of Fermi or Bose particles that live in a continuous phase space Ω. Then, we may describe the free energy of the quantum mechanical many-body system by means of

6.5 Examples and Applications

$$F[\rho] = U_L[\rho] - T^{FD,BE}S[\rho] \, , \quad (6.300)$$

where the quantum mechanical entropy $^{FD,BE}S$ is given by (6.107) and $U_L = \int_\Omega U_0(x)\rho(x)\,\mathrm{d}x$. Substituting (6.300) into the univariate version of (4.62), we obtain the evolution equation

$$\frac{\partial}{\partial t}\rho(x,t) = \frac{\partial}{\partial x}\left\{ M(\rho)\left[\frac{\mathrm{d}U_0(x)}{\mathrm{d}x}\rho(x,t) \mp T\frac{\partial}{\partial x}\ln\{1 \mp \rho(x,t)\}\right]\right\} \, , \quad (6.301)$$

where $M(\rho)$ denotes a density-dependent mobility coefficient. For $M = 1$ we get

$$\frac{\partial}{\partial t}\rho(x,t) = \frac{\partial}{\partial x}\frac{\mathrm{d}U_0(x)}{\mathrm{d}x}\rho(x,t) \mp T\frac{\partial^2}{\partial x^2}\ln\{1 \mp \rho(x,t)\} \, . \quad (6.302)$$

Equation (6.302) has been proposed as a classical stochastic description of fermions and bosons [165, 190, 193, 192, 323]. The upper sign is related to Fermi systems, the lower sign to Bose systems. Equation (6.302) can be regarded as the diffusion form of the entropy Fokker–Planck equation (6.301). In order to obtain the corresponding drift form, we first note that from (6.302) we conclude that the relation $\hat{L}_s(\rho) = \pm\ln[1 \pm \rho]$ holds. Next we note that for $^{FD,BE}S$ the mobility coefficient (6.48) reads

$$M(\rho) = \left[\frac{\mathrm{d}\hat{L}_s(\rho)}{\mathrm{d}\rho}\right]^{-1} = 1 \mp \rho \, . \quad (6.303)$$

Substituting this expression into (6.301), we obtain

$$\frac{\partial}{\partial t}\rho(x,t) = \frac{\partial}{\partial x}\frac{\mathrm{d}U_0(x)}{\mathrm{d}x}[1 \mp \rho(x,t)]\rho(x,t) + T\frac{\partial^2}{\partial x^2}\rho(x,t) \, . \quad (6.304)$$

As suggested by Kaniadakis and Quarati the drift form and its multivariate generalization can be derived using exclusion and inclusion principles related to Fermi and Bose statistics [327, 328]. Alternatively, the drift form (6.304) can also be derived using heuristic arguments [318]. In general, nonlinear Fokker-Planck equations for quantum particles can be derived from quantum mechanical Boltzmann equations [53, 154, 596] taking the so-called kinetic interaction principle into account [322, 323].

Stationary Solutions

Stationary distributions of (6.302) and (6.304) that obey the free energy principle $\delta F/\delta\rho_{st} = \mu$ can be derived from (6.108) and read

$$\rho_{st}(x) = \frac{1}{\exp\{[U_0(x) - \mu]/T\} \pm 1} \, , \quad (6.305)$$

see also Sect. 6.4.2. Equation (6.305) describes quantum mechanical distributions of particles or subsystems that behave like fermions (upper sign) or bosons (lower sign).

Approach to Stationary Solutions

Since the quantum mechanical entropy measures $^{\mathrm{FD}}S$ and $^{\mathrm{BE}}S$ are strictly concave, the corresponding free energy measures are bounded from below and the H-theorem of free energy Fokker–Planck equations applies, which implies that we have $\lim_{t\to\infty} \partial\rho/\partial t = 0$. This approach of transient solutions to stationary ones has also been illustrated by means of numerical studies [377].

Langevin Equations and Autocorrelations

Using $P(x,t;u) = \rho(x,t)/M_0$, we can transform (6.302) and (6.304) into

$$\frac{\partial}{\partial t} P(x,t;u) = \frac{\partial}{\partial x} P \frac{\mathrm{d}U_0(x)}{\mathrm{d}x} \mp \frac{T}{M_0} \frac{\partial^2}{\partial x^2} \ln\{1 \mp M_0 P\} \,, \qquad (6.306)$$

$$\frac{\partial}{\partial t} P(x,t;u) = \frac{\partial}{\partial x} \frac{\mathrm{d}U_0(x)}{\mathrm{d}x} [1 \mp M_0 P] P + T \frac{\partial^2}{\partial z^2} P \,. \qquad (6.307)$$

For the Fermi systems we require that $P < 1/M_0$ (which corresponds to the requirement $\rho < 1$). Note that (6.307) and (6.306) correspond to the entropy Fokker–Planck equations (6.46) and (6.50) involving the entropy

$$^{\mathrm{FD,BE}}S[P] = -\int_\Omega P(x) \ln P(x) \, \mathrm{d}x \mp \frac{1}{M_0} \int [1 \mp M_0 P(x)] \ln[1 \mp M_0 P(x)] \, \mathrm{d}x \,. \qquad (6.308)$$

Stationary distributions now read

$$P_{\mathrm{st}}(x) = \frac{1}{M_0 \left[\exp\{[U_0(x)-\mu]/T\} \mp 1\right]} \,. \qquad (6.309)$$

Equation (6.306) can be written in the form of a general nonlinear Fokker–Planck equation (2.10) with $D_2(P) = \mp T M_0^{-1} \ln(1 \mp P)/P$, which is not defined for $P = 0$ but is well-defined in the limit $P \to 0$ because of $\lim_{z\downarrow 0} \mp T M_0^{-1} \ln(1\mp z)/z = \lim_{z\downarrow 0} \pm T M_0^{-1} 1/(1 \mp z) = 1$ (rule of de l'Hopital). Furthermore, we have $D_2(P) > 0$ for $P > 0$ (and $P < 1/M_0$ in the Fermi case). Consequently, for initial distributions $u(x)$, for which (6.306) yields positive-definite C^∞-solutions $P(x,t;u)$, we can regard $D'_2(x,t;u) = D_2(P) = \mp T M_0^{-1} \ln(1\mp P)/P$ as the time and space-dependent second order Kramers–Moyal coefficient of a linear Fokker–Planck equation, which implies that (6.306) can be regarded as a strongly nonlinear Fokker–Planck equation. A complete description of the stochastic processes described by (6.306) can then be obtained from the solutions $P(x,t;u)$ and the transition probability density $P(x,t|x',t';u)$ defined by

$$\frac{\partial}{\partial t} P(x,t|x',t';u) = \frac{\partial}{\partial x} \frac{\mathrm{d}U_0(x)}{\mathrm{d}x} P(x,t|x',t';u)$$
$$\mp T \frac{\partial^2}{\partial x^2} \frac{\ln\{1 \mp M_0 P(x,t;u)\}}{M_0 P(x,t;u)} P(x,t|x',t';u) \,. \qquad (6.310)$$

6.5 Examples and Applications

The Ito–Langevin equation corresponding to (6.306) and (6.310) reads

$$\frac{\mathrm{d}}{\mathrm{d}t}X(t) = h(X) + \sqrt{T}\,\zeta(X(t),t)\,, \tag{6.311}$$

with

$$\zeta(y,t) = \sqrt{\frac{\pm\ln\{1 \pm M_0 P(y,t;u)\}}{M_0 P(y,t;u)}}\,\Gamma(t) \tag{6.312}$$

and $h(x)$ given by $h(x) = -\mathrm{d}U_0(x)/\mathrm{d}x$. Here, $\zeta(X,t)$ can be considered as a probability-dependent multiplicative fluctuating force. Likewise, for classical solutions $P(x,t;u)$ of (6.307) we can interpret the expression $D'_1(x,t;u) = D_1(x,P) = -\mathrm{d}U_0(x)/\mathrm{d}x[1 + M_0 P]$ as a time and space-dependent drift term of a linear Fokker–Planck equations, which means that (6.307) becomes a strongly nonlinear Fokker–Planck equation. In this case, the stochastic processes of interest can be computed from $P(x,t;u)$ defined by (6.307) and $P(x,t|x',t';u)$ described by

$$\frac{\partial}{\partial t}P(x,t|x',t';u) = \frac{\partial}{\partial x}\frac{\mathrm{d}U_0(x)}{\mathrm{d}x}[1 \mp M_0 P(x,t;u)]P(x,t|x',t';u)$$
$$+ T\frac{\partial^2}{\partial x^2}P(x,t|x',t';u)\,. \tag{6.313}$$

Alternatively, $P(x,t;u)$ and $P(x,t|x',t';u)$ can be obtained from

$$\frac{\mathrm{d}}{\mathrm{d}t}X(t) = h(X)\,[1 \pm M_0 P(x,t;u)]|_{x=X(t)} + \sqrt{T}\,\Gamma(t)\,, \tag{6.314}$$

for $h(x) = -\mathrm{d}U_0(x)/\mathrm{d}x$. In sum, both from the drift and diffusion forms the joint probability densities $P(x,t;x',t';u) = P(x,t|x',t';u)P(x',t';u)$, $P(x,t;x',t';x'',t'';u) = P(x,t|x',t';u)P(x',t'|x'',t'';u)P(x',t'';u)$ and so on can be derived. From these joint distributions, in turn, we obtain the density measures

$$\rho(x,t;x',t') = M_0 P(x,t;x',t';u) = P(x,t|x',t';u)\rho(x',t')\,,$$
$$\rho(x,t;x',t';x'',t'') = M_0 P(x,t;x',t';x'',t'';u)$$
$$= P(x,t|x',t';u)P(x',t'|x'',t'';u)\rho(x'',t'')$$
$$\ldots\,, \tag{6.315}$$

see also (3.23). These density functions provide us with the information about all possible autocorrelation functions of systems described by the quantum mechanical Fokker–Planck equations (6.302) and (6.304).

Nonlinearities and Distributions

Let us dwell on the drift form of the Fermi and Bose Fokker–Planck equation (6.307). For (6.307) related to the Bose statistics we may introduce the effective force $D_1(x,P) = h(x)[1 + M_0 P]$. We see that the force $h(x)$ is weighted by an expression involving the probability density in such a way that at positions where the probability density is large, the amount of $h(x)$ is increased. In consequence, if $h(x)$ is attractive at a particular position x^*, the probability density is increased at x^* and, in turn, the amount of the effective force $h_{\text{eff}}(x^*)$ is increased. Thus, the stationary probability densities of (6.307) in the case of the plus sign are more pronounced at the attractive points of $h(x)$ than those of the corresponding linear Fokker–Planck equations. This conclusion is in accordance with the notion that bosons have the tendency to gather to a greater extent at the states of minimal energy than particles described by the Maxwell–Boltzmann statistics. Furthermore, as far as (6.307) for fermions is concerned, we may infer by analogy that the stationary probability density is less pronounced at the attractive points of the force h in comparison with the stationary probability density of the corresponding linear Fokker–Planck equation – an inference that is in agreement with the fact that fermions repulse each other and avoid aggregation at a particular state.

Interpretation of the Univariate Case: Overdamped Motion

We may interpret the univariate equations as descriptions for the overdamped motion of fermions and bosons, where x is the position of the particles (Smoluchowski limit). Alternatively, we may consider x as the particle velocity of fermions and bosons: $x \to v$. Then, for the parabolic potential $U_0(v) = \gamma v^2/2$ with $\gamma > 0$ we would deal with particles that are subjected to a linear damping force: $h(v) = -\gamma v$.

6.5.5 Multivariate Generalizations

We discuss here three multivariate entropy Fokker–Planck equations that may be of particular interest [193]. They correspond to special cases of the Kramers equation (4.107) for vanishing mean field terms.

Generalized Kramers Equation for Linear Internal Energy Functionals

For $H_{\text{MF}} = 0$ the evolution equation (4.107) reduces to

$$\frac{\partial}{\partial t} P(\mathbf{p}, \mathbf{q}, t) = -\sum_{j=1}^{3} \left\{ \frac{p_j}{m} \frac{\partial}{\partial q_j} - \frac{\partial}{\partial p_j} \left(\frac{\mathrm{d}V(\mathbf{q})}{\mathrm{d}q_j} + \frac{\gamma}{m} p_j \right) \right\} P(\mathbf{p}, \mathbf{q}, t)$$

$$+ \gamma Q \left. \frac{\mathrm{d}B}{\mathrm{d}z} \right|_{\int_\Omega s[P]\,\mathrm{d}^3p\,\mathrm{d}^3q} \sum_{j=1}^{3} \frac{\partial^2}{\partial p_j^2} \hat{L}_s \left[P(\mathbf{p}, \mathbf{q}, t) \right] . \quad (6.316)$$

There might be solutions of (6.316) for which the corresponding Fokker–Planck equation for $P(\mathbf{p}, \mathbf{q}, t | \mathbf{p}', \mathbf{q}', t')$ yields well-defined Markov transition probability densities. These solutions of (6.316) can alternatively be computed from the Ito–Langevin equation

$$\frac{d}{dt}q_j(t) = \frac{p_j}{m},$$

$$\frac{d}{dt}p_j(t) = -\frac{\gamma}{m}p_j - \frac{d}{dq_j}V(\mathbf{q}) + \sqrt{\gamma Q \left.\frac{dB}{dz}\right|_{\int_\Omega s[P]d^3p\,d^3q}} \frac{\hat{L}_s[P]}{P}\Gamma_j(t),$$

(6.317)

for $j = 1, 2, 3$.

Kramers Equations for Classical Fermions and Bosons

Next, we consider ensembles of fermions and bosons. The multivariate counterparts of the quantum mechanical entropies (6.308) read

$$^{BE}S = -\int \left\{P\ln P + \frac{(1-M_0P)}{M_0}\ln(1-M_0P)\right\}d^3p_i\,d^3q_i, \quad (6.318)$$

$$^{FD}S = -\int \left\{P\ln P - \frac{(1+M_0P)}{M_0}\ln(1+M_0P)\right\}d^3p_i\,d^3q_i \quad (6.319)$$

for the Bose–Einstein statistics and the Fermi–Dirac statistics, respectively. Substituting these entropies with $Q = T$ into (6.316), we obtain

$$\frac{\partial}{\partial t}P(\mathbf{p},\mathbf{q},t) = -\sum_{j=1}^{3}\left\{\frac{p_j}{m}\frac{\partial}{\partial q_j} - \frac{\partial}{\partial p_j}\left(\frac{dV(\mathbf{q})}{dq_j} + \frac{\gamma}{m}p_j\right)\right\}P(\mathbf{p},\mathbf{q},t)$$

$$\mp \frac{\gamma T}{M_0}\sum_{j=1}^{3}\frac{\partial^2}{\partial p_j^2}\ln\{1 \mp M_0 P(\mathbf{p},\mathbf{q},t)\}, \quad (6.320)$$

where the upper sign holds for fermions and the lower sign for bosons. One can also show that the drift form of the multivariate fermions and bosons Fokker–Planck equations is given by [165, 318, 321, 327, 328]

$$\frac{\partial}{\partial t}P(\mathbf{p},\mathbf{q},t) = -\sum_{j=1}^{3}\left\{\frac{p_j}{m}\frac{\partial}{\partial q_j} - \frac{\partial}{\partial p_j}\left(\frac{dV}{dq_j} + \frac{\gamma}{m}p_j\right)\right\}[1 \mp M_0 P(\mathbf{p},\mathbf{q},t)]P(\mathbf{p},\mathbf{q},t)$$

$$+ \gamma T \sum_{j=1}^{3}\frac{\partial^2}{\partial p_j^2}P(\mathbf{p},\mathbf{q},t). \quad (6.321)$$

Again, the upper sign refers to particles with Fermi–Dirac statistics, while the lower sign refers to particles with Bose–Einstein statistics. Stationary solutions of (6.320) and (6.321) describe statistics similar to the Fermi–Dirac and the Einstein-Bose statistics:

$$P_{st}(\mathbf{p}, \mathbf{q}) = \frac{1}{M_0 \left[\exp\left\{\dfrac{\mathbf{p}^2/(2m) + V(\mathbf{q}) - \mu}{T}\right\} \pm 1\right]} . \tag{6.322}$$

If we consider solutions $P(\mathbf{p}, \mathbf{q}, t)$ of (6.320) for which (6.320) is a strongly nonlinear Fokker–Planck equation, Markov diffusion processes can be defined by means of the transient solutions $P(\mathbf{p}, \mathbf{q}, t)$ of (6.320) and the transition probability densities $P(\mathbf{p}, \mathbf{q}, t | \mathbf{p}', \mathbf{q}', t')$ satisfying

$$\frac{\partial}{\partial t} P(\mathbf{p}, \mathbf{q}, t | \mathbf{p}', \mathbf{q}', t') =$$

$$-\sum_{j=1}^{3} \left\{ \frac{p_j}{m} \frac{\partial}{\partial q_j} - \frac{\partial}{\partial p_j} \left(\frac{dV(\mathbf{q})}{dq_j} + \frac{\gamma}{m} p_j \right) \right\} P(\mathbf{p}, \mathbf{q}, t | \mathbf{p}', \mathbf{q}', t')$$

$$\pm \gamma T \sum_{j=1}^{3} \frac{\partial^2}{\partial p_j^2} \frac{\ln\{1 \pm M_0 P(\mathbf{p}, \mathbf{q}, t)\}}{M_0 P(\mathbf{p}, \mathbf{q}, t)} P(\mathbf{p}, \mathbf{q}, t | \mathbf{p}', \mathbf{q}', t') . \tag{6.323}$$

Then, the stochastic trajectories of fermions and bosons can be computed from the Ito–Langevin equation

$$\frac{d}{dt} q_j(t) = \frac{p_j}{m} ,$$

$$\frac{d}{dt} p_j(t) = -\frac{\gamma}{m} p_j - \frac{d}{dq_j} V(\mathbf{q}) + \sqrt{\pm \gamma T \frac{\ln\{1 \pm M_0 P(\mathbf{p}, \mathbf{q}, t)\}}{M_0 P(\mathbf{p}, \mathbf{q}, t)}} \Gamma_j(t) ,$$
$$\tag{6.324}$$

for $j = 1, 2, 3$. Likewise, if (6.321) can be regarded as a strongly nonlinear Fokker–Planck equation, Markov diffusion processes can be defined by means of the transient solutions $P(\mathbf{p}, \mathbf{q}, t)$ of (6.321) and the transition probabilities given by

$$\frac{\partial}{\partial t} P(\mathbf{p}, \mathbf{q}, t | \mathbf{p}', \mathbf{q}', t') =$$

$$-\sum_{j=1}^{3} \left\{ \frac{p_j}{m} \frac{\partial}{\partial q_j} - \frac{\partial}{\partial p_j} \left(\frac{dV}{dq_j} + \frac{\gamma}{m} p_j \right) \right\} [1 \pm M_0 P(\mathbf{p}, \mathbf{q}, t)] P(\mathbf{p}, \mathbf{q}, t | \mathbf{p}', \mathbf{q}', t')$$

$$+ \gamma Q \sum_{j=1}^{3} \frac{\partial^2}{\partial p_j^2} P(\mathbf{p}, \mathbf{q}, t | \mathbf{p}', \mathbf{q}', t') . \tag{6.325}$$

Alternatively, the processes can be obtained from the Langevin equation

$$\frac{\mathrm{d}}{\mathrm{d}t}q_j(t) = \frac{p_j}{m}[1 \pm M_0 P(\mathbf{p},\mathbf{q},t)] \;,$$
$$\frac{\mathrm{d}}{\mathrm{d}t}p_j(t) = -\left[\frac{\gamma}{m}p_j + \frac{\mathrm{d}}{\mathrm{d}q_j}V(\mathbf{q})\right][1 \pm M_0 P(\mathbf{p},\mathbf{q},t)] + \sqrt{\gamma T}\,\Gamma_j(t) \;.$$
(6.326)

The density functions $\rho(\mathbf{p},\mathbf{q},t)$, $\rho(\mathbf{p},\mathbf{q},t;\mathbf{p}',\mathbf{q}',t'),\ldots$ can be derived by means of

$$\rho(\mathbf{p},\mathbf{q},t) = M_0 P(\mathbf{p},\mathbf{q},t) \;,$$
$$\rho(\mathbf{p},\mathbf{q},t;\mathbf{p}',\mathbf{q}',t') = M_0 P(\mathbf{p},\mathbf{q},t;\mathbf{p}',\mathbf{q}',t')P(\mathbf{p}',\mathbf{q}',t')$$
$$\ldots$$
(6.327)

Alternatively, they may be computed from the stochastic trajectories $\mathbf{X}(t) = (\mathbf{p}(t),\mathbf{q}(t))$:

$$\rho(\mathbf{p},\mathbf{q},t) = M_0 \langle \delta(\mathbf{p}-\mathbf{p}(t))\,\delta(\mathbf{q}-\mathbf{q}(t))\rangle \;,$$
$$\rho(\mathbf{p},\mathbf{q},t;\mathbf{p}',\mathbf{q}',t') = M_0 \langle \delta(\mathbf{p}-\mathbf{p}(t))\,\delta(\mathbf{q}-\mathbf{q}(t))\,\delta(\mathbf{p}'-\mathbf{p}(t'))\,\delta(\mathbf{q}'-\mathbf{q}(t'))\rangle$$
$$\ldots$$
(6.328)

Quantum mechanical energy distributions

Finally, we may defined stochastic paths of quantum particles in energy space by means of the Kramers equations (6.320) and (6.321) and the Langevin equations (6.324) and (6.326), respectively. In order to illustrate the main point here, let us consider a spatially homogenous and isotropic system with $V(\mathbf{q}) = 0$. In this case the energy $\epsilon(t)$ of a particle is given by the kinetical energy $\epsilon(t) = \mathbf{p}^2(t)/[2m]$. First, we note that realization of $\epsilon(t)$ can be computed from the Langevin equations (6.324) and (6.326) if we solve them numerically for $\mathbf{p}(t)$ and subsequently put $\epsilon(t) = \mathbf{p}^2(t)/[2m]$. The energy distribution can then be obtained from the realizations of $\epsilon(t)$ like

$$P(\epsilon,t) = \langle \delta(\epsilon - \epsilon(t))\rangle \;. \tag{6.329}$$

Alternatively, we introduce the distribution $P(p,t)$ of the magnitude of the momentum $p = |\mathbf{p}|$: $P(p,t) = \langle \delta(p - p(t))\rangle$. Since in the isotropic system the probability to find a particle with $p < |\mathbf{p}| < p + \mathrm{d}p$ is given by $4\pi P(\mathbf{p},t)p^2\mathrm{d}p$, on the one hand, and $P(p,t)\mathrm{d}p$, on the other, we obtain

$$P(p,t) = 4\pi p^2 P(\mathbf{p},t)\big|_{|\mathbf{p}|=p} \;. \tag{6.330}$$

The distribution of particle energies $P(\epsilon,t)$ satisfies $P(\epsilon,t)\mathrm{d}\epsilon = P(p,t)\mathrm{d}p$ [498]. Using $\mathrm{d}\epsilon = 2m^{-1}p\,\mathrm{d}p$ and (6.330), we get $P(\epsilon,t) = 2\pi m p P(\mathbf{p},t)$ and

$$P(\epsilon,t) = [2m]^{3/2}\pi\sqrt{\epsilon}\,P(\mathbf{p},t)\big|_{|\mathbf{p}|^2/[2m]=\epsilon} \;. \tag{6.331}$$

That is, the energy distributions $P(\epsilon,t)$ can be obtained from the isotropic solutions $P(\mathbf{p},t)$ of the aforementioned Kramers equation and the Langevin equations for quantum particles. For example, in the stationary case from (6.322) and (6.331) with $V(\mathbf{q}) = 0$ we obtain

$$P_{\text{st}}(\epsilon) = \frac{a\sqrt{\epsilon}}{\left[\exp\left\{\frac{\epsilon - \mu}{T}\right\} \pm 1\right]}, \qquad (6.332)$$

with $a = [2m]^{3/2}\pi M_0^{-1}$. In fact, $P_{\text{st}}(\epsilon)$ describes the energy distribution of Fermi and Bose systems that are composed of particles with finite mass m, where the factor $\sqrt{\epsilon}$ corresponds to the density of quantum states. We will return the issue of quantum energy distributions in Sect. 6.5.6 and present there an alternative approach.

Kramers Equations for the Entropy by Tsallis

Substituting the entropy functional ^TS_q given by (6.121) into (6.316), we obtain

$$\frac{\partial}{\partial t} P(\mathbf{p},\mathbf{q},t) = -\sum_{j=1}^{3}\left\{\frac{p_j}{m}\frac{\partial}{\partial q_j} - \frac{\partial}{\partial p_j}\left(\frac{dV}{dq_j} + \frac{\gamma}{m}p_j\right)\right\}P + \gamma Q \sum_{j=1}^{3}\frac{\partial^2}{\partial p_j^2}P^q. \qquad (6.333)$$

For $q \in (0,1)$ stationary probability densities can be derived from (4.108) and are found as

$$P_{\text{st}}(\mathbf{p},\mathbf{q}) = \left\{\left(\frac{1-q}{q}\right)\left(\frac{1}{1-q} + \frac{[\mathbf{p}^2/(2m) + V(\mathbf{q})]}{Q} - \mu\right)\right\}^{1/(q-1)}. \qquad (6.334)$$

Note that for $q = 1$ the generalized Kramers equation (6.333) recovers the classical Kramers equation. Similarly, in the limit $q \to 1$ the stationary solution (6.334) reduces to a Boltzmann distribution of the Hamiltonian $\mathbf{p}^2/(2m) + V$. For $q > 1$ we are dealing with cut-off distributions, which usually require more detailed analysis. Multivariate nonlinear Fokker–Planck equations similar to (6.333) have been studied by Compte et al. [99, 100] and Borland et al. [68]. Equation (6.333) can be considered as a possible modification of the univariate entropy Fokker–Planck equation proposed by Plastino and Plastino that has been discussed in Sect. 6.5.2.

6.5.6 Metal Electron Model, Black Body Radiation Model, and Planck's Radiation Formula

In Sects. 6.5.4 and 6.5.5 we have considered quantum mechanical many-body systems in phase spaces spanned by a one-dimensional coordinate x and by

6.5 Examples and Applications

momentum and position vectors **p** and **q**. In this section, we extend the scope of our previous considerations in order to describe the evolution of quantum systems in the energy phase space. In doing so, we will be able to derive nonlinear Fokker–Planck equations for stochastic processes that are consistent with the experimentally observed energy distributions of electrons in metals and photons in black body cavities.

Let $g_i > 0$ describe the number of different quantum states that belong to the same energy level ϵ_i, that is, the degeneration of the energy level ϵ_i. Then the quantum entropy (6.98) becomes [14, 302, 373]

$$^{\text{FD,BE}}S(\rho_1,\ldots,\rho_N) = -\sum_{i=1}^{N} \rho_i \ln \rho_i + \sum_{i=1}^{N} g_i \ln g_i \mp \sum_{i=1}^{N} (g_i \mp \rho_i) \ln(g_i \mp \rho_i) \,. \quad (6.335)$$

Note that here and in what follows the upper sign refers to Fermi systems, while the lower sign refers to Bose systems. In the special case of a vanishing degeneration (i.e., for $g_i = 1$) the entropy (6.335) recovers the entropy measure (6.98) of the nondegenerated case. In the case of Fermi systems we require that $\rho_i < g_i$ for temperatures $T > 0$. For $T = 0$ we have $\rho_i = g_i$ up to the Fermi energy ϵ_F and $\rho_i = 0$ for energy states i with $\epsilon_i > \epsilon_F$. Studying systems with continuous energy levels $\epsilon \in \Omega = [0,\infty)$, we replace g_i by a function $g(\epsilon) \geq 0$. The expression $g(\epsilon)\,d\epsilon$ describes the number of states in an energy range between ϵ and $\epsilon + d\epsilon$. That is, $g(\epsilon)$ describes the density of states with respect to the energy scale ϵ [14, 410, 490, 619]. Using $g(\epsilon)$, we may modify (6.99) in order to obtain

$$^{\text{FD,BE}}S[\rho] = -\int_\Omega \rho(\epsilon) \ln \rho(\epsilon)\,d\epsilon + \int_\Omega g(\epsilon) \ln g(\epsilon)\,d\epsilon$$
$$\mp \int_\Omega [g(\epsilon) \mp \rho(\epsilon)] \ln[g(\epsilon) \mp \rho(\epsilon)]\,d\epsilon \,. \quad (6.336)$$

In the case of Fermi systems the constraint $\rho(\epsilon) < g(\epsilon)$ for $T > 0$ holds. Now, let us consider the free energy

$$F[\rho] = U[\rho] - T\,^{\text{FD,BE}}S[\rho] \quad (6.337)$$

with

$$U[\rho] = \int_\Omega \epsilon \rho(\epsilon)\,d\epsilon \,. \quad (6.338)$$

In order to derive stationary distributions ρ_{st} from the free energy principle $\delta F/\delta \rho = \mu$, we need to compute the variational derivatives of U and $^{\text{FD,BE}}S$. They read

$$\frac{\delta U}{\delta \rho} = \epsilon \quad (6.339)$$

and

$$\frac{\delta}{\delta\rho}{}^{\text{FD,BE}}S = -\ln\left(\frac{\rho}{g\mp\rho}\right). \tag{6.340}$$

From $\delta F/\delta\rho = \mu$ and by means of (6.337), (6.339), and (6.340), we obtain the Fermi–Dirac and Bose–Einstein distributions for quantum systems with degenerated energy levels:

$$\rho_{\text{st}}(\epsilon) = \frac{g(\epsilon)}{\exp\{(\epsilon-\mu)/T\}\pm 1}. \tag{6.341}$$

The free energy Fokker–Planck equation for the density $\rho(\epsilon,t)$ can be obtained from (4.13) by replacing $\mathbf{x}$ with ϵ. Thus, we get

$$\frac{\partial}{\partial t}\rho(\epsilon,t) = \frac{\partial}{\partial\epsilon}\rho\frac{\partial}{\partial\epsilon}\frac{\delta F}{\delta\rho}. \tag{6.342}$$

Using (6.337), (6.339), and (6.340), we obtain

$$\frac{\partial}{\partial t}\rho(\epsilon,t) = \frac{\partial\rho}{\partial\epsilon} + T\underbrace{\frac{\partial}{\partial\epsilon}\rho\frac{\partial}{\partial\epsilon}\frac{\rho}{g\mp\rho}}_{Y}. \tag{6.343}$$

Let us evaluate the expression Y. A detailed calculation shows that

$$\rho\frac{\partial}{\partial\epsilon}\frac{\rho}{g\mp\rho} = \pm\frac{dg}{d\epsilon}[1+\ln(g\mp\rho)] \mp \frac{d}{d\epsilon}[g\ln(g\mp\rho)]. \tag{6.344}$$

With this result at hand, (6.343) can be written as

$$\frac{\partial}{\partial t}\rho(\epsilon,t) = \frac{\partial\rho}{\partial\epsilon} \pm T\frac{\partial}{\partial\epsilon}\left\{\frac{dg}{d\epsilon}[1+\ln(g\mp\rho)]\right\} \mp T\frac{\partial^2}{\partial\epsilon^2}g\ln(g\mp\rho). \tag{6.345}$$

Equation (6.345) is well-defined in the limit $\rho \to 0$ because (6.345) is homogenous with respect to ρ. That is, $\rho = 0$ is a solution of (6.345) (which can be verified by substituting $\rho = 0$ into (6.345)). In order to obtain a semi-positive definite diffusion coefficient, we write the term $g\ln(g\mp\rho)$ in (6.345) as $g\ln(g\mp\rho) = g\ln(1\mp\rho/g) + g\ln(g)$, which gives us

$$\frac{\partial}{\partial t}\rho(\epsilon,t) = -\frac{\partial}{\partial\epsilon}\left\{\left[-1\mp\frac{T}{\rho}\frac{dg}{d\epsilon}\ln\left(1\mp\frac{\rho}{g}\right)\right]\rho\right\} \mp T\frac{\partial^2}{\partial\epsilon^2}g\ln\left(1\mp\frac{\rho}{g}\right). \tag{6.346}$$

Note that we have a reflective boundary at $\epsilon = 0$ with $\rho(0,t) = 0$. Furthermore, we will consider only functions $g(\epsilon)$ with $g(0) = 0$. In view of the term $\ln(1\mp\rho/g)$ occurring in (6.346), we restrict our considerations to solutions $\rho(\epsilon,t)$ of (6.346) for which in the limit $\epsilon \to 0$ the ratio ρ/g is finite for all $t \geq t_0$. The drift and diffusion coefficients read

$$d_1(\epsilon,\rho) = -1\mp\frac{T}{\rho}\frac{dg(\epsilon)}{d\epsilon}\ln\left(1\mp\frac{\rho}{g(\epsilon)}\right), \tag{6.347}$$

$$d_2(\epsilon,\rho) = \mp T\frac{g(\epsilon)}{\rho}\ln\left(1\mp\frac{\rho}{g(\epsilon)}\right). \tag{6.348}$$

The diffusion coefficient d_2 is positive definite, which can be seen if we distinguish explictly between Fermi and Bose systems. For Fermi systems we obtain $d_2 = Tg\ln(1/(1-\rho/g)) > 0$ for $0 < \rho(\epsilon) < g(\epsilon)$. For Bose systems we have $d_2 = Tg\ln(1+\rho/g)) > 0$ for $\rho(\epsilon) > 0$. By similar reasonings, we see that the drift term is composed of an attractive and repulsive part: $d_1 = d_1(-) + d_1(+)$. The attractive part is given by $d_1(-) = -1$ and drives the particles to states of zero energy. The repulsive part reads $d_1(+) = \mp T\rho^{-1}\mathrm{d}g/\mathrm{d}\epsilon \ln(1 \mp \rho/g) > 0$ and drives the quantum particles away from the ground state energy provided that the density of energy states increases with the energy (i.e., we have $\mathrm{d}g/\mathrm{d}\epsilon > 0$). Due to the interplay of these two forces, stable stationary distributions can be established. In the limit $\rho/g \to 0$ the drift and diffusion coefficients become independent of ρ:

$$\lim_{\rho/g \to 0} d_1(\epsilon, \rho) = -1 + T\frac{\mathrm{d}}{\mathrm{d}\epsilon}\ln g(\epsilon) , \qquad (6.349)$$

$$\lim_{\rho/g \to 0} d_2(\epsilon, \rho) = T . \qquad (6.350)$$

If the occupation number density ρ is normalized to M_0 with $M_0 = \int_\Omega \rho(\epsilon)\,\mathrm{d}\epsilon$, we can substitute $P(\epsilon, t; u) = \rho(\epsilon, t)/M_0$ into (6.346) and thus obtain a nonlinear Fokker–Planck equation for the probability density P. We proceed now under the hypothesis that the Fokker–Planck equation thus obtained is a strongly nonlinear Fokker–Planck equation. Then, the transition probability density $P(\epsilon, t|\epsilon', t'; u)$ of the stochastic process related to (6.346) satisfies

$$P(\epsilon, t|\epsilon', t'; u) = \frac{\partial}{\partial \epsilon}d_1(\epsilon, \rho(\epsilon, t))P + \frac{\partial^2}{\partial \epsilon^2}d_2(\epsilon, \rho(\epsilon, t))P , \qquad (6.351)$$

with $u(\epsilon) = \rho(\epsilon, t_0)$, and higher order density distributions can be obtained from

$$\rho(\epsilon, t; \epsilon', t') = P(\epsilon, t|\epsilon', t'; u)\rho(\epsilon', t') ,$$
$$\rho(\epsilon, t; \epsilon', t'; \epsilon'', t'') = P(\epsilon, t|\epsilon', t'; u)P(\epsilon', t'|\epsilon'', t''; u)\rho(\epsilon'', t'')$$
$$\cdots , \qquad (6.352)$$

see also Sect. 3.2. Alternatively, higher order density functions can be computed from the ensemble averages

$$\rho(\epsilon, t; \epsilon', t') = M_0 \langle \delta(\epsilon - \epsilon_L(t))\delta(\epsilon' - \epsilon_L(t')) \rangle ,$$
$$\rho(\epsilon, t; \epsilon', t', \epsilon'', t'') = M_0 \langle \delta(\epsilon - \epsilon_L(t))\delta(\epsilon' - \epsilon_L(t'))\delta(\epsilon'' - \epsilon_L(t'')) \rangle$$
$$\cdots , \qquad (6.353)$$

where $\epsilon_L(t)$ is given by the two-layered Ito–Langevin equation (see Sect. 3.4.1)

$$\frac{\mathrm{d}}{\mathrm{d}t}\epsilon_L(t) = d_1(\epsilon_L, \rho(\epsilon_L, t)) + \sqrt{d_2(\epsilon_L, \rho(\epsilon_L, t))}\Gamma(t) \qquad (6.354)$$

and $\rho(\epsilon, t)$ is a solution of (6.346). Both $\rho(\epsilon, t) = M_0 \langle \delta(\epsilon - \epsilon_L(t)) \rangle$ and higher order density functions (6.353) can be computed from the self-consistent Ito–Langevin equation (see Sect. 3.4.2)

$$\frac{d}{dt}\epsilon_L(t) = -1 \mp \frac{T}{M_0 \langle \delta(\epsilon - \epsilon_L(t)) \rangle} \frac{dg(\epsilon)}{d\epsilon} \ln\left(1 \mp \frac{M_0 \langle \delta(\epsilon - \epsilon_L(t)) \rangle}{g(\epsilon)}\right)\Bigg|_{\epsilon = \epsilon_L(t)}$$
$$+ \sqrt{\mp T \frac{g(\epsilon)}{M_0 \langle \delta(\epsilon - \epsilon_L(t)) \rangle} \ln\left(1 \mp \frac{M_0 \langle \delta(\epsilon - \epsilon_L(t)) \rangle}{g(\epsilon)}\right)\Bigg|_{\epsilon = \epsilon_L(t)}} \Gamma(t), \quad (6.355)$$

see Sect. 3.4.2. Let us apply now the Ito–Langevin equation (6.355) to describe the quantum statistics of electron gases and black body radiation.

Metal Electron Model

The electrons of the conduction band of metals can be regarded as a gas of fermions distributed over a continuous energy scale. In what follows we consider a free electron gas for which contributions of the potential energy to the total energy of the gas can be neglected. For the free electron gas the density of states is given by $g(\epsilon) = a\sqrt{\epsilon}$ with $a > 0$ [14, 410, 490, 619]. Consequently, for the free electron gas we have $dg/d\epsilon = a/(2\sqrt{\epsilon})$ and the Fermi Fokker–Planck equation (6.346) (upper sign) becomes

$$\frac{\partial}{\partial t}\rho(\epsilon, t) = -\frac{\partial}{\partial \epsilon}\left\{\left[-1 - \frac{aT}{2\sqrt{\epsilon}\rho}\ln\left(1 - \frac{\rho}{a\sqrt{\epsilon}}\right)\right]\rho\right\}$$
$$- aT\frac{\partial^2}{\partial \epsilon^2}\sqrt{\epsilon}\ln\left(1 - \frac{\rho}{a\sqrt{\epsilon}}\right). \quad (6.356)$$

The corresponding Ito–Langevin equation (6.355) reads

$$\frac{d}{dt}\epsilon_L(t) = -1 - \frac{aT}{2M_0\sqrt{\epsilon}\langle \delta(\epsilon - \epsilon_L(t)) \rangle}\ln\left(1 - \frac{M_0 \langle \delta(\epsilon - \epsilon_L(t)) \rangle}{a\sqrt{\epsilon}}\right)\Bigg|_{\epsilon = \epsilon_L(t)}$$
$$+ \sqrt{-T\frac{a\sqrt{\epsilon}}{M_0 \langle \delta(\epsilon - \epsilon_L(t)) \rangle}\ln\left(1 - \frac{M_0 \langle \delta(\epsilon - \epsilon_L(t)) \rangle}{a\sqrt{\epsilon}}\right)\Bigg|_{\epsilon = \epsilon_L(t)}} \Gamma(t). \quad (6.357)$$

Equation (6.357) is subjected to the constraint $\rho(\epsilon) < a\sqrt{\epsilon}$ for $T > 0$. From (6.349) and (6.350) we read off that the limiting cases

$$\lim_{\rho/g \to 0} d_1(\epsilon, \rho) = -1 + \frac{T}{2\epsilon}, \quad (6.358)$$

$$\lim_{\rho/g \to 0} d_2(\epsilon, \rho) = T \quad (6.359)$$

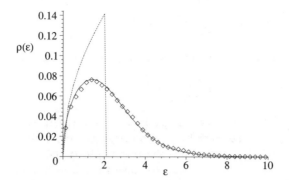

Fig. 6.17. Stationary solution of the Fermi Fokker–Planck equation (6.356). *Solid line*: analytical result (6.360). *Diamonds*: numerical results obtained by solving (6.357) using the averaging method for self-consistent Ito–Langevin equations. *Dashed line*: stationary solution in the limiting case $T \to 0$. Parameters: $\mu = \epsilon_F = 2$, $T = 1.0$, $a = 0.1$ ($L = 5000$, $\Delta t = 0.1$, $2[\Delta x]^2 = 0.1$, w_n via Box–Muller). In order to use a moderately small single time step Δt, a reflective boundary at $\epsilon_b = 0.1$ was used. In order to simulate a reflective boundary in the limit $\epsilon_b \to 0$, one needs to use a single time step $\Delta t \to 0$ such that $\Delta t \times d_1$ with d_1 given by (6.358) yields not too large values for $\epsilon = \epsilon_b$ [180]

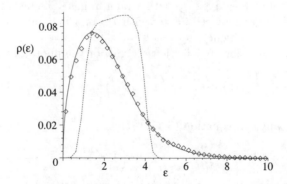

Fig. 6.18. Stationary solution as in Fig. 6.17 (*solid line and diamonds*). In addition, the initial distribution used to solve the Ito–Langevin equation numerically is shown. The initial distribution corresponds approximately to a uniform distribution in the range between $\epsilon = 1$ and $\epsilon = 4$ and satisfies the constraint $\rho(\epsilon) < g(\epsilon)$ [180]

hold. In the case of the free electron gas, the stationary solution (6.341) reads

$$\rho_{st}(\epsilon) = \frac{a\sqrt{\epsilon}}{\exp\{(\epsilon-\mu)/T\}+1} \,, \quad (6.360)$$

for $T > 0$, where μ is referred to as the Fermi energy ϵ_F. In the limit $T \to 0$ we obtain

$$\rho_{st}(\epsilon) = \begin{cases} a\sqrt{\epsilon} & \text{for } \epsilon \leq \mu = \epsilon_F \\ 0 & \text{for } \epsilon > \mu = \epsilon_F \end{cases}. \quad (6.361)$$

In order to simulate the electron gas by means of the Ito–Langevin equation (6.357) for a particular Fermi energy ϵ_F, we first compute M_0 by means of $M_0 = \int_\Omega \rho_{st}(\epsilon)\,d\epsilon$ and substitute the result into (6.357). Figure 6.17 shows the stationary distribution of the free electron gas as obtained from (6.360) and from a simulation of the Ito–Langevin equation (6.357). The dashed line describes the Fermi distribution at $T = 0$. In order to solve the Ito–Langevin equation (6.357) numerically it is important to choose an initial distribution that satisfies the constraint $\rho(\epsilon) < g(\epsilon) = a\sqrt{\epsilon}$, see Fig. 6.18.

Black Body Radiation Model and Planck's Radiation Formula

The electromagnetic radiation in a cavity with walls that have a particular temperature T exhibits a frequency distribution $\rho(\nu)$ given by Bose–Einstein statistics. The notion is that the radiation field consists of photons. A photon is a Bose particle that describes an electromagnetic wave with a frequency ν and energy $\epsilon = h\nu$, where h is Planck's constant [14, 269, 490]. We say the black body container is filled up with a photon gas. The density of states of the photon gas is given by $g(\nu) = a\nu^2$ with $a > 0$, which implies $dg/d\nu = 2a\nu$. For the sake of conveniency, we put $h = 1$ such that $\epsilon = \nu$. Then, the Bose Fokker–Planck equation (6.346) (lower sign) for the photon gas is given by

$$\frac{\partial}{\partial t}\rho(\nu,t) = -\frac{\partial}{\partial \nu}\left\{\left[-1 + \frac{2aT\nu}{\rho}\ln\left(1+\frac{\rho}{a\nu^2}\right)\right]\rho\right\} + aT\frac{\partial^2}{\partial \nu^2}\nu^2\ln\left(1+\frac{\rho}{a\nu^2}\right). \quad (6.362)$$

and the Ito–Langevin equation (6.355) reads

$$\frac{d}{dt}\nu_L(t) = -1 + \frac{2aT\nu}{M_0\langle\delta(\nu-\nu_L(t))\rangle}\ln\left(1+\frac{M_0\langle\delta(\nu-\nu_L(t))\rangle}{a\nu^2}\right)\bigg|_{\nu=\nu_L(t)}$$

$$+ \sqrt{\frac{a\nu^2 T}{M_0\langle\delta(\nu-\nu_L(t))\rangle}\ln\left(1+\frac{M_0\langle\delta(\nu-\nu_L(t))\rangle}{a\nu^2}\right)}\bigg|_{\nu=\nu_L(t)} \Gamma(t). \quad (6.363)$$

Moreover, for the photon gas (6.349) and (6.350) become

$$\lim_{\rho/g \to 0} d_1(\nu,\rho) = -1 + \frac{2T}{\nu}, \quad (6.364)$$

$$\lim_{\rho/g \to 0} d_2(\nu,\rho) = T. \quad (6.365)$$

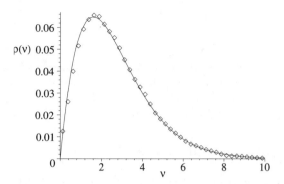

Fig. 6.19. Stationary solution of the Bose Fokker–Planck equation (6.362). *Solid line*: analytical result (6.366). *Diamonds*: numerical results obtained by solving (6.363) using the averaging method for self-consistent Ito–Langevin equations. Parameters: $T = 1.0$, $a = 0.1$ ($L = 20000$, $\Delta t = 0.1$, $2[\Delta x]^2 = 0.1$, w_n via Box–Muller). A reflective boundary at $\nu_b = 0.1$ was used (see also Fig. 6.17) [180]

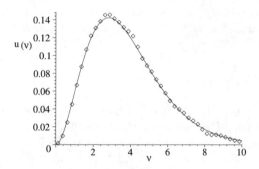

Fig. 6.20. Illustration of Planck's radiation formula. Planck's formula as obtained from (6.367) (*solid line*) and as obtained by solving numerically the Ito–Langevin equation (6.363) (*diamonds*). Parameters as in Fig. 6.19 [180]

Since the chemical potential μ of photons equals zero, the stationary solution (6.341) reads

$$\rho_{\text{st}}(\nu) = \frac{a\nu^2}{\exp\{\nu/T\} - 1} \, . \tag{6.366}$$

The spectral energy density $u(\nu)$ of a cavity with volume $V = 1$ is defined by $u(\nu) = \nu \rho_{\text{st}}(\nu)$ [14, 269, 293]. Using (6.366), we obtain Planck's radiation formula

$$u(\nu) = \frac{a\nu^3}{\exp\{\nu/T\} - 1} \ . \tag{6.367}$$

From (6.366) and $\nu \in \Omega = [0, \infty)$ it follows that the total mass M_0 of the photon gas for a particular temperature T is given by $M_0 = \int_0^\infty \rho_{\text{st}}(\nu) \, d\nu$. If we substitute this M_0-value into (6.363), we obtain a closed description of the stochastic evolution of the photon frequencies ν. In particular, $\rho(\nu, t)$ and $u(\nu)$ can be computed from $\rho(\nu, t) = M_0 \langle \delta(\nu - \nu_L(t)) \rangle$ and $u(\nu) = M_0 \nu \langle \delta(\nu - \nu_L(t)) \rangle_{\text{st}}$, see Figs. 6.19 and 6.20.

Relativistic Fermions and Bosons

In the case of quantum particles with high energies we need to account for relativistic effects. To this end, we may substitute in the quantum Fokker–Planck equation (6.346) and the quantum Langevin equations (6.354) and (6.355) the relativistic formula for the density of states of fermions and bosons given by [238, 373]

$$g(\epsilon) = a(\epsilon + mc^2)\sqrt{\epsilon(\epsilon + 2mc^2)} \ , \tag{6.368}$$

where c is the speed of light. In fact, it is clear that in the non-relativistic limit $\epsilon \ll mc^2$ the density of states function reduces to $g \propto \sqrt{\epsilon}$, whereas in the so-called ultrarelativistic limit $\epsilon \gg mc^2$ quantum particles can be treated as particle waves that satisfy $g \propto \epsilon^2$. For details see [184].

6.5.7 Population Dynamics

Entropy Fokker–Planck equations are in particularly tailored to describe the spread of populations. The reason for this is twofold.

First, linear diffusion equations require that the diffusion velocities of individuals can become arbitrarily large [247]. For example, if we assume that $\mathbf{v}(\mathbf{x}, t) = -c\rho^{-1}(\mathbf{x}, t)\nabla\rho(\mathbf{x}, t)$ holds (where $c > 0$ is a constant) and substitute this relation into (6.159), we obtain a linear diffusion equation but $|\mathbf{v}|$ can tend to infinity for $\rho \to 0$. Since the dispersion velocities of humans and animals are bounded, we may assume that the relationship between $\mathbf{v}$ and ρ differs from $\mathbf{v}(\mathbf{x}, t) = -c\rho^{-1}(\mathbf{x}, t)\nabla\rho(\mathbf{x}, t)$. In this case, the continuity equation yields a nonlinear evolution equation for ρ, For example, if we assume that $\mathbf{v}(\mathbf{x}, t) = -c\nabla\rho^\nu$ with $\nu \geq 1$ holds, then we have $\mathbf{v} = 0$ for $\rho \to 0$ and $\nabla\rho \to 0$. As we have seen in Sect. 6.5.1, the assumption $\mathbf{v} \propto -\nabla\rho^\nu$ leads to the porous medium equation (6.162), which in turn can be interpreted as an entropy Fokker–Planck equation.

Second, solutions of the linear diffusion equation typically exhibit non-vanishing tails (see, e.g., the Gaussian solutions of Wiener processes). The population densities observed in experiments, however, resemble cut-off distributions. As shown in Sects. 6.5.2 and 6.5.3, such cut-off distributions can be obtained as stationary solutions of entropy Fokker–Planck equations. Therefore, the evolution of the populations under investigation may appropriately be described by means of nonlinear Fokker–Planck equations involving general entropy and information measures.

For example, Okubo et al. have suggested that the dispersal of insects can be described by means of the evolution equation

$$\frac{\partial}{\partial t}\rho(x,t) = \frac{\partial}{\partial x}\gamma\,\text{sgn}(x)\rho(x,t) + Q\frac{\partial}{\partial x}\left[\rho(x,t)^\nu \frac{\partial}{\partial x}\rho(x,t)\right], \quad (6.369)$$

for $\gamma \geq 0$ and $\nu \geq 0$, where $\text{sgn}(x)$ is the function defined by $\text{sgn}(x) = 1$ for $x \geq 0$ and $\text{sgn}(x) = -1$ otherwise [455, 456]. In particular, the spatial distribution of mosquitoes within a mosquito swarm has been described by the cut-off distribution $\rho(x) = a\{1 - b|x|\}_+^2$ for $x \in \Omega = \mathbb{R}$ with $a, b > 0$ [456, Sect. 6.2.2]. This distribution corresponds to a stationary solution of (6.369) for $\nu = 1/2$. If M_0 describes the mass $M_0 = \int_\Omega \rho(x)\,dx$ and if we put $P = \rho/M_0$, then we can cast (6.369) into the entropy Fokker–Planck equation (6.170) with $q = \nu + 1$. Note that also other ν-values for (6.369) have been considered in population dynamics [438, 635]. Furthermore, Shigesada and Teramoto [456, Sect. 5.6] and Schweitzer [515, Sect. 9.1.2] proposed nonlinear Fokker–Planck equations of the form

$$\frac{\partial}{\partial t}\rho(x,t) = \frac{\partial}{\partial x}\frac{dU_0(x)}{dx}\rho(x,t) + \frac{\partial^2}{\partial x^2}\left[\alpha(x) + \beta(x)\rho(x,t)\right]\rho(x,t) \quad (6.370)$$

to describe the diffusion of populations and migration processes. For $\alpha(x) = \alpha$, $\beta(x) = \beta$, and $P = \rho/M_0$ the evolution equation (6.370) corresponds to the entropy Fokker–Planck equation (6.1) involving the free energy functional

$$F = \langle U_0 \rangle + \alpha\,{}^B S + \frac{\beta}{M_0}\,{}^T S_q \quad (6.371)$$

with $q = 2$.

7 General Nonlinear Fokker–Planck Equations

This chapter is concerned with general nonlinear Fokker–Planck equations that can be cast into the form (2.4). In Sect. 7.1 we will discuss the stability of systems with one-dimensional nonlinearities. In Sect. 7.2 we will show how to treat general nonlinear Fokker–Planck equations by means of analytical tools that have been developed for free energy Fokker–Planck equations. Sects. 7.3 and 7.4 are about examples and applications. Finally, we will give some references on the topic of nonlinear Fokker–Planck equations in Sect. 7.5. Note that in this chapter we will focus on the univariate case (2.10) with coefficients that do not depend explicitly on t:

$$\frac{\partial}{\partial t}P(x,t;u) = -\frac{\partial}{\partial x}D_1(x,P)P + \frac{\partial^2}{\partial x^2}D_2(x,P)P \ . \tag{7.1}$$

7.1 Linear Stability Analysis

7.1.1 Stationary Solutions

For one-dimensional nonlinearities (7.1) becomes

$$\frac{\partial}{\partial t}P(x,t;u) = -\frac{\partial}{\partial x}D_1(x,\langle A\rangle)P + \frac{\partial^2}{\partial x^2}D_2(x,\langle A\rangle)P \tag{7.2}$$

and involves the probability current

$$J = D_1(x,\langle A\rangle)P - \frac{\partial}{\partial x}D_2(x,\langle A\rangle)P \ , \tag{7.3}$$

see Sect. 2.2.4. We consider boundary conditions and coefficients D_1, D_2 for which stationary distributions can be computed from $J = 0$ and, consequently, are given by

$$P_{\mathrm{st}}(x) = \frac{1}{ZD_2(x,\langle A\rangle_{\mathrm{st}})}\exp\left\{\int^x \frac{D_1(x',\langle A\rangle_{\mathrm{st}})}{D_2(x',\langle A\rangle_{\mathrm{st}})}\,\mathrm{d}x'\right\} \ . \tag{7.4}$$

The order parameter $\langle A\rangle_{\mathrm{st}}$ can be determined by means of the self-consistency equation

$$\langle A \rangle_{\rm st} = R(\langle A \rangle_{\rm st}) \,, \tag{7.5}$$

where $R(m)$ is defined by

$$P(x,m) = \frac{1}{ZD_2(x,m)} \exp\left\{\int^x \frac{D_1(x',m)}{D_2(x',m)}\,dx'\right\}, \tag{7.6}$$

with $Z(m) = \int_\Omega D_2^{-1}(x,m)\exp\{\int^x D_1(x',m)/D_2(x',m)\,dx'\}\,dx$ and $R(m) = \int A(x)P(x,m)\,dx$. That is, $R(m)$ reads explicitly

$$R(m) = \frac{\int_\Omega \frac{A(x)}{D_2(x,m)} \exp\left\{\int^x \frac{D_1(x',m)}{D_2(x',m)}\,dx'\right\} dx}{\int_\Omega \frac{1}{D_2(x,m)} \exp\left\{\int^x \frac{D_1(x',m)}{D_2(x',m)}\,dx'\right\} dx}. \tag{7.7}$$

7.1.2 Stability of Stationary Solutions (Globally Nonoscillatory Instabilities)

Often, the hypothesis is made that the stability of stationary distributions can be read off from the slope of $R(m)$ at intersection points $m = R(m)$. That is, the hypothesis is made that the implications (5.32) derived in Sect. 5.1 for several generic free energy Fokker–Planck equations also hold in the general case. In fact, one can find support for this hypothesis [186, 187]. In order to elucidate this point, we confine ourselves to studying systems with diffusion coefficients $D_2(x,\langle A \rangle) = D_2(x)$, for which (7.2) reads

$$\frac{\partial}{\partial t} P(x,t;u) = -\frac{\partial}{\partial x} D_1(x,\langle A\rangle) P + \frac{\partial^2}{\partial x^2} D_2(x) P \,. \tag{7.8}$$

Likewise, stationary distributions are now given by

$$P_{\rm st}(x) = \frac{1}{ZD_2(x)} \exp\left\{\int^x \frac{D_1(x',\langle A\rangle_{\rm st})}{D_2(x')}\,dx'\right\} \tag{7.9}$$

(see (7.4)) and lead to self-consistency equations $\langle A \rangle_{\rm st} = R(\langle A \rangle_{\rm st})$ with

$$R(m) = \frac{\int_\Omega \frac{A(x)}{D_2(x)} \exp\left\{\int^x \frac{D_1(x',m)}{D_2(x')}\,dx'\right\} dx}{\int_\Omega \frac{1}{D_2(x)} \exp\left\{\int^x \frac{D_1(x',m)}{D_2(x')}\,dx'\right\} dx}, \tag{7.10}$$

see also (7.7). From (7.10) it follows that

$$\left.\frac{dR(m)}{dm}\right|_{\langle A \rangle_{\rm st}} = C\left[A, \frac{\partial}{\partial \langle A \rangle_{\rm st}}\int^X \frac{D_1(x',\langle A\rangle_{\rm st})}{D_2(x')}\,dx'\right], \tag{7.11}$$

where C describes the cross-correlation function

7.1 Linear Stability Analysis

$$C(f,g) = \langle fg \rangle_{st} - \langle f \rangle_{st} \langle g \rangle_{st} = \langle (f - \langle f \rangle_{st})(g - \langle g \rangle_{st}) \rangle_{st} \quad (7.12)$$

for functions $f(x)$ and $g(x)$. In order to apply linear stability analysis, we consider perturbations $\epsilon = P - P_{st}$ with $\int_\Omega \epsilon(x)\,\mathrm{d}x = 0$ of stationary solutions P_{st}. Substituting $P = P_{st} + \epsilon$ into (7.8) and neglecting nonlinear terms with respect to ϵ, we obtain the linear evolution equation

$$\frac{\partial}{\partial t}\epsilon(x,t) = -\frac{\partial}{\partial x}D_1(x,\langle A \rangle_{st})\epsilon + \frac{\partial^2}{\partial x^2}D_2(x)\epsilon$$
$$- \langle A \rangle_\epsilon \frac{\partial}{\partial x}\left[P_{st}(x)\frac{\partial D_1(x,\langle A \rangle_{st})}{\partial \langle A \rangle_{st}}\right], \quad (7.13)$$

with $\langle A \rangle_\epsilon = \int_\Omega A(x)\,\epsilon(x,t)\,\mathrm{d}x$. Equation (7.13) can be written as

$$\frac{\partial}{\partial t}\epsilon(x,t) = \frac{\partial}{\partial x}D_2 P_{st}\frac{\partial}{\partial x}\frac{\epsilon}{P_{st}} - \langle A \rangle_\epsilon \frac{\partial}{\partial x}\left[P_{st}(x)\frac{\partial D_1(x,\langle A \rangle_{st})}{\partial \langle A \rangle_{st}}\right] \quad (7.14)$$

(hint: use (7.14) as starting point and derive (7.13), see [176]).

Let us apply Shiino's decomposition to the perturbation ϵ. That is, we first write $\epsilon(x,t) = \sqrt{P_{st}}\epsilon'(x,t)$ with $\epsilon'(x,t) = \beta(t)\sqrt{P_{st}}\phi(x) + \chi_\perp(x,t)$. The function χ is assumed to be orthogonal to the first function $\varphi(x) = \sqrt{P_{st}(x)}\phi(x)$: $\int_\Omega \chi_\perp(x,t)\sqrt{P_{st}}(x)\phi(x)\,\mathrm{d}x = 0$. We also require that $\int_\Omega \sqrt{P_{st}}\chi_\perp(x,t)\,\mathrm{d}x = 0$ and $\int_\Omega P_{st}(x)\phi(x)\,\mathrm{d}x = 0$, which implies that $\int_\Omega \epsilon(x)\,\mathrm{d}x = 0$. We define $\phi(x)$ by

$$\phi(x,\langle A \rangle_{st}) = \frac{\partial}{\partial \langle A \rangle_{st}}\int^x \frac{D_1(x',\langle A \rangle_{st})}{D_2(x')}\,\mathrm{d}x'$$
$$-\left\langle \frac{\partial}{\partial \langle A \rangle_{st}}\int^X \frac{D_1(x',\langle A \rangle_{st})}{D_2(x')}\,\mathrm{d}x' \right\rangle_{st}, \quad (7.15)$$

which gives us $\epsilon(x,t)$ in the form of

$$\epsilon(x,t) = \beta(t)P_{st}(x)\left[\frac{\partial}{\partial \langle A \rangle_{st}}\int^x \frac{D_1(x',\langle A \rangle_{st})}{D_2(x')}\,\mathrm{d}x'\right.$$
$$\left.-\left\langle \frac{\partial}{\partial \langle A \rangle_{st}}\int^X \frac{D_1(x',\langle A \rangle_{st})}{D_2(x')}\,\mathrm{d}x' \right\rangle_{st}\right] + \sqrt{P_{st}(x)}\chi_\perp(x,t).$$
$$(7.16)$$

In what follows we assume that $\phi \neq 0$ holds. From our definitions it follows that $\int_\Omega A\phi P_{st}\,\mathrm{d}x = C$ and $\langle A \rangle_\epsilon = \beta C + E(\chi_\perp)$, where $E(\chi_\perp)$ denotes an expression that depends on $\chi_\perp$ and C depends on the arguments as in (7.11). Consequently, substituting (7.16) into the right-hand side of (7.14), we obtain

$$\frac{\partial}{\partial t}\epsilon(x,t) = \beta(t)(1-C)\frac{\partial}{\partial x}\left[P_{st}(x)\frac{\partial D_1(x,\langle A \rangle_{st})}{\partial \langle A \rangle_{st}}\right] + E'(\chi_\perp), \quad (7.17)$$

where C takes the arguments as in (7.11) and $E'(\chi_\perp)$ describes terms that depend on $\chi_\perp$ and vanish for vanishing functions $\chi_\perp$. Multiplying (7.17) with ϵ/P_{st}, and performing the integration $\int_\Omega \ldots dx$, we obtain

$$\frac{d}{dt} \int_\Omega \frac{\epsilon^2(x,t)}{2P_{st}(x)} dx = -\beta^2(1-C) \left\langle \frac{1}{D_2} \left[\frac{\partial D_1}{\partial \langle A \rangle_{st}}\right]^2 \right\rangle_{st} + E''(\chi_\perp), \quad (7.18)$$

where $E''(\chi_\perp)$ describes terms containing $\chi_\perp$ that satisfy $E''(\chi_\perp) = 0$ for $\chi_\perp = 0$. Due to the orthogonality of $\chi_\perp$ and $\sqrt{P_{st}}\phi$ it follows that the equivalence $\int_\Omega \epsilon^2(x,t)/P_{st}(x)\, dx = \beta^2 \langle \phi^2 \rangle_{st} + \int_\Omega \chi_\perp^2(x)\, dx$ holds, which leads to

$$\langle \phi^2 \rangle_{st} \frac{d}{dt} \frac{\beta^2(t)}{2} + \frac{d}{dt} \int_\Omega \frac{\chi_\perp^2(x,t)}{2} dx = -\beta^2(1-C) \left\langle \frac{1}{D_2} \left[\frac{\partial D_1}{\partial \langle A \rangle_{st}}\right]^2 \right\rangle_{st} + E''(\chi_\perp).$$
(7.19)

Note that for perturbations with $\chi_\perp(x,t_0) = 0$ we have $d[\int_\Omega \chi_\perp^2(x,t)\, dx]/dt = 0$ at $t = t_0$. Consequently, taking (7.11) and $E''(\chi_\perp = 0) = 0$ into account, for perturbations $\epsilon(x,t)$ that satisfy $\chi_\perp = 0$ and $\beta \neq 0$ at $t = t_0$ we obtain

$$\frac{d}{dt} \int_\Omega \frac{\epsilon^2(x,t_0)}{2P_{st}(x)} dx \bigg|_{\beta \neq 0, \chi_\perp = 0} = \frac{1}{2} \langle \phi^2 \rangle_{st} \frac{d}{dt} \beta^2(t_0)$$

$$= -\beta^2 \left(1 - \frac{dR}{dm}\bigg|_{\langle A \rangle_{st}}\right) \underbrace{\left\langle \frac{1}{D_2} \left[\frac{\partial D_1}{\partial \langle A \rangle_{st}}\right]^2 \right\rangle_{st}}_{Y}.$$
(7.20)

By assumption we have $\phi \neq 0 \Rightarrow \langle \phi^2 \rangle > 0$. In addition, in what follows we will consider the case in which the expression Y does not vanish. Then, if the inequality $dR(\langle A \rangle_{st})/dm > 1 \; (< 1)$ holds, the amplitude $|\beta|$ of the perturbation (7.16) increases (decreases) at $t = t_0$. Note that on account of the relation

$$\frac{d}{dt} \int_\Omega \frac{\epsilon^2(x,t)}{2P_{st}(x)} dx = \|\epsilon\| \frac{d}{dt} \|\epsilon\|, \quad (7.21)$$

for the norm (see also Sect. 5.1.2)

$$\|f\| = \sqrt{\int_\Omega \frac{f^2}{P_{st}} dx} \quad (7.22)$$

we have

$$\frac{d}{dt} \|\epsilon\| \bigg|_{\beta \neq 0, \chi_\perp = 0} = -\frac{\beta^2}{\|\epsilon\|} \left(1 - \frac{dR}{dm}\bigg|_{\langle A \rangle_{st}}\right) \left\langle \frac{1}{D_2} \left[\frac{\partial D_1}{\partial \langle A \rangle_{st}}\right]^2 \right\rangle. \quad (7.23)$$

Consequently, for $dR(\langle A \rangle_{st})/dm > 1 \; (< 1)$ the norm $\|\epsilon\|$ increases (decreases) at $t = t_0$ for perturbations (7.16) with $\beta \neq 0$ and $\chi_\perp = 0$. We consider now a fundamental class of stationary distributions, namely, distributions that involve simple fixed point instabilities with respect to the norm (7.22). That is, we assume that for all possible small deviations ϵ the norm either increases or decreases. In the former case, we have an unstable fixed point distribution. In the latter case, we have an asymptotically stable fixed point distribution. Note that this requirement implies that we exclude oscillatory instabilities and saddle point instabilities from our considerations. For simple fixed point distributions the sign of $1 - \mathrm{d}R(\langle A\rangle_\mathrm{st})/\mathrm{d}m$ tells us not only how a particular perturbation evolves with time but it tells us how all possible perturbations evolve. Consequently, we conclude that

$$\left.\frac{\mathrm{d}R(m)}{\mathrm{d}m}\right|_{\langle A\rangle_\mathrm{st}} < 1 \Rightarrow P_\mathrm{st} \text{ asymptotically stable,}$$

$$\left.\frac{\mathrm{d}R(m)}{\mathrm{d}m}\right|_{\langle A\rangle_\mathrm{st}} > 1 \Rightarrow P_\mathrm{st} \text{ unstable.} \qquad (7.24)$$

By similar reasoning, for stationary distributions that either describe an asymptotically stable fixed point, an unstable fixed point, or a saddle point with respect to the function space of probability densities and the norm (7.22), we conclude that

$$P_\mathrm{st} \text{ asymptotically stable} \Rightarrow \left.\frac{\mathrm{d}R(m)}{\mathrm{d}m}\right|_{\langle A\rangle_\mathrm{st}} < 1, \qquad (7.25)$$

$$\left.\frac{\mathrm{d}R(m)}{\mathrm{d}m}\right|_{\langle A\rangle_\mathrm{st}} > 1 \Rightarrow P_\mathrm{st} \text{ unstable} . \qquad (7.26)$$

That is, for stationary distributions related to saddle point instabilities we can derived necessary (but not sufficient) conditions for asymptotically stable stationary probability densities. Comparing the result obtained so far with (5.32), we realize that by means of self-consistency equation analysis we can determine to a certain extent the stability of stationary solutions of general nonlinear Fokker-Planck equations.

Globally Nonoscillatory Instabilities

Let us illustrate this issue by considering a stationary distribution $P_\mathrm{st}(x)$ defined on $\Omega = \mathbb{R}$. We assume that perturbations $\epsilon(x)$ of this distribution can be written by means of their Fourier transforms as $\epsilon(x,t) = \sqrt{P_\mathrm{st}(x)} \int a(k,t) \exp\{ikx\}\,\mathrm{d}k$ with $a(-k,t) = a^*(k,t)$ complex. We require that the amplitudes $a(k,t)$ of all Fourier modes involve real Lyapunov exponents $\lambda(k)$. That is, we use $a(k,t) = a_0(k)\exp\{\lambda(k)\,t\}$ with $\lambda(k)$ real and $a_0(-k) = a_0^*(k)$ complex. Consequently, at a bifurcation point the system exhibits a nonoscillatory behavior with respect to all Fourier modes (globally nonoscillatory instability). In this case, we find that the norm of ϵ reads

$\|\epsilon\|^2 = \pi \int |a_0(k)|^2 \exp\{2\lambda(k)t\} \, \mathrm{d}k$. In addition, a stationary distribution corresponds to an asymptotically stable fixed point if $\lambda(k) < 0$ for all k, a saddle point if $\lambda(k) > 0$ for some k and $\lambda(k) \leq 0$ for others, and an unstable fixed point if $\lambda(k) > 0$ for all k. Using the aforementioned relation $\|\epsilon\|^2 = \pi \int |a_0(k)|^2 \exp\{2\lambda(k)t\} \, \mathrm{d}k$, we can illustrate explicitly that the conclusions drawn in (7.24) hold if we know that we are dealing either with a stable or an unstable fixed point and that the conclusions drawn in (7.25) and (7.26) hold if the distribution $P_{\mathrm{st}}(x)$ may also correspond to a saddle point. For example, if $\mathrm{d}R(\langle A \rangle_{\mathrm{st}})/\mathrm{d}m > 1$ holds, then from (7.23) it follows that $\mathrm{d}\|\epsilon\|/\mathrm{d}t > 0$ which implies that there is at least one positive Lyapunov exponent. This, in turn, implies that $P_{\mathrm{st}}(x)$ corresponds to a saddle point or an unstable fixed point. In contrast, if $P_{\mathrm{st}}(x)$ corresponds to a stable fixed point, then we have $\lambda(k) < 0$ for all k and $\mathrm{d}\|\epsilon\|/\mathrm{d}t < 0$ for all kinds of perturbations. Consequently, we have $\mathrm{d}\|\epsilon\|/\mathrm{d}t < 0$ for the particular perturbation ϵ with $\chi_\perp = 0$ and $\beta \neq 0$. From (7.23) it then follows that $\mathrm{d}R(\langle A \rangle_{\mathrm{st}})/\mathrm{d}m < 1$ holds.

The Role of $\delta^2 S_B$

Using the relation $\delta^2 S_B[P_{\mathrm{st}}](\epsilon^2) = -\int_\Omega \epsilon^2(x)/P_{\mathrm{st}}(x) \, \mathrm{d}x$, which holds for the Boltzmann entropy (6.96), we can write (7.21) and (7.22) as

$$\|f\| = \sqrt{\int_\Omega \frac{f^2}{P_{\mathrm{st}}} \, \mathrm{d}x} = \sqrt{-\delta^2 S_B[P_{\mathrm{st}}](f)} \tag{7.27}$$

and

$$-\frac{\mathrm{d}}{\mathrm{d}t} \delta^2 S_B = \frac{\mathrm{d}}{\mathrm{d}t} \|\epsilon\|^2 . \tag{7.28}$$

From (7.20), (7.24), and (7.26) we conclude that

$$\frac{\mathrm{d}}{\mathrm{d}t} \left. \frac{\delta^2 S_B[P_{\mathrm{st}}](\epsilon)}{2} \right|_{\beta \neq 0, \chi_\perp = 0, t = t_0} < 0 \Leftrightarrow \left. \frac{\mathrm{d}R(m)}{\mathrm{d}m} \right|_{\langle A \rangle_{\mathrm{st}}} > 1 \Rightarrow P_{\mathrm{st}} \text{ unstable.} \tag{7.29}$$

The expression $\mathrm{d}\delta^2 S_B[P]/\mathrm{d}t$ is known as the excess entropy production [228, 353, 449]. From (7.29) it is clear that if the excess entropy production $\mathrm{d}\delta^2 S_B[P]/\mathrm{d}t$ at a stationary solution P_{st} is negative for perturbations of the form (7.16), then the stationary distribution is an unstable one. Indeed, in nonlinear nonequilibrium thermodynamics it has been shown that for many systems $\mathrm{d}\delta^2 S_B/\mathrm{d}t < 0$ is a necessary condition that a stationary state becomes unstable and a nonequilibrium phase transition occurs [228, 353, 449]. For systems that can be described by means of the nonlinear Fokker–Planck equation (7.8), we can draw the conclusions that the condition $\mathrm{d}\delta^2 S_B/\mathrm{d}t < 0$ is a sufficient condition for instability (provided that we are dealing with a globally nonoscillatory instability).

7.1.3 On an Additional Stability Coefficient*

From (7.17) we have derived the evolution equation (7.20) that completely describes the evolution of $\beta(t)$ at $t = t_0$ in combination with the stability coefficient $\tilde\lambda = 1 - \mathrm{d}R(\langle A\rangle_{\mathrm{st}})/\mathrm{d}m$. For $\tilde\lambda > 0$ the amplitude $|\beta(t)|$ decreases at $t = t_0$, whereas for $\tilde\lambda < 0$ the amplitude $|\beta(t)|$ increases at $t = t_0$. From (7.17) we may obtain alternative stability coefficients. These coefficients will yield the same result but they may reveal some additional information about the stability problem. Multiplying (7.17) with $A(x)$, integrating with respect to x and using partial integration (under the assumption that the surface term vanishes that occurs due to partial integration), we obtain

$$C\frac{\mathrm{d}}{\mathrm{d}t}\beta(t_0) + \underbrace{\int_\Omega \sqrt{P_{\mathrm{st}}(x)}A(x)\frac{\partial\chi_\perp(x,t_0)}{\partial t}\,\mathrm{d}x}_{Y}$$
$$= \left\langle \frac{\mathrm{d}A}{\mathrm{d}X}\frac{\partial D_1(X,\langle A\rangle_{\mathrm{st}})}{\partial\langle A\rangle_{\mathrm{st}}}\right\rangle_{\mathrm{st}} (C-1)\beta(t_0)\,, \qquad (7.30)$$

for perturbations with $\chi_\perp(x,t) = 0$ and $\beta \neq 0$ at $t = t_0$. If there are perturbations (7.16) for which the contribution $\chi_\perp(x,t)$ evolves in a subspace of the function space, which is orthogonal to $A(x)\sqrt{P_{\mathrm{st}}}$ (and orthogonal to $\phi(x)\sqrt{P_{\mathrm{st}}}$; see the preceding), then we have

$$\int_\Omega \chi_\perp(x,t)A(x)\sqrt{P_{\mathrm{st}}(x)}\,\mathrm{d}x = 0\,, \qquad (7.31)$$

for $t \geq t_0$ and, consequently, the term Y in (7.30) vanishes. Considering $C = \mathrm{d}R(\langle A\rangle_{\mathrm{st}})/\mathrm{d}m \neq 0$ and using (7.11), we obtain

$$\frac{\mathrm{d}}{\mathrm{d}t}\beta(t_0) = \left\langle \frac{\mathrm{d}A}{\mathrm{d}X}\frac{\partial D_1(X,\langle A\rangle_{\mathrm{st}})}{\partial\langle A\rangle_{\mathrm{st}}}\right\rangle_{\mathrm{st}} \left(1 - \left[\frac{\mathrm{d}R(m)}{\mathrm{d}m}\bigg|_{\langle A\rangle_{\mathrm{st}}}\right]^{-1}\right)\beta(t_0)\,, \qquad (7.32)$$

which can be written as

$$\frac{\mathrm{d}}{\mathrm{d}t}\beta(t) = -\frac{\delta}{\mathrm{d}R(\langle A\rangle_{\mathrm{st}})/\mathrm{d}m}\left(1 - \frac{\mathrm{d}R(m)}{\mathrm{d}m}\bigg|_{\langle A\rangle_{\mathrm{st}}}\right)\beta(t)$$
$$= \delta\left(1 - \left[\frac{\mathrm{d}R(m)}{\mathrm{d}m}\bigg|_{\langle A\rangle_{\mathrm{st}}}\right]^{-1}\right)\beta(t)\,, \qquad (7.33)$$

by introducing the stability coefficient δ defined by [187]

$$\delta = \left\langle \frac{\mathrm{d}A}{\mathrm{d}X}\frac{\partial D_1(X,\langle A\rangle_{\mathrm{st}})}{\partial\langle A\rangle_{\mathrm{st}}}\right\rangle_{\mathrm{st}}\,. \qquad (7.34)$$

Multiplying (7.33) with $\beta(t)$ and comparing the result with (7.20), we find

$$\delta = \underbrace{\frac{1}{\langle\phi^2\rangle_{\text{st}}} \left\langle \frac{1}{D_2} \left[\frac{\partial D_1}{\partial \langle A\rangle_{\text{st}}}\right]^2 \right\rangle_{\text{st}}}_{>0} \left.\frac{\mathrm{d}R(m)}{\mathrm{d}m}\right|_{\langle A\rangle_{\text{st}}}, \qquad (7.35)$$

which implies that δ and $\mathrm{d}R(\langle A\rangle_{\text{st}})/\mathrm{d}m$ may differ in magnitude but they have the same sign. We may evaluate δ in order to determine the stability of stationary solutions. However, as argued earlier, in doing so, we cannot gain information about the stability problem at hand that goes beyond the information contained in (7.25) and (7.26). That is, we still cannot say more than that the implications (7.25) and (7.26) hold.

An issue that seems to be of interest with regard to the coefficient δ is the conditions under which δ becomes relevant at all. For example, let us assume that there is a function $f(z)$ such that

$$A(x)f(\langle A\rangle_{\text{st}}) = \frac{\partial}{\partial \langle A\rangle_{\text{st}}} \int^x \frac{D_1(x', \langle A\rangle_{\text{st}})}{D_2(x')} \, \mathrm{d}x' \qquad (7.36)$$

holds. Then, (7.31) is satisfied because $\chi_\perp$ satisfies by definition the relation $\int_\Omega \chi_\perp(x,t)\sqrt{P_{\text{st}}}(x)\phi(x)\,\mathrm{d}x = 0$ and we have $\phi(x) \propto A(x)$. As we will see in Sect. 7.2.3, the relation (7.36) describes a matching condition for which local Lyapunov functionals can be derived.

7.2 Free Energy and Lyapunov Functional Analysis

In some cases, nonlinear Fokker–Planck equations of the form (7.1) can be studied by means of methods that have been developed for free energy Fokker–Planck equations. The idea is to transform a given nonlinear Fokker–Planck equation into a nonlinear Fokker–Planck equation that can be studied by means of free energy measures and Lyapunov functionals. Table 7.1 summarizes the problems and remedies that we will discuss in what follows.

Note that there is no guarantee that these transformations lead to success. That is, we present here a toolbox containing techniques that lead in some cases to the desired result but may fail in other cases. In addition, we would like to point out that nonlinear Fokker–Planck equations with multiplicative noise terms may be regarded as free energy Fokker–Planck equations with state-dependent mobility coefficients as demonstrated in Sect. 5.1.7.

7.2.1 Moving Frame Transformations

We consider here (2.10) for $D_1(x,t,P) = -\gamma x + F(t) - \mathrm{d}[\int_\Omega U_{\text{MF}}(x - x')P(x')\,\mathrm{d}x']/\mathrm{d}x$ and $D_2(x,t,P)P = Q\hat{L}_s(P)$, $S = \int_\Omega s(P)\,\mathrm{d}x$, $\gamma \geq 0$, and $U_{\text{MF}}(z) = U_{\text{MF}}(-z)$, which reads

7.2 Free Energy and Lyapunov Functional Analysis

Table 7.1. Treatment of nonlinear Fokker–Planck equations (NLFPEs) using free energy measures and Lyapunov functionals

Problem	Free energy FPE with F not bounded	NLFPE with $D_2(P)$	NLFPE with $D_1(x, \langle A \rangle)$	NLFPE with $D_1(x, P), D_2(P)$
Remedy	moving frame transformation	derivation of entropy/ information measures	derivation of local Lyapunov functional	derivation of Lyapunov functional
Result	free energy FPE with F bounded	free energy FPE	stability analysis by means of Lyapunov's direct method	NLFPE involving a Lyapunov functional

$$\frac{\partial}{\partial t} P(x,t;u) = \frac{\partial}{\partial x}\left[\gamma x - F(t) + \frac{\partial}{\partial x}\int_\Omega U_{\mathrm{MF}}(x-x')P(x',t;u)\,\mathrm{d}x'\right]P$$
$$+ Q\frac{\partial^2}{\partial x^2}\hat{L}_s(P), \qquad (7.37)$$

where $\hat{L}_s$ is defined by (4.6) and acts on the entropy kernel $s(z)$. Equation (7.37) can be regarded as a special case of the nonlinear Fokker–Planck equation (4.7) supplemented with an explicitly time-dependent drift force $F(t)$. Using (4.55), the nonlinear Fokker–Planck equation (7.37) can be written as

$$\frac{\partial}{\partial t} P(x,t;u) = \frac{\partial}{\partial x}\left[\gamma x - F(t) + \frac{\partial}{\partial x}\int_\Omega U_{\mathrm{MF}}(x-x')P(x',t;u)\,\mathrm{d}x'\right]P$$
$$- Q\frac{\partial}{\partial x}P\frac{\partial}{\partial x}\frac{\delta S}{\delta P}. \qquad (7.38)$$

We consider probability densities $P(x,t;u)$ subjected to periodic boundary conditions (with $\gamma = 0$, $x \in \Omega = [a,b]$ and $U_{\mathrm{MF}}(z) = U_{\mathrm{MF}}(z+T)$ with $b - a = T > 0$) and natural boundary conditions (with $x \in \Omega = \mathbb{R}$). Solutions $P(x,t;u)$ of (7.38) can be studied in a moving frame described by the coordinate y satisfying

$$y = x - g(t). \qquad (7.39)$$

The probability density P' in the moving frame (or moving coordinate system) is given by $P'(y,t)\mathrm{d}y = P(x,t;u)\mathrm{d}x$ and

$$P'(y,t) = P(y+g(t),t;u), \qquad (7.40)$$

because we have $\mathrm{d}x = \mathrm{d}y$ [498]. Let us choose $g(t_0) = 0$ such that

$$P'(x, t_0) = P(x, t_0; u) = u(x) \, . \tag{7.41}$$

Then, we have

$$\frac{\partial}{\partial t} P'(y, t) = \frac{\partial}{\partial x} P(x, t; u)\bigg|_{y+g(t)} \frac{\mathrm{d}}{\mathrm{d}t} g(t) + \frac{\partial}{\partial t} P(x, t; u)\bigg|_{y+g(t)} \, , \tag{7.42}$$

$$\frac{\partial}{\partial y} P'(y, t) = \frac{\partial}{\partial x} P(x, t; u)\bigg|_{y+g(t)} \, , \tag{7.43}$$

$$\frac{\partial^2}{\partial y^2} P'(y, t) = \frac{\partial^2}{\partial x^2} P(x, t; u)\bigg|_{y+g(t)} \, . \tag{7.44}$$

From (4.55) the relation

$$\frac{\partial}{\partial x} P \frac{\partial}{\partial x} \frac{\delta S}{\delta P} = \frac{\partial^2}{\partial x^2} \hat{L}_s(P) = \frac{\mathrm{d}\hat{L}_s}{\mathrm{d}z}\bigg|_P \frac{\partial^2 P}{\partial x^2} + \frac{\mathrm{d}^2 \hat{L}_s}{\mathrm{d}z^2}\bigg|_P \left[\frac{\partial P}{\partial x}\right]^2 \tag{7.45}$$

can be derived which, in turn, can be evaluated at $x = y + g(t)$:

$$\frac{\partial}{\partial x} P \frac{\partial}{\partial x} \frac{\delta S}{\delta P}\bigg|_{y+g(t)} = \left\{ \frac{\mathrm{d}\hat{L}_s}{\mathrm{d}z}\bigg|_P \frac{\partial^2 P}{\partial x^2} + \frac{\mathrm{d}^2 \hat{L}_s}{\mathrm{d}z^2}\bigg|_P \left[\frac{\partial P}{\partial x}\right]^2 \right\}_{y+g(t)}$$

$$= \frac{\partial}{\partial y} P' \frac{\partial}{\partial y} \frac{\delta S}{\delta P'} \, . \tag{7.46}$$

Let us write (7.38) as

$$\frac{\partial}{\partial t} P(x, t; u) = \gamma P + [\gamma x - F(t)] \frac{\partial}{\partial x} P$$

$$+ \frac{\partial}{\partial x} \left(P \frac{\partial}{\partial x} \int_\Omega U_{\mathrm{MF}}(x - x') P(x', t; u) \, \mathrm{d}x' \right) - Q \frac{\partial}{\partial x} P \frac{\partial}{\partial x} \frac{\delta S}{\delta P} \, . \tag{7.47}$$

If we substitute (7.47) into the right-hand side of (7.42), we obtain

$$\frac{\partial}{\partial t} P'(y, t) = \left\{ \gamma P + \left[\gamma x - F(t) + \frac{\mathrm{d}g(t)}{\mathrm{d}t}\right] \frac{\partial}{\partial x} P \right.$$

$$\left. + \frac{\partial}{\partial x} \left(P \frac{\partial}{\partial x} \int_\Omega U_{\mathrm{MF}}(x - x') P(x', t; u) \, \mathrm{d}x' \right) - Q \frac{\partial}{\partial x} P \frac{\partial}{\partial x} \frac{\delta S}{\delta P} \right\}_{y+g(t)} \, . \tag{7.48}$$

Replacing x by $y + g(t)$ and x' by $y' + g(t)$ and using (7.43) and (7.44), we find

$$\frac{\partial}{\partial t} P'(y, t) = \left[\frac{\mathrm{d}g(t)}{\mathrm{d}t} - F(t) + \gamma g(t)\right] \frac{\partial}{\partial y} P'(y, t) + \frac{\partial}{\partial y} \gamma y P'(y, t)$$

$$+ \frac{\partial}{\partial y} \left(P' \frac{\mathrm{d}}{\mathrm{d}y} \int_\Omega U_{\mathrm{MF}}(y - y') P'(y, t) \, \mathrm{d}y' \right) - Q \frac{\partial}{\partial y} P' \frac{\partial}{\partial y} \frac{\delta S}{\delta P'} \, . \tag{7.49}$$

7.2 Free Energy and Lyapunov Functional Analysis

Finally, let us put

$$g(t) = \int_{t_0}^{t} \exp\{-\gamma(t-z)\} F(z) \, \mathrm{d}z , \qquad (7.50)$$

which implies that the probability densities P' and P satisfy the relation

$$P'(y,t) = P\left(y + \int_{t_0}^{t} e^{-\gamma(t-z)} F(z) \, \mathrm{d}z, t; u\right) \qquad (7.51)$$

and that $\mathrm{d}g/\mathrm{d}t = -\gamma g + F(t)$ holds. Then, (7.49) reduces to

$$\frac{\partial}{\partial t} P'(y,t) = \frac{\partial}{\partial y}\left[\gamma y + \frac{\mathrm{d}}{\mathrm{d}y}\int_\Omega U_{\mathrm{MF}}(y - y') P'(y,t)\, \mathrm{d}y'\right] P'$$
$$-Q\frac{\partial}{\partial y} P' \frac{\partial}{\partial y}\frac{\delta S}{\delta P'} . \qquad (7.52)$$

Note that for natural boundary conditions the phase space Ω in the moving frame description is the same as in the original problem described by (7.38). The reason for this is that for natural boundary conditions the transformation $x \to y = x - g(t)$ for finite $g(t)$ does not affect $\Omega = \mathbb{R}$. For periodic boundary conditions we require that $\gamma = 0$ holds and that $U_{\mathrm{MF}}(z)$ be a periodic function. Due to the moving frame transformation the phase space $\Omega = [a,b]$ is shifted to $\Omega'(t) = [a - g(t), b - g(t)]$. However, on account of our requirement all quantities involved in (7.52) are periodic functions, which implies that at every time point t we can shift Ω' by an arbitrary constant without changing the behavior of the solution. Therefore, we may shift $\Omega'(t)$ back to Ω. In fact the transformation $\Omega'(t) \to \Omega$ has tacitly been carried out to obtain (7.49-7.52).

Using the free energy functional

$$F[P] = \frac{\gamma}{2}\int_\Omega y^2 P(y)\, \mathrm{d}y + \frac{1}{2}\int_\Omega\int_\Omega U_{\mathrm{MF}}(y - y')\, \mathrm{d}y\, \mathrm{d}y' - QS[P] , \qquad (7.53)$$

the evolution equation (7.52) can be cast into the free energy Fokker–Planck equation

$$\frac{\partial}{\partial t} P'(y,t) = \frac{\partial}{\partial y} P'(y,t) \frac{\partial}{\partial y}\frac{\delta F}{\delta P'} . \qquad (7.54)$$

In sum, by means of the moving frame transformation (7.51), the nonlinear Fokker–Planck equation (7.38) becomes a free energy Fokker–Planck equation defined by (7.53) and (7.54).

Example

Let us illustrate the moving frame transformation for the special case $F(t) = \omega = \text{constant}$ that is frequently discussed in the literature. For example, for a periodic random variable $X \in \Omega \in [0, 2\pi]$ and $t \geq 0$ we may consider the nonlinear Fokker–Planck equation

$$\frac{\partial}{\partial t} P(x, t; u) = -\frac{\partial}{\partial x}\left[\left(\omega - \frac{\mathrm{d}}{\mathrm{d}x}\int_\Omega U_{\mathrm{MF}}(x - x') P(x', t; u)\, \mathrm{d}x'\right) P\right] + Q\frac{\partial}{\partial x^2} P. \tag{7.55}$$

From (7.51), $\gamma = 0$, and $F(t) = \omega$ it follows that

$$P'(y, t) = P(y + \omega(t - t_0), t; u). \tag{7.56}$$

Using (7.56), the evolution equation (7.55) can be written as [365]

$$\frac{\partial}{\partial t} P'(y, t) = \frac{\partial}{\partial y}\left[P' \frac{\mathrm{d}}{\mathrm{d}y}\int_\Omega U_{\mathrm{MF}}(y - y') P'(y, t)\, \mathrm{d}y'\right] + Q \frac{\partial^2}{\partial y^2} P'. \tag{7.57}$$

7.2.2 Derivation of Entropy and Information Measures

In Sect. 4.5.2 we have shown that by means of the transformation (4.58), we can write the free energy Fokker–Planck equation (4.49),

$$\frac{\partial}{\partial t} P(\mathbf{x}, t; u) = -\nabla \cdot [\mathbf{I}(\mathbf{x}, t, P)\, P] + \nabla \cdot \left[\mathcal{M}(\mathbf{x}, t, P) P \cdot \nabla \frac{\delta F}{\delta P}\right], \tag{7.58}$$

in terms of the nonlinear Fokker–Planck equation (2.4),

$$\frac{\partial}{\partial t} P(\mathbf{x}, t; u) = -\sum_{i=1}^{M} \frac{\partial}{\partial x_i} D_i(\mathbf{x}, t, P) P(\mathbf{x}, t; u)$$

$$+ \sum_{i,k=1}^{M} \frac{\partial^2}{\partial x_i \partial x_k} D_{ik}(\mathbf{x}, t, P) P(\mathbf{x}, t; u). \tag{7.59}$$

The question now arises: to what extent is this transformation invertible? In other words, at issue is to find a mapping that transforms the nonlinear Fokker–Planck equation (7.59) into the free energy Fokker–Planck equation (7.58). In general, such a mapping will not exist. However, in some cases, we can indeed associate entropy and energy functionals to nonlinear Fokker–Planck equations and, in doing so, cast them into the form of free energy Fokker–Planck equations [190].

In what follows, we confine ourselves to discussing the univariate nonlinear Fokker–Planck equation (2.10) for the special case

$$\frac{\partial}{\partial t} P(x, t; u) = -\frac{\partial}{\partial x} D_1(x, P) P(x, t; u) + \frac{\partial^2}{\partial x^2} D_2(P) P(x, t; u), \tag{7.60}$$

where D_1 can be derived from an internal energy functional

$$D_1(x, P) = -\frac{\mathrm{d}}{\mathrm{d}x} \frac{\delta U}{\delta P} \tag{7.61}$$

7.2 Free Energy and Lyapunov Functional Analysis

and $D_2(z)$ is a function of z. That is, $D_2(P)$ does not denote a functional of P. We assume that $D_2(P)$ involves a proportional constant $Q > 0$ that can be regarded as an overall noise amplitude and decompose $D_2(P)$ as follows

$$D_2(P) = Q D'_2(P) \,. \tag{7.62}$$

Our objective is to transform (7.60) into (4.7) for an entropy and information measure given by $S[P] = \int_\Omega s(P)\, dx$. Requiring the equivalence between (7.60) and the free energy Fokker–Planck equation (4.7) for a univariate random variable, we need to determine s such that

$$D'_2(P) P = \hat{L}_s(P) \tag{7.63}$$

holds. Since $\hat{L}_s$ defined by (4.6) satisfies

$$\hat{L}_s(z) = s - z \frac{ds}{dz} = -z^2 \frac{d}{dz} \frac{s(z)}{z} \,, \tag{7.64}$$

the relation (7.63) becomes

$$\frac{D'_2(P)}{P} = -\frac{d}{dP} \frac{s(P)}{P} \,, \tag{7.65}$$

which gives us our final result

$$s(P) = -P \int_{P_0}^{P} \frac{D'_2(z)}{z}\, dz + s_0 \frac{P}{P_0} \,. \tag{7.66}$$

Since $\hat{L}_f(z)$ vanishes for linear functions (i.e., we have $f(z) = az \Rightarrow \hat{L}_f(z) = 0$), the lower boundary P_0 is irrelevant for the expression $\hat{L}_s(P)$. Therefore, we write (7.60) as

$$\frac{\partial}{\partial t} P(x,t;u) = -\frac{\partial}{\partial x} D_1(x,P) P + Q \frac{\partial^2}{\partial x^2} \hat{L}_s(P) \,,$$
$$s(P) = -P \int^P \frac{D'_2(z)}{z}\, dz \,. \tag{7.67}$$

Alternatively, we may write (7.67) as

$$\frac{\partial}{\partial t} P(x,t;u) = -\frac{\partial}{\partial x} D_1(x,P) P + Q \frac{\partial}{\partial x} P \frac{\partial}{\partial x} \frac{\delta S}{\delta P} \,,$$
$$S[P] = S_0 - \int_\Omega P(x) \int^{P(x)} \frac{D'_2(z)}{z}\, dz\, dx \,. \tag{7.68}$$

In deriving this result we have made use of (4.55) and have exploited s_0 and P_0 to add an offset S_0 to the entropy $S[P] = \int_\Omega s(P)\, dx$. Taking the drift term (7.61) into account, we realize that (7.68) can be written as

$$\frac{\partial}{\partial t} P(x,t;u) = \frac{\partial}{\partial x} P \frac{\partial}{\partial x} \frac{\delta F}{\delta P} \qquad (7.69)$$

with the free energy functional $F = U[P] - QS[P]$. We will return to the relationship between the evolution equations (7.60), (7.68), (7.69) in Sect. 7.2.4.

Let us conclude this section with an example. Let us consider the nonlinear Fokker–Planck equation

$$\frac{\partial}{\partial t} P(x,t;u) = -\frac{\partial}{\partial x} h(x) P(x,t;u) + Q \frac{\partial^2}{\partial x^2} \frac{P(x,t;u)}{1+\alpha P(x,t;u)}, \qquad (7.70)$$

for $\alpha \geq 0$. From $D_2'(P) = 1/(1+\alpha P)$ and (7.66) we obtain

$$s(P) = -P \int^P \frac{1}{z(1+\alpha z)} dz = -P \ln \frac{P}{1+\alpha P} \qquad (7.71)$$

and the entropy

$$^{\mathrm{NSF}} S_\alpha[P] = -\int_\Omega P(x) \ln \frac{P(x)}{1+\alpha P(x)} dx + S_0 \qquad (7.72)$$

addressed in Sect. 6.4.2. The relevance of this entropy and information measure will be discussed in Sect. 7.4.4.

7.2.3 Derivation of Local Lyapunov Functionals

Ito Case

We study next (7.8) for drift terms that are linear with respect to the order parameter $\langle A \rangle$ such that we have

$$D_1(x, \langle A \rangle) = h(x) - \kappa \left(A(x) - \langle A(X) \rangle \right) \qquad (7.73)$$

and

$$\frac{\partial}{\partial t} P(x,t;u) = -\frac{\partial}{\partial x} \left[h(x) - \kappa \left(A(x) - \langle A(X) \rangle \right) \right] P + \frac{\partial^2}{\partial x^2} D_2(x) P \qquad (7.74)$$

with $\kappa > 0$. We refer to (7.74) as the Ito case of a nonlinear Fokker–Planck equation. The Stratonovich case will be defined below, see also Sect. 2.2.2. We will analyze the stability of stationary solutions. To this end, we will proceed as in Sect. 7.1.2. While in Sect. 7.1.2, we have excluded oscillatory instabilities from our considerations, we will derive here some more rigorous results.

By means of (7.4), stationary solutions of (7.74) can be written as

$$P_{\mathrm{st}}(x) = \frac{1}{ZD_2(x)} \exp \left\{ \int^x \frac{h(x') - \kappa A(x')}{D_2(x')} dx' + \kappa \langle A \rangle_{\mathrm{st}} \int^x \frac{1}{D_2(x')} dx' \right\}, \qquad (7.75)$$

where Z denotes a normalization constant. In line with our remark in Sect. 7.1.1 note that this result holds, for example, for natural and periodic boundary conditions. In the latter case, however, we need to require that h, A, D be defined in such as way that $P(x+T) = P(x)$ holds, where T is the period of the process under consideration. By virtue of

$$P(x,m) = \frac{1}{Z(m)D_2(x)} \exp\left\{ \int^x \frac{h(x') - \kappa[A(x') - m]}{D_2(x')} \, dx' \right\},$$
$$R(m) = \int_\Omega A(x) P(x,m) \, dx, \tag{7.76}$$

we define the self-consistency equation $m = R(m)$, whose solutions determine the stationary order parameter values $\langle A_{\text{st}} \rangle$ occurring in (7.75). Differentiation of $R(m)$ with respect to m gives us

$$\left. \frac{dR(m)}{dm} \right|_{\langle A \rangle_{\text{st}}} = \kappa C\left(A(x), \int^x \frac{1}{D_2(x')} \, dx' \right), \tag{7.77}$$

where $C(f,g)$ is defined by (7.12). Substituting $P = P_{\text{st}} + \epsilon$ into (7.74), the evolution equation for the perturbation ϵ is found as

$$\frac{\partial}{\partial t} \epsilon(x,t) = -\frac{\partial}{\partial x} [h(x) - \kappa (A(x) - \langle A \rangle_{\text{st}})] \epsilon + \frac{\partial^2}{\partial x^2} D_2(x) \epsilon$$
$$- \kappa \int_\Omega A(x') \epsilon(x') \, dx' \frac{dP_{\text{st}}}{dx}. \tag{7.78}$$

Using (7.75), we can transform (7.78) into

$$\frac{\partial}{\partial t} \epsilon(x,t) = \frac{\partial}{\partial x} \left[D_2(x) P_{\text{st}} \frac{\partial}{\partial x} \frac{\epsilon}{P_{\text{st}}} \right] - \kappa \int_\Omega A(x') \epsilon(x',t) \, dx' \frac{dP_{\text{st}}}{dx}, \tag{7.79}$$

which corresponds to (7.14) for systems with drift coefficients (7.73). Let us write (7.79) as

$$\frac{\partial}{\partial t} \epsilon(x,t) = \frac{\partial}{\partial x} \left[D_2(x) P_{\text{st}} \frac{\partial}{\partial x} \left\{ \frac{\epsilon}{P_{\text{st}}} - \kappa \int_\Omega A(x') \, \epsilon(x',t) \, dx' \int^x \frac{1}{D_2(x')} dx' \right\} \right]. \tag{7.80}$$

If we consider now functions $A(x)$ and diffusion coefficients $D_2(x)$ that satisfy

$$A(x) = c \int^x \frac{1}{D_2(x')} \, dx' \tag{7.81}$$

(where c is a constant) and introduce the functional

$$\Psi[P] = \frac{\kappa}{2c} K_A(X) - {}^{\text{B}}S[P], \tag{7.82}$$

or, alternatively, the functional

$$\Psi[P] = -\frac{\kappa}{2c}\langle A\rangle^2 - {}^B S[P] \, , \qquad (7.83)$$

then (7.80) can equivalently be expressed as

$$\frac{\partial}{\partial t}\epsilon(x,t) = \frac{\partial}{\partial x}D_2(x)P_{\mathrm{st}}\frac{\partial}{\partial x}\int_\Omega \frac{\delta^2\Psi[P_{\mathrm{st}}]}{\delta P(x)\delta P(y)}\epsilon(y,t)\,\mathrm{d}y \, . \qquad (7.84)$$

Having obtained (7.84), we can proceed as in Sect. 5.1.2. Analogously to the derivation of (5.23) and (5.25), we can show that for $L = \delta^2\Psi[P_{\mathrm{st}}](\epsilon)/2$ the relations

$$\frac{\mathrm{d}}{\mathrm{d}t}\delta^2\Psi \leq 0 \, , \quad \frac{\mathrm{d}}{\mathrm{d}t}\delta^2\Psi = 0 \Leftrightarrow \delta^2\Psi[P_{\mathrm{st}}](\epsilon) = 0 \qquad (7.85)$$

hold. The second variation $\delta^2\Psi$ can be evaluated as demonstrated in Sect. 5.1.5 for the K_A-model. Thus, we obtain a stability condition similar to the one listed in the second column of Table 5.1. If

$$\tilde\lambda = 1 - \frac{\kappa}{c}K_{A,\mathrm{st}}(X) \qquad (7.86)$$

is positive, then we have $\delta^2\Psi > 0$ for $\epsilon \neq 0$ and $\delta^2\Psi$ corresponds to a Lyapunov functional that decreases with time and converges to zero in the long time limit. Consequently, perturbations vanish in the long time limit and P_{st} corresponds to an asymptotically stable distribution. Conversely, if $\tilde\lambda < 0$, then there exists a perturbation $\epsilon(x)$ such that $\delta^2\Psi[P_{\mathrm{st}}](\epsilon) < 0$ and, consequently, P_{st} is unstable. Note that Ψ does not determine the global behavior of (7.74), that is, the solutions $P(x,t;u)$. Rather $\delta^2\Psi$ determines the local behavior of (7.74). That is, it determines the evolution of perturbations of stationary distributions P_{st}. Therefore, we may say that $\delta^2\Psi$ describes a local Lyapunov functional. Note also that $\delta^2\Psi$ can be regarded as a modification of the excess entropy term $\delta^2 S_B$ discussed in Sect. 7.1.2.

Using the matching condition (7.81), the stationary distribution (7.75) and the self-consistency equation (7.76) read

$$P_{\mathrm{st}}(x) = \frac{1}{cZ}\frac{\mathrm{d}A}{\mathrm{d}x}\exp\left\{\frac{1}{c}\int^x h(x')\frac{\mathrm{d}A}{\mathrm{d}x'}\,\mathrm{d}x' - \frac{\kappa}{2c}\Big[A(x) - \langle A\rangle_{\mathrm{st}}\Big]^2\right\} \qquad (7.87)$$

and

$$P(x,m) = \frac{1}{cZ(m)}\frac{\mathrm{d}A}{\mathrm{d}x}\exp\left\{\int^x h(x')\frac{\mathrm{d}A}{\mathrm{d}x'}\,\mathrm{d}x' - \frac{\kappa}{2c}[A(x)-m]^2\right\} \, ,$$

$$R(m) = \int_\Omega A(x)P(x,m)\,\mathrm{d}x \, , \qquad (7.88)$$

which eventually leads to $\mathrm{d}R/\mathrm{d}m = \kappa K_{A,\mathrm{st}}(X)/c$ at $m = \langle A\rangle_{\mathrm{st}}$. As a result, we conclude that for $\mathrm{d}R(\langle A\rangle_{\mathrm{st}})/\mathrm{d}m < 1$ stationary distributions are asymptotically stable and for $\mathrm{d}R(\langle A\rangle_{\mathrm{st}})/\mathrm{d}m > 1$ stationary distributions are unstable. Note that this is in line with the stability analysis by means of self-consistency equations, see Sect. 5.1.3.

Stratonovich Case

In order to study noise induced phenomena, it is often more appropriate to consider Fokker–Planck equations such as

$$\frac{\partial}{\partial t}P(x,t;u) = -\frac{\partial}{\partial x}\left[h(x) - \kappa\left(A(x) - \langle A(X)\rangle_P\right)\right]P + \frac{\partial}{\partial x}g(x)\frac{\partial}{\partial x}g(x)P, \tag{7.89}$$

with $g(x) > 0$. For $\kappa = 0$ the nonlinear Fokker–Planck equation (7.89) describes a stochastic process defined by the Stratonovich–Langevin equation [498]

$$\frac{\mathrm{d}}{\mathrm{d}t}X(t) = h(X) + \underbrace{g(X)\Gamma(t)}_{\text{Stratonovich}}. \tag{7.90}$$

Using $D_2(x) = g^2(x)$, we can cast (7.89) into the form of (7.74):

$$\frac{\partial}{\partial t}P(x,t;u) = -\frac{\partial}{\partial x}\left[h(x) + \frac{1}{2}\frac{\mathrm{d}D_2(x)}{\mathrm{d}x} - \kappa\left(A(x) - \langle A\rangle_P\right)\right]P + \frac{\partial^2}{\partial x^2}D_2(x)P. \tag{7.91}$$

If (7.91) is a strongly nonlinear Fokker–Planck equation, then the Ito–Langevin equation corresponding to (7.91) reads

$$\frac{\mathrm{d}}{\mathrm{d}t}X(t) = h(X) + \frac{1}{2}\frac{\mathrm{d}D_2(X)}{\mathrm{d}x} - \kappa\left[A(X) - \langle A\rangle\right] + \underbrace{\sqrt{D_2(X)}\Gamma(t)}_{\text{Ito}} \tag{7.92}$$

and transition probability densities are defined by (see Chap. 3):

$$\frac{\partial}{\partial t}P(x,t|x',t';u) =$$
$$-\frac{\partial}{\partial x}\left[h(x) + \frac{1}{2}\frac{\mathrm{d}D_2(x)}{\mathrm{d}x} - \kappa\left(A(x) - \langle A\rangle_{P(x,t;u)}\right)\right]P(x,t|x',t';u)$$
$$+\frac{\partial^2}{\partial x^2}D_2(x)P(x,t|x',t';u). \tag{7.93}$$

Consequently, (7.92) is the Ito–Langevin equation of the Stratonovich Fokker–Planck equation (7.89). Comparing (7.91) with (7.74), we find that we need to replace h by $h + 0.5\mathrm{d}D_2(x)/\mathrm{d}x$, that is, we need to add the so-called spurious drift term $0.5\mathrm{d}D_2(x)/\mathrm{d}x$. Using $h \to h + 0.5\mathrm{d}D_2(x)/\mathrm{d}x$ and the matching condition (7.81), the stationary distribution (7.75) and the self-consistency equation (7.76) become

$$P_{\mathrm{st}}(x) = \frac{1}{Z}\sqrt{\frac{1}{c}\frac{\mathrm{d}A}{\mathrm{d}x}}\exp\left\{\int^x h(x')\frac{\mathrm{d}A}{\mathrm{d}x'}\,\mathrm{d}x' - \frac{\kappa}{2c}\left[A(x) - \langle A\rangle_{\mathrm{st}}\right]^2\right\} \tag{7.94}$$

and

$$P(x, m) = \frac{1}{Z(m)} \sqrt{\frac{1}{c} \frac{dA}{dx}} \exp\left\{ \int^x h(x') \frac{dA}{dx'} dx' - \frac{\kappa}{2c} [A(x) - m]^2 \right\},$$
$$R(m) = \int_\Omega A(x) P(x, m) \, dx \, . \tag{7.95}$$

Comparing (7.87) and (7.94), we realize that the factor $1/D_2 = c^{-1} dA/dx$ in front of the exponential function is replaced by $1/\sqrt{D_2}$. An example of a Stratonovich nonlinear Fokker–Planck equation involving a local Lyapunov functional will be given in Sect. 7.3.2.

Generalization

The results obtained so far can be generalized to the nonlinear Fokker–Planck equation (7.8). As shown in Sect. 7.1.2 perturbations ϵ of stationary states satisfy (7.14). Equation (7.14), in turn, can be written as

$$\frac{\partial}{\partial t} \epsilon(x, t) = \frac{\partial}{\partial x} D_2 P_{st} \frac{\partial}{\partial x} \left[\frac{\epsilon}{P_{st}} - \langle A \rangle_\epsilon \int^x \frac{\partial}{\partial \langle A \rangle_{st}} \frac{D_1(x', \langle A \rangle_{st})}{D_2(x')} dx' \right], \tag{7.96}$$

which generalizes (7.80). Assuming that there is a function $f(z)$ such that the matching condition (7.36) holds, we write (7.96) as

$$\frac{\partial}{\partial t} \epsilon(x, t) = \frac{\partial}{\partial x} D_2 P_{st} \frac{\partial}{\partial x} \left[\frac{\epsilon}{P_{st}} - A(x) f(\langle A \rangle_{st}) \langle A \rangle_\epsilon \right]. \tag{7.97}$$

Using $B(z)$ with

$$f(z) = -\frac{d^2 B}{dz^2} \tag{7.98}$$

and

$$\Psi[P] = B(\langle A \rangle) - {}^B S[P], \tag{7.99}$$

we can transform (7.97) into (7.84). Consequently, (7.85) holds and the stability coefficient can be obtained in analogy to the $B(\langle A \rangle)$-model discussed in Sect. 5.1.5. We have

$$\tilde{\lambda} = 1 + K_{A,st}(X) \left. \frac{d^2 B}{dm^2} \right|_{\langle A \rangle_{st}}. \tag{7.100}$$

For stationary solutions with $\tilde{\lambda} > 0$ we find that $\delta^2 \Psi[P_{st}](\epsilon)$ is a local Lyapunov functional and that the distributions correspond to asymptotically stable ones. In contrast, for a stationary solution P_{st} with $\tilde{\lambda} < 0$, we have $\delta^2 \Psi < 0$ for $\epsilon \neq 0$. Since we have $d\delta^2 \Psi/dt \leq 0$ the function $|\delta^2 \Psi|$ increases as long as ϵ is small, which indicates that P_{st} describes an unstable stationary solution.

Note that in the special case (7.73), the matching condition (7.36) reads $A(x)f(\langle A\rangle_{\text{st}}) = \kappa \int^x D_2^{-1}(x')\,\mathrm{d}x'$, which implies that f is independent of $\langle A\rangle$ and corresponds to a constant. If we put $f = \kappa/c$, we reobtain (7.81). From $f = \kappa/c$ it follows that $B(z) = -\kappa z^2/(2c)$ and the Lyapunov functional (7.99) reduces to the special case (7.83). Likewise, the stability coefficient (7.100) recovers the special case given by (7.86).

7.2.4 Derivation of Lyapunov Functionals

We discuss now how to derive Lyapunov functionals for nonlinear Fokker–Planck equations [166]. Such Lyapunov functionals are monotonically decreasing functions for the solutions of nonlinear Fokker–Planck equations. They are bounded from below and become stationary if and only if we are dealing with stationary solutions of the nonlinear Fokker–Planck equations, see (4.3). They can be used to show that an H-theorem holds and that transient solutions converge to stationary ones, see Sect. 4.3. In addition, they may be used to determine the stability of stationary distributions, see Sect. 5.1.

We consider a nonlinear Fokker–Planck equation of the form

$$\frac{\partial}{\partial t}P(x,t;u) = -\frac{\partial}{\partial x}M(x,P)\underbrace{\left\{(h(x) + k[x,P])P - \frac{\partial}{\partial x}L(P)\right\}}_{j}. \qquad (7.101)$$

Equation (7.101) involves a probability-dependent diffusion coefficient $L(P)$, a probability-independent drift force $h(x)$, and a probability-dependent drift coefficient $k[x, P]$. Here, $L(z)$ is a function of z, whereas $k[x, P]$ is a functional of P and may also depend explicitly on the state variable x. Here, M denotes a probability-dependent mobility and we require $M > 0$. Note also that we have introduced above a mobility-independent probability current j.

Equation (7.101) can be cast into the form (2.10) using

$$D_1(x, P)P = M(x, P)(h(x) + k[x, P])P - L(P)\frac{\partial}{\partial x}M(x, P),$$
$$D_2(x, P)P = M(x, P)L(P). \qquad (7.102)$$

In general, Fokker–Planck equations such as given by (7.101) have multiple stationary solutions, see Chap. 2.4. Here, we require the existence of at least one stationary solution P_{st} and that in the stationary case the mobility-independent probability current j vanishes. That is, P_{st} satisfies

$$(h(x) + k[x, P_{\text{st}}])P_{\text{st}}(x) = \frac{\partial}{\partial x}L(P_{\text{st}}). \qquad (7.103)$$

Note that from (7.103) it follows that the integral relation $\int^x h(x') + k[x'.P_{\text{st}}]\,\mathrm{d}x' = \int^P [z^{-1}\mathrm{d}L/\mathrm{d}z]\,\mathrm{d}z$ holds, which implies that for periodic boundary conditions we need to require that the integral $\int^x h(x')+k[x'.P_{st}]\,dx'$ be a periodic function with respect to x. For $M(z) = 1$, $k[x,P] = \tilde{k}(x)$, and $L(P) = QP$ with $Q > 0$, the nonlinear Fokker–Planck equation (7.101) recovers the linear Fokker–Planck equation

$$\frac{\partial}{\partial t}P(x,t;u) = -\frac{\partial}{\partial x}\left(h(x)+\tilde{k}(x)\right)P + Q\frac{\partial^2}{\partial x^2}P \ . \tag{7.104}$$

In what follows, we assume that there exists a limiting case in which (7.101) reduces to a linear evolution equation of the form (7.104). We denote this limiting case as "lin-lim". For example, the Plastino–Plastino Fokker–Planck equation related to the entropy ^TS_q becomes linear for $q \to 1$, see Sect. 6.5.2. We can then define $\tilde{k}$ and the fluctuation strength Q in (7.103) by

$$\tilde{k}(x) = \text{lin-lim } k[x,P] \ , \quad Q = \frac{1}{P}\text{lin-lim } L(P) \ . \tag{7.105}$$

By means of the functional

$$N[x,P] = -\int^x k[x',P]P(x',t;u)\,dx' + L(P) \ , \tag{7.106}$$

the nonlinear Fokker–Planck equation (7.101) can be expressed as

$$\frac{\partial}{\partial t}P(x,t;u) = -\frac{\partial}{\partial x}M(x,P)\left\{h(x)P - \frac{\partial}{\partial x}N[P]\right\} \ . \tag{7.107}$$

Introducing the functional

$$f[w] = \exp\left\{\frac{1}{Q}\int^x \frac{1}{w(x')}\frac{dN[x',w]}{dx'}\,dx'\right\} \tag{7.108}$$

that acts on a probability density $w(x)$, we obtain

$$\frac{\partial \ln f[P]}{\partial x} = \frac{1}{QP}\frac{\partial N[x,P]}{\partial x} \ , \tag{7.109}$$

$$\frac{d \ln f[P_{st}]}{dx} = \frac{1}{QP_{st}}\frac{dN[x,P_{st}]}{dx} = \frac{h(x)}{Q} \ , \tag{7.110}$$

and

$$\frac{\partial}{\partial t}P(x,t;u) = \frac{\partial}{\partial x}\left[P(x,t;u)M(x,P)\frac{\partial}{\partial x}\ln\left\{\frac{f[P]}{f[P_{st}]}\right\}\right] \ . \tag{7.111}$$

Let us dwell on the functional f for a moment. From (7.110) it follows that

$$f[P_{st}] = \frac{1}{Z}W(x) \tag{7.112}$$

7.2 Free Energy and Lyapunov Functional Analysis

holds, where $Z > 0$ is an integration constant and W describes the Boltzmann distribution of $V = -\int^x h(x') \, dx'$ defined by

$$W(x) = \frac{\exp\{-V(x)/Q\}}{\int_\Omega \exp\{-V(x')/Q\} \, dx'} . \tag{7.113}$$

For the univariate case of the free energy Fokker–Planck equation (4.7), we have $k[x, P] = -d\delta U_{\mathrm{NL}}[P]/\delta P dx$ and $\partial L(P)/\partial x = P^{-1}\partial \delta S[P]/\delta P \partial x$ and have

$$G[w] = \frac{1}{e} f[w] , \tag{7.114}$$

where G is defined by (4.112). That is, $f[w]$ is a generalization of the distortion functional $G[w]$ discussed in Sect. 4.7.1.

The form of (7.111) suggests that the functional

$$DI[P, P_{\mathrm{st}}] = \int_\Omega \ln\left\{\frac{f[P]}{f[P_{\mathrm{st}}]}\right\} \frac{\partial P}{\partial t} \, dx \tag{7.115}$$

is smaller than or equal to zero for solutions of (7.111). In fact, substituting (7.111) into (7.115), integrating by parts, and assuming that the surface term thus obtained vanishes, we obtain

$$DI[P, P_{\mathrm{st}}] = -\int_\Omega P(x, t; u) M(x, P) \left[\frac{\partial}{\partial x} \ln\left\{\frac{f[P]}{f[P_{\mathrm{st}}]}\right\}\right]^2 dx \leq 0 . \tag{7.116}$$

Note that for periodic boundary conditions the aforementioned surface term vanishes provided that $k[x, P]$ is defined such that f is a periodic function with respect to x. On the basis of (7.115), we can make an attempt to construct a monotonically decreasing functional given by $I = \int^t DI(t') \, dt'$:

$$I[P, P_{\mathrm{st}}] = \int^t dt' \left[\int_\Omega dx \ln\left\{\frac{f[P]}{f[P_{\mathrm{st}}]}\right\} \frac{\partial P}{\partial t'}\right] \tag{7.117}$$

$$\Rightarrow \frac{d}{dt} I[P, P_{\mathrm{st}}] = DI[P, P_{\mathrm{st}}] \leq 0 . \tag{7.118}$$

In the limiting case in which the nonlinear Fokker–Planck equation (7.101) becomes linear (i.e., if the limits $M(z) \to 1$, $k[x, P] \to \tilde{k}(x)$, and $L(z) \to Qz$ hold) the functional (7.108) reduces to

$$\text{lin-lim} f[w(x)] = w(x) \exp\left\{-\frac{1}{Q}\int^x \tilde{k}(x') \, dx'\right\} \tag{7.119}$$

and (7.118) yields

$$\text{lin-lim} \frac{d}{dt} I[P, P_{\mathrm{st}}] = \int_\Omega \ln\left\{\frac{P}{P_{\mathrm{st}}}\right\} \frac{\partial P}{\partial t} \, dx = \frac{d}{dt} \int_\Omega P \ln\left\{\frac{P}{P_{\mathrm{st}}}\right\} dx$$

$$\Rightarrow \text{lin-lim} I[P, P_{\mathrm{st}}] = I_0 + \int_\Omega P \ln\left\{\frac{P}{P_{\mathrm{st}}}\right\} dx , \tag{7.120}$$

where I_0 denotes an arbitrary integration constant and the normalization condition was taken into account (i.e., $\mathrm{d} \int_\Omega P \mathrm{d}x / \mathrm{d}t = 0$). We realize that the functional I recovers in the linear limit the Boltzmann–Kullback measure (4.123). Moreover, from (7.113) and (7.119) we obtain $P_{\mathrm{st}}(x) = \exp\{[-V(x) + \int^x \tilde{k}(x')\, \mathrm{d}x']/Q\}/Z'$, where Z' is a normalization constant.

The question now arises whether or not the functional (7.117) satisfies the conditions listed in (4.29) and can be used to show that transient solutions $P(x,t;u)$ become stationary in the long time limit. In any case, I is constant in the stationary case, that is, $\partial P/\partial t = 0 \Rightarrow \mathrm{d}I/\mathrm{d}t = 0$. In order to verify the boundedness of I and the implication $\mathrm{d}I/\mathrm{d}t = 0 \Rightarrow \partial P/\partial t = 0$, we need to eliminate the expression $\partial P/\partial t$ in (7.117). To this end, we write (7.118) as

$$\frac{\mathrm{d}}{\mathrm{d}t} I[P, P_{\mathrm{st}}] = \int_\Omega \ln f[P_{\mathrm{st}}] \frac{\partial P}{\partial t} \mathrm{d}x - \int_\Omega \ln f[P] \frac{\partial P}{\partial t} \mathrm{d}x \ . \tag{7.121}$$

Next, we integrate (7.109) and (7.110) with respect to x and substitute the results into (7.121). Thus, we obtain $I[P, P_{\mathrm{st}}] = I[P]$ with

$$I[P] = \frac{1}{Q} \langle V(X) \rangle + I_{\mathrm{NL}}[P] \ , \tag{7.122}$$

$$\frac{\mathrm{d}}{\mathrm{d}t} I_{\mathrm{NL}}[P] = \frac{1}{Q} \int_\Omega \mathrm{d}x \frac{\partial P}{\partial t} \left[\int^x \mathrm{d}x' \frac{1}{P(x',t;u)} \frac{\partial}{\partial x'} N[x', P] \right] \ . \tag{7.123}$$

In the following, we will illustrate how to derive explicit expressions for the nonlinear part I_{NL}. To this end, we will consider free energy and entropy Fokker–Planck equations for which we know that Lyapunov functionals in terms of free energy measures exist.

Free Energy Fokker–Planck Equations with Boltzmann Statistics

In order to study systems with linear diffusion terms, we set $M(x,P) = 1$ and $L(P) = QP$, which implies $N[x,P] = -\int^x k[x',P]P(x')\,\mathrm{d}x' + QP$. Then, (7.122) and (7.123) read

$$I[P] = \frac{1}{Q} \langle V(X) \rangle - {}^{\mathrm{B}}S[P] + I_1[P] \ , \tag{7.124}$$

$$\frac{\mathrm{d}}{\mathrm{d}t} I_1[P] = -\frac{1}{Q} \int_\Omega \mathrm{d}x \frac{\partial P}{\partial t} \left[\int^x \mathrm{d}x' \, k[x', P] \right] \ , \tag{7.125}$$

where ${}^{\mathrm{B}}S$ denotes the Boltzmann entropy (6.96) in the univariate case: ${}^{\mathrm{B}}S[P] = -\int_\Omega P \ln P \, \mathrm{d}x$.

Bounded Potential Model

In the case of the bounded potential model (5.83), we have

7.2 Free Energy and Lyapunov Functional Analysis

$$k = k_{\mathrm{MF}}[x, P] = \frac{\mathrm{d}B(\langle A \rangle)}{\mathrm{d}\langle A \rangle} \frac{\mathrm{d}A(x)}{\mathrm{d}x} . \tag{7.126}$$

Consequently, the right hand side of (7.125) reads

$$\mathrm{RHS} = -\frac{1}{Q} \left. \frac{\mathrm{d}B(z)}{\mathrm{d}z} \right|_{z=\langle A \rangle} \int_\Omega A(x) \frac{\partial P}{\partial t} \mathrm{d}x = \frac{1}{Q} \frac{\mathrm{d}}{\mathrm{d}t} B(\langle A \rangle) . \tag{7.127}$$

Then, the functional I reads

$$I[P] = \frac{\langle V(X) \rangle + B(\langle X \rangle)}{Q} - {}^{\mathrm{B}}S[P] . \tag{7.128}$$

The expression $\langle V(x) \rangle + B(\langle X \rangle)$ may be interpreted as the internal energy U of the system under consideration. Then, $F = QI = U - Q^{\mathrm{B}}S$ can be regarded as a free energy measure, which is indeed the Lyapunov functional of the bounded potential model (5.83), see Sect. 5.1.5.

Kuramoto–Shinomoto–Sakaguchi Model

For the KSS model (5.210), we have $\Omega = [0, 2\pi]$, $V = 0$, $k[x, P] = -\mathrm{d}[\int_\Omega U_{\mathrm{MF}}(x-x')P(x')\,\mathrm{d}x']/\mathrm{d}x$ and $U_{\mathrm{MF}}(z) = U_{\mathrm{MF}}(-z)$. Therefore, we can cast (7.125) into the form

$$\frac{\mathrm{d}}{\mathrm{d}t} I_1[P] = \frac{1}{Q} \int_\Omega \int_\Omega P(y, t) \frac{\partial P(x, t; u)}{\partial t} U_{\mathrm{MF}}(x-x')\,\mathrm{d}x\,\mathrm{d}y . \tag{7.129}$$

Due to the symmetry of U_{MF}, we obtain

$$I[P] = \frac{1}{Q} \left[\frac{1}{2} \int_\Omega \int_\Omega P(x) P(x') U_{\mathrm{MF}}(x-x')\,\mathrm{d}x\,\mathrm{d}x' \right] - {}^{\mathrm{B}}S[P] . \tag{7.130}$$

That is, we reobtain the result derived in Sect. 5.3.2, namely, that the KSS model involves the Lyapunov functional $F = QI = U - Q^{\mathrm{B}}S$ with $U = 0.5 \int_\Omega \int_\Omega P(x)P(x')U_{\mathrm{MF}}(x-x')\,\mathrm{d}x\,\mathrm{d}x'$.

General Case

For $k[x, P] = -\partial \delta U_{\mathrm{NL}}[P]/\delta P \partial x$ the integral relation (7.125) reads

$$\frac{\mathrm{d}}{\mathrm{d}t} I_1[P] = \frac{1}{Q} \int_\Omega \frac{\partial P}{\partial t} \frac{\delta U_{\mathrm{NL}}[P]}{\delta P}\,\mathrm{d}x , \tag{7.131}$$

which implies that $I_1 = U_{\mathrm{NL}}/Q$ and

$$I[P] = \frac{\langle V(X) \rangle + U_{\mathrm{NL}}}{Q} - {}^{\mathrm{B}}S[P] . \tag{7.132}$$

Consequently, the functional (7.131) is a possible candidate for a Lyapunov functional of the nonlinear Fokker–Planck equation

$$\frac{\partial}{\partial t}P(x,t;u) = -\frac{\partial}{\partial x}M(x,P)\left\{\left(h(x) - \frac{\partial}{\partial x}\frac{\delta U_{\mathrm{NL}}[P]}{\delta P}\right)P - Q\frac{\partial}{\partial x}P\right\}. \tag{7.133}$$

In fact, (7.133) is a free energy Fokker–Planck equation with $F = QI$, which implies that I and F correspond to Lyapunov functionals provided that they are bounded from below.

Entropy Fokker–Planck Equations

We address here two closely intertwined approaches to entropy Fokker–Planck equations.

Entropy Fokker–Planck Equations
Derived from Linear Nonequilibrium Thermodynamics

Let us consider the univariate entropy Fokker–Planck equation (6.4) for measures $S = \int_\Omega s(P)\,\mathrm{d}x$, which reads

$$\frac{\partial}{\partial t}P(x,t;u) = -\frac{\partial}{\partial x}M(x,P)\left\{h(x)P - Q\frac{\partial}{\partial x}\hat{L}_s(P)\right\}. \tag{7.134}$$

By comparison with (7.101), we obtain $k[x,P] = 0$ and $L(P) = Q\hat{L}_s[P]$, which implies $N[P] = Q\hat{L}_s[P]$, see (7.106). In this case, on account of the relation

$$Q\int^x \frac{1}{P(x',t)}\frac{\partial}{\partial x'}\hat{L}_s[P(x',t)]\,\mathrm{d}x' = -Q\left.\frac{\mathrm{d}s(z)}{\mathrm{d}z}\right|_{P(x,t;u)}, \tag{7.135}$$

we find that (7.123) reads

$$\frac{\mathrm{d}}{\mathrm{d}t}I_{\mathrm{NL}}[P](t) = \int_\Omega \frac{\partial P}{\partial t}\left.\frac{\mathrm{d}s(z)}{\mathrm{d}z}\right|_{P(x,t;u)}\mathrm{d}x, \tag{7.136}$$

which implies

$$I_{\mathrm{NL}}[P] = S[P] + S_0, \tag{7.137}$$

where S_0 is an arbitrary integration constant. From (7.122) it then follows that

$$I[P] = \frac{1}{Q}\langle V(X)\rangle - S[P] - S_0. \tag{7.138}$$

That is, $I[P]$ is a promising candidate for a Lyapunov functional of (7.134). In fact, in Sect. 4.5.7, we have shown that for $S_0 = 0$ the expression $F = QI = \langle V(X)\rangle - QS[P]$ corresponds to a Lyapunov functional of the nonlinear Fokker–Planck equation (7.134).

Entropy Fokker–Planck Equations
Derived from the Kinetical Interaction Principle

Let us consider stochastic processes described by the generalized Fokker–Planck equation

$$\frac{\partial}{\partial t} P(x,t;u) = -\frac{\partial}{\partial x} P\, M(x,P) \left\{ h(x) - Q \frac{\partial}{\partial x} \ln \kappa(P) \right\}, \qquad (7.139)$$

where $\kappa(z)$ is a monotonically increasing function. Equation (7.139) may be regarded as the Smoluchowski limit [498] of the generalized Fokker–Planck equation derived by Kaniadakis [323] by means of the kinetical interaction principle. According to the kinetical interaction principle, the function κ describes how transition probability densities depend on process probability densities P. In the absence of such a dependency we obtain $\kappa(z) = z$. Then, for $M = 1$ the entropy Fokker–Planck equation (7.139) becomes linear with respect to P. Comparing (7.139) with the Fokker–Planck equation (7.101), we can make the following identifications: $k[x,P] = 0$ and $L(P) = Q \int^x dx' P(x',t)\, \partial \ln \kappa(P(x',t))/\partial x'$. From (7.106) and (7.108) we obtain $f[w] = \kappa(w(x))$. That is, the distortion functional (7.108) corresponds to the deformation function κ. Inserting this result into (7.117) and changing the order of the integrations, we obtain

$$I[P, P_{\mathrm{st}}] = \int_\Omega dx \left[\int^t dt' \ln\left\{ \frac{\kappa(P)}{\kappa(P_{\mathrm{st}})} \right\} \frac{\partial P}{\partial t'} \right]. \qquad (7.140)$$

Since $\kappa(z)$ is a function of z, the integral (7.140) can be written as

$$I[P, P_{\mathrm{st}}] = \int_\Omega dx \left[\int^{P(x)} dz \ln\left\{ \frac{\kappa(z)}{\kappa(P_{\mathrm{st}}(x))} \right\} \right]. \qquad (7.141)$$

Using the entropy measure

$$S_\kappa[P] = -\int_\Omega dx \int^P dz \ln \kappa(z) \qquad (7.142)$$

and taking into account that stationary distributions satisfy $\ln[\kappa(P_{\mathrm{st}})] = -V(x)/Q$ with $V(x) = -\int^x h(x')\, dx'$ (see (7.139)), we can write I as

$$I = -S_\kappa[P] + \frac{1}{Q} \langle V(X) \rangle. \qquad (7.143)$$

The functional I is a good candidate for a Lyapunov functional. By means of the explicit expression for I, we can now examine whether or not I is bounded from below. Furthermore, we arrive once again at an interpretation of I in terms of a free energy measure, $F = QI = \langle V \rangle - Q\, S_\kappa[P]$, and we see that (7.139) is a free energy Fokker–Planck equation (4.4) for F.

Equation (7.60) Discussed in Sect. 7.2.2

Let us consider (7.60) as a special case of (7.101). Comparing (7.60) and (7.101), we obtain $M = 1$, $h(x) + k[x, P] = -d\delta U/\delta P dx$, and $L(P) = QD'_2(P)P$. Let us put $h(x) = -d\delta U_L/\delta P dx = -dU_0/dx$ and $k[x, P] = -d\delta U_{NL}/\delta P dx$. Then, (7.122) and (7.123) read

$$I[P] = \frac{U_L}{Q} + I_{NL}[P] , \tag{7.144}$$

$$\frac{d}{dt} I_{NL}[P] = \frac{d}{dt}\frac{U_{NL}}{Q} + \underbrace{\int_\Omega dx \frac{\partial P}{\partial t}\left[\int^x dx' \frac{1}{P(x',t;u)}\frac{\partial}{\partial x'}D'_2(P)P\right]}_{Y} . \tag{7.145}$$

The integral Y can be transformed into

$$Y = \int_\Omega dx \frac{\partial P}{\partial t}\left[D'_2(P) + \int^P dz \frac{D'_2(z)}{z}\right]$$

$$= \frac{d}{dt}\int_\Omega dx P \int^P dz \frac{D'_2(z)}{z} . \tag{7.146}$$

Introducing the entropy

$$S[P] = S_0 - \int_\Omega P(x) \int^{P(x)} \frac{D'_2(z)}{z} dz\, dx , \tag{7.147}$$

we can write the functional I_{NL} as $I_{NL} = U_{NL}/Q - S$, which implies that $F = QI$ is given by $F = U - QS$. That is, we have reobtained the free energy and entropy measures derived in Sect. 7.2.2.

7.3 Examples

7.3.1 Traveling Waves

In Sect. 5.3.2, the KSS model was studied from the perspective of free energy Fokker–Planck equations. In general, the KSS model is defined by the nonlinear Fokker–Planck equation

$$\frac{\partial}{\partial t}P(x,t;u) = -\frac{\partial}{\partial x}\left[\omega - \frac{\partial}{\partial x}\int_\Omega U_{MF}(x-x')P(x',t;u)\,dx'\right]P + Q\frac{\partial^2}{\partial x^2}P , \tag{7.148}$$

for a random variable $X \in [0, 2\pi]$ and $t \geq t_0$. Here, $U_{MF}(z)$ describes an arbitrary coupling potential with a Fourier expansion given by

$$U_{\mathrm{MF}}(z) = -\sum_{n=1}^{\infty}[c_n \cos(nz) + c'_n \sin(nz)] \qquad (7.149)$$

and $c_n, c'_n \in \mathbb{R}$. Just as in Sect. 5.3.2, we have put a minus sign in (7.149), which implies that for $c_n > 0$ and $c'_n = 0$ the potential $U_{\mathrm{MF}}(z)$ has a minimum at $z = 0$. Note also that the very first Fourier coefficient, which describes a constant offset, has been neglected because the corresponding term in the Fokker–Planck equation vanishes (i.e., for $U_{\mathrm{MF}} = c_0$ we have $c_0 \partial \int_\Omega P(x', t; u)\,\mathrm{d}x']/\partial x = 0$). By means of the moving frame transformation (7.56),

$$P'(y,t) = P(y + \omega(t - t_0), t; u) , \qquad (7.150)$$

we transform (7.148) into

$$\frac{\partial}{\partial t}P'(y,t) = \frac{\partial}{\partial y}\left[P'(y,t)\int_\Omega \frac{\mathrm{d}U_{\mathrm{MF}}(y-y')}{\mathrm{d}y}P'(y',t)\,\mathrm{d}y'\right] + Q\frac{\partial^2}{\partial y^2}P'(y,t) , \qquad (7.151)$$

and, in doing so, eliminate the constant drift force described by ω.

Stationary solutions

It is clear that the uniform distributions

$$P_{\mathrm{st}}(x) = \frac{1}{2\pi} \,,\quad P'_{\mathrm{st}}(y) = \frac{1}{2\pi} \qquad (7.152)$$

are stationary solutions of (7.148) and (7.151). With respect to the moving frame, further stationary distributions satisfy

$$P'_{\mathrm{st}}(y) = \frac{1}{Z} e^{-\frac{\int_\Omega U_{\mathrm{MF}}(y-y')P'_{\mathrm{st}}(y')\,\mathrm{d}y'}{Q}} , \qquad (7.153)$$

which can be verified by substituting (7.153) into (7.151), see also (5.213). Note that the implicit equation (7.153) is also solved by the uniform distribution $P'_{\mathrm{st}}(y) = 1/(2\pi)$ for $Z = 1/(2\pi)$. Whereas the uniform stationary distribution exists for all coupling potentials U_{MF} and parameters ω and Q, nonuniform stationary distributions of the form (7.153) do not necessarily exist. If a nonuniform stationary distribution exists, then there exist infinitely many of them due to the invariance of the system with respect to shifts of the phase space. That is, if $P'_{\mathrm{st}}(y) \neq 1/(2\pi)$ is a stationary solution, then $P'_{\mathrm{st}}(y + \Delta y)$ for arbitrary Δy is a stationary solution as well.

Traveling Wave Solutions

If nonuniform stationary distributions (7.153) exist, then they give rise to traveling wave solutions. Traveling wave solutions are solutions $P(x, t; u)$ that exhibit a fixed overall shape, but this shape is shifted with time along the x-axis, see Fig. 7.1. From (7.150) it follows that $P(x,t;u) = P'(x-\omega(t-t_0),t)$. Using (7.153) and the inverse transformation of (7.150) given by

$$P_{\text{st}}(x,t;u) = P'_{\text{st}}(x - \omega(t-t_0)) , \qquad (7.154)$$

we conclude that traveling wave solutions of (7.148) are described by

$$P_{\text{st}}(x,t;P'_{\text{st}}) = \frac{1}{Z} e^{-\frac{\int_\Omega U_{\text{MF}}(x - \omega(t-t_0) - y') P'_{\text{st}}(y') \, dy'}{Q}} , \qquad (7.155)$$

where $u(x)$ is given by $P'_{\text{st}}(x)$ (i.e., we have $P_{\text{st}}(x,t_0;P'_{\text{st}}) = P'_{\text{st}}(x)$). The moving frame transformation (7.150) can be exploited to draw further conclusions.

- If an H-theorem holds for the moving frame system, that is, if $P'(y,t) \to P'_{\text{st}}(y)$ for $t \to \infty$ holds, then the solutions of (7.148) either converge to the uniform distribution or to traveling wave solutions (7.155).
- If the uniform stationary distribution in the moving frame is asymptotically stable (unstable), then the uniform stationary distribution in the fixed frame is asymptotically stable (unstable) as well.
- If a nonuniform stationary distribution $P'_{\text{st}}(y)$ in the moving frame is stable (unstable), then the corresponding traveling wave solutions (7.155) are stable (unstable). Here, stable means that for all kinds of initial distributions $u(x) = P'_{\text{st}}(x) + \epsilon(x)$, where $\epsilon(x)$ describes a small perturbation, the solution $P_{\text{st}}(x,t;u)$ converges to a traveling wave solution $P_{\text{st}}(x,t;u') = P'_{\text{st}}(x - \omega(t-t_0) - \Delta x)$ that has the same shape as the unperturbed solution but in general differs with respect to the phase offset (i.e, we have $u'(x) = P'_{\text{st}}(x - \Delta x)$).

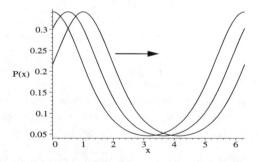

Fig. 7.1. Illustration of a traveling wave solution of (7.148)

General Case

In general, the coupling potential U_{MF} exhibits nonvanishing Fourier coefficients c'_n. Let us examine the stability of the uniform stationary distribution in this case by means of linear stability analysis [365]. We put

$$P'(y,t) = P'_{\text{st}}(y) + \epsilon(y,t) = \frac{1}{2\pi} + \epsilon(y,t), \quad (7.156)$$

where ϵ denotes a small perturbation. Substituting (7.156) into (7.151) and taking only terms linear in ϵ into account, we obtain

$$\frac{\partial}{\partial t}\epsilon(y,t) = \frac{1}{2\pi}\frac{\partial^2}{\partial y^2}\int_\Omega U_{\text{MF}}(y-y')\epsilon(y',t)\,\mathrm{d}y' + Q\frac{\partial^2}{\partial y^2}\epsilon(y,t). \quad (7.157)$$

The Fokker–Planck operator $\hat{F}$ defined by

$$\hat{F}\epsilon(y,t) = Q\frac{\partial^2}{\partial y^2}\epsilon(y,t) \quad (7.158)$$

has complex valued eigenfunctions given by $\epsilon_n(y) = \exp\{iny\}/\sqrt{2\pi}$ that satisfy $\partial^2\epsilon_n/\partial y^2 = -n^2\epsilon_n$. Consequently, we express $\epsilon(x,t)$ in terms of $\epsilon(y,t) = \sum_{-\infty}^\infty a_n(t)\exp\{iny\}/\sqrt{2\pi}$ with $a_0 = 0$ because of $\int_\Omega \epsilon(y,t)\,\mathrm{d}y = 0$. Note that the coefficients a_n satisfy $a_n = a^*_{-n}$ (where a^*_{-n} denotes the complex conjugate of a_{-n}). Likewise, we describe U_{MF} by means of the Fourier expansion $U_{\text{MF}}(z) = -\sum_{n=-\infty}^\infty \tilde{c}_n \exp\{inz\}/2$ with complex coefficients $\tilde{c}_n = c_n + ic'_n$. Then, we get

$$\int_\Omega U_{\text{MF}}(y-y')\epsilon(y',t)\,\mathrm{d}y' = -\frac{1}{2\sqrt{2\pi}}\sum_n \tilde{c}_n e^{iyn}\sum_k a_k \underbrace{\int_\Omega e^{-iy'n}e^{iy'k}\,\mathrm{d}y'}_{2\pi\delta_{nk}}$$

$$= -\frac{\pi}{\sqrt{2\pi}}\sum_n \tilde{c}_n a_n e^{iyn} \quad (7.159)$$

and

$$\frac{\partial}{\partial t}\epsilon(y,t) = \frac{1}{\sqrt{2\pi}}\frac{\partial^2}{\partial y^2}\sum_n\left(Q - \frac{\tilde{c}_n}{2}\right)a_n(t)e^{iyn}$$

$$= -\frac{1}{\sqrt{2\pi}}\sum_n\left(Q - \frac{\tilde{c}_n}{2}\right)n^2 a_n(t)e^{iyn}. \quad (7.160)$$

Substituting $\epsilon(y,t) = \sum_{-\infty}^\infty a_n(t)\exp\{iny\}/\sqrt{2\pi}$ into (7.160) and exploiting the orthogonality of the eigenfunctions, we finally obtain

$$\frac{\mathrm{d}}{\mathrm{d}t}a_n(t) = -n^2\left(Q - \frac{\tilde{c}_n}{2}\right)a_n(t). \quad (7.161)$$

From (7.161) it follows that for $Q - \text{Re}(\tilde{c}_n)/2 > 0$ the Fourier amplitudes $a_n(t)$ vanish and $P'_{\text{st}} = 1/(2\pi)$ is asymptotically stable. In contrast, if there is an index n such that $Q - \text{Re}(\tilde{c}_n)/2 < 0$, then the amplitude $a_n(t)$ increases as a function of time and $P'_{\text{st}} = 1/(2\pi)$ is unstable. In sum, the stability of the uniform distribution is determined by the signs of the coefficients

$$\tilde{\lambda}_n = Q - \frac{c_n}{2}, \qquad (7.162)$$

where c_n describe the real part of $\tilde{c}_n$.

Nonstationary Solutions and Traveling Wave Solutions

The previously derived result carries over to the stationary solution $P_{\text{st}}(x) = 1/(2\pi)$ of (7.148): $P_{\text{st}}(x) = 1/(2\pi)$ is asymptotically stable if the inequality $\tilde{\lambda}_n > 0$ holds for all n and unstable if there is at least one $\tilde{\lambda}_n$ with $\tilde{\lambda}_n < 0$. As a result, if the uniform distribution becomes unstable, it might be the case that the nonlinear Fokker–Planck equation (7.148) has only solutions that are nonstationary in the limit $t \to \infty$. The shape of these nonstationary solutions may vary with time. However, if $P'(y,t)$ converges to a nonuniform stationary probability density $P'_{\text{st}}(y)$ in the limit $t \to \infty$, then $P(x,t;u)$ converges to a traveling wave solution described by $P(x,t;P'_{\text{st}}) = P'_{\text{st}}(x - \omega(t - t_0))$. Unfortunately, at this stage we cannot determine the asymptotic behavior of $P'(y,t)$, which implies that we cannot determine whether or not $P(x,t;u)$ converges to traveling wave solutions if the uniform distribution becomes unstable.

Free Energy Case with $\omega \neq 0$

If $c'_n = 0$ holds for every n, then U_{MF} defined by (7.149) corresponds to a symmetric potential. In this case, (7.151) is equivalent to the KSS model discussed in Sect. 5.3.2. That is, with respect to the moving frame we are dealing with a free energy Fokker–Planck equation. For this Fokker–Planck equation the H-theorem of free energy Fokker–Planck equations applies and, consequently, the limiting case $\partial P'/\partial t = 0$ for $t \to \infty$ holds. This, in turn, implies that the limiting case

$$\lim_{t \to \infty} P(x,t;u) = P(x,t;P'_{\text{st}}) = P'_{\text{st}}(x - \omega(t - t_0)) \qquad (7.163)$$

holds. Equation (7.163) states that transient solutions of (7.148) converge either to traveling wave solutions (if P'_{st} differs from the uniform distribution) or to the uniform distribution (if P'_{st} corresponds to the uniform distribution). Furthermore, from the stability analysis carried out in Sect. 5.3.2, we conclude that if $\tilde{\lambda} = Q - c_n/2 > 0$ holds for all n, then $P'_{\text{st}} = 1/(2\pi)$ and, consequently, the uniform probability density $P_{\text{st}}(x) = 1/(2\pi)$ is asymptotically stable. If there is at least one index n with $\tilde{\lambda} = Q - c_n/2 < 0$, then $P'_{\text{st}}(y) = 1/(2\pi)$

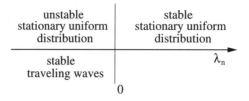

Fig. 7.2. Bifurcation diagram for solutions of (7.148) for symmetric potentials $U_{\mathrm{MF}}(z)$. Traveling wave solutions emerge from the uniform distribution when coefficients $\tilde{\lambda}_n$ become negative

as well as $P_{\mathrm{st}}(x) = 1/(2\pi)$ are unstable (see also (5.227)). Note that this result is consistent with the stability analysis derived for the general case (see (7.162)).

Taking the implications of the H-theorem and the stability analysis together, we obtain a clear picture about the behavior of the uniform distribution at an instability point, see Fig. 7.2. If the distributions $P_{\mathrm{st}}(x) = P'_{\mathrm{st}}(y) = 1/(2\pi)$ become unstable due to a change of the control parameters, then $P'(y,t)$ converges to a stable nonuniform distribution and $P(x,t;u)$ approaches a stable traveling wave solution of the form (7.155). Finally, note that in the parameter regime with the stable uniform distribution we do not have traveling wave solutions if we deal with a sine coupling-force, because in this case the uniform distribution is the only possible stationary solution (see Sect. 5.3.2).

7.3.2 Reentrant Noise-Induced Phase Transitions

Reentrant phase transitions have bifurcation lines that are qualitatively different from the bifurcation line shown in Fig. 5.18 [601]. For example, a system with a monostable and a bistable parameter regime and a control parameter $\alpha \in [0, \infty)$ exhibits a reentrant phase transition if it is bistable for a control parameter $\alpha \in [c_1, c_2]$ and monostable for $\alpha \notin [c_1, c_2]$ with $0 < c_1 < c_2 < \infty$. In this case, by increasing gradually the control parameter α we enter the bistable regime at $\alpha = c_1$ and leave the bistable regime at $\alpha = c_2$. For examples, see [219, 224, 434, 466, 601, 602].

In this section we will study a reentrant phase transition, which arises due to the impact of a multiplicative noise source that, in turn, induces a drift term via the Stratonovich interpretation [187]. To this end, we return to our discussion in Sect. 7.2.3 and consider a system described by the nonlinear Fokker–Planck equation (7.91) and the matching condition (7.81). We choose the drift and diffusion coefficients in line with coefficients that have frequently been used in the context of reentrant bifurcations [219, 224, 434, 466, 601, 602]. We use $g(x) = \sqrt{Q}(1 + bx^2)$ and $D_2(x) = Q(1 + bx^2)^2$ with $b > 0$, which means that there is a mean field force involving $A(x) = c \int_0^x D_2^{-1}(x') \, dx' = c[\arctan(\sqrt{b}x)/\sqrt{b} + x/(1+bx^2)]/(2Q)$. Since the expression $\arctan(x) + x/(1+x^2)$ behaves like an arctan-function[1], we introduce the modified arctan-function

$$\arctan'(x) = \arctan(x) + \frac{x}{1+x^2} \ . \qquad (7.164)$$

For $c = 2Q$ we then obtain from (7.91) the mean field model

$$\frac{\partial}{\partial t} P(x,t;u) = -\frac{\partial}{\partial x} \left[h(x) - \kappa \left(\frac{\arctan'(\sqrt{b}x)}{\sqrt{b}} - \left\langle \frac{\arctan'(\sqrt{b}X)}{\sqrt{b}} \right\rangle \right) \right] P$$

$$+ Q \frac{\partial}{\partial x} (1+bx^2) \frac{\partial}{\partial x} (1+bx^2) P \ . \qquad (7.165)$$

Furthermore, we put $h(x) = -ax(1+bx^2)^2$ with $a > 0$ [601]. then, stationary distributions of (7.165) satisfy the implicit equation

$$P_{\text{st}}(x) =$$

$$\frac{1}{Z(1+bx^2)} \exp\left\{ -\frac{1}{2Q} \left(ax^2 + \frac{\kappa}{2} \left[\frac{\arctan'(\sqrt{b}x)}{\sqrt{b}} - \left\langle \frac{\arctan'(\sqrt{b}X)}{\sqrt{b}} \right\rangle_{\text{st}} \right]^2 \right) \right\}$$

$$(7.166)$$

(see (7.94) and replace $Z\sqrt{Q}$ by Z). Accordingly, the self-consistency equation $m = R(m)$ for $m = \langle \arctan'(X) \rangle$ involves the functions

$$P(x,m) = \frac{1}{Z(m)} \frac{1}{1+bx^2} \exp\left\{ -\frac{1}{2Q} \left(ax^2 + \frac{\kappa}{2} \left[\frac{\arctan'(\sqrt{b}x)}{\sqrt{b}} - m \right]^2 \right) \right\} ,$$

$$R(m) = \int_\Omega \arctan'(x) P(x,m) \, dx \ . \qquad (7.167)$$

Symmetric Solution

Since $\arctan'(x)$ describes an antisymmetric function, the relation $R(0) = 0$ holds, which in turn implies that a symmetric solution P_{st} with $\langle X \rangle = 0$ and $\left\langle \arctan'(\sqrt{b}X)/\sqrt{b} \right\rangle = 0$ exists for all parameters κ and Q. Note that for stationary distributions of the form (7.166), the implication $\langle X \rangle_{\text{st}} \neq 0 \Leftrightarrow \left\langle \arctan'(\sqrt{b}X)/\sqrt{b} \right\rangle_{\text{st}} \neq 0$ holds.

Asymmetric Solutions and Reentrant Phase Transitions

Let us examine the emergence of asymmetric solutions with nonvanishing first moments. To this end, we first determine the qualitative behavior of

---

[1] Both $\arctan(x)$ and $\arctan(x) + x/(1+x^2)$ are symmetric and bounded functions that increase strictly monotonically: $d[\arctan(x)+x/(1+x^2)]/dx = 2/(1+x^2)^2 > 0$.

$R'(m) = R(m) - m$. It is clear that the relation $R'(m) = 0$ defines solutions of the self-consistency equation $m = R(m)$. Moreover, if the inequality $\mathrm{d}R'(m)/\mathrm{d}m < 0 \, (> 0)$ holds for $R'(m) = 0$, we are dealing with solutions of $m = R(m)$ with $\mathrm{d}R/\mathrm{d}m < 1 \, (> 1)$ and asymptotically stable (unstable) stationary distributions (see Sect. 7.2.3). The function $R'(m)$ is shown in Fig. 7.3 for three values of Q. We realize that when asymmetric solutions with $\langle \arctan'(\sqrt{b}X) \rangle / \sqrt{b} \neq 0$ and $\langle X \rangle \neq 0$ emerge, then the symmetric solution becomes unstable (because the slope $\mathrm{d}R'(0)/\mathrm{d}m$ becomes positive). In addition, we see that the bifurcation is reentrant. Bifurcation lines can be computed from $\tilde{\lambda} = 0 \Rightarrow \mathrm{d}R(m)/\mathrm{d}m = \kappa K_{A,\mathrm{st}}/c = 1$ with $c = 2Q$ (see Sect. 7.2.3) and are depicted in Fig. 7.4. The bifurcation diagram for the order parameter $m = \langle A \rangle_{\mathrm{st}}$ as a function of Q is given in Fig. 7.5. Note that in Fig. 7.5 only the stable solutions are shown. Assuming that (7.165) describes a strongly nonlinear Fokker–Planck equation we have determined the stationary solutions of (7.165) by solving numerically the corresponding Ito–Langevin equation, see Fig. 7.5. In line with (7.92), this Ito–Langevin equation reads

$$\frac{\mathrm{d}}{\mathrm{d}t}X(t) = h(X) - \frac{\kappa}{\sqrt{b}}\left[\arctan'(\sqrt{b}X) - \left\langle\arctan'(\sqrt{b}X)\right\rangle\right]$$
$$+ g(X)\frac{\mathrm{d}g(X)}{\mathrm{d}x} + g(X)\Gamma(t) , \quad (7.168)$$

with drift and diffusion functions given by $h(x) = -ax(1 + bx^2)^2$ and $g(x) = \sqrt{Q}(1 + bx^2)$ (see above). Finally, note that the reentrant phase transitions discussed here may describe the emergence and the vanishing of group behavior in populations of interacting members as argued in Sect. 5.5.5.

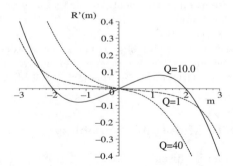

Fig. 7.3. Illustration of the self-consistency equation $R'(m) = R(m) - m = 0$ for the nonlinear Fokker–Planck equation (7.165). The function $R'(m)$ for $\kappa = 20$ and $a = 1$ is depicted for several values of Q: $Q = 1$, $Q = 10$, $Q = 40$. Solutions of $R'(m) = 0$ describe stationary distributions. Solutions of $R'(m) = 0$ with negative (positive) slopes correspond to asymptotically stable (unstable) distributions; reprinted from [187], © 2004, with permission from Elsevier

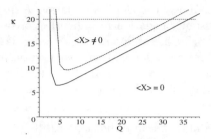

Fig. 7.4. Bifurcation lines of reentrant phase transitions described by the mean field model (7.165). Bifurcation lines are computed from $\kappa K_{A,\text{st}}/(2Q) = 1$ for $a = 1$ (*solid line*) and $a = 1.5$ (*thick dashed line*) and for $b = 0.2$; reprinted from [187], © 2004, with permission from Elsevier

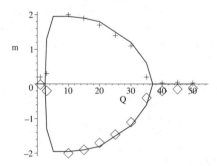

Fig. 7.5. Order parameter $m = \langle \arctan'(\sqrt{b}X)/\sqrt{b} \rangle_{\text{st}}$ of a reentrant bifurcation described by (7.165). Solid lines are computed from $m = R(m)$ and (7.167). Crosses and diamonds are obtained from simulations of Langevin (7.168) using the averaging method for self-consistent Langevin equations. The noise amplitude Q is gradually increased along the horizontal line depicted in Fig. 7.4. Parameters: $a = 1$, $b = 0.2$, $\kappa = 20$ ($L = 20000$, $\Delta t = 10^{-4}$, w_n via Box–Muller. The control parameter Q is gradually increased starting with $Q = 1$. For every Q value 2×10^5 iterations were carried out. Random variables for upper and lower branches are initially delta distributed at 0.01 and -0.01); reprinted from [187], © 2004, with permission from Elsevier

7.3.3 Systems with Multistable Variability

So far, we have primarily discussed Fokker–Planck equations that either exhibit finite dimensional nonlinearities and are nonlinear with respect to their drift terms (see Chap. 5), or exhibit infinite dimensional nonlinearities and are nonlinear with respect to their diffusion terms (see Chap. 6). In contrast, we consider now the Fokker–Planck equation

$$\frac{\partial}{\partial t}P(x,t;u) = \frac{\partial}{\partial x}\gamma x P + \frac{\partial^2}{\partial x^2}D_2(\langle A(X)\rangle)P \qquad (7.169)$$

that is nonlinear with respect to the diffusion term and features a diffusion coefficient that depends on a single order parameter $\langle A \rangle$ [194]. We assume that $\gamma > 0$ and $X \in \Omega = \mathbb{R}$ and consider solutions P subjected to natural boundary conditions. Note that D_2 does not depend explicitly on x. Therefore, the diffusion term can also be written as $D_2(\langle A(X) \rangle) \partial^2 P / \partial x^2$. Furthermore, we require $D_2(z) > 0$. Substituting $D_1(x) = -\gamma x$ and $D_2(x, \langle A \rangle) = D_2(\langle A \rangle)$ into (7.4), we obtain

$$P_{\text{st}}(x) = \sqrt{\frac{\gamma}{2\pi D_2(\langle A \rangle_{\text{st}})}} \exp\left\{-\frac{\gamma x^2}{2 D_2(\langle A \rangle_{\text{st}})}\right\} . \tag{7.170}$$

Note that we have determined the normalization constant of (7.170) by means of the Gaussian distribution (3.110). In a similar vein, from (7.6) and (7.7) it follows that

$$P(x,m) = \sqrt{\frac{\gamma}{2\pi D_2(m)}} \exp\left\{-\frac{\gamma x^2}{2 D_2(m)}\right\} \tag{7.171}$$

and

$$R(m) = \int_\Omega A(x) P(x,m) \, \mathrm{d}x , \tag{7.172}$$

which gives us the self-consistency equation $m = R(m)$. The stationary order parameter values $\langle A \rangle_{\text{st}}$ that satisfy (7.170) correspond to the solutions m of the self-consistency equation $m = R(m)$. Alternatively, we may determine $\langle A \rangle_{\text{st}}$ in a two-step approach. Accordingly, the first step is to determine $D_2(\langle A \rangle_{\text{st}})$. To this end, we note that the relation

$$D_2(\langle A \rangle_{\text{st}}) = D_2\left(\int_\Omega A(x) \sqrt{\frac{\gamma}{2\pi D_2(\langle A \rangle_{\text{st}})}} \exp\left\{-\frac{\gamma x^2}{2 D_2(\langle A \rangle_{\text{st}})}\right\} \mathrm{d}x\right) \tag{7.173}$$

holds. Therefore, we define

$$\tilde{P}(x,m) = \sqrt{\frac{\gamma}{2\pi m}} \exp\left\{-\frac{\gamma x^2}{2m}\right\} , \tag{7.174}$$

$$\tilde{R}(m) = D_2\left(\int_\Omega A(x) \tilde{P}(x,m) \, \mathrm{d}x\right) , \tag{7.175}$$

and the self-consistency equation

$$m = \tilde{R}(m) . \tag{7.176}$$

Solutions of (7.176) give us the stationary parameters values $D(\langle A \rangle_{\text{st}})$. In a second step, we substitute $D(\langle A \rangle_{\text{st}})$ into (7.170) and compute $\langle A \rangle_{\text{st}} = \int_\Omega A(x) P_{\text{st}}(x) \, \mathrm{d}x$.

In order to determine the stability of the distribution (7.170), one may examine the stability of cumulants M_1 and K. Multiplying (7.169) by x and

x^2, respectively, integrating with respect to x, and using partial integration, we obtain the evolution equations

$$\frac{d}{dt}M_1(t) = -\gamma M_1(t) \tag{7.177}$$

and

$$\frac{d}{dt}M_2(t) = -2\gamma M_2(t) + 2D_2(\langle A \rangle) . \tag{7.178}$$

In deriving these results, we have assumed that the surface terms vanish that arise from the integrations by parts. It is clear from (7.177) and (7.178) that the variance $K = M_2 - (M_1)^2$ evolves like

$$\frac{d}{dt}K(t) = -2\gamma K(t) + 2D_2(\langle A \rangle) . \tag{7.179}$$

From (7.177) we read off that the first cumulant M_1 is asymptotically stable and vanishes in the long time limit, which is consistent with (7.170). Therefore, at issue is to study the evolution of $K(t)$. To this end, we first derive an exact time-dependent solution of (7.169). Substituting

$$P_G(x, t; u) = \sqrt{\frac{1}{2\pi K(t)}} e^{-\frac{[x - x_0 M_1(t)]^2}{2K(t)}} \tag{7.180}$$

for $u = \sqrt{2\pi K(t_0)}^{-1} \exp\{-[x-x_0]^2/2K(t_0)\}$ into (7.169), we obtain

$$\frac{d}{dt}M_1(t) = -\gamma M_1(t) ,$$
$$\frac{d}{dt}K(t) = -2\gamma K(t) + 2D_2(\langle A \rangle_{P_G}) . \tag{7.181}$$

By means of this time-dependent solution, we can study the behavior of perturbations of the stationary variance $K_{st} = D_2(\langle A \rangle_{st})$. To this end, we consider the initial distributions

$$u(x) = \sqrt{\frac{1}{2\pi(K_{st} + v_0)}} e^{-\frac{[x-w_0]^2}{2(K_{st}+v_0)}} , \tag{7.182}$$

where v_0 and w_0 denote small quantities. For $M_1(t) = M_1(t_0) + w(t) = w(t)$ and $K(t) = K_{st} + v(t)$ with $w(t_0) = w_0$ and $v(t_0) = v_0$, the distribution (7.180) and the evolution equations (7.181) become

$$P_G(x, w(t), v(t)) = \sqrt{\frac{1}{2\pi[K_{st} + v(t)]}} e^{-\frac{[x - w(t)]^2}{2[K_{st} + v(t)]}} \tag{7.183}$$

and
$$\frac{d}{dt}w(t) = -\gamma w(t) , \qquad (7.184)$$
$$\frac{d}{dt}v(t) = -2\gamma(v(t) + K_{st}) + 2D_2(\langle A \rangle_{P_G(x,w,v)}) . \qquad (7.185)$$

From (7.184) it follows that perturbations w vanish in any case. In order to gain some insight into the evolution of $v(t)$, we consider an interval $I = [t_0, t_0 + \Delta t]$ for which $v(t)$ and $w(t)$ with $t \in I$ can be regarded as small quantities and linearize (7.185) with respect to v and w. Accordingly, we find

$$\langle A \rangle_{P_G(x,w,v)} = \langle A \rangle_{st} + \frac{\partial}{\partial v} \int_\Omega A(x) P_G(x,w,v) \, dx \bigg|_{v=w=0} v$$
$$+ \underbrace{\frac{\partial}{\partial w} \int_\Omega A(x) P_G(x,w,v) \, dx \bigg|_{v=w=0} w}_{Y} + O(vw, v^2, w^2) . \qquad (7.186)$$

In what follows, we assume that $A(x) = A(-x)$ holds. Then, we have $Y = 0$ and, taking only linear terms into account, (7.186) becomes

$$\langle A \rangle_{P_G(x,w,v)} = \langle A \rangle_{st} + \frac{\partial}{\partial v} \int_\Omega A(x) P_G(x,w,v) \, dx \bigg|_{v=w=0} v , \qquad (7.187)$$

which implies that

$$D_2\left(\langle A \rangle_{P_G(x,w,v)}\right) = D_2\left(\langle A \rangle_{st}\right)$$
$$+ \frac{dD_2(z)}{dz} \bigg|_{z=\langle A \rangle_{st}} \frac{\partial}{\partial v} \int_\Omega A(x) P_G(x,w,v) \, dx \bigg|_{v=w=0} v . \qquad (7.188)$$

Note that in (7.188) we have again taken into account only terms that are linear in v. Using $-2\gamma K_{st} + 2D_2(\langle A \rangle_{st}) = 0$, from (7.185) and (7.188) we finally obtain

$$\frac{d}{dt}v(t) = -\left[\gamma - \frac{dD_2(z)}{dz}\bigg|_{z=\langle A \rangle_{st}} \frac{\partial}{\partial v} \int_\Omega A(x) P_G(x,w,v) \, dx \bigg|_{v=w=0}\right] 2v . \qquad (7.189)$$

Let us define the stability coefficient

$$\tilde{\lambda} = \gamma - \frac{dD_2(z)}{dz}\bigg|_{z=\langle A \rangle_{st}} \frac{\partial}{\partial v} \int_\Omega A(x) P_G(x,w,v) \, dx \bigg|_{v=w=0} . \qquad (7.190)$$

For $\tilde{\lambda} < 0$ the variance is unstable, which implies that P_{st} is unstable. For $\tilde{\lambda} > 0$ the variance is asymptotically stable, which is a necessary condition for

P_{st} being asymptotically stable. As we will show next, there is a geometrical interpretation of $\tilde{\lambda}$. Using

$$\frac{\partial}{\partial v}\int_\Omega A(x)P_G(x,w,v)\,\mathrm{d}x\bigg|_{v=w=0} = \gamma\frac{\partial}{\partial m}\int_\Omega A(x)\tilde{P}(x,m)\,\mathrm{d}x\bigg|_{D_2(\langle A\rangle_{\text{st}})} \tag{7.191}$$

and

$$\frac{\mathrm{d}\tilde{R}}{\mathrm{d}m}\bigg|_{\langle A\rangle_{\text{st}}} = \frac{\mathrm{d}D_2(z)}{\mathrm{d}z}\bigg|_{z=\langle A\rangle_{\text{st}}}\frac{\mathrm{d}}{\mathrm{d}m}\int_\Omega A(x)\tilde{P}(x,m)\,\mathrm{d}x\bigg|_{D_2(\langle A\rangle_{\text{st}})}, \tag{7.192}$$

we get

$$\frac{\mathrm{d}}{\mathrm{d}t}v(t) = -2\gamma\left[1 - \frac{\mathrm{d}\tilde{R}(m)}{\mathrm{d}m}\bigg|_{D_2(\langle A\rangle_{\text{st}})}\right]v \tag{7.193}$$

and

$$\tilde{\lambda} = 1 - \frac{\mathrm{d}\tilde{R}(m)}{\mathrm{d}m}\bigg|_{D_2(\langle A\rangle_{\text{st}})}. \tag{7.194}$$

That is, $\tilde{\lambda}$ is related to the slope of $\tilde{R}(m)$ at intersection points $m = \tilde{R}(m)$. Therefore, the linear stability analysis that we have carried out above is consistent with a stability analysis based on the self-consistency equation (7.176).

H-Theorem

We consider now a diffusion coefficient $D_2(\langle A\rangle)$ for which the Gaussian solution (7.180) converges to its stationary solution (7.170) in the long time limit. As we will show next, in this case, transient solutions in general converge to stationary ones [194]. Let us consider the solution $W(x,t;u)$ of the linear Fokker–Planck equation

$$\frac{\partial}{\partial t}W(x,t;u') = \frac{\partial}{\partial x}\gamma x W(x,t) + \frac{\partial^2}{\partial x^2}D_2(\langle A(X)\rangle_P)W, \tag{7.195}$$

where $\langle A(X)\rangle_P = \int_\Omega A(x)P(x,t;u)\,\mathrm{d}x$ is the expectation value related to a solution of the nonlinear Fokker–Planck equation (7.169). We consider the initial distribution

$$u'(x) = \frac{1}{\sqrt{2\pi\varpi(t)}}\exp\left\{-\frac{[x-\mu(t_0)]^2}{2\varpi(t_0)}\right\}, \tag{7.196}$$

with

$$\mu(t_0) = \langle X\rangle_{P(x,t_0;u)}, \quad \varpi(t_0) = \langle [x-\mu(t_0)]^2\rangle_{P(x,t_0;u)}. \tag{7.197}$$

Then, by analogy to (7.180) and (7.181) we realize that

$$W(x,t;u') = \frac{1}{\sqrt{2\pi\varpi(t)}} \exp\left\{-\frac{[x-\mu(t)]^2}{2\varpi(t)}\right\} \qquad (7.198)$$

is a solution of (7.195) for

$$\frac{d}{dt}\mu(t) = -\gamma\mu ,$$
$$\frac{d}{dt}\varpi(t) = -2\gamma\varpi + 2D_2(\langle A\rangle_P) . \qquad (7.199)$$

In general, the initial distribution $u(x)$ of P does not correspond to $u'(x)$ of W and, consequently, the distributions W and P are not equivalent.

Next, we exploit the H-theorem for linear Fokker–Planck equations [498]. We use the Boltzmann–Kullback measure

$$^BK[P,W] = \int_\Omega P(x,t;u) \ln\left[\frac{P(x,t;u)}{W(x,t;u')}\right] dx \geq 0 , \qquad (7.200)$$

see (4.123). The key issue for the proof of the H-theorem is to realize that by definition the evolution equations for W and P have common drift and diffusion coefficients. In particular, the diffusion coefficient D_2 describes for both evolution equations the same time-dependent function $D'_2(t) = D_2(\langle A\rangle)$. On account of this property, we can calculate the derivative of K by means of standard techniques developed for linear Fokker–Planck equations [498] and thus obtain

$$\frac{d}{dt} {}^BK[P,W] = -2D_2(\langle A\rangle) \int_\Omega P(x,t;u) \left\{\frac{\partial}{\partial x} \ln\left[\frac{P(x,t;u)}{W(x,t;u')}\right]\right\}^2 dx \leq 0 . \qquad (7.201)$$

From (7.200) and (7.201) it follows that in the limit $t \to \infty$ the derivative of K vanishes. From $D_2 > 0$ and $\lim_{t\to\infty} dK/dt = 0$ it follows that

$$\lim_{t\to\infty} [P(x,t;u) - W(x,t;u')] = 0 . \qquad (7.202)$$

Equation (7.202) implies that in the limit $t \to \infty$ the expectation values $\langle A\rangle$ of $P(x,t;u)$ and $W(x,t;u')$ become equivalent: $\lim_{t\to\infty}[\langle A\rangle_P - \langle A\rangle_W] = 0$. For $\langle A\rangle_P = \langle A\rangle_W$ the distribution W corresponds to the Gaussian solution P_G described by (7.180). Consequently, we obtain

$$P(x,t;u) = W(x,t;u') = P_G(x,t;u'') \qquad (7.203)$$

for $t \to \infty$. For diffusion coefficients D_2, for which P_G converges in the limit $t \to \infty$ to a stationary solution of the form (7.170), we conclude that for arbitrary initial distributions $u(x)$ the general solution $P(x,t;u)$ converges to a stationary distribution of the form (7.170).

Pitchfork Bifurcation Model

As an example, we will study a many-body system with a diffusion coefficient D_2 that is composed of two parts: $D_2 = Q + Q_{\mathrm{MF}}$. Here, Q corresponds to a constant fluctuation strength, whereas Q_{MF} couples the subsystems. For $Q_{\mathrm{MF}} = 0$ the stationary second moment $M_{2,\mathrm{st}}$ is given by $M_{2,\mathrm{st}} = Q/\gamma$. We assume that Q_{MF} measures the deviation of the second moment $M_2(t)$ from Q/γ. We choose

$$D_2(\langle X^2 \rangle) = Q + \alpha \left(\langle X^2 \rangle - \frac{Q}{\gamma} \right) - \beta \left(\langle X^2 \rangle - \frac{Q}{\gamma} \right)^3, \qquad (7.204)$$

with $\alpha > 0$ and $\beta > 0$. Substituting D_2 into (7.169), we get

$$\frac{\partial}{\partial t} P(x,t;u) = \frac{\partial}{\partial x} \gamma x P + \left\{ Q + \alpha \left(\langle X^2 \rangle - \frac{Q}{\gamma} \right) - \beta \left(\langle X^2 \rangle - \frac{Q}{\gamma} \right)^3 \right\} \frac{\partial^2}{\partial x^2} P. \qquad (7.205)$$

An evolution equation for M_2 can be obtained by multiplying (7.205) with x^2, integrating with respect to x, and integrating by parts:

$$\frac{d}{dt} M_2(t) = -2(\gamma - \alpha) \left\{ M_2(t) - \frac{Q}{\gamma} \right\} - 2\beta \left\{ M_2(t) - \frac{Q}{\gamma} \right\}^3. \qquad (7.206)$$

Equation (7.206) describes a pitchfork bifurcation of the variable $q(t) = M_2(t) - Q/\gamma$ [246, 255, 448]. The stationary values become $M_{2,\mathrm{st}}^{(a)} = Q/\gamma$ for $\alpha > 0$ and $M_{2,\mathrm{st}}^{(b,\pm)} = Q/\gamma \pm \sqrt{(\alpha - \gamma)/\beta}$ for $\alpha > \gamma$, see Fig. 7.6. Using conventional linear stability analysis, we find that $M_{2,\mathrm{st}}^{(a)}$ describes an asymptotically stable stationary solution of (7.206) for $\alpha < \gamma$ and an unstable one

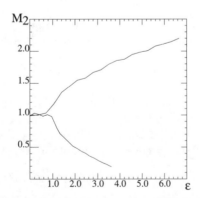

Fig. 7.6. Bifurcation diagram: $M_{2,\mathrm{st}}$ as function of $\epsilon = \alpha/\gamma$ (stable solutions only). Both curves were computed from the Langevin equation (7.210) using the averaging method for self-consistent Langevin equations. Parameters: $\gamma = Q = 1$, $\beta = 4$ ($L = 10^4$, $\Delta t = 0.01$, 3000 iterations for every ϵ, w_n via Box–Muller) [194]

for $\alpha > \gamma$, whereas $M_{2,\text{st}}^{(b,\pm)}$ describe stable stationary solutions for $\alpha > \gamma$. The stability of M_2 may carry over to the stability of the corresponding stationary probability densities. In fact, we obtain the very same results from the stability analysis based on (7.189) and (7.190). To see this, we substitute (7.204) with $\mathrm{d}D_2(z)/\mathrm{d}z = \alpha + 3\beta(z - Q/\gamma)^2$ and $\int_\Omega A(x) P_G(x,w,v) \, \mathrm{d}x = M_{2,\text{st}} + v$ for $w = 0$ into (7.190) and thus obtain $\tilde{\lambda} = \gamma - \alpha$ for $M_{2,\text{st}}^{(a)}$. Likewise, for $M_{2,\text{st}}^{(b,\pm)}$ we get $\tilde{\lambda} = -2(\gamma - \alpha)$. These results indicate that for $\alpha < \gamma$ (i.e. for $\epsilon = \alpha/\gamma < 1$) the stationary probability density (7.170) with $\langle A \rangle_{\text{st}} = M_{2,\text{st}}^{(a)}$ is stable, whereas for $\alpha > \gamma$ (i.e. for $\epsilon = \alpha/\gamma > 1$) it is unstable and there are two stable stationary probability densities (7.170) with $\langle A \rangle_{\text{st}} = M_{2,\text{st}}^{(b,\pm)}$. In fact, numerical simulations confirm this hypothesis, see Figs. 7.6 and 7.7.

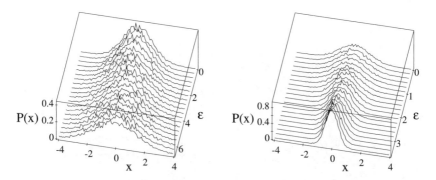

Fig. 7.7. *Left*: stationary stable probability densities corresponding to the upper branch in Fig. 7.6. Beyond the critical value $\epsilon = 1$ the probability density evolves towards the uniform distribution when ϵ is increased. *Right*: stationary stable probability densities corresponding to the lower branch in Fig. 7.6. Beyond the critical value $\epsilon = 1$ the probability density evolves towards a δ-distribution for increasing ϵ [194]

Basins of Attraction

We consider the bistable regime with $\alpha > \gamma$. Equation (7.206) can be written as

$$\frac{\mathrm{d}}{\mathrm{d}t} M_2(t) = -\frac{\mathrm{d}V}{\mathrm{d}M_2} , \qquad (7.207)$$

with

$$V(z) = (\gamma - \alpha)\left[z - \frac{Q}{\gamma}\right]^2 + \frac{\beta}{2}\left[z - \frac{Q}{\gamma}\right]^4 . \qquad (7.208)$$

From (7.208) we see that the potential V is bistable and symmetric with respect to $z_0 = Q/\gamma$ (i.e., we have $V(Q/\gamma + z) = V(Q/\gamma - z)$). Therefore, for initial distributions $u(x)$ with $M_2(t_0) > Q/\gamma$ the second moment converges to the potential minimum at $z > Q/\gamma$, which implies that $P(x,t;u)$ converges to the Gaussian stationary distribution (7.170) with $M_{2,\text{st}}^{(b,+)} = Q/\gamma + \sqrt{(\alpha-\gamma)/\beta}$. In contrast, for initial distributions with $M_2(t_0) < Q/\gamma$ we find that $M_2(t)$ converges to the potential minimum at $z < Q/\gamma$, which implies that $P(x,t;u)$ converges to P_{st} with $M_{2,\text{st}}^{(b,-)} = Q/\gamma - \sqrt{(\alpha-\gamma)/\beta}$, see Fig. 7.8. We conclude that there are two basins of attraction in the function space of probability densities. These basins are separated by initial distributions $u(x)$ with vanishing second moment. For these initial distributions solutions $P(x,t;u)$ converge to the unstable stationary solution with $M_{\text{st}}^{(a)} = Q/\gamma$. That is, initial distributions $u(x)$ with $M_2(t_0) = 0$ describe a separatrix in the function space of probability densities that separates the basins of attraction.

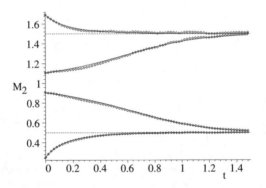

Fig. 7.8. Order parameter M_2 as a function of t as obtained from a simulation of the Langevin equation (7.210) using the averaging method (*diamonds*) and from solving numerically (7.207) (*solid lines*) for initial distribution $u(x)$ with different $M_2(0)$ ($t_0 = 0$). The initial distributions were symmetric uniform distributions $u(x) = 1/(2a)$ for $x \in [-a,a]$ and $u(x) = 0$ for $x \notin [a,a]$. We considered the bistable regime $\epsilon = \alpha/\gamma > 1$. For initial distributions with $M_2(0) > Q/\gamma$ stochastic processes converge to the wide stationary solution with $M_{2,\text{st}} > Q/\gamma$ (*dashed upper line*). For initial distributions with $M_2(0) < Q/\gamma$ stochastic processes converge to the narrow stationary solution with $M_{2,\text{st}} < Q/\gamma$ (*dashed lower line*). Initial values: $M_2(0) = 1.7$, $M_2(0) = 1.1$, $M_2(0) = 0.9$, and $M_2(0) = 0.25$ (*from top to bottom*). Parameters: $Q = 1$, $\gamma = 1$, $\alpha = 2$, $\beta = 4$ ($L = 5 \times 10^5$, $\Delta t = 0.01$, w_n via Box–Muller)

Langevin Equation

For every initial distribution $u(x)$ the second moment M_2 defined by (7.207) corresponds to a function of time which describes the overdamped motion of a particle in the globally attractive potential (7.208). Consequently, by substituting $M_2(t, u)$ as a function of t into (7.205), we can write (7.205) as a linear Fokker–Planck equation involving a diffusion coefficient $D'_2(t, u)$ that depends on time t and the initial distribution u. We consider now only initial distributions $u(x)$ with second moments $M_2(t_0, u)$, for which the inequalities

$$M_2(t, u) > 0 , \quad D'_2(t, u) > 0 \tag{7.209}$$

hold, where $M_2(t, u)$ is the solution of (7.207). Then, the linear Fokker–Planck equation related to $D'_2(t, u)$ is well-defined, which implies that (7.205) is a strongly nonlinear Fokker–Planck equation (see Chap. 3) that can be studied numerically by means of its corresponding Ito–Langevin equation. Introducing the control parameter $\epsilon = \alpha/\gamma$, the Ito–Langevin equation of (7.205) reads

$$\frac{\mathrm{d}}{\mathrm{d}t} X(t) = -\gamma X(t) + \sqrt{Q + \epsilon \gamma \left(\langle X^2 \rangle - \frac{Q}{\gamma} \right) - \beta \left(\langle X^2 \rangle - \frac{Q}{\gamma} \right)^3} \, \Gamma(t) \, . \tag{7.210}$$

From our previous discussion of (7.206) it follows that the pitchfork bifurcation occurs if $\alpha = \gamma$ holds, which means that the critical point is at $\epsilon = 1$. Note that $D_2 > 0$ is always satisfied by the stationary values $M_{2,\mathrm{st}}^{(a)}$ and $M_{2,\mathrm{st}}^{(b,+)}$. However, the admissible range of ϵ is restricted in the case of $M_{2,\mathrm{st}}^{(b,-)}$. Substituting $M_{2,\mathrm{st}}^{(b,-)}$ into (7.204), we see that the requirement $D_2 > 0$ implies that ϵ must satisfy $\epsilon < Q^2 \beta/\gamma^3 + 1$. Figure 7.6 shows the bifurcation diagram obtained by simulating (7.210). The simulation of the lower branch ($M_{2,\mathrm{st}}^{(b,-)}$) indicates the convergence of the stationary probability density to a delta distribution when ϵ approaches its maximal admissible value $\epsilon_{\max} = Q^2 \beta/\gamma^3 + 1$. In our simulation we have $\gamma = Q = 1$, $\beta = 4$, and $\epsilon_{\max} = 5$. Figure 7.7 illustrates the stationary probability densities $P_{\mathrm{st}}(x)$ for different values of ϵ corresponding to the upper branch (left panel) and the lower branch (right panel) of the diagram in Fig. 7.6. Finally, Fig. 7.8 shows the evolution of the order parameter $\langle X^2 \rangle$ as computed from the Langevin equation (7.210) for several initial values and illustrates the basins of attraction of the two stationary solutions that exist for $\epsilon > 1$.

7.4 Applications

7.4.1 Landau Form and Plasma Particles

The evolution of the velocity distribution of plasma particles can be described by means of a nonlinear Fokker–Planck equation. Let us consider a plasma composed of one species of charged particles. The particles are described by their positions and velocities. Considering a spatially homogeneous plasma, we are left with the density function for the particle velocities $\rho(\mathbf{v}, t)$ with $\mathbf{v} \in \Omega = \mathbb{R}^3$ normalized to $M_0 = \int_\Omega \rho(\mathbf{v}, t)\,\mathrm{d}^3 v$. Alternatively, we may consider the probability distribution of particle velocities $P = \rho/M_0$. Then, $P(\mathbf{v}, t)$ satisfies the evolution equation

$$\frac{\partial}{\partial t} P(\mathbf{v}, t) = -\sum_{i=1}^{3} \frac{\partial}{\partial v_i} D_i(\mathbf{v}, P) P + \sum_{i,k=1}^{3} \frac{\partial^2}{\partial v_i \partial v_k} D_{ik}(\mathbf{v}, P) P \,, \qquad (7.211)$$

with

$$D_i(\mathbf{v}, P) = a \frac{\partial}{\partial v_i} \int_\Omega \frac{P(\mathbf{v}', t)}{|\mathbf{v} - \mathbf{v}'|} \,\mathrm{d}^3 v' \,,$$

$$D_{ik}(\mathbf{v}, P) = b \frac{\partial^2}{\partial v_i \partial v_k} \int_\Omega |\mathbf{v} - \mathbf{v}'|\, P(\mathbf{v}', t)\,\mathrm{d}^3 v' \,. \qquad (7.212)$$

Equation (7.211) is referred to as the Landau form of the Fokker–Planck equation [36, 85, 134, 138, 292, 347, 447]. The drift and diffusion terms involve the Rosenbluth potentials

$$\Phi_1(\mathbf{v}) = \int_\Omega \frac{P(\mathbf{v}', t)}{|\mathbf{v} - \mathbf{v}'|}\,\mathrm{d}^3 v' \,,$$

$$\Phi_2(\mathbf{v}) = \int_\Omega |\mathbf{v} - \mathbf{v}'|\, P(\mathbf{v}', t)\,\mathrm{d}^3 v' \,. \qquad (7.213)$$

The Landau form accounts for collisions of particles and Coulomb interactions between particles. Both transient solutions [13, 398, 447, 500, 560] and stationary solutions [540] have been studied. The Landau form can also be supplemented with additional drift terms related to the collisionless Boltzmann equation [52, 83, 84, 152]. Note that the Landau form of the Fokker–Planck equation also has applications in astrophysics [58, 370].

7.4.2 Bunch-Particle Distributions of Particle Beams

Particle beams of electron storage rings can be bunched. A particle bunch is a group of particles that move together and have roughly the same energy. Charged bunch-particles produce electromagnetic fields. These fields act back on the bunch-particles and are called wakefields [91, 628]. In line with the Vlasov theory for the interaction between charged particles and their fields, there is a mean field theory of the wakefields, which leads to Vlasov–Fokker–Planck equations for particle distributions [125, 126, 283, 284, 285, 286, 534, 535, 558, 609]. Our objective is to relate the solutions of these Fokker–Planck equations to the random motions of single beam particles [177].

Vlasov–Fokker–Planck Equations and Canonical-Dissipative Systems

Let us describe the particles in bunched longitudinal beams traveling through electron storage rings in terms of their relative positions and rescaled energy deviations. Relative positions are defined with respect to moving frames related to the traveling bunches and will be denoted by q. Rescaled energy deviations will be denoted by p and are typically measured in terms of appropriately rescaled energy deviations from beam design energies. It then turns out that the bunch-particle distribution $\rho(p,q,t)$ with $(p,q) \in \Omega = \mathbb{R}^2$ and $M_0 = \int_\Omega \rho(p,q,t)\,dp\,dq$ satisfies the evolution equation

$$\frac{\partial}{\partial t}\rho(p,q,t) + \frac{\partial}{\partial q}\frac{\partial H}{\partial p}\rho - \frac{\partial}{\partial p}\frac{\partial H}{\partial q}\rho = \gamma\frac{\partial}{\partial p}(p\rho) + D\frac{\partial^2}{\partial p^2}\rho, \qquad (7.214)$$

with H defined by

$$H = H_0(p,q) + \int_\Omega H_{\mathrm{W}}(q-q')\rho(p',q')\,dp'\,dq'. \qquad (7.215)$$

The first expression $H_0(p,q)$ describes a single particle Hamiltonian and usually simply reads $H_0(p,q) = p^2/2 + q^2/2$. The integral in (7.215) reflects particle-particle interactions in a self-consistent mean field fashion (Vlasov theory). The function H_{W} determines the details of the wakefield that reflects the particle-particle couplings [125, 126, 283, 284, 285, 286, 534, 535, 558, 609]. The terms on the left-hand side of (7.214) describe damping and diffusion of the particles due to synchrotron radiation. Here, $\gamma > 0$ is a damping constant and $D > 0$ is a diffusion coefficient. The right-hand side of (7.214) is also called the Fokker–Planck collision operator. The left-hand side of (7.214) is of type of a Vlasov equation [85, 138, 347, 442, 447]. Therefore, we refer to (7.214) as a Vlasov–Fokker–Planck equation. Aiming at descriptions for single particle motions, we transform (7.214) into an evolution equation for a probability density $P(p,q,t) = \rho(p,q,t)/M_0$ normalized to unity:

$$\frac{\partial}{\partial t}P(p,q,t) + \frac{\partial}{\partial q}\frac{\partial H'}{\partial p}P - \frac{\partial}{\partial p}\frac{\partial H'}{\partial q}P = \gamma\frac{\partial}{\partial p}pP + D\frac{\partial^2}{\partial p^2}P, \qquad (7.216)$$

with

$$H' = H_0(p,q) + M_0\int_\Omega H_{\mathrm{W}}(q-q')P(p',q')\,dp'\,dq'. \qquad (7.217)$$

In general, $H_{\mathrm{W}}(z)$ is composed of a symmetric and an antisymmetric part. If $H_{\mathrm{W}}(z)$ describes a symmetric function, then there is a free energy approach to the problem at hand.

Langevin Equation

Let us write (7.216) as

$$\frac{\partial}{\partial t}P(p,q,t) = -\nabla \cdot (\mathbf{h}P) + D\frac{\partial^2}{\partial p^2}P, \qquad (7.218)$$

with the probability-dependent drift vector

$$\mathbf{h}(p,q,P) = \mathbf{I}(p,q,P) + \gamma p \begin{pmatrix} 1 \\ 0 \end{pmatrix}, \qquad (7.219)$$

where $\mathbf{I} = (I_p, I_q)$ is given by $I_p = -\partial H'/\partial q$ and $I_q = \partial H'/\partial p$ and ∇ is defined by $\nabla = (\partial/\partial p, \partial/\partial q)$. We assume that for solutions of (7.218), the time-dependent drift vector $\mathbf{h}'$ defined by $\mathbf{h}'(p,q,t) = \mathbf{h}(p,q,P)$ corresponds to the first Kramers–Moyal coefficient of a Markov diffusion process. Then, (7.218) is a strongly nonlinear Fokker–Planck equation and the transition probability density $P(p,q,t|p',q',t')$ of the Markov diffusion process satisfies

$$\frac{\partial}{\partial t}P(p,q,t|p',q',t) =$$
$$-\nabla \cdot [\mathbf{h}(p,q,P(p,q,t))P(p,q,t|p',q',t)] + D\frac{\partial^2}{\partial p^2}P(p,q,t|p',q',t), \qquad (7.220)$$

where $P(p,q,t)$ is a solution of (7.218). Equations (7.218) and (7.220) define a Markov process in terms of the joint probability densities

$$P(p_n, q_n, t_n; \ldots; p_1, q_1, t_1) =$$
$$P(p_n, q_n, t_n | p_{n-1}, q_{n-1}, t_{n-1}) \cdots P(p_2, q_2, t_2 | p_1, q_1, t_1) P(p_1, q_1, t_1), \qquad (7.221)$$

see Chap. 3. Let $p(t)$ and $q(t)$ define the random variables related to $P(p,q,t) = \langle \delta(p-p(t))\delta(q-q(t))\rangle$. Then, the joint probability densities (7.221) in general and $P(p,q,t)$ in particular can also be obtained from the Langevin equation

$$\frac{d}{dt}q(t) = \frac{\partial H_0(p,q)}{\partial p},$$
$$\frac{d}{dt}p(t) = -\frac{\partial H_0(p,q)}{\partial q} - M_0\frac{\partial}{\partial q}\int_\Omega H_W(q-q')P(p',q',t)\,dp'\,dq'$$
$$-\gamma p + \sqrt{D}\,\Gamma(t). \qquad (7.222)$$

Free Energy Case

If $H_W(z)$ describes a symmetric function, then (7.216) can be interpreted as a free energy Fokker–Planck equation of a canonical-dissipative system. To see this, for $H_W(z) = H_W(-z)$ we define the internal energy $U[P]$ as

$$U[P] = \langle H_0 \rangle + \frac{M_0}{2} \int_\Omega \int_\Omega H_W(q-q') P(p,q) P(p',q') \, dp \, dq \, dp' \, dq' \ . \quad (7.223)$$

Furthermore, we define the conservative drift forces

$$I_p = -\frac{\partial}{\partial q}\frac{\delta U}{\delta P}, \quad I_q = \frac{\partial}{\partial p}\frac{\delta U}{\delta P} \quad (7.224)$$

related to the conservative drift vector $\mathbf{I} = (I_p, I_q)$. Using $\nabla = (\partial/\partial p, \partial/\partial q)$, the Vlasov–Fokker–Planck equation (7.216) can be written as

$$\frac{\partial}{\partial t} P(p,q,t) = -\nabla \cdot (\mathbf{I}P) + \gamma \frac{\partial}{\partial p} P \frac{\partial}{\partial p}\frac{\delta F}{\delta P}, \quad (7.225)$$

where F is the free energy $F = U - Q\,^{\mathrm{B}}S$ involving the Boltzmann entropy $^{\mathrm{B}}S = -\int P \ln P \, dp \, dq$ and a noise amplitude $Q = D/\gamma$. Equation (7.225) corresponds to a free energy Fokker–Planck equation of a canonical-dissipative system (see Sect. 4.6) and for $H_0(p,q) = p^2/2 + q^2/2$ it can be regarded as a special case of the generalized Kramers equation (4.106) for systems with Boltzmann statistics. Consequently, stationary distributions can be obtained from the free energy principle $\delta F/\delta P = \mu$. From (5.5) we obtain the implicit equation

$$P_{\mathrm{st}}(p,q) = \frac{1}{Z} \exp\left\{-\frac{H'(p,q,P_{\mathrm{st}})}{Q}\right\}, \quad (7.226)$$

with H' given by (7.217). From solutions of (7.226) we then obtain the stationary particle density distributions $\rho_{\mathrm{st}}(p,q) = M_0 P_{\mathrm{st}}(p,q)$.

Haissinski Distributions of Bunched Particle Beams

Following [608], we confine ourselves to model the wakefield H_W by means of a delta function: $H_{\mathrm{MF}} = H_W = \kappa M_0^{-1} \delta(q-q')$. In doing so, we will be able to derive analytical expressions and to study at least qualitatively possible impacts of wakefields. Then, for $H_0(p,q) = p^2/2 + q^2/2$ the Vlasov–Fokker–Planck equation (7.216) reads

$$\frac{\partial}{\partial t} P(p,q,t) + p \frac{\partial P(p,q,t)}{\partial q} - \frac{\partial P(p,q,t)}{\partial p}\left[q + \kappa \frac{\partial}{\partial q} P(q,t)\right]$$

$$= \gamma \frac{\partial}{\partial p}[pP(p,q,t)] + D \frac{\partial^2}{\partial p^2} P(p,q,t) \quad (7.227)$$

and (7.222) becomes

$$\frac{d}{dt} q(t) = p \ ,$$

$$\frac{d}{dt} p(t) = -q - \kappa \frac{\partial P(q,t)}{\partial q} - \gamma p + \sqrt{D}\, \Gamma(t) \ , \quad (7.228)$$

with $P(q,t) = \int P(p,q,t)\,\mathrm{d}p = \langle \delta(q-q(t)) \rangle$. Stationary distributions can be found in the form of $P_{\mathrm{st}}(p,q) = P_{\mathrm{st}}(p)P_{\mathrm{st}}(q)$ with

$$P_{\mathrm{st}}(p) = \frac{1}{\sqrt{2\pi Q}} \exp\left\{-\frac{p^2}{2Q}\right\} \qquad (7.229)$$

and $P_{\mathrm{st}}(q)$ given by

$$[Q + \kappa P_{\mathrm{st}}(q)]\frac{\mathrm{d}P_{\mathrm{st}}(q)}{\mathrm{d}q} = -qP_{\mathrm{st}}(q)\ . \qquad (7.230)$$

This can be verified by substituting $P_{\mathrm{st}}(p,q)$ into (7.227) (see also [608]). Note also that we have $Q = D/\gamma$ as stated earlier. Stationary solutions satisfying $P_{\mathrm{st}}(p,q) = P_{\mathrm{st}}(p)P_{\mathrm{st}}(q)$ with $P_{\mathrm{st}}(p)$ and $P_{\mathrm{st}}(q)$ described by (7.229) and (7.230) are called Haissinski solutions [91, 250]. Using a stability analysis similar to the one developed in Sect. 5.1, we can show that the Haissinski solutions are stable provided that they exist [608]. One may determine $P_{\mathrm{st}}(q)$ by solving the differential equation (7.230). Alternatively, from the free energy principle we obtain (7.226) in the form of

$$P_{\mathrm{st}}(p,q) = \frac{1}{Z} \exp\left\{-\frac{p^2}{2Q}\right\} \exp\left\{-\frac{q^2/2 + \kappa P_{\mathrm{st}}(q)}{Q}\right\}\ , \qquad (7.231)$$

which gives us $P_{\mathrm{st}}(p,q) = P_{\mathrm{st}}(p)P_{\mathrm{st}}(q)$ as well as the distribution (7.229) and

$$P_{\mathrm{st}}(q) = \frac{1}{Z'} \exp\left\{-\frac{q^2/2 + \kappa P_{\mathrm{st}}(q)}{Q}\right\}\ , \qquad (7.232)$$

where Z' is a normalization constant. Note that differentiating (7.232) with respect to q, we obtain (7.230) again. Equation (7.232) can be evaluated using the concept of distortion functions (see Sects. 4.7.1 and 7.2.4). To this end, we transform (7.232) into

$$P_{\mathrm{st}}(q)\exp\left\{\frac{\kappa}{Q}P_{\mathrm{st}}(q)\right\} = \frac{1}{Z''}W(q)\ , \quad W(q) = \frac{1}{\sqrt{2\pi Q}}\exp\left\{-\frac{q^2}{2Q}\right\}\ , \qquad (7.233)$$

where Z'' is another normalization constant. Next, we recall that the inverse of the function f given by $f(\mathrm{LW}) = \mathrm{LW}\exp\{\mathrm{LW}\}$ is Lambert's W-function denoted here by $\mathrm{LW}(f)$ [627]. Then, we introduce the distortion function $G(z) = z\exp\{\kappa z/Q\}$ and its inverse function $G^{-1}(z)$ given by $G^{-1}(z) = Q\mathrm{LW}(\kappa z/Q)/\kappa$. Thus, we get

$$P_{\mathrm{st}}(q) = \frac{Q}{\kappa}\mathrm{LW}\left(\frac{\kappa}{Z''Q}W(q)\right)\ , \qquad (7.234)$$

where Z'' is determined by the requirement $\int P_{\mathrm{st}}(q)\,\mathrm{d}q = 1$. In sum, (7.229) and (7.234) define the stationary Haissinski solutions of the Vlasov–Fokker–Planck equation (7.227). Note that an expression similar to (7.234) can also be derived using alternative reasoning [577].

7.4 Applications 347

For $\kappa = 0$ we have $LW(z) = z$, $Z'' = 1$ and $P_{\mathrm{st}}(q) = W(q)$. That is, we deal with a Gaussian distribution. For $\kappa > 0$ we can read off from (7.228) that particles are driven away from regions of high density (because $P(q,t)$ acts as a potential) and, consequently, the distributions $P_{\mathrm{st}}(q)$ are wider than the Gaussian distribution $W(q)$. For $\kappa < 0$ particles are attracted by regions of high density (because now $-P(q,t)$ acts as a potential) and, consequently, we deal with squeezed distributions and the distributions $P_{\mathrm{st}}(q)$ are smaller than the Gaussian distribution $W(q)$, see Fig. 7.9. Since the parameter κ reflects the impact of the wakefield and the interactions between beam particles, we see that the particle-particle interactions can result in general in a distortion of the Gaussian distribution $W(q)$ and in particular in a squeezing or widening of the beam particle distribution $P_{\mathrm{st}}(q)$.

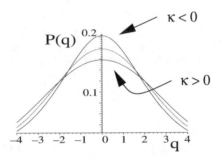

Fig. 7.9. Haissinski distributions given by (7.234) for $Q = 5$ and several parameters of κ: broadened distribution for $\kappa = 15$ (*solid line*), Gaussian distribution for $\kappa = 0$ (*dashed line*), and squeezed distribution for $\kappa = -10$ (*solid line*); reprinted from [177], © 2003, with permission from Elsevier

Now let us simulate the Langevin equation (7.228). To this end, we interpret (7.228) as a self-consistent Langevin equation and solve it by means of the averaging method. That is, we discretize (7.228) by means of the Euler forward scheme

$$q_{n+1}^l = q_n^l + \Delta t\, p_n^l ,$$
$$p_{n+1}^l = p_n^l$$
$$-\Delta t \left[q_n^l - \frac{\kappa}{(L-1)\sqrt{2\pi}[\Delta x]^2} \sum_{k=1}^{L} \left([q_n^l - q_n^k] \exp\left\{ -\frac{1}{2} \left[\frac{q_n^l - q_n^k}{\Delta x} \right]^2 \right\} \right) + \gamma p_n \right]$$
$$+\sqrt{D\Delta t}\, w_n^l , \qquad (7.235)$$

where q_n^l and p_n^l are realizations of the random variables $q(t_n)$ and $p(t_n)$ at time points $t_n = n\Delta t$. For details, see Sect. 3.4.4. Figure 7.10 (left panel)

348   7 General Nonlinear Fokker–Planck Equations

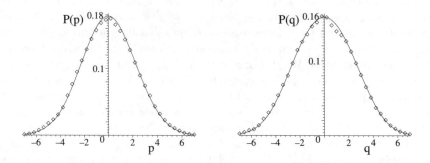

Fig. 7.10. *Left panel*: Gaussian stationary distribution $P_{\rm st}(p)$ of the Vlasov–Fokker–Planck equation (7.227) as obtained from the analytical expression (7.229) (*solid line*) and the simulation of the Langevin equation (7.228) by means of (7.235) (*diamonds*). *Right panel*: Haissinski distribution $P_{\rm st}(q)$ computed from (7.234) (*solid line*) and the Langevin equation (7.228) (*diamonds*). Parameters: $Q = 5$, $\kappa = 15$, $\gamma = 1$ ($L = 25000$, $\Delta t = 0.03$, $2[\Delta x]^2 = 0.1$, $p_0^l = q_0^l = 0$, evaluation of the stationary case p_n, q_n at $n = 200$, w_n via Box-Muller); reprinted from [177], © 2003, with permission from Elsevier

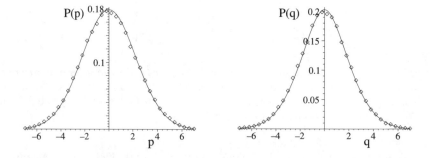

Fig. 7.11. *Left panel*: Gaussian distribution $P_{\rm st}(p)$ for $\kappa = -10$ computed from (7.229) (*solid line*) and the Langevin equation (7.228) (*diamonds*). *Right panel*: Haissinski distribution $P_{\rm st}(q)$ for $\kappa = -10$ computed from (7.234) (*solid line*) and the Langevin equation (7.228) (*diamonds*) (parameters other than κ as in Fig. 7.10); reprinted from [177], © 2003, with permission from Elsevier

shows $P_{\rm st}(p)$ computed from the random walk model (7.235) for $\kappa > 0$ and shows for the sake of comparison the exact result given by the Gaussian distribution (7.229). Figure 7.10 (right panel) shows the Haissinski distribution $P_{\rm st}(q)$ as obtained from the simulation of (7.235) for the same parameter value κ and as computed from (7.234). Figure 7.11 (right panel) shows the Haissinski distribution $P_{\rm st}(q)$ in the case of $\kappa < 0$ as obtained from the simulation of the Langevin equation (7.228) by means of (7.235) and as computed from (7.234). We have also computed the distribution $P_{st}(p)$ for this parameter value. The result is shown in Fig. 7.11 (left panel) and is equivalent to the distribution $P_{st}(p)$ shown in Fig. 7.10 (left panel) – as predicted by the theory. Figures 7.10 and 7.11 illustrate that we can describe the single particle dynamics of a particle ensemble satisfying the Vlasov–Fokker–Planck equations (7.227) by means of the corresponding Langevin equation (7.228) and its discretization (7.235).

Haissinski Distributions – General Case

Our considerations can easily be generalized to systems with wakefields $H_W = \kappa M_0^{-1} \delta(q - q')$ and single particle Hamiltonians $H_0 = p^2/2 + V(q)$, where $V(q)$ describes a potential with respect to the generalized coordinate q. Then, (7.227) is replaced by

$$\frac{\partial}{\partial t} P(p,q,t) + p \frac{\partial P(p,q,t)}{\partial q} - \frac{\partial P(p,q,t)}{\partial p} \left[\frac{dV}{dq} + \kappa \frac{\partial}{\partial q} P(q,t) \right]$$
$$= \gamma \frac{\partial}{\partial p} [p P(p,q,t)] + D \frac{\partial^2}{\partial p^2} P(p,q,t) \ . \tag{7.236}$$

Likewise, we need to replace (7.228) by

$$\frac{d}{dt} q(t) = p \ ,$$
$$\frac{d}{dt} p(t) = -\frac{\partial}{\partial q} [V(q) + \kappa P(q,t)] - \gamma p + \sqrt{D} \, \Gamma(t) \ . \tag{7.237}$$

Stationary solutions are given by $P_{st}(p,q) = P_{st}(p) P_{st}(q)$ with $P_{st}(p)$ defined by (7.229) and $P_{st}(q)$ given by (7.234), where $W(q)$ now corresponds to the Boltzmann distribution

$$W(q) = \frac{\exp\{-V(q)/Q\}}{\int \exp\{-V(q)/Q\} \, dq} \ , \tag{7.238}$$

with $Q = D/\gamma$.

7.4.3 Noise Generator

As pointed out by Primak and colleagues, Langevin equations involving probability-dependent diffusion terms can be used to model noise sources [488, 489]. We will briefly dwell on this issue and discuss this idea from the perspective of nonlinear Fokker–Planck equations. In particular, we will show that Primak's noise generators can be described in terms of probability density maintaining nonlinear Fokker–Planck equations [165]. Our departure point is the trivial nonlinear Fokker–Planck equation

$$\frac{\partial}{\partial t}P(x,t;u) = 0 ,\tag{7.239}$$

for a random variable $X(t) \in \Omega = \mathbb{R}$. Equation (7.239) states that $P(x,t;u)$ is equivalent to the initial distribution $u(x)$ for all $t \geq t_0$. We assume now that there exists a constant C such that $u(x)$ satisfies the relation

$$\forall x : 0 < \frac{1}{u(x)}\left[C - \int_{-\infty}^{x} y\,u(y)\,dy\right] < \infty .\tag{7.240}$$

For example, for symmetric probability densities ($u(x) = u(-x)$) we can choose $C = 0$. Next, we interpret the right-hand side of (7.239) according to

$$\frac{\partial}{\partial t}P(x,t;u) = k_1 \frac{\partial}{\partial x} xP(x,t;u) - k_1 \frac{\partial}{\partial x} xP(x,t;u)$$

$$= k_1 \frac{\partial}{\partial x} xP(x,t;u) + \frac{\partial^2}{\partial x^2} \underbrace{k_1 \frac{1}{P(x,t;u)}\left[C - \int_{-\infty}^{x} y\,P(y,t;u)\,dy\right] P(x,t;u)}_{D_2(x,P)} ,$$

(7.241)

with $k_1 > 0$. Note that on account of the inequality (7.240) and the observation $P(x,t;u) = u(x)$ the diffusion coefficient $D_2(x,P)$ is well-defined and positive. By means of (7.241) we can now define a Markov diffusion process that satisfies (7.239). To this end, we consider initial distributions $P(x,t,u) = P_{\mathrm{st}}(x) = u(x)$ for which the conditional probability densities

$$\frac{\partial}{\partial t}P(x,t|x',t';u) = k_1 \frac{\partial}{\partial x} xP(x,t|x',t';u) + \frac{\partial^2}{\partial x^2} D_2'(x,u)P(x,t|x',t';u) ,\tag{7.242}$$

with

$$D_2'(x,u) = D_2(x,P) = k_1 \frac{1}{u(x)}\left[C - \int_{-\infty}^{x} y\,u(y)\,dy\right]\tag{7.243}$$

correspond to Markov transition probability density. In this case, (7.239) is a strongly nonlinear Fokker–Planck equation and, consequently, the Ito–Langevin equation that corresponds to (7.239) (or, more precisely, to the interpretation of (7.239) in terms of (7.242)) is given by

$$\frac{d}{dt}X(t) = -k_1 X(t) + \sqrt{\frac{k_1}{P(x;u)}\left[C - \int_{-\infty}^{x} y\,P_{\mathrm{st}}(y)\,dx\right]}\bigg|_{x=X(t)} \Gamma(t) .\tag{7.244}$$

The probability density $P_{\mathrm{st}}(x)$ occurring in (7.244) may be computed from $P_{\mathrm{st}}(x) = u(x)$ (two-layered Langevin equation) or from $P_{\mathrm{st}}(x) = \langle \delta(x - X(t))\rangle$

(self-consistent Langevin equation). For details, see Chap. 3. Equation (7.244) provides us with a convenient tool to generate a stationary stochastic process with a particular distribution $P_{st}(x) = u(x)$ and a particular time scale k_1. That is, we can supplement a stochastic process that is given by a distribution $P_{st}(x)$ with a process time scale k_1. Since both properties (P_{st} and k_1) can be chosen independently, it has been suggested to use the Ito–Langevin equation (7.244) as a noise generator in engineering problems [488].

7.4.4 Accuracy-Flexibility Trade-Off

At first sight, one may think that for biological systems flexibility and accuracy are an antagonistic pair. The more accurately tasks are carried out, the slower the performance. That is, it seems to be plausible that flexibility can only be increased at the cost of accuracy. In fact, there is a reciprocal relationship between performance time and performance accuracy of humans, which is known as Fitts' law [160, 433, 445, 499]. Fitts' law states that the faster movements are performed, the less accurate they are. Therefore, it is safe to say that biological systems have to deal with a trade-off between accuracy and flexibility. Having said that we realize, however, that biological systems can usually handle that problem. Many biological systems possess the ability to adapt quickly and adequately to environmental and internal changes and in this sense show both accuracy and flexibility.

For example, humans and animals match their gaits to locomotory speeds in order to minimize energy consumption. Consequently, they switch from one gait to another at particular velocities [12, 299, 426, 427]. Similarly, when performance speed is increased, humans tend to abandon movement patterns requiring a high degree of coordination and show involuntary switches to simpler and more stable coordination patterns. For example, this has been observed for rhythmic isofrequency finger and wrist movements [109, 110, 216, 336, 339, 461], paced unimanual arm movements [471], and polyrhythmic tapping [50, 133, 475] (see also Sect. 5.5.6). Furthermore, perceptual patterns of apparent motion change qualitatively when the frequency of stroboscopically presented objects is increased [195, 266, 357, 521]. In the regions preceding and following transition regions, the performed patterns usually exhibit a high degree of accuracy [339, 475]. Similarly, in the pre- and posttransition regions of perceived apparent motions, humans describe their visual experiences in terms of meaningful perceptual patterns. In sum, we have given examples of qualitative changes in locomotion, coordinated movements, and perception under critical task conditions. In these examples, transitions between well-defined (and in this sense accurate) spatio-temporal motoric and perceptual patterns are observed and the perception-action systems of interest are characterized by both flexibility and accuracy.

Therefore, at issue is to clarify how flexibility and accuracy can be implemented in a system at the same time. In this context, it has been shown that stochastic processes that involve a particular kind of stochastic feedback, namely, negative stochastic feedback, exhibit a relatively small accuracy-flexibility trade-off [190, 196].

Negative Stochastic Feedback Model

Our departure point is the Langevin equation

$$\frac{d}{dt}X(t) = h(X) + \sqrt{Q}\ g\left(P(x,t;u)\right)|_{x=X(t)}\ \Gamma(t)\ , \qquad (7.245)$$

for $t \geq t_0$ and $X \in \Omega = \mathbb{R}$ with $X(t)$ distributed according to $u(x)$ at $t = t_0$. We assume that this Langevin equation is related to a strongly nonlinear Fokker–Planck equation. Then, in line with Sect. 3.4, the probability density P is either computed in a self-consistent fashion from $X(t)$ or is taken as the solution of this nonlinear Fokker–Planck equation. Interpreting (7.245) as an Ito–Langevin equation, as shown in Sect. 3.4, the corresponding strongly nonlinear Fokker–Planck equation reads

$$\frac{\partial}{\partial t}P(x,t;u) = -\frac{\partial}{\partial x}hP + Q\frac{\partial^2}{\partial x^2}g^2(P)P\ . \qquad (7.246)$$

If we put $g'(x,t) = g(P(x,t;u))$ and interpret (7.245) as the Stratonovich–Langevin equation

$$\frac{d}{dt}X(t) = h(X) + \sqrt{Q}\ \underbrace{g'(X,t)\Gamma(t)}_{\text{Stratonovich}}\ , \qquad (7.247)$$

then from the equivalence

$$-\frac{Q}{2}\frac{\partial [g']^2}{\partial X} + \sqrt{Q}\ \underbrace{g'(X,t)\Gamma(t)}_{\text{Stratonovich}} = \sqrt{Q}\ \underbrace{g'(X,t)\Gamma(t)}_{\text{Ito}} \qquad (7.248)$$

(see e.g. [298, Sect. 5.4.2]) it follows that (7.247) is related to the nonlinear Fokker–Planck equation

$$\frac{\partial}{\partial t}P(x,t;u) = -\frac{\partial}{\partial x}\left[h(x) + Q\,g(P)\frac{\partial g(P)}{\partial x}\right]P + Q\frac{\partial^2}{\partial x^2}g^2(P)P\ . \qquad (7.249)$$

We will deal with both cases simultaneously by studying the evolution equation

$$\frac{\partial}{\partial t}P(x,t;u) = -\frac{\partial}{\partial x}\left[\left\{h(x) + 2Q(1-\varpi)\,g(P)\frac{\partial g(P)}{\partial x}\right\}P - Q\frac{\partial}{\partial x}g^2(P)P\right]\ . \qquad (7.250)$$

That is, $P(x,t;u)$ of (7.250) is equivalent to $P(x,t;u) = \langle \delta(x - X(t)) \rangle$ derived from the Langevin equation (7.245) if we interpret (7.245) according to Ito calculus and put $\varpi = 1$ or if we interpret (7.245) according to Stratonovich calculus and put $\varpi = 1/2$. Furthermore, we assume that h is the force of a globally attractive potential. That is, we have $V(x) = -\int^x h(x')\,\mathrm{d}x$ and $V(x \to \pm\infty) \to \infty$. Finally, we require that $g(z)$ satisfy the conditions

$$g \geq 0 \,, \quad \frac{\partial g}{\partial z} \leq 0 \,, \tag{7.251}$$

for $z \geq 0$.

Qualitative Discussion: Stationary Case

Let us discuss qualitatively the impact of the function g on the stationary behavior of a statistical ensemble described by (7.250) and (7.251). For $g = 1$, the Fokker–Planck equation is linear with respect to P and the stationary probability density is given by the Boltzmann distribution

$$W(x) = \frac{1}{Z_B} \exp\left\{-\frac{V(x)}{Q}\right\} \,, \tag{7.252}$$

with $Z_B = \int_\Omega \exp\{-V(x)/Q\}\mathrm{d}x$. The positions of the maxima of the stationary probability density (7.252) correspond to the positions of the minima of the potential V. For $\partial g/\partial P < 0$, however, the maxima of the probability density P describe minima of the noise amplitude $\sqrt{Q}g(P)$. Consequently, in the stationary case, members of a statistical ensemble that are located in close vicinity of the minima of V are less affected by noise than members at other positions. By comparison with the trivial case $g = \text{const}$, this mechanism results in a more pronounced aggregation of ensemble members at the minima of the potential V. We can also look at this effect in a different way. We may say that the nonlinearity $g(P)$ "distorts" the probability density W such that large values are enhanced whereas small values are weakened. Roughly speaking, we anticipate that due to the inequality $\partial g/\partial P < 0$ we have the following circular causality: the more probable a stable state related to a potential minimum is, the less it will be affected by noise and, conversely, the less the stable state is affected by noise, the more probable it is. Therefore, we may say that (7.245-7.251) describe processes with negative stochastic feedback [190, 196]. Next, let us examine in a more quantitative manner stationary solutions of the negative stochastic feedback model.

Stationary Solutions

Equation (7.250) corresponds to the nonlinear Fokker–Planck equation (2.10) with $D_1(x, P) = h + 2Q(1 - \varpi)\, g(P)\partial g(P)/\partial x$ and $D_2(P) = Qg^2$. Consequently, stationary distributions of (7.250) satisfy (2.23). Since we are looking for solutions with $P(x \to \pm\infty) = 0$ (natural boundary conditions), the probability current J vanishes and we obtain

$$\frac{h(x)}{Q} = \varpi \frac{\partial}{\partial x} g^2(P_{\text{st}}) + \frac{g^2(P_{\text{st}})}{P_{\text{st}}} \frac{\partial P_{\text{st}}}{\partial x} . \qquad (7.253)$$

By virtue of (7.252) and (7.253), the implicit equation

$$\frac{1}{Z} W(x) = \exp\left\{ \varpi g^2(P_{\text{st}}) + \int^{P_{\text{st}}} \frac{g^2(z)}{z} dz \right\} = f(P_{\text{st}}) \qquad (7.254)$$

can be found. The variable Z guarantees the normalization of P_{st}. The function f describes the mapping of the probability density $P_{\text{st}}(x)$ to $W(x)$ and can be regarded as a distortion function similar to the distortion functionals discussed in Sects. 4.7.1 and 7.2.4. Assuming that f is invertible with $F = f^{-1}$, we can compute P_{st} from W by means of

$$P_{\text{st}}(x) = F\left(\frac{1}{Z} W(x) \right) . \qquad (7.255)$$

In order to illustrate this procedure, we put $g(P) = g_{n,\alpha}$ with

$$g_{n,\alpha} = \left(\frac{1}{1+\alpha P} \right)^{n/2}, \quad \alpha \geq 0, \; n = 1, 2, 3, \ldots \qquad (7.256)$$

Note that for $\alpha = 0$ we have $g_{n,\alpha} = 1$ and the nonlinear Fokker–Planck equation becomes linear. Substituting (7.256) into (7.254), we obtain [144]

$$f_{n,\varpi,\alpha}(P_{\text{st}}) =$$
$$\frac{P_{\text{st}}}{1+\alpha P_{\text{st}}} \exp\left\{ \frac{\varpi}{(1+\alpha P_{\text{st}})^n} + \sum_{k=1}^{n-1} \binom{n-1}{k} \frac{(-\alpha P_{\text{st}})^k}{k(1+\alpha P_{\text{st}})^k} + C_n \right\} . \qquad (7.257)$$

The constants C_n can assume arbitrary values and can be compensated by the normalization Z. Focusing on weak nonlinearities ($n = 1, 2$), we choose $C_1 = 0, C_2 = 1$ and obtain

$$f_{1,\varpi,\alpha}(P_{\text{st}}) = \frac{P_{\text{st}}}{1+\alpha P_{\text{st}}} \exp\left\{ \frac{\varpi}{1+\alpha P_{\text{st}}} \right\} ,$$
$$f_{2,\varpi,\alpha}(P_{\text{st}}) = \frac{P_{\text{st}}}{1+\alpha P_{\text{st}}} \exp\left\{ \frac{\varpi}{(1+\alpha P_{\text{st}})^2} + \frac{1}{1+\alpha P_{\text{st}}} \right\} . \qquad (7.258)$$

Properties of the Distortion Functions

Figure 7.12 shows the distortion functions $f_{n=(1,2),\varpi,\alpha}$ for both the Ito and the Stratonovich case. In the limit $P_{\text{st}} \to \infty$ the distortion functions $f_{n=(1,2),\varpi,\alpha}(P_{\text{st}})$ converge to a finite value $1/\alpha$. Consequently, all the functions $F_{1,\varpi=1/2,\alpha}$, $F_{1,\varpi=1,\alpha}$, and $F_{2,\varpi=1/2,\alpha}$ have a singularity at $1/\alpha$. This implies that in order to map W onto P_{st} (see (7.255)) the normalization constant Z must satisfy the inequality $1/\alpha < \max\{W(x)\}/Z$. Figure 7.12 demonstrates that the functions F describe mappings $W \to P_{\text{st}}$ such that large values of W are amplified and small values of W are weakened. Finally, note that a detailed analysis reveals that $f_{2,\varpi=1/2,\alpha}(z)$ (Stratonovich case) as a function of z is a strictly concave function, whereas $f_{2,\varpi=1,\alpha}(z)$ (Ito case) has a maximum at $P_{\text{st}} = 1/\alpha$ with a value $1/(2\alpha)\exp\{3/4\} > 1/\alpha$. For this reason, we only consider the lower branch of $F_{2,1,\alpha}$, that is, the lower branch of the curve in the inlet of Fig. 7.12 corresponding to graph IV.

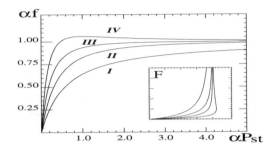

Fig. 7.12. Distortion function $f_{n,\varpi,\alpha}$ as a function of P_{st} for $n = 1$ (I, II) and $n = 2$ (III, IV): the Ito case ($\varpi = 1 : II, IV$) results in stronger distortions than the Stratonovich calculus ($\varpi = 1/2 : I, III$); the inlet shows the inverse graphs $F_{n,\varpi,\alpha} = f^{-1}n, \varpi, \alpha$; reprinted from [196], © 2002 Kluwer Academic Publishers, with kind permission of Kluwer Academic Publishers

Double-Well Potential

In order to analyze further properties of the mapping $W \to P_{\text{st}}$, we choose a particular deterministic force h. We consider $h(x) = -dV/dx$ with

$$V(x) = -\frac{k_1}{2}x^2 + \frac{k_2}{3}x^3 + \frac{k_3}{4}x^4 \qquad (7.259)$$

and $k_1 > 0$, $k_2 > 0$, and $k_3 > 0$, which has a maximum at the origin and both a local and a global minimum, see Fig. 7.13. The Boltzmann distribution W of this potential can be computed by means of (7.252). To obtain a normalized solution of (7.255), we generate a family of distribution functions $P_i(x; Z_i)$ that are regarded as candidates for P_{st} by applying $F_{n,\varpi,\alpha}$ to W for various Z_i. Explicitly, we initially estimate Z_0, integrate over $P_0(x; Z_0)$, and use the resulting value to estimate a new approximation of Z_1, which gives us $P_1(x; Z_1)$. This procedure is repeated iteratively until convergence. Final estimates of the probability densities are given in Fig. 7.13. The undeformed probability density ($g_{n,\alpha} = 1$, curve (i)) shows a much lower probability at the global minimum of the potential compared to the cases including nonlinear feedback (see curves (ii) and (iii)).

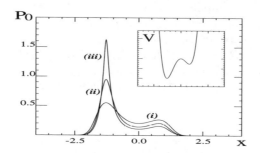

Fig. 7.13. Stationary probability density $P_{\rm st}(x)$ of (7.250) with g and h given by (7.256) and (7.259), respectively. Parameter values are (i) $g=1$, (ii) $n=1, \varpi=1$, and (iii) $n=2, \varpi=1/2$; $\alpha=1, Q=50, k_1=50, k_2=k_3=100$. The inlet shows the corresponding potential $V(x)$; reprinted from [196], © 2002 Kluwer Academic Publishers, with kind permission of Kluwer Academic Publishers

Transient Solutions

Let us discuss now the transient behavior of solutions of the Langevin equation (7.245). To this end, we consider the Ito case for a weak nonlinearity and choose $n=1$ in (7.256). Then, (7.245) reads

$$\frac{\partial}{\partial t}X(t) = h(X) + \sqrt{\frac{Q}{1+\alpha P(x,t)}}\bigg|_{x=X(t)} \Gamma(t) \ . \tag{7.260}$$

Likewise, (7.250) reads

$$\frac{\partial}{\partial t}P(x,t;u) = -\frac{\partial}{\partial x}h(x)P + Q\frac{\partial^2}{\partial x^2}\frac{P}{1+\alpha P} \tag{7.261}$$

and is equivalent to (7.70) discussed in Sect. 7.2.2. In Sect. 7.2.2 we have shown that (7.261) is an entropy Fokker–Planck equation of the form (6.1) involving the entropy measure $^{\rm NSF}S_\alpha$ defined by (7.72). Since $^{\rm NSF}S_\alpha$ is a strictly concave measure (see Sect. 6.4.2), the free energy $F = \langle V \rangle - Q\,^{\rm NSF}S_\alpha$ is bounded from below (case A in Table 4.2) and the H-theorem for free energy Fokker–Planck equations applies. Thus, we conclude that transient solutions of (7.261) converge to stationary ones in the long time limit.

The Langevin equation (7.260) was solved numerically using the histogram method for self-consistent Langevin equations, see Sect. 3.4.4. Figures 7.14 and 7.15 show the solutions for two different parameter values of

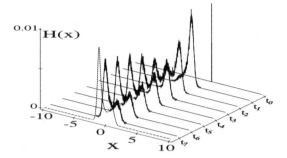

Fig. 7.14. Simulation of the Langevin equation (7.260) with g and h given by (7.256) and (7.259), respectively. Numerical results are plotted at different times t_i with $t_{i+1} > t_i$. The dashed line in front corresponds to the analytical result, see Fig. 7.13 (ii). The histogram method for self-consistent Langevin equations was used. Parameters: $\alpha = 1$, $Q = 50$, $k_1 = 50, k_2 = k_3 = 100$ ($L = 40000$, $\Delta t = 1/1000$, $\Delta x = 1/100$, w_n via Box–Muller); reprinted from [196], © 2002 Kluwer Academic Publishers, with kind permission of Kluwer Academic Publishers

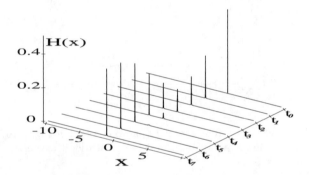

Fig. 7.15. Simulation results for $\alpha = 100$ (other parameters as in Fig. 7.14); reprinted from [196], © 2002 Kluwer Academic Publishers, with kind permission of Kluwer Academic Publishers

α. In the stationary case, the maximum p_m of P is located at the global minimum x_m of the potential (7.259). The numerical simulation for a weak nonlinearity is in good agreement with the analytical solution given above, see Fig. 7.14. If we increase the nonlinearity, the maximum value of the probability density $p_m = P(x_m)$ increases and the variance K decreases, which results in sharply peaked distributions as shown in Fig. 7.15. In this case, the transient probability densities are essentially described by two peaks that correspond to the two minima of the potential. Therefore, the stochastic process primarily consists of discrete jumps between the minima of the potential. The probability to find realizations of the stochastic process outside close vicinities of the potential minima becomes negligible.

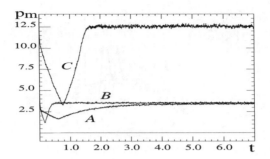

Fig. 7.16. Maximal probability p_m as function of time t. Stochastic processes with negative stochastic feedback ($\alpha = 5$: curve B and $\alpha = 20$: curve C; $Q = 50$) show shorter transient parts compared to the reference process without stochastic feedback ($\alpha = 0, Q = 4.5$: curve A); reprinted from [196], © 2002 Kluwer Academic Publishers, with kind permission of Kluwer Academic Publishers

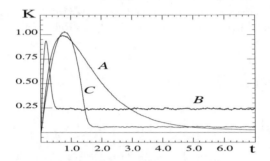

Fig. 7.17. Variance K as a function of t (parameters as in Fig. 7.16); reprinted from [196], © 2002 Kluwer Academic Publishers, with kind permission of Kluwer Academic Publishers

In order to compare transition times for systems with and without stochastic feedback, we adjusted the parameters α and Q in such a way that the stationary probability densities have approximately either the same peak probability density p_m or the same variance K. Figures 7.16 and 7.17 show the results of these simulations. The graphs A, B, and C were computed from three different processes, which will be referred to as $A-$, $B-$, and $C-$processes, respectively. The $A-$process corresponds to the conventional stochastic process given by (7.245) with $g = 1$. The $B-$ and $C-$processes correspond to processes with stochastic feedback. The parameters α and Q are chosen such that the stationary peak probabilities p_m of the $A-$ and $B-$processes and the stationary variances K of the $A-$ and $C-$processes assume the same values. In comparison to the relaxation processes defined by the ordinary Langevin equation (curves A in Figs. 7.16 and 7.17), the impact of the negative stochastic feedback results in a much faster relaxation to stationary values of p_m (see Fig. 7.16). Stronger nonlinearities again increase the probability p_m at x_m (curve C in Fig. 7.16). The stationary variance decreases with increasing α (compare tails of curves B and C of Fig. 7.17), because increasing α implies that the effective diffusion coefficient $Q/(1+\alpha P(x))$ decreases for states x with $P(x) > 0$. From Fig. 7.17 it is clear that the variances of systems with stochastic feedback converge to their stationary values much faster than the variances of systems without stochastic feedback.

Towards a Minimization of the Accuracy-Flexibility Trade-Off

We have illustrated that the aggregation of the subsystems of a many-body system in the minimum of a potential (related to some kind of energy or effort) can be amplified by means of negative stochastic feedback and that the transition times are significantly shorter than those of comparable systems without feedback. Consequently, a biological system under the impact of negative stochastic feedback can detect the position of the minimum value of a potential force with a large probability and within a short transient period. In other words, such a system exhibits the ability to generate sharply peaked probability densities and to respond quickly to changes in both the environment and its internal parameters. Moreover, the form of (7.245) does not require any a priori knowledge about the underlying potential force. Thus, the system can be considered to be self-organized.

Bose Systems and Negative Stochastic Feedback

In closing this section, we would like to point out that Bose systems can be considered as another example of systems with negative stochastic feedback. To this end, let us consider for $\varpi = 1$ the nonlinearity

$$g_{BE}(P) = \sqrt{\frac{\ln[1+P]}{P}} , \qquad (7.262)$$

which leads to the Fokker–Planck equation

$$\frac{\partial}{\partial t}P(x,t;u) = -\frac{\partial}{\partial x}h(x)P + Q\frac{\partial}{\partial x}\ln[1+P] . \qquad (7.263)$$

This evolution equation, in turn, corresponds to the Fermi Fokker–Planck equation (6.306) if we put $M_0 = 1$ and $Q = T$. The stationary probability density of (7.263) reads

$$P_{\text{st}}(x) = \frac{1}{\exp\{(U_0(x) - \mu)/Q\} - 1} , \qquad (7.264)$$

and describes a Bose–Einstein statistics. Just as in our first example (see (7.256)), in the limit $P \to 0$ the nonlinearity g_{BE} converges to unity (which can be shown by applying the rule of de l'Hopital to $[g_{BE}]^2$). Moreover, for low density distributions, the diffusion term of (7.263) is approximately given by $\ln[1 + P] \approx P/[1 + 0.5P]^{-1} + O(P^3)$. That is, up to terms of the order of two, the nonlinear Fokker–Planck equation for Bose particles (7.263) is equivalent to (7.261) with $\alpha = 1/2$. Furthermore, we obtain

$$\frac{d}{dP} g_{BE}(P) = \frac{1}{2P^2(1+P)} \sqrt{\frac{P}{\ln[1+P]}} Z(1+P) , \qquad (7.265)$$

where $Z(y)$ is defined by $Z(y) = y - 1 - y \ln y$ for $y > 1$ ($P > 0$) and $Z(1) = 0$. Since the inequality $\ln z < z - 1$ holds for $z > 0$ and $z \neq 1$, it follows that $Z(y) < 0$ for $y > 1$ (hint: substitute z by $1/y$). Consequently, g_{BE} satisfies (7.251) and Bose systems can be regarded within the Fokker–Planck approach as systems with negative stochastic feedback.

7.5 Bibliographic Notes

In this section, we will address some studies that are concerned with free energy Fokker–Planck equations, entropy Fokker–Planck equations, and nonlinear Fokker–Planck equations in general and have not been mentioned so far. In doing so, we will not give a review on the subject of nonlinear Fokker–Planck equations but we will illustrate the variety of topics that have been discussed in the context of nonlinear Fokker–Planck equations.

Natural Boundary Conditions

K-Model with Additive Noise

Pioneering works on the K-model discussed in Sect. 5.1.5 have been carried out by Desai and Zwanzig [135] and Kometani and Shimizu [352]. The free energy approach to the K-model is due to Shiino [522, 523]. The K-model and various modifications concerning the drift coefficient of the model have been discussed in the literature [30, 32, 77, 88, 127, 128, 141, 142, 295, 393, 431, 435].

K-Model with Multiplicative Noise

The K-model with multiplicative noise has also been studied [20, 59, 392, 434, 644, 645]. In this context, discontinuous phase transitions [434], discontinuous phase transitions related to reentrant bifurcations [644], the impact of colored noise on reentrant bifurcations [392], and propagation of signals in cellular media [645] have been examined.

K-Model and Spatially Extended Systems

The K-model with multiplicative noise has been used as a mean field approximation of spatially distributed systems with diffusive coupling (i.e., coupling described by the Laplacian operator or next neighbor coupling) [219, 220, 224, 288, 601, 600, 602].

Pattern Formation

The formation of spatial patterns in spatially distributed systems with diffusive coupling and multiplicative noise sources has been studied by means of nonlinear Fokker–Planck equations [220, 466, 646].

Multiple Species and Relaxation Oscillators

Multivariate nonlinear Fokker–Planck equations involving nonlinear drift terms like in the K-model have been studied in several studies. In these studies the components of the respective random vectors correspond to different kinds of subsystems or species [123, 218, 649]. For example, Yamaguchi and co-workers have studied the collective behavior of coupled relaxation oscillators [638, 639]. Schuster and Wagner have considered a Wilson–Cowan model that comes with two dynamical variables for each oscillatory subunit [513, 514]. In order to study mean field coupled networks of oscillatory neurons, single neurons have been modeled by means of the two-dimensional Morris–Lecar system [272] and the FitzHugh–Nagumo equation [280, 320, 462]. Kawai and colleagues have studied a two-component mean field model that describes self-sustained oscillations in terms of a time-dependent probability density and a nonstationary oscillating order parameter [330].

Inertia Effects

Multivariate nonlinear Fokker–Planck equations and related models taking inertia effects into account have also been considered. In these studies the focus was on the formation and breakup of swarms [420, 516], reentrant bifurcations [330], and the impact of parametric driving forces [54]. In addition, changes of the crystallographic structure of matter (so-called structural phase transitions) have been addressed in the context of the independent-site approximation, which is some kind of mean field approximation [80, 161].

Surface Physics

Solutions of the Kadar–Parisi–Zhang equation have been examined by means of nonlinear Fokker–Planck equations [225, 404]. In doing so, the growth of surfaces and roughening phenomena have been studied. In addition wetting processes [132] have been described by nonlinear Fokker–Planck equations.

Particle Sedimentation

Particle sedimentation in fluids has been modeled by means of a multivariate Langevin equation with probability-dependent coefficients [290]. Roughly speaking, the model is given by the Langevin equation

$$\frac{d}{dt}X(t) = V(t), \quad (7.266)$$

$$\frac{d}{dt}V(t) = -\gamma X + a(\langle X \rangle) + \sqrt{b(K(X))}\Gamma(t), \quad (7.267)$$

where $a(z)$ and $b(z)$ are functions of z.

Batchelor and Turbulent Diffusion

As early as 1952, Batchelor proposed the diffusion equation

$$\frac{\partial}{\partial t}P(x,t;u) = Q\frac{\partial^2}{\partial x^2}\langle X^2 \rangle P(x,t;u) \quad (7.268)$$

in order to describe the displacement of particles in turbulent fields [40]. Equation (7.268) has a structure similar to the model discussed in Sect. 7.3.3.

Stochastics of Ion Channels

Ion channels of neurons can be regarded as two-state systems. Accordingly, at an ion channel one may find an ion either inside or outside the membrane of a neuron. Let $X_1 \geq 0$ and $X_2 \geq 0$ denote the occupation numbers of these states. In line with the quantum mechanical description of lasers [256], it has been suggested that $\mathbf{X} = (X_1, X_2)$ satisfies the Langevin equation

$$\frac{d}{dt}\mathbf{X}(t) = \begin{pmatrix} -w_{21} & w_{12} \\ w_{21} & -w_{12} \end{pmatrix}\mathbf{X} + G(\langle X_1 \rangle, \langle X_2 \rangle) \cdot \mathbf{\Gamma}(t), \quad (7.269)$$

where G describes a 2×2 matrix that depends on the mean values $\langle X_1 \rangle$ and $\langle X_2 \rangle$ [262]. It is clear from (7.269) that the corresponding Fokker–Planck equation involves a diffusion coefficient that depends on $\langle X_1 \rangle$ and $\langle X_2 \rangle$ as well. That is, the system is described by means of a nonlinear Fokker–Planck equation.

Periodic Boundary Conditions

Winfree and the Theory of Synchronization

Much of our interest in the behavioral synchronization of populations was fueled by Winfree's seminal works on that topic [632, 633]. The original model proposed by Winfree differs slightly from the KSS model studied in this chapter and in Sect. 5.3.2. While the KSS model involves a mean field coupling term $\sum_k U_{\mathrm{MF}}(X_i - X_k)$, the Winfree model exhibits a coupling term given by $\sum_k A(X_k)$. Models of this kind have also been considered in the more recent literature [27, 391].

KSS Model with Distributed Eigenfrequencies

The KSS model discussed in Sect. 7.3.1 describes an ensemble of phase oscillators with a common eigenfrequency ω. Taking a more general point of view, we may think of ensembles of coupled oscillators with different eigenfrequencies. In this context, the question arises as to what extent oscillators with different eigenfrequencies synchronize their behavior. For systems that are not subjected to fluctuating forces, this question and related issues have been studied in [115, 363, 365, 367, 368, 369, 443, 504, 505, 513, 514, 554, 553, 564, 565, 563, 603, 620, 640, 647] for time-continuous systems and in [112, 113, 114, 575] for time-discrete systems. Oscillator populations with some kind of distributed state-dependent eigenfrequencies have been considered in [599]. Taking fluctuating forces into account and using nonlinear Fokker–Planck equations, the behavior of oscillator populations with distributed eigenfrequencies has been studied in [4, 26, 60, 61, 63, 62, 96, 102, 103, 364, 365, 366, 477, 503, 542, 555, 643]. For reviews, see [552, 557]. Such coupled oscillator models have been used to describe associative memories [369, 640, 643]. A coupled oscillator population that is subjected to white noise fluctuating forces, on the one hand, and exhibits eigenfrequencies that fluctuate according to a dichotomic Markov process, on the other hand, has been studied as well [355].

Multiplicative Noise

The KSS model with sine-coupling and the mean field HKB model for systems with multiplicative noise have been studied in [343, 341, 344, 463, 491, 492].

Delays

Mean field models including delays have also been investigated in the context of nonlinear Fokker–Planck equations [342, 450, 642]. In particular, the phenomenon of stochastic resonance has been addressed [342].

Higher Order Coupling Functions

Models similar to the KSS model that involve higher order coupling functions such as $\sin(2x)$, $\sin(3x)$, ... instead of the sine-coupling function have been studied in [25, 117, 118, 119, 275, 365].

Inertia Effects

The dynamics of mean field coupled phase oscillators under the impact of inertia effects and related models have attracted considerable attention [5, 22, 23, 24, 159, 226, 380, 381, 382, 383, 384, 385, 477, 562, 581]. For example, Daido has studied the circadian rhythms using arrays of two-component oscillators [120]. Coupled phase oscillators that take some kind of inertia effects into account have been examine by Hadley et al. in the context of phase locking in arrays of Josephson-junctions [248]. In a similar vein, Acebron et al. have focused on the synchronization of coupled phase oscillators with inertia terms [6].

Parkinson's Disease

Several authors have suggested that Parkinsonian tremor is caused by populations of neurons that synchronize their oscillatory activity [390]. In a series of papers, Tass [569, 570, 571, 572, 573, 574] proposed a model of coupled neural oscillators that are putatively involved in Parkinsonian tremor (see also [568]). According to this oscillator model, the couplings between neural oscillators in healthy people are so weak that they do not lead to the degree of synchronization necessary to create Parkinsonian tremor. The oscillators are said to be de-synchronized. In contrast, in patients with Parkinson's disease, the oscillators are assumed to be strongly connected with each other. Due to these connections, a mean field force emerges and results in a self-organized synchronization of the oscillator population. This synchronization, however, can be destroyed by stimulating the synchronized neurons with an appropriate external signal. In other words, it is believed that an external signal can be used to de-couple neural oscillators. The hope is that this de-coupling mechanism can be applied to help patients suffering from Parkinson's disease.

Miscellaneous Topics

In addition to the aforementioned applications, mean field models for periodic random variables have been used to describe Josephson arrays [629, 630], Landau damping [556], arrays of semi-conductor lasers [356], charge density waves [60], and have been proven to be useful tools for the study of coupled neurons described by Hodgkin–Huxley equations [272, 276].

Nonlinear Diffusion Equations

Nonlinear diffusion equations and their applications are discussed in Crank's book [101], whereas mathematical aspects of nonlinear diffusion equations are addressed in the Proceedings by Ni et al. [446]. As illustrated in Chap. 6, such nonlinear diffusion equations are closely related to entropy Fokker–Planck equations involving entropy measures different from the Boltzmann entropy. A discussion on the renormalization group approach to the Barenblatt–Pattle solutions of the porous medium equation can be found in [229].

Mathematical Literature

In the mathematical literature, nonlinear Fokker–Planck equations have been investigated in line with a seminal study by McKean Jr. [411] (see Sect. 3.1) and are sometimes referred to as McKean–Vlasov equations [73, 217, 222, 414,

454]. In this context, the propagation of molecular chaos (see Sect. 3.8) has been addressed [414, 415]. Stochastic processes related to nonlinear Fokker–Planck equations have also been called processes depending on Brownian paths [452]. The convergence of transient solutions of nonlinear Fokker–Planck equations to stationary ones has been examined by means of functionals that are similar to the Lyapunov functionals discussed in Chap. 4 and in this chapter, see, for example, [30, 31, 32, 87]. The relationship between Langevin equations similar to the ones presented in Sect. 3.4 and nonlinear Fokker–Planck equations has been studied in the context of so-called semilinear parabolic partial differential equations by Milstein and colleagues [423, 424, 425]. In a similar vein, it might be helpful to study nonlinear Fokker–Planck equations in the context of the theory of nonlinear parabolic partial differential equations [205, 206, 395].

8 Epilogue

One may raise the question: what is at the heart of this book? In reply, it may be said: it is the nonlinearity of nature. This book is concerned with nonlinear science in general and synergetic systems in particular. Nonlinear science tries to meet the demand of the modern society for analyzing and dealing with highly complex many-body systems. It is a rapidly growing field of science that lives in a triangle with corners at the theory, the numerics, and the experiment. In the preceding chapters we have shown that nonlinear Fokker–Planck equations are powerful tools to describe and understand various phenomena that are observed in the inanimate and animate world and arise due to the nonlinearity of systems. In recognition of the aforementioned triangle of nonlinear science, we have discussed nonlinear Fokker–Planck equations with an eye on theoretical physics, numerical solutions by means of Langevin equations, and top-down modeling of experimental findings.

In this final chapter we would like to discuss two issues. The first issue concerns the top-down modeling by means of nonlinear Fokker–Planck equations and possible ways of its verification and falsification. The second issue is about generalizations. Both issues are somewhat of a speculative character. That is, we will not give final answers. Rather we will make some hypotheses.

Top-Down Modeling and Its Verification and Falsification

Although nonlinear Fokker–Planck equations can describe a variety of frequently observed phenomena, this does not imply that the underlying mechanisms are adequately described in terms of nonlinear Fokker–Planck equations. As usual, a particular phenomenon can be described by means of several models involving different notions about the fundamental mechanisms that give rise to the observed phenomenon. In this context, the problem is to select a particular model from a set of possible models that all do their job equally well. For example, if a system exhibits a power law distribution, we can describe the system by means of a nonlinear Fokker–Planck equation as shown in Chap. 6. Alternatively, one can describe power law distribution by means of linear Fokker–Planck equations with state-dependent diffusion coefficients. In the context of nonlinear Fokker–Planck equations, the establishment of states with power law distributions is regarded as a collective phenomenon. The notion is that a power law distribution arises from the interactions between the subsystems of a many-body system. In the context of linear Fokker–Planck equations, power law distributions describe a single system that is subjected to a multiplicative noise source [65, 325, 436, 547] or to some kind of temperature fluctuations [42]. From this example we learn that we can make a choice of a particular model if there is additional information available that comes, for example, in terms of the microscopic structure of a system. Having said that, it would be beneficial if we could verify or falsify irrespective of such additional information a particular model just on the basis of the observed phenomenon, that is, on the basis of experimental data. In fact, there are data analysis techniques available that can be used to extract the model equations of systems described by Markov diffusion processes from experimental data [208, 209, 210, 494, 536]. In principle, such data analysis techniques could also be applied to extract the evolution equations of nonlinear families of Markov processes and, consequently, of systems described by strongly nonlinear Fokker–Planck equations. Accordingly, based on experimental observations one could determine whether the system under consideration is adequately described by means of a linear or nonlinear Fokker–Planck equation. In addition, one could determine the dependencies of the drift and diffusion terms on state variables and density measures. Let us illustrate the latter point. For example, let us consider a system that can be described by a univariate random variable $X(t)$. Let us further assume that we have reason to believe that the system can be described by means of a strongly nonlinear Fokker–Planck equation with a diffusion coefficient $D_2(x, P)$ that does not explicitly depend on the state variable x but is a function (not a functional) of P, that is, we have $D_2(x, P) = D_2(P)$. Then, we can compute $D_2'(x, t) = D_2(P(x, t))$ by means of

$$D_2'(x,t) = D_2(P(x,t)) = \lim_{\Delta t \to 0} \frac{1}{2\Delta t} \langle [X(t - \Delta t) - X(t)]^2 \rangle |_{X(t)=x}, \quad (8.1)$$

see (3.81) and the references cited above. Likewise, we can compute the probability density P from

$$P(x,t) = \langle \delta(x - X(t)) \rangle = \lim_{\Delta x \to 0} \frac{1}{\sqrt{2\pi}\Delta x} \left\langle \exp\left\{ -\frac{1}{2} \left[\frac{x - X(t)}{\Delta x} \right]^2 \right\} \right\rangle, \quad (8.2)$$

see (3.49). If we regard now for a fixed time t and several x-values. the pairs $(P(x,t), D_2'(x,t))$ as points of a two-dimensional plane and plot these points on the plane, then they will represent a graph that corresponds to the function $P \to D_2(P)$. For example, for systems described by the nonlinear Fokker–Planck equation proposed by Plastino and Plastino that has been addressed in Chap. 6, this procedure will give us the graph $D_2(P) = P^{q-1}$ [200].

Generalizations

As stated in Chap. 2, the nonlinear Fokker–Planck equation (2.4) does not describe the most general nonlinear Fokker–Planck equation that we can think of. We may think of generalizations concerning the drift and diffusion coefficients of (2.4), the transition probability densities related to solutions of (2.4), and the Fokker–Planck operator involved in (2.4). For example, a more general class of nonlinear Fokker–Planck equations may be given by nonlinear parabolic partial differential equations that have, roughly speaking, the same Fokker–Planck operator as the nonlinear Fokker–Planck equation (2.4) but have a more general type of drift and diffusion coefficients [205, 206, 395, 446]. Furthermore, in Chap. 3, we confined ourselves to discussing nonlinear Fokker–Planck equations in the context of linear evolution equations for transition probability densities. Alternatively, one may consider nonlinear Fokker–Planck equations related to nonlinear evolution equations for transition probability densities [67]. Finally, some of the concepts and methods discussed in this book may be applied to other types of evolution equations for probability distributions and density measures. In this context, we may think of Liouville equations, master equations, Boltzmann equations, fractional linear Fokker–Planck equations, and fractional nonlinear Fokker–Planck equations. For example, two concepts that have been frequently used in this book, namely, distortion functionals and Kullback measures related to concavity inequalities, have also found applications in the theory of nonlinear master equations, see (4.122) and [185]. Furthermore, the notion of Fokker-Planck equations determined by free energy measures may be applied to nonlinear reaction-diffusion equations as well [178].

References

1. S. Abe. A note on the q-deformation-theoretic aspect of the generalized entropies in nonextensive physics. *Phys. Lett. A*, 224:326–330, 1997.
2. S. Abe. Remark on the escort distribution representation of nonextensive statistical mechanics. *Phys. Lett. A*, 275:250–253, 2000.
3. S. Abe and Y. Okamoto. *Nonextensive statistical mechanics and its applications*. Springer, Berlin, 2001.
4. J. A. Acebron, L. L. Bonilla, S. De Leo, and R. Spigler. Breaking the symmetry in bimodal frequency distributions of globally coupled oscillators. *Phys. Rev. E*, 57:5287–5290, 1998.
5. J. A. Acebron, L. L. Bonilla, and R. Spigler. Synchronization in populations of globally coupled oscillators with inertia effects. *Phys. Rev. E*, 62:3437–3454, 2000.
6. J. A. Acebron and R. Spigler. Adaptive frequency model for phase-frequency synchronization in large populations of globally coupled nonlinear oscillators. *Phys. Rev. Lett.*, 81:2229–2232, 1998.
7. M. Acharyya. Comparison of mean-field and Monte Carlo approaches to dynamic hysteresis in Ising ferromagnets. *Physica A*, 253:199–204, 1998.
8. J. Aczel and Z. Daroczy. *On measures of information and their characterizations*. Academic Press, New York, 1975.
9. G. Adam and O. Hittmair. *Wärmetheorie*. Vieweg, Braunschweig, 1970, in German.
10. G. S. Agarwal. Fluctuation-dissipation theorems for systems in non-thermal equilibrium and applications. *Z. Physik*, 252:25–38, 1972.
11. A. Aharoni. *Introduction to the theory of ferromagnetism*. Clarendon Press, Oxford, 1996.
12. R. McN. Alexander. Optimization and gaits in the locomotion of vertebrates. *Physiol. Rev.*, 69:1199–1227, 1989.
13. E. J. Allen and H. D. Victory (Jr.). A computational investigation of the random particle method for numerical solution of the kinetic Vlasov-Poisson-Fokker-Planck equations. Physica A 209:318–346, (1994).
14. M. Alonso and E. J. Finn. *Fundamental university phyiscs, Vol III*. Addison-Wesley Publishing Company, Reading, 1968.
15. E. L. Amazeen, P. G. Amazeen, P. J. Treffner, and M. T. Turvey. Attention and handedness in bimanual coordination dynamics. *J. Exp. Psychol. - Hum. Percept. Perform.*, 23:1552–1560, 1997.
16. P. G. Amazeen, E.L. Amazeen, and M. T. Turvey. Dynamics of human intersegmental coordination: theory and research. In D. A. Rosenbaum and C. E. Collyer, editors, *Timing of behavior*, pages 237–259, Cambridge, 1998. MIT Press.

17. O. Amir and S. Neuman. Gaussian closure of transient unsaturated flow in random soils. *Transp. Porous Media*, 54:55–77, 2004.
18. D. J. Amit. *Modeling brain function*. Cambridge University Press, Cambridge, 1989.
19. D. J. Amit and Y. Verbin. *Statistical physics*. World Scientific, Singapore, 1999.
20. V. S. Anishchenko, V. V. Astakhov, A. B. Neiman, T.E. Vadivasova, and L. Schimansky-Geier. *Nonlinear dynamics of chaotic and stochastic systems*. Springer, Berlin, 2002. Sec 3.1.6.
21. C. Anteneodo and A. R. Plastino. Maximum entropy approach to stretched exponential probability distributions. *J. Phys. A: Math. Gen.*, 32:1089–1097, 1999.
22. C. Anteneodo and C. Tsallis. Breakdown of exponential sensitivity to initial conditions: role of the range of interctions. *Phys. Rev. Lett.*, 80:5313–5316, 1998.
23. M. Antoni, H. Hinrichsen, and S. Ruffo. On the microcanonical solution of a system of fully coupled particles. *Chaos, solitons & fractals*, 13:393–399, 2002.
24. M. Antoni and S. Ruffo. Clustering and relaxation in Hamiltonian long-range dynamics. *Phys. Rev. E*, 52:2361–2374, 1995.
25. T. Aonishi and M. Okada. Multibranch entrainment and slow evolution among branches in coupled oscillators. *Phys. Rev. Lett.*, 88:024102, 2002.
26. A. Arenas and C. J. Perez-Vincente. Exact long-time behavior of a network of phase oscillators under random fields. *Phys. Rev. E*, 50:949–956, 1994.
27. J. T. Ariaratnam and S. H. Strogatz. Phase diagram for the Winfree model of coupled nonlinear oscillators. *Phys. Rev. Lett.*, 86:4278–4281, 2001.
28. T. Arimitsu and N. Arimitsu. Analysis of turbulence by statistics based on generalized entropies. *Physica A*, 295:177–194, 2001.
29. S. Arimoto. Information-theoretical considerations on estimation problems. *Inf. and Control*, 19:181–194, 1971.
30. A. Arnold, L. L. Bonilla, and P. Markowich. Liapunov functionals and large-time-asymptotics of mean-field nonlinear Fokker-Planck equations. *Transp. Theor. Stat. Phys.*, 25:733–751, 1996.
31. A. Arnold, P. Markowich, and G. Toscani. On large time asymptotics for Drift-Diffusion-Possion Systems. *Transp. Theor. Stat. Phys.*, 29:571–581, 2000.
32. A. Arnold, P. Markowich, G. Toscani, and A. Unterreiter. On convex Sobolev inequalities and the rate of convergence to equilibrium for Fokker-Planck type equations. *Commun. Part. Diff. Eq.*, 26:43–100, 2001.
33. L. Arnold. *Stochastische Differentialgleichungen*. Oldenbourg Verlag, München, 1973, in German.
34. D. G. Aronson. The porous medium equation. In A. Dobb and B. Eckmann, editors, *Nonlinear diffusion problems - Lecture notes in mathematics, Vol. 1224*, pages 1–46, Berlin, 1986. Springer.
35. P. Bak. *How Nature Works: the science of self-organized criticality*. Springer, New York, 1996.
36. R. Balescu. *Equilibrium and nonequilibrium statistical mechanics*. John Wiley and Sons, New York, 1975.
37. F. Bardou, J. P. Bouchaud, A. Aspect, and C. Cohen-Tannoudij. *Lévy statistics and laser cooling*. Cambridge University Press, Cambridge, 2002.
38. G. I. Barenblatt, V. M. Entov, and V. M. Ryzhik. *Theory of fluid flows through natural rocks*. Kluwer Academic Publisher, Dordrecht, 1990.

39. A. G. Bashkirov and A. V. Vityazev. Information entropy and power-law distributions for chaotic systems. *Physica A*, 277:136–145, 2000.
40. G. K. Batchelor. Diffusion in a field of homogeneous turbulence II. The relative motion of particle. *Proc. Camb. Phil. Soc*, 48:345–362, 1952.
41. C. Beck. Application of generalized thermostatistics to fully developed turbulence. *Physica A*, 277:115–123, 2000.
42. C. Beck. Dynamical foundations of nonextensive statistical mechanics. *Phys. Rev. Lett.*, 87:180601, 2001.
43. C. Beck. Scaling exponents in fully developed turbulence from nonextensive statistical mechanics. *Physica A*, 295:195–200, 2001.
44. C. Beck. Nonextensive methods in turbulence and particle physics. *Physica A*, 305:209–217, 2002.
45. C. Beck. Superstatistics in hydrodynamic turbulence. *Physica D*, 193:195–207, 2004.
46. C. Beck and E. G. D. Cohen. Superstatistics. *Physica A*, 322:267–275, 2003.
47. C. Beck, G. L. Lewis, and H. L. Swinney. Measuring nonextensivity parameters in a turbulent Couette-Taylor flow. *Phys. Rev. E*, 63:035303, 2001.
48. C. Beck and F. Schlögl. *Thermodynamics of chaotic systems*. Cambridge University Press, Cambridge, 1993.
49. P. J. Beek, C. E. Peper, and D. F. Stegeman. Dynamical models of movement coordination. *Hum. Movement Sci.*, 14:573–608, 1995.
50. P. J. Beek, C. E. Peper, and P. C. W. van Wieringen. Frequency locking, frequency modulation, and bifurcations in dynamic movement systems. In G. E. Stelmach and J. Requin, editors, *Tutorials in motor behavior II*, pages 599–622, Amsterdam, 1992. North-Holland Publ. Company.
51. P. J. Beek, W. E. I. Rikkert, and P. C. W. van Wieringen. Limit cycle properties of rhythmic forearm movements. *J. Exp. Psychol. - Hum. Percept. Perform.*, 22:1077–1093, 1996.
52. A. R. Bell, R. G. Evans, and D. J. Nicholas. Electron energy transport in steep temperature gradients in laser-produced plasma. *Phys. Rev. Lett.*, 46:243–246, 1981.
53. N. Bellomo and M. Lo Schiavo. From the Boltzmann equation to generalized kinetic models in applied sciences. *Math. Comput. Model.*, 26:43–76, 1997.
54. I. Bena, C. van den Broeck, R. Kawai, M. Copelli, and K. Lindenberg. Collective behavior of parametric oscillators. *Phys. Rev. E*, 65:036611, 2002.
55. M. Bestehorn and R. Friedrich. Rotationally invariant order parameter equations for natural patterns in nonequilibrium systems. *Phys. Rev. E*, 59:2642–2652, 1999.
56. J. Bhattacharya and H. Petsche. Enhanced phase synchrony in the electroencephalography γ band for musicians while listening to music. *Phys. Rev. E*, 64:012902, 2001.
57. J. J. Binnen, N. J. Dowrick, A. J. Fisher, and M. E. J. Newman. *The theory of critical phenomena*. Oxford University Press, New York, 1992.
58. J. Binney and S. Tremaine. *Galatic dynamics*. Princeton University Press, Princeton, New Jersey, 1987.
59. T. Birner, K. Lippert, R. Müller, A. Kühnel, and U. Behn. Critical behavior of nonequilibrium phase transitions to magnetically ordered states. *Phys. Rev. E*, 65:046110, 2002.

60. L. L. Bonilla. Stable nonequilibrium probability densities and phase transitions for mean-field models in the thermodynamic limit. *J. Stat. Phys.*, 46:659–678, 1987.
61. L. L. Bonilla, J. C. Neu, and R. Spigler. Nonlinear stability of incoherence and collective synchronization in a population of coupled oscillators. *J. Stat. Phys.*, 67:313–330, 1992.
62. L. L. Bonilla, C. J. Perez-Vicente, F. Ritort, and J. Soler. Exactly solvable phase oscillator models with synchronization dynamics. *Phys. Rev. Lett.*, 81:3643–3646, 1998.
63. L. L. Bonilla, C. J. P. Vicente, and J. M. Rubi. Glassy synchronization in a population of coupled oscillators. *J. Stat. Phys.*, 70:921–937, 1993.
64. E. P. Borges and I. Roditi. A family of nonextensive entropies. *Phys. Lett. A*, 246:399–402, 1998.
65. L. Borland. Ito-Langevin equations with generalized thermostatistics. *Phys. Lett. A*, 245:67–72, 1998.
66. L. Borland. Microscopic dynamics of the nonlinear Fokker-Planck equation. *Phys. Rev. E*, 57:6634–6642, 1998.
67. L. Borland. Option pricing formulas based on a non-Gaussian stock price model. *Phys. Rev. Lett.*, 89:098701, 2002.
68. L. Borland, F. Pennini, A. R. Plastino, and A. Plastino. The nonlinear Fokker-Planck equation with state-dependent diffusion - A nonextensive maximum entropy approach. *Eur. Phys. J. B*, 12:285–297, 1999.
69. L. Borland, A. R. Plastino, and C. Tsallis. Information gain within nonextensive thermostatistics. *J. Math. Phys.*, 39:6490–6501, 1998. Erratum: 40 (1999) 2196-2196.
70. J.-P. Bouchaud and A. Georges. Anomalous diffusion in disordered media: statistical mechanics, models and physical applications. *Phys. Rep.*, 195:127–293, 1990.
71. J. P. Bouchaud and M. Potters. *Theory of financial risks*. Cambridge University Press, Cambridge, 2000.
72. A. C. Branka and D. M. Heyes. Algorithms for Brownian dynamics simulation. *Phys. Rev. E*, 58:2611–2615, 1998.
73. W. Braun and K. Hepp. The Vlasov dynamics and its fluctuations in the $1/n$ limit of interacting classical particles. *Commun. Math. Phys.*, 56:101–113, 1977.
74. J. Brenner and P. Lesky. *Mathematik für Ingenieure und Naturwissenschaftler*. Aula-Verlay, Wiesbaden, 1989, in German.
75. S. L. Bressler, R. Coppola, and R. Nakamura. Episodic multiregional cortical coherence at multiple frequencies during visual task performance. *Nature*, 366:153–156, 1993.
76. L. Brevdo, R. Helmig, M. Haragus-Courcelle, and K. Kirchgässner. Permanent fronts in two-phase flows in a porous medium. *Transp. Porous Media*, 44:507–537, 2001.
77. J. J. Brey, J. M. Casado, and M. Morillo. On the dynamics of a stochastic nonlinear mean-field model. *Physica A*, 128:497–508, 1984.
78. J. C. Bronski, L. D. Carr, B. Deconinck, J. N. Kutz, and K. Promislow. Stability of repulsive Bose-Einstein condensates in a periodic potential. *Phys. Rev. E*, 63:036612, 2001.
79. I. N. Bronstein and K. A. Semendjajew. *Taschenbuch der Mathematik*. Harri Deutsch, Thun, 1989, in German.

80. A. D. Bruce. Structural phase transitions. *Adv. Phys.*, 29:111–217, 1980.
81. J. Buck and E. Buck. Mechanism of rhythmic synchronous flashing of fireflies. *Science*, 159:1319–1327, 1968.
82. J. Buck and E. Buck. Synchronous fireflies. *Sci. American*, 234(5):74–85, 1976.
83. V. Y. Bychenkov, J. Myatt, W. Rozmus, and V. T. Tikhonchuk. Ion acoustic waves in plasmas with collisional electrons. *Phys. Rev. E*, 50:5134–5137, 1994.
84. V. Y. Bychenkov, W. Rozmus, and V. T. Tikhonchuk. Nonlocal electron transport in a plasma. *Phys. Rev. Lett.*, 75:4405–4408, 1995.
85. R. A. Cairns. *Plasma physics*. Blackie, Philadelphia, 1985.
86. H. B. Callen. *Thermodynamics and an introduction to thermostatistics*. John Wiley and Sons, New York, 1985.
87. J. A. Carillo, A. Jüngel, P. A. Markowich, G. Toscani, and A. Unterreiter. Entropy dissipation methods for degenerate parabolic problems and generalized Sobolev inequalities. *Monatshefte für Mathematik*, 113:1–82, 2001.
88. J. M. Casado and M. Morillo. Phase transitions in a nonlinear stochastic model: A numerical simulation study. *Phys. Rev. A*, 42:1875–1879, 1990.
89. D. Chandler. *Introduction to modern statistical mechanics*. Oxford University Press, New York, 1987.
90. S. Chandrasekhar. *Liquid crystals*. Cambridge University Press, Cambridge, 1977.
91. A. Chao. *Physics of collective instabilities in high energy accelerators*. John Wiley and Sons, New York, 1993.
92. P. H. Chavanis. Generalized thermodynamics and Fokker-Planck equations: applications to stellar dynamics in two-dimensional turbulence. *Phys. Rev. E*, 68:036108, 2003.
93. P. H. Chavanis and J. Sommeria. Statistical mechanics of the shallow water system. *Phys. Rev. E*, 65:026302, 2002.
94. P. H. Chavanis, J. Sommeria, and R. Robert. Statistical mechanics of two-dimensional vortices and collisionless stellar systems. *Astrophys. J.*, 471:385–399, 1996.
95. B. L. Cheng, R. I. Epstein, R. A. Guyer, and A. C. Young. Earthquake-like behaviour of soft gamma-ray repeaters. *Nature*, 382:518–520, 1996.
96. M. Y. Choi, Y. W. Kim, and D. C. Hong. Periodic synchronization in a driven system of coupled oscillators. *Phys. Rev. E*, 49:3825–3832, 1994.
97. E. G. D. Cohen. The kinetic theory of moderately dense gases. In H. J. M. Hanley, editor, *Transport phenomena in fluids*, pages 119–207, New York, 1969. Marcel Dekker.
98. E. G. D. Cohen. Superstatistics. *Physica D*, 193:35–52, 2004.
99. A. Compte and D. Jou. Non-equilibrium thermodynamics and anomalous diffusion. *J. Phys. A: Math. Gen.*, 29:4321–4329, 1996.
100. A. Compte, D. Jou, and Y. Katayama. Anomalous diffusion in linear shear flows. *J. Phys. A: Math. Gen.*, 30:1023–1030, 1997.
101. J. Crank. *The mathematics of diffusion*. Clarendon Press, Oxford, 1975.
102. J. D. Crawford. Scaling and singularities in the entrainment of globally coupled oscillators. *Phys. Rev. Lett.*, 74:4341–4344, 1995.
103. J. D. Crawford and K. T. R. Davies. Synchronization of globally coupled phase oscillators: singularities and scaling for general couplings. *Physica D*, 125:1–46, 1999.

104. M. C. Cross and P. C. Hohenberg. Pattern formation outside of equilibrium. *Rev. Mod. Phys.*, 65:851–1112, 1993.
105. E. M. F. Curado and F. D. Nobre. Derivation of nonlinear fokker-planck equations by means of approximations to the master equation. *Phys. Rev. E*, 67:021107, 2003.
106. E. M. F. Curado and C. Tsallis. Generalized statistical mechanics: connection with thermodynamics. *J. Phys. A: Math. Gen.*, 24:L69–L72, 1991. Errata: 24: 3187, 25: 1019.
107. M. G. E. da Luz, S. V. Buldyrev, S. Havlin, E. P. Rapsos, H. E. Stanley, and G. M. Viswanathan. Improvements in the statistical approach to random Lévy flight searches. *Physica A*, 295:89–92, 2001.
108. A. Daffertshofer. Effects of noise on the phase dynamics of nonlinear oscillators. *Phys. Rev. E*, 58:327–338, 1998.
109. A. Daffertshofer, C. E. Peper, and P. J. Beek. Spectral analyses of event-related encephalographic signals. *Phys. Lett. A*, 266:290–302, 2000.
110. A. Daffertshofer, C. E. Peper, T. D. Frank, and P. J. Beek. Spatio-temporal patterns of encephalographic signals during polyrhythmic tapping. *Hum. Movement Sci.*, 19:475–498, 2000.
111. A. Daffertshofer, C. van den Berg, and P. J. Beek. A dynamical model for mirror movements. *Physica D*, 132:243–266, 1999.
112. H. Daido. Discrete-time population dynamics of interacting self-oscillators. *Prog. Theor. Phys.*, 75:1460–1463, 1986.
113. H. Daido. Population dynamics of randomly interacting self-oscillators: I tractable models without frustration. *Prog. Theor. Phys.*, 77:622–634, 1987.
114. H. Daido. Scaling behavior at the onset of mutual entrainment in a population of interacting oscillators. *J. Phys. A: Math. Gen.*, 20:L629–L636, 1987.
115. H. Daido. Quasientrainment and slow relaxation in a population of oscillators with random and frustrated interactions. *Phys. Rev. Lett.*, 68:1073–1076, 1992.
116. H. Daido. Order function and macroscopic entrainment in uniformly coupled limit-cycle oscillators. *Prog. Theor. Phys.*, 88:1213–1218, 1993.
117. H. Daido. Generic scaling at the onset of macroscopic mutual entrainment in limit-cycle oscillators with uniform all-to-all coupling. *Phys. Rev. Lett.*, 73:760–763, 1994.
118. H. Daido. Multibranch entrainment and scaling in large populations of coupled oscillators. *Phys. Rev. Lett.*, 77:1406–1409, 1996.
119. H. Daido. Onset of cooperative entrainment in limit-cycle oscillators with uniform all-to-all interactions: bifurcation of the order function. *Physica D*, 91:24–66, 1996.
120. H. Daido. Why circadian rhythms are circadian: competitive population dynamics of biological oscillators. *Phys. Rev. Lett.*, 87:048101, 2001.
121. F. Dalfovo, S. Giorgini, L. P. Pitaevskii, and S. Stringari. Theory of Bose-Einstein condensation in trapped gases. *Rev. Mod. Phys.*, 71:463–512, 1999.
122. H. Damasio. The brain binds entities and events by multiregional activation from convergence zones. *Neural Comput.*, 1:123–132, 1989.
123. S. Dano, F. Hynne, S. De Monte, F. d'Ovidio, P. G. Sorensen, and H. Westerhoff. Synchronization of glycotic oscillations in a yeast cell population. *Faraday Discuss.*, 120:261–276, 2001.
124. Z. Daroczy. Generalized information functions. *Inf. and Control*, 16:36–51, 1970.

125. R. C. Davidson, H. Qin, and P. J. Channell. Periodically-focused solutions to the nonlinear Vlasov-Maxwell equations for intense charged particle beams. *Phys. Lett. A*, 258:297–304, 1999.
126. R. C. Davidson, H. Qin, and T. S. F. Wang. Vlasov-Maxwell description of electron-ion two-stream instability in high-intensity linacs and storage rings. *Phys. Lett. A*, 252:213–221, 1999.
127. D. A. Dawson. Critical dynamics and fluctuations for a mean-field model of cooperative behavior. *J. Stat. Phys.*, 31:29–85, 1983.
128. D. A. Dawson and J. Gärtner. *Large deviations, free energy functional and quasi-potential for a mean field model of interacting diffusions*. American Mathematical Society, Providence, 1989. Memories of the American Mathematical Society, Vol. 78, No. 398.
129. P. de Gennes. *The physics of liquid crystals*. Clarendon Press, Oxford, 1974.
130. S. R. de Groot and P. Mazur. *Non-equilibrium thermodynamics*. North-Holland Publ. Company, Amsterdam, 1962.
131. W. H. de Jeu. *Physical properties of liquid crystalline materials*. Gordon and Beach, New York, 1980.
132. F. de los Santos, M. M. Telo da Gamma, and M. A. Munoz. Nonequilibrium wetting transitions with short range forces. *Phys. Rev. E*, 67:021607, 2003.
133. G. C. DeGuzman and J. A. S. Kelso. Multifrequency behavioral patterns and the phase attractive circle map. *Biol. Cybern.*, 64:485–495, 1991.
134. J. L. Delcroix. *Introduction to the theory of ionized gases*. Interscience publishers, London, 1960.
135. R. C. Desai and R. Zwanzig. Statistical mechanics of a nonlinear stochastic model. *J. Stat. Phys.*, 19:1–24, 1978.
136. R. P. Di Sisto, S. Martinez, R. B. Orellana, A. R. Plastino, and A. Plastino. General thermostatistical formalisms, invariance under uniform spectrum translations, and Tsallis q-additivity. *Physica A*, 265:590–613, 1999.
137. D. Diakonov, L. M. Jensen, C. J. Pethick, and H. Smith. Loop structure of the lowest Bloch band for a Bose-Einstein condensate. *Phys. Rev. A*, 66:013604, 2002.
138. Y. N. Dnestrovskii and D. P. Kostomarov. *Numerical simulations of plasmas*. Springer, Berlin, 1986.
139. M. Doi and S. F. Edwards. *The theory of polymer dynamics*. Clarendon Press, Oxford, 1988.
140. G. Drazer, H. S. Wio, and C. Tsallis. Anomalous diffusion with absorption: exact time-dependent solutions. *Phys. Rev. E*, 61:1417–1422, 2000.
141. A. N. Drozdov and M. Morillo. Solution of nonlinear Fokker-Planck equations. *Phys. Rev. E*, 54:931–937, 1996.
142. A. N. Drozdov and M. Morillo. Validity of basic concepts in nonlinear cooperative Fokker-Planck models. *Phys. Rev. E*, 54:3304–3313, 1996.
143. I. Dukovski, J. Machta, and L. V. Chayes. Invaded cluster simulations of the XY model in two and three dimensions. *Phys. Rev. E*, 65:026702, 2002.
144. H. Dwight. *Tables of integrals and other mathematical data*. MacMillan, New York, 1957.
145. E. B. Dynkin. *Markov processes*, volume I. Springer, Berlin, 1965.
146. W. Ebeling. Problems of a statistical ensemble theory for systems far from equilibrium. In J. A. Freund and T. Pöschel, editors, *Stochastic processes in physics, chemistry and biology*, pages 392–399, Berlin, 2000. Springer.

147. W. Ebeling and H. Engel-Herbert. The influence of external fluctuations on self-sustained temporal oscillations. *Physica A*, 104:378–396, 1980.
148. W. Ebeling and R. Feistel. *Physik der Selbstorganisation und Evolution*. Akademie-Verlag, Berlin, 1982, in German.
149. W. Ebeling and L. Schimansky-Geier. Transition phenomena in multidimensional systems - models of evolution. In F. Moss and P. V. E. McClintock, editors, *Noise in nonlinear dynamical systems, Vol. 1*, pages 279–306, Cambridge, 1989. Cambridge University Press.
150. R. Eckhorn, O.-J. Grüsser, U. Kröller, K. Pellnitz, and B. Pöpe. Efficiency of different neuronal codes: information transfer calculations for three different neuronal systems. *Biol. Cybern.*, 22:49–60, 1976.
151. A. K. Engel, P. König, A. K. Kreiter, and W. Singer. Interhemispheric synchronization of oscillatory neuronal responses in cat visual cortex. *Science*, 252:1177–1179, 1991.
152. E. M. Epperlein, G. J. Rickard, and A. R. Bell. Two-dimensional nonlocal electron transport in laser-produced plasma. *Phys. Rev. Lett.*, 61:2453–2456, 1988.
153. P. Espanol, M. Serrano, and H. C. Öttinger. Thermodynamically admissible form for discrete hydrodynamics. *Phys. Rev. Lett.*, 83:4542–4545, 1999.
154. B. C. Eu. *Kinetic theory and irreversible thermodynamics*. John Wiley and Sons, New York, 1992.
155. B. U. Felderhof. Orientational relaxation in the Maier-Saupe model of nematic liquid crystals. *Physica A*, 323:88–106, 2003.
156. B. U. Felderhof and R. B. Jones. Mean field theory of the nonlinear response of an interacting dipolar system with rotational diffusion to an oscillating field. *J. Phys. C*, 15:4011–4024, 2003.
157. M. Fialkowiski. Viscous properties of biaxial nematic liquid crystals: the method of calculation of the Leslie viscosity coefficients. *Phys. Rev. E*, 55:2902–2915, 1997.
158. M. Fialkowiski and S. Hess. Orientational phenomena in a plastic flow of a two-dimensional square crystal. *Physica A*, 282:65–76, 2000.
159. M. C. Fibro. Analytic estimation of the Lyapunov exponent in a mean-field model undergoing a phase transition. *Phys. Rev. E*, 57:6599–6603, 1998.
160. P. M. Fitts. The information capacity of the human motor system in controlling the amplitude of movement. *J. Exp. Psychol.*, 47:381–391, 1954.
161. S. Flach. Long-time correlations in a model of structural phase transitions with infinite range interactions. *Z. Physik B*, 82:419–424, 1991.
162. A. D. Fokker. *Annalen der Physik*, 43:810, 1914.
163. T. D. Frank. On nonlinear and nonextensive diffusion and the second law of thermodynamics. *Phys. Lett. A*, 267:298–304, 2000.
164. T. D. Frank. H-theorem for Fokker-Planck equations with drifts depending on process mean values. *Phys. Lett. A*, 280:91–96, 2001.
165. T. D. Frank. A Langevin approach for the microscopic dynamics of nonlinear Fokker-Planck equations. *Physica A*, 301:52–62, 2001.
166. T. D. Frank. Lyapunov and free energy functionals of generalized Fokker-Planck equations. *Phys. Lett. A*, 290:93–100, 2001.
167. T. D. Frank. Generalized Fokker-Planck equations derived from generalized linear nonequilibrium thermodynamics. *Physica A*, 310:397–412, 2002.

168. T. D. Frank. Generalized multivariate Fokker-Planck equations derived from kinetic transport theory and linear nonequilibrium thermodynamics. *Phys. Lett. A*, 305:150–159, 2002.
169. T. D. Frank. Interpretation of Lagrange multipliers of generalized maximum-entropy distributions. *Phys. Lett. A*, 299:153–158, 2002.
170. T. D. Frank. On a general link between anomalous diffusion and nonextensivity. *J. Math. Phys.*, 43:344–350, 2002.
171. T. D. Frank. On a mean field Haken-Kelso-Bunz model and a free energy approach to relaxation processes. *Nonlin. Phenom. Complex Syst.*, 5(4):332–341, 2002.
172. T. D. Frank. Stability analysis of mean field models described by Fokker-Planck equations. *Annalen der Physik*, 11:707–716, 2002.
173. T. D. Frank. Stochastic processes described by nonlinear parabolic partial differential equations — Common properties and applications to human movement sciences. In *Computational mathematics and modeling: CMM 2002*, pages 257–263. East-West Journal of Mathematics, Special Issue 2002, 2002.
174. T. D. Frank. A note on the Markov property of stochastic processes described by nonlinear Fokker-Planck equations. *Physica A*, 320:204–210, 2003.
175. T. D. Frank. On the boundedness of free energy functionals. *Nonlin. Phenom Complex Syst.*, 6(3):696–704, 2003.
176. T. D. Frank. On the second variation of free energies and nonlinear Fokker-Planck equations involving periodic variables. *Prog. Theor. Phys. Supp.*, 150:48–56, 2003.
177. T. D. Frank. Single particle dynamics of many-body systems described by Vlasov-Fokker-Planck equations. *Phys. Lett. A*, 319:173–180, 2003.
178. T. D. Frank. Asymptotic properties of nonlinear diffusion, nonlinear drift-diffusion, and nonlinear reaction-diffusion equations. *Annalen der Physik*, 13:461–469, 2004.
179. T. D. Frank. Autocorrelation functions of nonlinear Fokker-Planck equations. *Eur. Phys. J. B*, 37:139–142, 2004.
180. T. D. Frank. Classical Langevin equations for the free electron gas and blackbody radiation. *J. Phys. A*, 37:3561–3567, 2004.
181. T. D. Frank. Complete description of a generalized Ornstein-Uhlenbeck process related to the nonextensive Gaussian entropy. *Physica A*, 340:251–256, 2004.
182. T. D. Frank. Dynamic mean field models: H-theorem for stochastic processes and basins of attraction of stationary processes. *Physica D*, 195:229–243, 2004.
183. T. D. Frank. Fluctuation-dissipation theorems for nonlinear Fokker-Planck equations of the Desai-Zwanzig type and Vlasov-Fokker-Planck equations. *Phys. Lett. A*, 329:475–485, 2004.
184. T. D. Frank. Modelling the stochastic single particle dynamics of relativistic fermions and bosons using nonlinear drift-diffusion equations. *Mathematical and Computer Modelling*, 2004/05, in press.
185. T. D. Frank. On a nonlinear master equation and the Haken-Kelso-Bunz model. *J. Biol. Phys.*, 30:139–159, 2004.
186. T. D. Frank. Stability analysis of nonequilibrium mean field models by means of self-consistency equations. *Phys. Lett. A*, 327:146–151, 2004.
187. T. D. Frank. Stability analysis of stationary states of mean field models described by Fokker-Planck equations. *Physica D*, 189:199–218, 2004.

188. T. D. Frank. Stochastic feedback, nonlinear families of Markov processes, and nonlinear Fokker-Planck equations. *Physica A*, 331:391–408, 2004.
189. T. D. Frank and P. J. Beek. A mean field approach to self-organization in spatially extended perception-action and psychological systems. In W. Tschacher and J. P. Dauwalder, editors, *The dynamical systems approach to cognition*, pages 159–179, Singapore, 2003. World Scientific.
190. T. D. Frank and A. Daffertshofer. Nonlinear Fokker-Planck equations whose stationary solutions make entropy-like functionals stationary. *Physica A*, 272:497–508, 1999.
191. T. D. Frank and A. Daffertshofer. Exact time-dependent solutions of the Renyi Fokker-Planck equation and the Fokker-Planck equations related to the entropies proposed by Sharma and Mittal. *Physica A*, 285:351–366, 2000.
192. T. D. Frank and A. Daffertshofer. H-theorem for nonlinear Fokker-Planck equations related to generalized thermostatistics. *Physica A*, 295:455–474, 2001.
193. T. D. Frank and A. Daffertshofer. Multivariate nonlinear Fokker-Planck equations and generalized thermostatistics. *Physica A*, 292:392–410, 2001.
194. T. D. Frank, A. Daffertshofer, and P. J. Beek. Multivariate Ornstein-Uhlenbeck processes with mean field dependent coefficients — Application to postural sway. *Phys. Rev. E*, 63:011905, 2001.
195. T. D. Frank, A. Daffertshofer, and P. J. Beek. On an interpretation of apparent motion as a self-organizing process. *Behav. Brain Sci.*, 24:668–669, 2001.
196. T. D. Frank, A. Daffertshofer, and P. J. Beek. Impacts of statistical feedback on the flexibility-accuracy trade-off of biological systems. *J. Biol. Phys.*, 28:39–54, 2002.
197. T. D. Frank, A. Daffertshofer, P. J. Beek, and H. Haken. Impacts of noise on a field theoretical model of the human brain. *Physica D*, 127:233–249, 1999.
198. T. D. Frank, A. Daffertshofer, C. E. Peper, P. J. Beek, and H. Haken. Towards a comprehensive theory of brain activity: coupled oscillator systems under external forces. *Physica D*, 144:62–86, 2000.
199. T. D. Frank, A. Daffertshofer, C. E. Peper, P. J. Beek, and H. Haken. H-theorem for a mean field model describing coupled oscillator systems under external forces. *Physica D*, 150:219–236, 2001.
200. T. D. Frank and R. Friedrich. Estimating the nonextensivity of systems from experimental data: a nonlinear diffusion equation approach. *Physica A*, 2004/05, in press.
201. T. D. Frank, C. E. Peper, A. Daffertshofer, and P. J. Beek. Variability of brain activity during rhythmic unimanual finger movements, In K. Davids, S. Bennett, and K. Newell, editors, *Variability in the Movement system: a multi-disciplinary perspective*, Champaign, Human Kinetics, 2004/05, in press.
202. T. D. Frank and A. R. Plastino. Generalized thermostatistics based on the Sharma-Mittal entropy and escort mean values. *Eur. Phys. J. B*, 30:543–549, 2002.
203. B. R. Frieden. *Physics from Fisher information*. Cambridge University Press, Cambridge, 1998.
204. B. R. Frieden, A. Plastino, A. R. Plastino, and B. H. Soffer. Fisher-based thermodynamics: Its Legendre transform and concavity properties. *Phys. Rev. E*, 60:48–53, 1999.
205. A. Friedman. *Partial differential equations of parabolic type*. Prentice-Hall, Inc., Englewood Cliffs, N.J., 1964.

206. A. Friedman. *Partial differential equations*. Holt, Rinehart and Winston, Inc., New York, 1969.
207. R. Friedrich. Phase diffusion equation for roll patterns of systems lacking reflectional symmetry. *Z. Physik B*, 92:129–131, 1993.
208. R. Friedrich and J. Peinke. Description of a turbulent cascade by a Fokker-Planck equation. *Phys. Rev. Lett.*, 78:863–866, 1997.
209. R. Friedrich and J. Peinke. Statistical properties of a turbulent cascade. *Physica D*, 102:147–155, 1997.
210. R. Friedrich, J. Peinke, and Ch. Renner. How to quantify deterministic and random influences on the statistics of the foreign exchange market. *Phys. Rev. Lett.*, 84:5224–5227, 2000.
211. R. Friedrich, G. Radons, T. Ditzinger, and A Henning. Ripple formation through an interface instability from moving growth and erosion sources. *Phys. Rev. Lett.*, 85:4884–4887, 2000.
212. R. Friedrich and C. Uhl. Spatio-temporal analysis of human electroencephalograms: Petit-mal epilepsy. *Physica D*, 98:171–182, 1996.
213. U. Frisch. *Turbulence*. Cambridge University Press, Cambridge, 1995.
214. A. Fuchs and V. K. Jirsa. The HKB model revisited: how varying the degree of symmetry controls dynamics. *Hum. Movement Sci.*, 19:425–449, 2000.
215. A. Fuchs, V. K. Jirsa, and J. A. S. Kelso. Theory of the relation between human brain activity (MEG) and hand movements. *Neuroimage*, 11:359–369, 2000.
216. A. Fuchs, J. A. S. Kelso, and H. Haken. Phase transitions in the human brain: spatial mode dynamics. *Int. J. Bif. and Chaos*, 2:917–939, 1992.
217. T. Funaki. A certain class of diffusion processes associated with nonlinear parabolic equations. *Z. Wahrscheinlichkeitstheory verw. Gebiete*, 67:331–348, 1984.
218. H. Gang, H. Haken, and X. Fagen. Stochastic resonance with sensitive frequency dependence in globally coupled continuous systems. *Phys. Rev. Lett.*, 77:1925–1928, 1996.
219. J. Garcia-Ojalvo, J. M. R. Parrondo, J. M. Sancho, and C. van den Broeck. Reentrant transition induced by multiplicative noise in the time-dependent Ginzburg-Landau model. *Phys. Rev. E*, 54:6918–6921, 1996.
220. J. Garcia-Ojalvo and J. M. Sancho. *Noise in spatially extended systems*. Springer, New York, 1999.
221. C. W. Gardiner. *Handbook of stochastic methods*. Springer, Berlin, 2 edition, 1997.
222. J. Gärtner. On the McKean–Vlasov limit for interacting diffusions. *Math. Nachr.*, 137:197–248, 1988.
223. T. Geisel, J. Nierwetberg, and A. Zacherl. Accelerated diffusion in Josephson junctions and related chaotic systems. *Phys. Rev. Lett.*, 54:616–619, 1985.
224. W. Genovese, M. A. Munoz, and J. M. Sancho. Nonequilibrium transitions induced by multiplicative noise. *Phys. Rev. E*, 57:R2495–R2498, 1998.
225. L. Giada and M. Marsili. First-order phase transition in a nonequilibrium growth process. *Phys. Rev. E*, 62:6015–6020, 2000.
226. A. Giansanti, D. Moroni, and A. Campa. Universal behavior in the static and dynamic properties of the α-XY model. *Chaos, solitons & fractals*, 13:407–416, 2002.

227. C. Giordano, A. R. Plastino, M. Casas, and A. Plastino. Nonlinear diffusion under a time dependent external force: q-maximum entropy solutions. *Eur. Phys. J. B*, 22:361–368, 2001.
228. P. Glansdorff and I. Prigogine. *Thermodynamic theory of structure, stability, and fluctuations*. John Wiley and Sons, New York, 1971.
229. N. Goldenfeld. *Lecture on phase transitions and the renormalization group*. Addison-Wesley Publishing Company, New York, 1992.
230. P. Gopikrishnan, M. Meyer, L. A. N. Amaral, and H. E. Stanley. Inverse cubic law for the distribution of stock price variations. *Eur. Phys. J. B*, 3:139–140, 1998.
231. P. Gopikrishnan, V. Plerou, L. A. N. Amaral, M. Meyer, , and H. E. Stanley. Scaling of the distribution of fluctuations of financial market indices. *Phys. Rev. E*, 60:5305–5316, 1999.
232. A. N. Gorban and I. V. Karlin. Family of additive entropy functions out of thermodynamic limit. *Phys. Rev. E*, 67:016104, 2003.
233. A. Goswami. *Quantum mechanics*. Wm. C. Brown Publishers, Dubuque (IA), 1992.
234. M. E. Gouvea and A. S. T. Pires. Optimized mean-field theory for the three-dimensional xy spin model. *Phys. Rev. B*, 54:14907–14909, 1996.
235. R. Graham. *Z. Physik B*, 26:281, 1976.
236. R. Graham and H. Haken. Fluctuations and stability of stationary non-equilibrium systems in detailed balance. *Z. Physik*, 245:141–153, 1971.
237. R. Graham and H. Haken. Generalized thermodynamic potential for Markov systems in detailed balance and far from thermal equilibrium. *Z. Physik*, 243:289–302, 1971.
238. W. T. Grandy (Jr.). *Foundations of statistical mechanics, Vol. I*. Kluwer Academic Publisher, Dordrecht, 1987.
239. C. M. Gray, P. König, A. K. Engel, and W. Singer. Oscillatory responses in cat visual cortex exhibit inter-columnar synchronization which reflects global stimulus properties. *Nature*, 338:334–337, 1989.
240. M. S. Green. *J. Chem. Phys.*, 19:1036, 1951.
241. M. S. Green. Markov random processes and the statistical mechanics of time-dependent phenomena. *J. Chem. Phys.*, 20:1281–1295, 1952.
242. A. Greiner, W. Strittmatter, and J. Honerkamp. Numerical integration of stochastic differential equations. *J. Stat. Phys.*, 51:95–108, 1988.
243. M. Grmela and H. C. Öttinger. Dynamics and thermodynamics of complex fluids. I. development of a general formalism. *Phys. Rev. E*, 56:6620–6632, 1997.
244. E. P. Gross. *Nuovo Cimento*, 20:454, 1961.
245. J. Gross, P. A. Tass, S. Salenius, R. Hari, H. J. Freund, and A. Schnitzler. Cortico-muscular synchronization during isometric muscle contraction in humans as revealed by magnetoencephalography. *J. Physiology*, 527:623–631, 2000.
246. J. Guckenheimer and P. Holmes. *Nonlinear oscillations, dynamical systems, and bifurcations of vector fields*. Springer, Berlin, 1984.
247. M. E. Gurtin and R. C. MacCamy. On the diffusion of biological populations. *Mathematical Biosciences*, 33:35–49, 1977.
248. P. Hadley, M. R. Beasley, and K. Wiesenfeld. Phase locking of Josephson-junction series arrays. *Phys. Rev. B*, 38:8712–8719, 1988.

249. W. Hahn. *Stability of motion.* Springer, Berlin, 1967.
250. J. Haissinski. *Nuovo Cimento Soc. Ital. Fis. B*, 18:72, 1973.
251. H. Haken. Distribution function for classical and quantum systems far from thermal equilibrium. *Z. Physik*, 263:267–282, 1973.
252. H. Haken. Cooperative phenomena in systems far from thermal equilibrium and in nonphysical systems. *Rev. Mod. Phys.*, 47:67–121, 1975.
253. H. Haken. Generalized Onsager-Machlup function and classes of path integral solutions of the Fokker-Planck equation and the master equation. *Z. Physik B*, 24:321, 1976.
254. H. Haken. *Synergetics. An introduction.* Springer, Berlin, 1977.
255. H. Haken. *Advanced synergetics.* Springer, Berlin, 1983.
256. H. Haken. *Light - Laser light dynamics.* North-Holland Publ. Company, Amsterdam, 1985.
257. H. Haken. *Information and self-organization.* Springer, Berlin, 1988.
258. H. Haken. Synergetics: an overview. *Rep. Prog. Phys.*, 52:515–553, 1989.
259. H. Haken. *Principles of brain functioning.* Springer, Berlin, 1996.
260. H. Haken. Slaving principle revisited. *Physica D*, 97:95–103, 1996.
261. H. Haken. *Information and self-organization.* Springer, Berlin, 2 edition, 2000.
262. H. Haken. *Brain dynamics.* Springer, Berlin, 2002.
263. H. Haken. *Synergetics: Itroduction and Advanced topics.* Springer, Berlin, 2004.
264. H. Haken, J. A. S. Kelso, and H. Bunz. A theoretical model of phase transitions in human hand movements. *Biol. Cybern.*, 51:347–356, 1985.
265. H. Haken, C. E. Peper, P. J. Beek, and A. Daffertshofer. A model for phase transitions in human hand movements during multifrequency tapping. *Physica D*, 90:179–196, 1996. Erratum: 92:260-260, 1996.
266. H. Haken and M. Stadler. *Synergetics of cognition.* Springer, Berlin, 1990.
267. H. Haken and M. Wagner (eds.). *Cooperative phenomena.* Springer, Berlin, 1973.
268. H. Haken and H. C. Wolf. *Molecular physics and elements of quantum chemistry.* Springer, Berlin, 1995.
269. H. Haken and H. C. Wolf. *The physics of atoms and quanta.* Springer, Berlin, 2000.
270. H. Haken (ed.). *Synergetics — Cooperative phenomena in multi-component systems.* Teubner, Stuttgart, 1973.
271. H. Haken (ed.). *Cooperative effects.* North-Holland Publ. Company, Amsterdam, 1974.
272. S. K. Han, C. Kurrer, and Y. Kuramoto. Dephasing and bursting in coupled neural oscillators. *Phys. Rev. Lett.*, 75:3190–3193, 1995.
273. P. Hänggi and H. Thomas. Time evolution, correlations, and linear response of non-Markov processes. *Z. Physik B*, 26:85–92, 1977.
274. P. Hänggi and H. Thomas. Stochastic processes: time evolution, symmetries and linear response. *Phys. Rep.*, 88:207–319, 1982.
275. D. Hansel, G. Mato, and C. Meunier. Clustering and slow switching in globally coupled phase oscillators. *Phys. Rev. E*, 48:3470–3477, 1993.
276. D. Hansel, G. Mato, and C. Meunier. Phase dynamics for weakly coupled Hodgkin-Huxley neurons. *Europhysics Letters*, 23:367–372, 1993.
277. F. E. Hanson, J. F. Case, E. Buck, and J. Buck. Synchrony and flash entrainment in a New Guinea firefly. *Science*, 174:161–164, 1971.

278. H. Hara. Path integrals for Fokker-Planck equation described by generalized random walks. *Z. Physik B*, 45:159–166, 1981.
279. W. A. Harrison. *Applied quantum mechanics*. World Scientific, Singapore, 2000.
280. H. Hasegawa. Dynamic mean-field theory of spiking neuron ensembles: response to a single spike with independent noise. *Phys. Rev. E*, 67:041903, 2003.
281. S. Havlin and D. Ben-Avraham. Diffusion in disordered media. *Adv. Phys.*, 36:695–798, 1987.
282. C. V. Heer. *Statistical mechanics, kinetic theory, and stochastic processes*. Academic Press, New York, 1972.
283. S. Heifets. Microwave instability beyond threshold. *Phys. Rev. E*, 54:2889–2898, 1996.
284. S. Heifets. Saturation of the coherent beam-beam instability. *Phys. Rev. ST-AB*, 4:044401, 2001.
285. S. Heifets. Single-mode coherent synchroton radiation instability of a bunched beam bunch. *Phys. Rev. ST-AB*, 6:080701, 2003.
286. S. Heifets and B. Podobedov. Single bunch stability to monopol excitation. *Phys. Rev. ST-AB*, 2:044402, 1999.
287. H. G. E. Hentschel and I. Procaccia. Relative diffusion in turbulent media: The fractal dimension of clouds. *Phys. Rev. A*, 29:1461–1470, 1984.
288. A. Hernandez-Machado. The effect of noise on spatio-temporal patterns. In P. E. Cladis and P. Palffy-Muhoray, editors, *Spatio-temporal patterns in nonequilibrium complex systems*, pages 521–528, New York, 1995. Addison-Wesley Publishing Company.
289. S. Hess. *Z. Naturforschung A*, 31:1034, 1976.
290. C. H. Hesse and E. Ramos. Dynamic simulation of a stochastic model for particle sedimentation in fluids. *Appl. Math. Modelling*, 18:437–445, 1994.
291. H. B. Hollinger and M. J. Zenzen. *The nature of irreversibility*. D. Reidel Publishing Company, Dordrecht, 1985.
292. E. H. Holt and R. E. Haskell. *Plasma dynamics*. MacMillan, New York, 1965.
293. J. Honerkamp. *Statistical physics. An advanced approach with applications*. Springer, Berlin, 1998.
294. D. C. Hong. Effect of excluded volume and anisotropy on granular statistics: Fermi statistics and condensation. In T. Pöschel and S. Luding, editors, *Granular gases*, pages 429–444, Springer, 2001. Berlin.
295. M.-O. Hongler and R. C. Desai. Study of a class of models for self-organization: equilibrium analysis. *J. Stat. Phys.*, 32:585–614, 1983.
296. M.-O. Hongler and D. M. Ryter. Hard mode stationary states generated by fluctuations. *Z. Physik B*, 31:333–337, 1978.
297. W. Horsthemke and A. Bach. *Z. Physik B*, 22:189, 1975.
298. W. Horsthemke and R. Lefever. *Noise-induced transitions*. Springer, Berlin, 1984.
299. D. F. Hoyt and C. R. Taylor. Gait and energetics of locomotion in horses. *Nature*, 292:239–240, 1981.
300. M. Hütter, I. V. Karlin, and H. C. Öttinger. Dynamic mean-field models from a nonequilibrium thermodynamics perspective. *Phys. Rev. E*, 68:016115, 2003.
301. M. Hütter and H. C. Öttinger. Modification of linear response theory for mean-field approximation. *Phys. Rev. E*, 54:2526–2530, 1996.

302. H. Ibach and H. Lüth. *Solid-state physics: an introduction to principles of material science*. Springer, Berlin, 2003.
303. P. Ilg and S. Hess. Nonequilibrium dynamics and magnetoviscosity of moderately concentrated magnetic liquids: a dynamic mean-field study. *Z. Naturforsch. A*, 58:589–600, 2003.
304. P. Ilg, I. V. Karlin, and Öttinger H. C. Generating moment equations in the Doi model of liquid-crystalline polymers. *Phys. Rev. E*, 60:5783–5787, 1999.
305. M. Iosifescu and P. Tautu. *Stochastic processes and its applications in biology and medicine*, volume I. Springer, Berlin, 1973.
306. R. Jain and S. Ramakumar. Stochastic dynamics modeling of the protein sequence length distribution in genomes: implications for microbial evolution. *Physica A*, 273:476–485, 1999.
307. S. Jain and A. P. Young. Monte Carlo simulations of XY spin glasses. *J. Phys. C: Solid State Phys.*, 19:3913–3923, 1986.
308. A. M. Jayannavar and M. C. Mahato. Macroscopic equation of motion in inhomogeneous media: a microscopic treatment. *Pramana - journal of physics*, 45:369–376, 1995.
309. E.T. Jaynes. Information theory and statistical mechanics. *Phys. Rev.*, 106:620–630, 1975.
310. E.T. Jaynes. Information theory and statistical mechanics II. *Phys. Rev.*, 108:171–190, 1975.
311. D. Z. Jin and H. E. Dubin. Regional maximum entropy theory of vortex crystal formation. *Phys. Rev. Lett.*, 80:4434–4437, 1998.
312. V. K. Jirsa. *Theoretical neurodynamics: from sub to supra network levels*, unpublished.
313. V. K. Jirsa, R. Friedrich, and H. Haken. Reconstruction of the spatio-temporal dynamics of a human magnetoencephalogram. *Physica D*, 89:100–122, 1995.
314. V. K. Jirsa and H. Haken. A field theory of electromagnetic brain activity. *Phys. Rev. Lett.*, 77:960–963, 1996.
315. V. K. Jirsa and H. Haken. A derivation of a macroscopic field theory of the brain from the quasi-microscopic neural dynamics. *Physica D*, 99:503–526, 1997.
316. J. V. Jose, L. P. Kadanoff, S. Kirkpatrick, and D. R. Nelson. Renormalization, vortices, and symmetry-breaking perturbations in the two-dimensional planar model. *Phys. Rev. B*, 16:1217–1241, 1977.
317. M. Kac. The physical background of Langevin equations. In A. K. Aziz, editor, *Lectures in differential equations*, volume II, pages 147–166, New York, 1969. Van Nostrand Reinhold Company.
318. L. P. Kadanoff. *Statistical physics: statics, dynamics and renormalization*. World Scientific, Singapore, 2000.
319. V. I. Kalikmanov. *Statistical physics of fluids*. Springer, Berlin, 2001.
320. T. Kanamaru, T. Horita, and Y. Okabe. Theoretical analysis of array-enhanced stochastic resonance in the diffusively coupled FitzHugh-Nagumo equation. *Phys. Rev. E*, 64:031908, 2001.
321. G. Kaniadakis. Generalized Boltzmann equation describing the dynamics of bosons and fermions. *Phys. Lett. A*, 203:229–234, 1995.
322. G. Kaniadakis. H-theorem and generalized entropies within the framework of nonlinear kinetics. *Phys. Lett. A*, 288:283–291, 2001.
323. G. Kaniadakis. Nonlinear kinetics underlying generalized statistics. *Physica A*, 296:405–425, 2001.

324. G. Kaniadakis. Statistical mechanics in the context of special relativity. *Phys. Rev. E*, 66:056125, 2003.
325. G. Kaniadakis and G. Lapenta. Microscopic dynamics underlying anomalous diffusion. *Phys. Rev. E*, 62:3246–3249, 2000.
326. G. Kaniadakis, A. Lavagno, and P. Quarati. Kinetic approach to fractional exclusion statistics. *Nuclear Physics B*, 466:527–537, 1996.
327. G. Kaniadakis and P. Quarati. Kinetic equation for classical particles obeying an exclusion principle. *Phys. Rev. E*, 48:4263–4270, 1993.
328. G. Kaniadakis and P. Quarati. Classical model of bosons and fermions. *Phys. Rev. E*, 49:5103–5116, 1994.
329. W. L. Kath. Waiting and propagating fronts in nonlinear diffusion. *Physica D*, 12:375–381, 1984.
330. R. Kawai, X. Sailer, and L. Schimansky-Geier. Macroscopic limit cycle via noise-induced phase transition. *Phys. Rev. E*, 69:051104, 2004.
331. H. Kawamura. Numerical studies of chiral ordering in three-dimensional XY spin glasses. *Phys. Rev. B*, 51:12398–12409, 1995.
332. J. Keizer. *Statistical thermodynamics of nonequilibrium processes*. Springer, New York, 1987.
333. D. C. Kelly. *Thermodynamics and statistical physics*. Academic Press, New York, 1973.
334. J. A. S. Kelso. Phase transitions and critical behavior in human bimanual coordination. *Am. J. Physiology: Regulatory, Integrative and Comparative Physiology*, 15:R1000–R1004, 1984.
335. J. A. S. Kelso. *Dynamic patterns - The self-organization of brain and behavior*. MIT Press, Cambridge, 1995.
336. J. A. S. Kelso, S. L. Bressler, S. Buchannan, G. C. DeGuzman, M. Ding, A. Fuchs, and T. Holroyd. A phase transition in human brain and behavior. *Phys. Lett. A*, 169:134–144, 1992.
337. J. A. S. Kelso, J. D. DelColle, and G. Schöner. Action-perception as a pattern formation process. In M. Jeannerod, editor, *Attention and performance XIII*, pages 139–169, Hillsdale, New Jersey, 1990. Erlbaum.
338. J. A. S. Kelso, A. Fuchs, R. Lancaster, D. Cheyne T. Holroyd, and H. Weinberg. Dynamic cortical activity in the human brain reveals motor equivalence. *Nature*, 392:814–818, 1998.
339. J. A. S. Kelso, J. P. Scholz, and G. Schöner. Non-equilibrium phase transitions in coordinated biological motion: Critical fluctuations. *Phys. Lett. A*, 118:279–284, 1986.
340. E. H. Kennard. *Kinetic theory of gases*. McGraw-Hill Book Company, New York, 1938.
341. S. Kim, S. H. Park, C. R. Doering, and C. S. Ryu. Reentrant transitions in globally coupled active rotators with multiplicative and additive noise. *Phys. Lett. A*, 224:147–153, 1997.
342. S. Kim, S. H. Park, and H.-B. Pyo. Stochastic resonance in coupled oscillator systems with time delay. *Phys. Rev. Lett.*, 82:1620–1623, 1999.
343. S. Kim, S. H. Park, and C. S. Ryu. Noise-induced transitions in coupled oscillator systems with a pinning force. *Phys. Rev. E*, 54:6042–6052, 1996.
344. S. Kim, S. H. Park, and C. S. Ryu. Noise-enhanced multistability in coupled oscillator systems. *Phys. Rev. Lett.*, 78:1616–1619, 1997.
345. S. Kirkpatrick and D. Sherrington. Infinite-ranged models of spin-glasses. *Phys. Rev. B*, 17:4384–4403, 1978.

346. C. Kittel. *Elementary statistical physics*. John Wiley and Sons, New York, 1958.
347. Y. L. Klimontovich. *Statistical physics*. Harwood Academic Publ., New York, 1986.
348. Y. L. Klimontovich. *Turbulent motion and the structure of chaos*. Kluwer Academic Publisher, Dordrecht, 1991.
349. P. J. Klinko and B. N. Miller. Mean field theory of spherical gravitating systems. *Phys. Rev. E*, 62:5783–5792, 2000.
350. P. E. Kloeden and E. Platen. *The numerical solution of stochastic differential equations*. Springer, Berlin, 1992.
351. L. Knopoff. Scale invariance of earthquakes. In B. Dubrulle, F. Graner, and D. Sornette, editors, *Scale invariance and beyond*, pages 159–172, Berlin, 1997. Springer.
352. K. Kometani and H. Shimizu. A study of self-organizing processes of nonlinear stochastic variables. *J. Stat. Phys.*, 13:473–490, 1975.
353. D. Kondepudi and I. Prigogine. *Modern thermodynamics*. John Wiley and Sons, New York, 1998.
354. V. N. Kondratyev. Statistics of magnetic noise in neutron star crusts. *Phys. Rev. Lett.*, 88:221101, 2002.
355. M. Kostur, J. Luczka, and L. Schimansky-Geier. Nonequilibrium coupled Brownian phase oscillators. *Phys. Rev. E*, 65:051115, 2002.
356. G. Kozyreff, A. G. Vladimirov, and P. Mandel. Global coupling with time delay in an array of semiconductor lasers. *Phys. Rev. Lett.*, 85:3809–3812, 2000.
357. P. Kruse, H.-O Carmesin, L. Pahlke, D. Strüber, and M. Stadler. Continuous phase transitions in the perception of multistabile visual patterns. *Biol. Cybern.*, 40:23–42, 1996.
358. R. Kubo. The fluctuation-dissipation theorem and Brownian motion. In R. Kubo, editor, *Many-body theory*, pages 1–16, Tokyo and New York, 1966. Syokabo and Benjamin.
359. R. Kubo. *Statistical Mechanics*. North-Holland Publ. Company, Amsterdam, 1967.
360. R. Kubo, M. Toda, and N. Hashitsume. *Statistical Physics II*. Springer, Berlin, 1985.
361. S. Kullback. *Information theory and statistics*. Dover Publications, New York, 1968.
362. S. Kullback and R. A. Leibler. On information and sufficiency. *Ann. Math. Stat.*, 22:79–86, 1951.
363. Y. Kuramoto. Self-entrainment of a population of coupled non-linear oscillators. In H. Araki, editor, *Int. symposium on mathematical problems in theoretical physics*, pages 420–422, Berlin, 1975. Springer.
364. Y. Kuramoto. Rhythms and turbulence in populations of chemical oscillators. *Physica A*, 106:128–143, 1981.
365. Y. Kuramoto. *Chemical oscillations, waves, and turbulence*. Springer, Berlin, 1984.
366. Y. Kuramoto. Cooperative dynamics of oscillator community - a study based on lattice of rings. *Prog. Theor. Phys. Suppl.*, 79:223–240, 1984.
367. Y. Kuramoto. Phase- and center-manifold reductions for large populations of coupled oscillators with application to non-locally coupled systems. *Int. J. Bif. and Chaos*, 7:789–805, 1997.

368. Y. Kuramoto and I. Nishikawa. Statistical macrodynamics of large dynamical systems. Case of a phase transition in oscillator communities. *J. Stat. Phys.*, 49:569–605, 1987.
369. M. G. Kuzmina, E. A. Manykin, and I. I. Surina. Oscillatory networks with Hebbian matrix of connections. In J. Mira and F. Sandoval, editors, *From natural to artificial neural computation*, pages 246–251, Berlin, 1995. Springer.
370. C. Lancellotti and M. Kiessling. Self-similar gravitational collapse in stellar dynamics. *Astrophys. J.*, 549:L93–L96, 2001.
371. L. D. Landau and E. M. Lifshitz. *Statistical Physics*. Pergramon Press, London, 1958.
372. L. D. Landau and E. M. Lifshitz. *Electrodynamics of continuous media*. Pergramon Press, London, 1960.
373. P. T. Landsberg. *Thermodynamics*. Interscience publishers, London, 1961.
374. P. T. Landsberg. Is equilibrium always an entropy maximum. *J. Stat. Phys.*, 35:159–169, 1984.
375. P. T. Landsberg. Entropies galore. *Braz. J. Phys.*, 29:46–49, 1999.
376. P. T. Landsberg and V. Vedral. Distributions and channel capacities in generalized statistical mechanics. *Phys. Lett. A*, 247:211–217, 1998.
377. G. Lapenta, G. Kaniadakis, and P. Quarati. Stochastic equation of systems of particles obeying an exclusion principle. *Physica A*, 225:323–335, 1996.
378. R. G. Larson and H. C. Öttinger. Effect of molecular elasticity on out-of-plane orientations in shearing flows of liquid-crystalline polymers. *Macromolecules*, 24:6270–6282, 1991.
379. J. LaSalle and S. Lefschetz. *Stability by Lyapunov's direct method*. Academic Press, New York, 1961.
380. V. Latora and A. Rapisarda. Dynamical quasi-stationary states in a system with long-range forces. *Chaos, solitons & fractals*, 13:401–406, 2002.
381. V. Latora, A. Rapisarda, and S. Ruffo. Lyapunov instability and finite size effects in a system with long-range forces. *Phys. Rev. Lett.*, 80:692–695, 1998.
382. V. Latora, A. Rapisarda, and S. Ruffo. Chaos and statistical mechanics in the Hamiltonian mean field model. *Physica D*, 131:38–54, 1999.
383. V. Latora, A. Rapisarda, and S. Ruffo. Superdiffusion and out-of-equilibrium chaotic dynamics with many degrees of freedoms. *Phys. Rev. Lett.*, 83:2104–2107, 1999.
384. V. Latora, A. Rapisarda, and C. Tsallis. Non-Gaussian equilibrium in a long-range Hamiltonian system. *Phys. Rev. E*, 64:056134, 2001.
385. V. Latora, A. Rapisarda, and C. Tsallis. Fingerprints of nonextensive thermodynamics in a long-range Hamiltonian system. *Physica A*, 305:129–136, 2002.
386. J. L. Lebowitz and P. G. Bergmann. Irreversible Gibbsian ensembles. *Annals of Physics*, 1:1–23, 1957.
387. E. K. Lenzi, C. Anteneodo, and L. Borland. Escape time in anomalous diffusive media. *Phys. Rev. E*, 63:051109, 2001.
388. E. K. Lenzi, M. K. Lenzi, H. Belich, and L. S. Lucena. Specific heat in the nonextensive statistics: effective temperature and Lagrange paramter β. *Phys. Lett. A*, 292:315–319, 2002.
389. E. K. Lenzi, R. S. Mendes, and L. R. da Silva. Statistical mechanics based on Renyi entropy. *Physica A*, 280:337–345, 2000.

390. R. Levy, W. D. Hutchison, A. M. Lozano, and J. O. Dostrovsky. High-frequency synchronization of neuronal activity in the subthalamic nucleus of parkinsonian patients with limb tremor. *J. Neurosci.*, 20:7766–7775, 2000.
391. J. H. Li and P. Hänggi. Spatially periodic stochastic system with infinite globally coupled oscillators. *Phys. Rev. E*, 64:011106, 2001.
392. J. H. Li and Z. Q. Huang. Nonequilibrium phase transition in the case of correlated noises. *Phys. Rev. E*, 53:3315–3318, 1996.
393. J. H. Li, Z. Q. Huang, and D. Y. Xing. Nonequilibrium transitions for a stochastic globally coupled model. *Phys. Rev. E*, 58:2838–2842, 1998.
394. J. A. S. Lima, R. Silva, and A. R. Plastino. Nonextensive thermostatistics and the H-Theorem. *Phys. Rev. Lett.*, 86:2938–2941, 2001.
395. J. D. Logan. *Transport modeling in hydrogeochemical systems*. Springer, Berlin, 2001.
396. R. Lovett and M. Baus. Van der Waals theory for the spatial distribution of the tension in an interface. I Density functional theory. *J. Chem. Phys.*, 111:5544–5554, 1999.
397. W. D. Luedtke and U. Landman. Slip diffusion and Levy flights of an adsorbed gold nanocluster. *Phys. Rev. Lett.*, 82:3835–3838, 1999. Erratum: 83 (1999) 1702.
398. W. M. MacDonald, M. N. Rosenbluth, and W. Chuck. Relaxation of a system of particles with Coulomb interaction. *Phys. Rev.*, 107:350–353, 1957.
399. W. Maier and A. Saupe. *Z. Naturforschung A*, 13:564, 1958.
400. W. Maier and A. Saupe. *Z. Naturforschung A*, 15:287, 1960.
401. L. C. Malacarne, R. S. Mendes, I. T. Pedron, and E. K. Lenzi. Nonlinear equation for anomalous diffusion: unified power-law and stretched exponential exact solution. *Phys. Rev. E*, 63:030101, 2001.
402. R. N. Mantegna and H. E. Stanley. Scaling behavior in the dynamics of an economic index. *Nature*, 376:46–49, 1995.
403. R. N. Mantegna and H. E. Stanley. *An introduction to econophysics*. Cambridge University Press, Cambridge, 2000.
404. M. Marsili and A. J. Bray. Soluble infinite-range model of kinetic roughening. *Phys. Rev. Lett.*, 76:2750–2753, 1996.
405. S. Martinez, A. R. Plastino, and A. Plastino. Nonlinear Fokker-Planck equations and generalized entropies. *Physica A*, 259:183–192, 1998.
406. F. Matasubara, T. Iyota, and S. Inawashiro. Effect of anisotropy on a short-range $\pm j$ Heisenberg spin glass in three dimensions. *Phys. Rev. Lett.*, 67:1458–1461, 1991.
407. J. Maucourt and D.R. Grempel. Lower critical dimension of the xy spin-glass model. *Phys. Rev. Lett.*, 80:770–773, 1998.
408. N. M. Maurits, B. A. C. van Vlimmeren, and J. G. E. M. Fraaije. Mesoscopic phase separation dynamics of compressible copolymer melts. *Phys. Rev. E*, 56:816–825, 1997.
409. R. M. May. Simple mathematical models with very complicated dynamics. *Nature*, 261:459–467, 1976.
410. G. F. Mazenko. *Equilibrium statistical mechanics*. John Wiley and Sons, Chichester, 2000.
411. H. P. McKean (Jr.). Propagation of chaos for a class of nonlinear parabolic equations. In A. K. Aziz, editor, *Lectures in differential equations*, volume II, pages 177–193, New York, 1969. Van Nostrand Reinhold Company.

412. F. Mechsner, D. Kerzel, G. Knoblich, and W. Prinz. Perceptual basis of bimanual coordination. *Nature*, 414:69–73, 2001.
413. A. Meibom and I. Balslev. Composite power laws in shock fragmentation. *Phys. Rev. Lett.*, 76:2492–2494, 1996.
414. S. Meleard. Asymptotic behavior of some interacting particle systems: McKean-Vlasov and Boltzmann models. In C. Graham, T. G. Kurtz, S. Meleard, P. E. Potter, M. Pulvirenti, and D. Talay, editors, *Probabilistic models for nonlinear partial differential equations*, pages 42–95, Berlin, 1996. Springer.
415. S. Meleard and S. R. Coppoletta. A propagation of chaos result for a system of particles with moderate interaction. *Stochastic processes and their applications*, 26:317–332, 1987.
416. R. S. Mendes. Some general relations in arbitrary thermostatistics. *Physica A*, 242:299–308, 1997.
417. F. G. Mertens and A. R. Bishop. Dynamics of vortices in two-dimensional magnets. In P. L. Christiansen, M. P. Sorenson, and A. C. Scott, editors, *Nonlinear Science at the down of the 21st century*, pages 137–170, Berlin, 2000. Springer.
418. R. Metzler and J. Klafter. The random walk's guide to anomalous diffusion: a fractional dynamics approach. *Phys. Rep.*, 339:1–77, 2000.
419. A. S. Mikhailov. *Foundations of synergetics, vol I*. Springer, Berlin, 1990.
420. A. S. Mikhailov and D. H. Zanette. Noise-induced breakdown of coherent collective motion in swarms. *Phys. Rev. E*, 60:4571–4575, 1999.
421. J. Miller. Statistical mechanics of Euler equations in two dimensions. *Phys. Rev. Lett.*, 65:2137–2140, 1990.
422. J. Miller, P. B. Weichman, and M. C. Cross. Statistical mechanics, Euler's equation, and Jupiter's red spot. *Phys. Rev. A*, 45:2328–2359, 1992.
423. G. N. Milstein. The probability approach to numerical solution of nonlinear parabolic equations. *Num. Methods Part. Diff. Eq.*, 18:490–522, 2002.
424. G. N. Milstein and M. V. Tretyakov. Numerical methods for nonlinear parabolic equations with small parameter based on probability approach. *Math. Comput.*, 60(229):237–267, 2000.
425. G. N. Milstein and M. V. Tretyakov. Numerical solution of Dirichlet problems for nonlinear parabolic equations by probability approach. *IMA J. Num. Analysis.*, 21:887–917, 2001.
426. A. E. Minetti. The biomechanics of skipping gaits: a third locomotion paradigm? *Proc. R. Soc. Lond. B*, 265:1227–1235, 1998.
427. A. E. Minetti, L. P. Ardigo, E. Reinach, and F. Saibene. The relationship between mechanical work and energy expenditure of locomotion in horses. *J. Exp. Biol.*, 202:2329–2338, 1999.
428. E. W. Montroll and B. J. West. Models of population growth, diffusion, competition and rearrangement. In H. Haken, editor, *Synergetics - Cooperative phenomena in multi-component systems*, pages 143–156, Berlin, 1973. Springer.
429. E. W. Montroll and B. J. West. On an enriched collection of stochastic processes. In E. W. Montroll and J. L. Lebowitz, editors, *Studies in statistical mechanics*, pages 61–175, Amsterdam, 1979. North-Holland Publ. Company.
430. E. N. Moore. *Theoretical mechanics*. John Wiley and Sons, New York, 1983.
431. M. Morillo, J. Gomez-Ordonez, and J. M. Casado. Stochastic resonance in a mean-field model of cooperative behavior. *Phys. Rev. E*, 52:316–320, 1995.

432. P. M. Morse. *Thermal physics.* W. A. Benjamin, Inc., New York, 1964.
433. D. Mottet and R. J. Bootsma. The dynamics of goal-directed rhythmic aiming. *Biol. Cybern.*, 80:235–245, 1999.
434. R. Müller, K. Lippert, A. Kühnel, and U. Behn. First-order nonequilibrium phase transition in a spatially extended system. *Phys. Rev. E*, 56:2658–2662, 1997.
435. T. Munakata. Liquid instability and freezing - reductive preturbation approach. *J. Phys. Soc. Japan*, 43:1723–1728, 1977.
436. T. Munakata and S. Mitsuoka. Langevin dynamics for generalized thermodynamics. *J. Phys. Soc. Japan*, 69:92–96, 2000.
437. A. Münster. *Statistical thermodynamics Vol 1.* Springer, Berlin, 1969.
438. J. D. Murray. *Mathematical biology.* Springer, Berlin, 1993.
439. V. N. Murthy and E. E. Fetz. Oscillatory activity in sensorimotor cortex of awake monkeys: Synchronization of local field potentials and relation to behavior. *J. Neurophysiology*, 76:3949–3967, 1996.
440. J. Naudts. Generalized thermostatistics and mean-field theory. *Physica A*, 332:279–300, 2004.
441. Z. Neda, E. Ravasz, T. Vicsek, Y. Brechet, and A. L. Barabasi. Physics of the rhythmic applause. *Phys. Rev. E*, 61:6987–6992, 2000.
442. F. M. Nekrasov, A. G. Elfimov, C. A. de Azevedo, and A. S. de Assis. Effect of background plasma nonlinearities on dissipation processes in plasma. *Phys. Lett. A*, 251:44–48, 1999.
443. J. C. Neu. Large populations of coupled chemical oscillators. *SIAM J. Appl. Math.*, 38:305–316, 1980.
444. A. C. Newell, T. Passot, and J. Lega. Order parameter equations for patterns. *Annu. Rev. Fluid Mech.*, 25:399–453, 1993.
445. K. M. Newell and D. M. Corcos. *Variability and motor control.* Human Kinetics Publishers, Champaign, 1993.
446. W. M. Ni, L. A. Peletier, and J. Serrin. *Nonlinear diffusion equations and their equilibrium states.* Springer, Berlin, 1988.
447. D. R. Nicholson. *Introduction to plasma theory.* John Wiley and Sons, New York, 1983.
448. G. Nicolis. *Introduction to nonlinear sciences.* Cambridge University Press, Cambridge, 1995.
449. G. Nicolis and I. Prigogine. *Self-organization in nonequilibrium system.* John Wiley and Sons, New York, 1977.
450. E. Niebur, H. G. Schuster, and D. M. Kammen. Collective frequencies and metastability in networks of limit-cycle oscillators with delays. *Phys. Rev. Lett.*, 67:2753–2756, 1991.
451. F. D. Nobre, E. M. F. Curado, and G. Rowlands. A procedure for obtaining general nonlinear Fokker-Planck equations. *Physica A*, 334:109–118, 2004.
452. D. Ocone and E. Pardoux. Non-adapted solutions of stochastic differential equations. In C. I. Byrnes, C. F. Martin, and R. E. Saeks, editors, *Analysis and control of nonlinear systems*, pages 591–598, Amsterdam, 1988. North-Holland Publ. Company.
453. L. Oddershede, P. Dimon, and J. Bohr. Self-organized criticality in fragmenting. *Phys. Rev. Lett.*, 71:3107–3110, 1993.
454. K. Oelschläger. A law of large numbers for moderately interacting diffusion processes. *Z. Wahrscheinlichkeitstheorie verw. Gebiete*, 69:279–322, 1985.

455. A. Okubo. *Diffusion and ecological problems: mathematical models*. Springer, Berlin, 1980.
456. A. Okubo and S. A. Levin. *Diffusion and ecological problems: modern perspectives*. Springer, Berlin, 2001.
457. A. Ott, J. P. Bouchard, D. Langevin, and W. Urbach. Anomalous diffusion in "living polymers": A genuine Levy flight? *Phys. Rev. Lett.*, 65:2201–2204, 1990.
458. H. C. Öttinger. *Stochastic processes in polymeric fluids*. Springer, Berlin, 1996.
459. H. C. Öttinger and M. Grmela. Dynamics and thermodynamics of complex fluids. II. illustration of a general formalism. *Phys. Rev. E*, 56:6633–6655, 1997.
460. A. R. R. Papa. On one-parameter-dependent generalizations of Boltzmann-Gibbs statistical mechanics. *J. Phys. A: Math. Gen.*, 31:5271–5276, 1998.
461. H. Park, D. R. Collins, and M. T. Turvey. Dissociation of muscular and spatial constraints on patterns of interlimb coordination. *J. Exp. Psychol. - Hum. Percept. Perform.*, 27:32–47, 2001.
462. K. Park, Y. C. Lai, Z. Liu, and A. Nachman. Aperiodic stochastic resonance and phase synchronization. *Phys. Lett. A*, 326:391–396, 2004.
463. S. H. Park and S. Kim. Noise-induced phase transitions in globally coupled active rotators. *Phys. Rev. E*, 53:3425–3430, 1996.
464. R. D. Parks. *Superconductivity*. Marcel Dekker, New York, 1969.
465. J. Y. Parlange, W. L. Hogarth, M. B. Parlange, R. Haverkamp, D.A. Barry, P.J. Ross, and T.S. Steenhuis. Approximate analytical solution of the nonlinear diffusion equation for arbitrary boundary conditions. *Transp. Porous Media*, 30:45–55, 1998.
466. J. M. R. Parrondo, C. van den Broeck, J. Buceta, and F. J. de la Rubia. Noise-induced spatial patterns. *Physica A*, 224:153–161, 1996.
467. R. E. Pattle. Diffusion from an instantaneous point source with a concentration-dependent coefficient. *Quart. J. Mech. Appl. Math.*, 12:407–409, 1959.
468. T. W. Patzek, D. B. Silin, S. M. Benson, and G. I. Barenblatt. On vertical diffusion of gases in a horizontal reservoir. *Transp. Porous Media*, 51:141–156, 2003.
469. I. T. Pedron, R. S. Mendes, L. C. Malacarne, and E. K. Lenzi. Nonlinear anomalous diffusion equation and fractal dimension: Exact generalized gaussian solution. *Phys. Rev. E*, 65:041108, 2002.
470. L. A. Peletier. The porous media equation. In H. Amann, N. Bazley, and K. Kirchgässner, editors, *Applications of nonlinear analysis in the physical science*, pages 229–241, Boston, 1981. Pitman Advanced Publishing Program.
471. C. E. Peper and P. J. Beek. Are frequency-induced transitions in rhythmic coordination mediated by a drop in amplitude? *Biol. Cybern.*, 79:291–300, 1998.
472. C. E. Peper and P. J. Beek. Distinguishing between the effects of frequency and amplitude on interlimb coupling in tapping a 2:3 polyrhythm. *Exp. Brain Res.*, 118:78–92, 1998.
473. C. E. Peper, P. J. Beek, and P. C. W. van Wieringen. Coupling strength in tapping a 2:3 polyrhythm. *Hum. Movement Sci.*, 14:217–245, 1995.
474. C. E. Peper, P. J. Beek, and P. C. W. van Wieringen. Frequency-induced transitions in bimanual tapping. *Biol. Cybern.*, 73:301–309, 1995.

475. C. E. Peper, P. J. Beek, and P. C. W. van Wieringen. Multifrequency coordination in bimanual tapping: Asymmetrical coupling and signs of supercriticality. *J. Exp. Psychol. - Hum. Percept. Perform.*, 21:1117–1138, 1995.
476. C. E. Peper and R. G. Carson. Bimanual coordination between isometric contractions and rhythmic movements: a symmetric coupling. *Exp. Brain Res.*, 129:417–432, 1999.
477. A. Pikovsky, M. Rosenblum, and J. Kurths. *Synchronization: a universal concept in nonlinear sciences*. Cambridge University Press, Cambridge, 2001.
478. L. P. Pitaevskii. *Sov. Phys. JETP*, 13:451, 1961.
479. M. Planck. *Sitzungsber. Preuss. Akad. Wissens.*, page 324, 1917.
480. A. Plastino and A. R. Plastino. On the universality of thermodynamics' Legendre transform structure. *Phys. Lett. A*, 226:257–263, 1997.
481. A. R. Plastino, M. Casas, and A. Plastino. A nonextensive maximum entropy approach to a family of nonlinear reaction-diffusion equations. *Physica A*, 280:289–303, 2000.
482. A. R. Plastino, H. G. Miller, A. R. Plastino, and G. D. Yen. The role of information measures in the determination of the maximum entropy-minimum norm solution of the generalized inverse problem. *J. Math. Phys.*, 38:6675–6682, 1997.
483. A. R. Plastino and A. Plastino. Stellar polytropes and Tsallis entropy. *Phys. Lett. A*, 174:384–386, 1993.
484. A. R. Plastino and A. Plastino. Non-extensive statistical mechanics and generalized Fokker-Planck equation. *Physica A*, 222:347–354, 1995.
485. M. Plischke and B. Bergersen. *Equilibrium statistical physics*. World Scientific, Singapor, 1994.
486. J. Portugali. *Self-organization and the city*. Springer, Berlin, 2000.
487. A. A. Post, C. E. Peper, A. Daffertshofer, and P. J. Beek. Relative phase dynamics in perturbed interlimb coordination: Stability and stochasticity. *Biol. Cybern.*, 83:443–459, 2000.
488. S. Primak. Generation of compound non-Gaussian random processes with a given correlation function. *Phys. Rev. E*, 61:100–103, 2000.
489. S. Primak, V. Lyandres, and V. Kontorovich. Markov models of non-Gaussian exponentially correlated processes and their applications. *Phys. Rev. E*, 63:061103, 2001.
490. F. Reif. *Fundamentals of statistical and thermal physics*. McGraw-Hill Book Company, New York, 1965.
491. P. Reimann, R. Kawai, C. Van den Broeck, and P. Hänggi. Coupled Brownian motors: anomalous hysteresis and zero-bias negative conductance. *Europhysics Letters*, 45:545–551, 1999.
492. P. Reimann, C. Van den Broeck, and R. Kawai. Nonequilibrium noise in coupled phase oscillators. *Phys. Rev. E*, 60:6402–6406, 1999.
493. C. Renner, J. Peinke, and R. Friedrich. Evidence of Markov properties of high frequency exchange rate data. *Physica A*, 298:499–520, 2001.
494. C. Renner, J. Peinke, R. Friedrich, O. Chanal, and B. Chabaud. Universality of small scale turbulence. *Phys. Rev. Lett.*, 89:124502, 2002.
495. A. Renyi. *Probability theory*. North-Holland Publ. Company, Amsterdam, 1970.
496. L. F. Richard. Atmospheric diffusion shown on a distance-neighbour group. *Proc. Roy. Soc. London A*, 110:709–737, 1926.

497. A. Rigo, A. R. Plastino, M. Carsas, and A. Plastino. Anomalous diffusion coupled with Verhulst-like growth dynamics: exact time-dependent solutions. *Phys. Lett. A*, 276:97–102, 2000.
498. H. Risken. *The Fokker-Planck equation — Methods of solution and applications*. Springer, Berlin, 1989.
499. D. A. Rosenbaum. *Human motor control*. Academic Press, New York, 1991.
500. M. Rosenbluth, W. M. MacDonald, and D. L. Judd. Fokker-Planck equation for an inverse square force. *Phys. Rev.*, 107:1–6, 1957.
501. J. M. Rubi and A. Perez-Madrid. Mesoscopic non-equilibrium thermodynamics approach to the dynamics of polymers. *Physica A*, 298:177–186, 2001.
502. A. Rubinowicz. *Quantum Mechanics*. Elsevier Publishing Company, Amsterdam, 1968.
503. H. Sakaguchi. Cooperative phenomena in coupled oscillator systems under external fields. *Prog. Theor. Phys.*, 79:39–46, 1988.
504. H. Sakaguchi and Y. Kuramoto. A soluble active rotator model showing phase transitions via mutual entrainment. *Prog. Theor. Phys.*, 76:576–581, 1986.
505. H. Sakaguchi, S. Shinomoto, and Y. Kuramoto. Local and global self-entrainment in oscillator lattices. *Prog. Theor. Phys.*, 77:1005–1010, 1987.
506. H. Sakaguchi, S. Shinomoto, and Y. Kuramoto. Phase transitions and their bifurcation analysis in a large population of active rotators with mean-field coupling. *Prog. Theor. Phys.*, 79:600–607, 1988.
507. N. A. Salingaros and B. J. West. A universal rule for the distribution of sizes. *Environ. Plann. B*, 26:909–923, 1999.
508. F. Schlögl. Small fluctuations in a steady non-equilibrium state. *Z. Physik B*, 33:199–204, 1979.
509. F. Schlögl. *Probability and heat*. Vieweg, Braunschweig, 1989.
510. E. Schöll and P. T. Landsberg. Generalised equal areas rules for spatially extended systems. *Z. Physik B*, 72:515–521, 1988.
511. G. S. Schöner and H. Haken. The slaving principle for Stratonovich stochastic differential equations. *Z. Physik B*, 63:493–504, 1986.
512. G. S. Schöner, H. Haken, and J. A. S. Kelso. A stochastic theory of phase transitions in human hand movement. *Biol. Cybern.*, 53:247–257, 1986.
513. H. G. Schuster and P. Wagner. A model for neural oscillators in the visual cortex. 1. Mean-field theory and derivation of the phase equations. *Biol. Cybern.*, 64:77–82, 1990.
514. H. G. Schuster and P. Wagner. A model for neural oscillators in the visual cortex. 2. Phase description of the feature dependent synchronization. *Biol. Cybern.*, 64:83–85, 1990.
515. F. Schweitzer. *Brownian agents and active particles*. Springer, Berlin, 2003.
516. F. Schweitzer, W. Ebeling, and B. Tilch. Statistical mechanics of canonical-dissipative systems and applications to swarm dynamics. *Phys. Rev. E*, 64:021110, 2001.
517. M. O. Scully and M. S. Zubairy. *Qunatum optics*. Cambridge University Press, Cambridge, 1997.
518. C. E. Shannon. A mathematical theory of communication. *Bell System Tech. J.*, 27:379-423,623-656, 1948. 379-423,623-656.
519. B. D. Sharma and A. Garg. Nonadditive measures of average charge for heterogeneous questionnaires. *Inf. and Control*, 41:232–242, 1979.
520. B. D. Sharma and D. P. Mittal. New non-additive measures of entropy for discrete probability distribution. *J. Math. Sci.*, 10:28–40, 1975.

521. R. N. Shepard. Perceptual-cognitive universals as reflections of the world. *Psychonomic Bulletin & Review*, 1:2–28, 1994.
522. M. Shiino. H-theorem and stability analysis for mean-field models of non-equilibrium phase transitions in stochastic systems. *Phys. Lett. A*, 112:302–306, 1985.
523. M. Shiino. Dynamical behavior of stochastic systems of infinitely many coupled nonlinear oscillators exhibiting phase transitions of mean-field type: H-theorem on asymptotic approach to equilibrium and critical slowing down of order-parameter fluctuations. *Phys. Rev. A*, 36:2393–2412, 1987.
524. M. Shiino. Free energies based on generalized entropies and H-theorems for nonlinear Fokker-Planck equations. *J. Math. Phys.*, 42:2540–2553, 2001.
525. M. Shiino. Nonlinear Fokker-Planck equation exhibiting bifurcation phenomena and generalized thermostatistics. *J. Math. Phys.*, 43:2654–2669, 2002.
526. M. Shiino. Nonlinear Fokker-Planck equations with and without bifurcations and generalized thermostatistics. *J. Korean Phys. Soc.*, 40:1037–1040, 2002.
527. M. Shiino. Stability analysis of mean-field-type nonlinear Fokker-Planck equations associated with a generalized entropy and its application to the self-gravitating system. *Phys. Rev. E*, 67:056118, 2003.
528. H. Shimizu. Muscular contraction mechanism as a hard mode instability. *Prog. Theor. Phys.*, 52:329–330, 1974.
529. H. Shimizu and T. Yamada. Phenomenological equations of motion of muscular contraction. *Prog. Theor. Phys.*, 47:350–351, 1972.
530. S. Shinomoto and Y. Kuramoto. Cooperative phenomena in two-dimensional active rotator systems. *Prog. Theor. Phys.*, 75:1319–1327, 1986.
531. S. Shinomoto and Y. Kuramoto. Phase transitions in active rotator systems. *Prog. Theor. Phys.*, 75:1105–1110, 1986.
532. M. F. Shlesinger, B. J. West, and J. Klafter. Lévy dynamics of enhanced diffusion: Application to turbulence. *Phys. Rev. Lett.*, 58:1100–1103, 1987.
533. M. F. Shlesinger, G. M. Zaslavsky, and U. Frisch. *Lévy flights and related topics in physics*. Springer, Berlin, 1995.
534. Y. Shobuda and K. Hirata. The existence of a static solution for the Haissinski equation with purely inductive wake force. *Part. Accel.*, 62:165–177, 1999.
535. Y. Shobuda and K. Hirata. Proof of the existence and uniqueness of a solution for the Haissinski equation with a capacitive wake function. *Phys. Rev. E*, 64:67501, 2001.
536. S. Siegert, R. Friedrich, and J. Peinke. Analysis of data sets of stochastic systems. *Phys. Lett. A*, 243:275–280, 1998.
537. R. B. Silberstein. Steady-state visually evoked potentials, brain resonances and cognitive processes. In P. L. Nunez, editor, *Neocortical dynamics and human EEG rhythms*, pages 272–303, New York, 1995. Oxford University Press.
538. W. Singer. The formation of cooperative cell assemblies in the visual cortex. In J. Krüger, editor, *Neural cooperativity*, pages 165–183, Berlin, 1991. Springer.
539. W. Singer. Synchronization of cortical activity and its putative role in information processing and learning. *Annu. Rev. Physiol.*, 55:349–374, 1993.
540. M. Soler, F. C. Martinez, and J. M. Donoso. Integral kinetic method for one dimension: the spherical case. *J. Stat. Phys.*, 69:813–835, 1992.
541. T. H. Solomon, E. R. Weeks, and H. L. Swinney. Observation of anomalous diffusion and levy flight in a two-dimensional rotating flow. *Phys. Rev. Lett.*, 71:3975–3978, 1993.

542. H. Sompolinsky, D. Golomb, and D. Kleinfeld. Cooperative dynamics in visual processing. *Phys. Rev. A*, 43:6990–7011, 1991.
543. O. Sontolongo-Costa, Y. Moreno-Vega, J. J. Lloveras-Gonzalez, and J. C. Antoranz. Criticality in droplet fragmentation. *Phys. Rev. Lett.*, 76:42–45, 1996.
544. V. Sorensen, R. P. Ingvaldsen, and H. T. A. Whiting. The application of coordination dynamics to the analysis of discrete movements using table-tennis as a paradigm skill. *Biol. Cybern.*, 85:27–38, 2001.
545. H. Spohn. Surface dynamics below the roughening transition. *J. Phys. I France*, 3:69–81, 1993.
546. H. E. Stanley. *Introduction to phase transitions and critical phenomena*. Oxford University Press, New York, 1971.
547. D. A. Stariolo. The Langevin and Fokker-Planck equations in the framework of a generalized statistical mechanics. *Phys. Lett. A*, 185:262–264, 1994.
548. M. Steiner, J. Villain, and C. G. Windsor. Theoretical and experimental studies on one-dimensional magnetic systems. *Advances in Physics*, 25:87–209, 1976.
549. D. Sternad, M. T. Turvey, and E. L. Saltzman. Dynamics of 1:2 coordination: generalizing relative phase to n:m rhythms, sources of symmetry breaking, temporal scaling, latent 1:1, and bistability. *J. Motor Behav.*, 31:207–247, 1999.
550. R. L. Stratonovich. *Topics in the theory of random noise, Vol. 1*. Gordon and Beach, New York, 1963.
551. G. Strobl. *Condensed matter physics — Crystals, liquids, liquid crystals, and polymers*. Springer, Berlin, 2004.
552. S. H. Strogatz. From Kuramoto to Crawford: exploring the onset of synchronization in populations of coupled oscillators. *Physica D*, 143:1–20, 2000.
553. S. H. Strogatz and R. E. Mirollo. Collective synchronization in lattices of nonlinear oscillators with randomness. *J. Phys. A: Math. Gen.*, 21:L699–L705, 1988.
554. S. H. Strogatz and R. E. Mirollo. Phase-locking and critical phenomena in lattices of coupled nonlinear oscillators with random intrinsic frequencies. *Physica D*, 31:143–168, 1991.
555. S. H. Strogatz and R. E. Mirollo. Stability of incoherence in a population of coupled oscillators. *J. Stat. Phys.*, 63:613–635, 1991.
556. S. H. Strogatz, R. E. Morillo, and P. C. Matthews. Coupled nonlinear oscillators below the synchronization threshold: relaxation by generalized Landau damping. *Phys. Rev. Lett.*, 68:2730–2733, 1992.
557. S. H. Strogatz and I. Stewart. Coupled oscillators and biological synchronization. *Sci. American*, 269(6):68–75, 1993.
558. G. V. Stupakov, B. N. Breizman, and M. S. Pekker. Nonlinear dynamics of microwave instability in accelerators. *Phys. Rev. E*, 55:5976–5984, 1997.
559. A. Taigbenu. Simulations of unsaturated flow in multiply zoned media by Green element models. *Transp. Porous Media*, 45:387–406, 2001.
560. M. Takai, H. Akiyama, and S. Takeda. Stabilization of drift-cyclotron loss-cone instability of plasma by high frequency field. *J. Phys. Soc. Japan*, 50:1716–1722, 1981.
561. M. Takatsuji. Information-theoretical approach to a system of interacting elements. *Biol. Cybern.*, 17:207–210, 1975.

562. F. Tamarit and C. Anteneodo. Rotator with long-range interactions: connection with the mean-field approximation. *Phys. Rev. Lett.*, 84:208–211, 2000.
563. P. Tass. Phase and frequency shifts in a population of phase oscillators. *Phys. Rev. E*, 56:2043–2060, 1997.
564. P. Tass and H. Haken. Synchronized in networks of limit cycle oscillators. *Z. Physik B*, 100:303–320, 1996.
565. P. Tass and H. Haken. Synchronized oscillations in the visual cortex - a synergetic model. *Biol. Cybern.*, 74:31–39, 1996.
566. P. Tass, J. Kurths, M. Rosenblum, J. Weule, A. Pikovsky, J. Volkmann, A. Schnitzler, and H. J. Freund. Complex phase synchronization in neurophysiological data. In C. Uhl, editor, *Analysis of neurophysiological brain functioning*, pages 252–273, Berlin, 1999. Springer.
567. P. Tass, M. G. Rosenblum, J. Weule, J. Kurths, A. Pikovsky, J. Volkmann, A. Schnitzler, and H. J. Freund. Detection of n:m phase locking from noisy data: application to magnetoencephalography. *Phys. Rev. Lett.*, 81:3291–3294, 1998.
568. P. A. Tass. *Phase resetting in medicine and biology - Stochastic modelling and data analysis*. Springer, Berlin, 1999.
569. P. A. Tass. Stochastic phase resetting: A theory for deep brain stimulation. *Prog. Theor. Phys. Suppl.*, 139:301–313, 2000.
570. P. A. Tass. Effective desynchronization by means of double-pulse phase resetting. *Europhysics Letters*, 53:15–21, 2001.
571. P. A. Tass. Effective desynchronization with a resetting pulse train followed by a single pulse. *Europhysics Letters*, 55:171–177, 2001.
572. P. A. Tass. Desynchronization of brain rhythms with soft phase-resetting techniques. *Biol. Cybern.*, 87:102–115, 2002.
573. P. A. Tass. Effective desynchronization with bipolar double-pulse stimulation. *Phys. Rev. E*, 66:036226, 2002.
574. P. A. Tass. A model of desynchronizing deep brain stimulation with a demand-controlled coordinated reset of neural subpopulations. *Biol. Cybern.*, 89:81–88, 2003.
575. J.-N. Teramae and Y. Kuramoto. Strong desynchronizing effects of weak noise in globally coupled systems. *Phys. Rev. E*, 63:036210, 2001.
576. Y. P. Terletskii. *Statistical Physics*. North-Holland Publ. Company, Amsterdam, 1971.
577. C. Thomas, R. Bartonlini, J. I. M. Botman, G. Dattoli, L. Mezi, and M. Migliorati. An analytical solution for the Haissinski equation with purely inductive wake fields. *Europhysics Letters*, 60:66–71, 2002.
578. R. F. Thompson. *The brain*. W. H. Freeman and Company, New York, 1985.
579. J. Tobochnik. Critical point phenomena and phase transitions. *Am. J. Phys.*, 69:255–263, 2001.
580. T. Tome and M. J. de Oliveira. Dynamic phase transition in the kinetic Ising model under a time-dependent oscillating field. *Phys. Rev. A*, 41:4251–4254, 1990.
581. A. Torcini and M. Antoni. Equilibrium and dynamical properties of two-dimensional n-body systems with long-range attractive interactions. *Phys. Rev. E*, 59:2746–2763, 1999.
582. C. Tsallis. Possible generalization of Boltzmann-Gibbs statistics. *J. Stat. Phys.*, 52:479–487, 1988.

583. C. Tsallis. Non-extensive thermostatistics: brief review and comment. *Physica A*, 221:227–290, 1995.
584. C. Tsallis. Nonextensive thermostatistics and fractals. *Fractals*, 3:541–547, 1995.
585. C. Tsallis. Some comments on Boltzmann-Gibbs statistical mechanics. *Chaos, solitons & fractals*, 6:539–559, 1995.
586. C. Tsallis. Lévy distributions. *Phys. World*, 10(7):42–45, 1997.
587. C. Tsallis. Nonextensive statistics: theoretical, experimental and computational evidences and connections. *Braz. J. Phys.*, 29:1–35, 1999.
588. C. Tsallis. Entropic nonextensivity: a possible measure of complexity. *Chaos, solitons and fractals*, 13:371–391, 2002.
589. C. Tsallis. What should a statistical mechanics satisfy to reflect nature? *Physica D*, 193:3–34, 2004.
590. C. Tsallis, C. Anteneodo, L. Borland, and R. Osorio. Nonextensive statistical mechanics and economics. *Physica A*, 324:89–100, 2003.
591. C. Tsallis and D. J. Bukman. Anomalous diffusion in the presence of external forces: Exact time-dependent solutions and their thermostatistical basis. *Phys. Rev. E*, 54:R2197–R2200, 1996.
592. C. Tsallis, R. S. Mendes, and A. R. Plastino. The role of constraints within generalized nonextensive statistics. *Physica A*, 261:534–554, 1998.
593. C. Tsallis and A. M. C. Souza. Constructing a statistical mechanics for Beck-Cohen superstatistics. *Phys. Rev. E*, 67:026106, 2003.
594. M. T. Turvey. Coordination. *Am. Psychol.*, 45:938–953, 1990.
595. B. Tyldesley and J. I. Grieve. *Muscles, nerves, and movement — Kinesiology in daily living*. Blackwell Science, Oxford, 1996.
596. E. A. Uhling and G. E. Uhlenbeck. Transport phenomena in Einstein-Bose and Fermi-Dirac gases. *Phys. Rev.*, 43:552–561, 1933.
597. A. Upadhyaya, J. P. Rieu, J. A. Glazier, and Y. Sawada. Anomalous diffusion and non-Gaussian velocity distribution of Hydra cells in cellular aggragates. *Physica A*, 293:549–558, 2001.
598. I. Vajda. Axioms for a-entropy of a generalized probability scheme. *Kybernetika*, 2:105–112, 1968. summary in English on pages 111 and 112.
599. C. van den Broeck, I. Bena, P. Reimann, and J. Lehmann. Coupled browian motors on a tilted washboard. *Annalen der Physik*, 9:713–720, 2000.
600. C. van den Broeck, J. M. R. Parrondo, J. Armero, and A. Hernandez-Machado. Mean field model for spatially extended systems in the presence of multiplicative noise. *Phys. Rev. E*, 49:2639–2643, 1994.
601. C. van den Broeck, J. M. R. Parrondo, and R. Toral. Noise-induced nonequilibrium phase transition. *Phys. Rev. Lett.*, 73:3395–3398, 1994.
602. C. van den Broeck, J. M. R. Parrondo, R. Toral, and R. Kawai. Nonequilibrium phase transitions induced by multiplicative noise. *Phys. Rev. E*, 55:4084–4094, 1997.
603. J. L. van Hemmen and W. F. Wreszinski. Lyapunov function for the Kuramoto model of nonlinearly coupled oscillators. *J. Stat. Phys.*, 72:145–166, 1993.
604. N. G. van Kampen. Derivation of the phenomenological equations from the master equations. i. even variables only. *Physica*, 23:707–719, 1957.
605. N. G. van Kampen. Derivation of the phenomenological equations from the master equations. ii. even and odd variables. *Physica*, 23:816–829, 1957.
606. N. G. van Kampen. Condensation of a classical gas with long-range attraction. *Phys. Rev*, 135:A362–369, 1964.

607. N. G. van Kampen. *Stochastic processes in physics and chemistry.* North-Holland Publ. Company, Amsterdam, 1981.
608. M. Venturini. Stability analysis of longitudinal beam dynamics using non-canonical Hamiltonian methods and energy principles. *Phys. Rev. ST-AB*, 5:054403, 2002.
609. M. Venturini and R. Warnock. Bursts of coherent synchrotron radiation in electron storage rings: a dynamical model. *Phys. Rev. Lett.*, 89:224802, 2002.
610. F. Verhulst. *Nonlinear differential equations and dynamical systems, 2nd ed.* Springer, Berlin, 1996.
611. J. Villian. Elementary excitations in magnetic chains. In T. Riste, editor, *Ordering in strongly fluctuating condensed matter systems*, pages 91–106, New York, 1979. Plenum Press.
612. G. M. Viswanathan, V. Afanasyev, S. V. Buldyrev, S. Havlin, M. G. E. da Luz, E. P. Raposo, and H. E. Stanley. Lévy flights search patterns of biological organisms. *Physica A*, 295:85–88, 2001.
613. E. Vives and A. Planes. Is Tsallis thermodynamics nonextensive? *Phys. Rev. Lett.*, 88:020601, 2002.
614. C. von der Malsburg and J. Buhmann. Sensory segmentation with coupled neural oscillators. *Biol. Cybern.*, 67:233–242, 1992.
615. E. V. Votyakov, A. DeMartino, and D. H. E. Gross. Thermodynamics of rotating self-gravitating systems. *Eur. Phys. J. B*, 29:593–603, 2002.
616. T. J. Walker. Acoustic synchrony: two mechanisms in the snowy tree cricket. *Science*, 166:891–894, 1969.
617. M. C. Wang and G. E. Uhlenbeck. On the theory of Brownian motion II. *Rev. Mod. Phys.*, 17:323–342, 1945.
618. Q. Wang, W. E, C. Liu, and P. Zhang. Kinetic theory for flows of nonhomogeneous rodlike liquid crystalline polymers with a nonlocal intermolecular potential. *Phys. Rev. E*, 65:051504, 2002.
619. G. H. Wannier. *Statistical physics.* Dover Publications, New York, 1966.
620. S. Watanabe and S. H. Strogatz. Integrability of a globally coupled oscillator array. *Phys. Rev. Lett.*, 70:2391–2394, 1993.
621. W. Weaver and C. E. Shannon. *The mathematical theory of communication.* University of Illinois Press, Illinois, 1949.
622. E. R. Weeks and H. L. Swinney. Anomalous diffusion resulting from strongly asymmetric random walks. *Phys. Rev. E*, 57:4915–4920, 1998.
623. M. F. Wehner and W. G. Wolfer. Numerical evaluation of path-integral solutions to Fokker-Planck equations. *Phys. Rev. A*, 27:2663–2670, 1983.
624. M. F. Wehner and W. G. Wolfer. Numerical evaluation of path-integral solutions to Fokker-Planck equations. III. Time and functionally dependent coefficients. *Phys. Rev. A*, 35:1795–1801, 1987.
625. A. Wehrl. General properties of entropy. *Rev. Mod. Phys.*, 50:221–250, 1978.
626. A. Wehrl. The many facets of entropy. *Rep. Math. Phys.*, 30:119–129, 1991.
627. E. W. Weisstein. *CRC Concise encyclopedia of mathematics.* Chapman and Hall/CRC, Boca Raton, 1998.
628. H. Wiedemann. *Particle accelerator physics II — Nonlinear and higher-order beam dynamics.* Springer, Berlin, 1993.
629. K. Wiesenfeld, P. Colet, and S. H. Strogatz. Synchronization transitions in a disordered Josephson series array. *Phys. Rev. Lett.*, 76:404–407, 1996.

630. K. Wiesenfeld, P. Colet, and S. H. Strogatz. Frequency locking in Josephson arrays: connection with the Kuramoto model. *Phys. Rev. E*, 57:1563–1569, 1998.
631. R. H. Wimmers, P. J. Beek, and P. C. W. van Wieringen. Phase transitions in rhythmic tracking movements: A case of unilateral coupling. *Hum. Movement Sci.*, 11:217–226, 1992.
632. A. T. Winfree. Biological rhythms and the behavior of populations of coupled oscillators. *J. Theor. Biol.*, 16:15–42, 1967.
633. A. T. Winfree. *The geometry of biological time*. Springer, Berlin, 1980.
634. T. P. Witelski. Perturbation analysis for wetting fronts in Richards' equation. *Transp. Porous Media*, 27:121–134, 1997.
635. T. P. Witelski. Segregation and mixing in degenerate diffusion in population dynamics. *J. Math. Biol.*, 35:695–712, 1997.
636. C. Wolf. Two-state paramagnetism induced by Tsallis and Renyi statistics. *Int. J. Theor. Phys.*, 37:2433–2438, 1998.
637. A. Wunderlin. On the slaving principle. In *Springer proceedings in physics*, pages 140–147, Berlin, 1987. Springer.
638. Y. Yamaguchi, K. Kometani, and H. Shimizu. Self-synchronization of nonlinear oscillators in the presence of fluctuations. *J. Stat. Phys.*, 26:719–743, 1981.
639. Y. Yamaguchi and H. Shimizu. Theory of self-synchronization in the presence of native frequency distribution and external noises. *Physica D*, 11:213–226, 1984.
640. M. Yamana, M. Shiino, and M. Yoshioka. Oscillator neural network model with distributed native frequencies. *J. Phys. A: Math. Gen.*, 32:3525–3533, 1999.
641. T. Yamano. On the robust thermodynamical structure against arbitrary entropy form and energy mean values. *Eur. Phys. J. B*, 18:103–106, 2000.
642. M. K. S. Yeung and S. H. Strogatz. Time delay in the Kuramoto model of coupled oscillators. *Phys. Rev. Lett.*, 82:648–651, 1999.
643. M. Yoshioka and M. Shiino. Associative memory storing an extensive number of patterns based on a network of oscillators with distributed natural frequencies in the presence of external white noise. *Phys. Rev. E*, 61:4732–4744, 2000.
644. A. A. Zaikin, J. Garcia-Ojalvo, and L. Schimansky-Geier. Nonequilibrium first-order phase transition induced by additive noise. *Phys. Rev. E*, 60:R6275–R6277, 1999.
645. A. A. Zaikin, J. Garcia-Ojalvo, L. Schimansky-Geier, and J. Kurths. Noise induced propagation in monostable media. *Phys. Rev. Lett.*, 88:010601, 2002.
646. A. A. Zaikin and L. Schimansky-Geier. Spatial patterns induced by additive noise. *Phys. Rev. E*, 58:4355–4360, 1998.
647. D. H. Zanette. Propagating structures in globally coupled systems with time delays. *Phys. Rev. E*, 62:3167–3172, 2000.
648. D. S. Zhang, G. W. Wei, D. J. Kouri, and D. K. Hoffman. Numerical method for the nonlinear Fokker-Planck equation. *Phys. Rev. E*, 56:1197–1206, 1997.
649. Y. Zhang, G. Hu, H. Liu, and J.H. Xiao. Collective behavior in globally coupled systems consisting of two kinds of competing cells. *Phys. Rev. E*, 57:2543–2548, 1998.

Index

Accuracy-flexibility trade-off, 351
Anomalous diffusion, 15
   and nonextensivity, 219
   Plastino–Plastino model, 251
   porous medium equation, 250
   Sharma–Mittal model, 274

Basins of attraction, 25
   dynamical Takatsuji model, 140
   multistable variability model, 339
BCS theory, 234
Bifurcation diagram
   bounded $B(M_1)$-model, 155
   Desai-Zwanzig model, 152
   dynamical Takatsuji model, 144
   KSS model with sine coupling, 162
   Maier–Saupe model, 200
   multistable variability model, 339
   reentrant phase transition, 331
Bistability, *see* Multistability
Black body radiation model, 294
Boltzmann equation, 28
Boundary conditions, 23
Brownian particle, 1

Canonical-dissipative systems, 90
Chapman–Kolmogorov equation, 47
Chemical potential, 82
Classifications, 26
Cluster amplitude and cluster phase, 156
Continuity equation, 22
Continuous phase transitions, 187
Control parameters, 4
Correlation functions
   dynamical Takatsuji model, 144
   Fluctuation–dissipation theorem, 70
   Plastino–Plastino model, 258

Shimizu–Yamada model, 69
strongly nonlinear case, 36

Damping constant, 1
Damping force, 1
Darcy's law, 248
Data analysis, 368
Discontinuous phase transitions, 187
Distortion functionals, 319
   free energy case, 97
   Lambert's W-function, 346
   negative stochastic feedback model, 354
Distributions
   Boltzmann distributions, 79
   bunch-particle distributions, 342
   cut-off distributions, 10
      Plastino–Plastino model, 253
      population dynamics, 297
      porous medium equation, 250
      Sharma–Mittal model, 264
   Gaussians, 57
   Haissinski distributions, 345
   power law, 10
      Plastino–Plastino model, 252
      Sharma–Mittal model, 264
Drift and diffusion coefficients, 20
Drift and diffusion terms, 20
Drift- and diffusion forms, 222
   Fermi and Bose systems, 280

Entropy and information measures, 224
   derivation of entropies, 310
   entropy increase, 218
   entropy production, 81, 87
   examples, 231
   κ-entropy, 247
   average information measure, 244

Boltzmann entropy, 232
Fermi and Bose systems, 233, 289
Fisher information, 247
negative stochastic feedback entropy, 245
Renyi entropy, 236
Sharma–Mittal entropy, 239
Tsallis entropy, 238
vorticity entropy, 235
excess entropy, 304
properties, 224
concavity, 226
extensivity, 228
nonextensivity, 228

Families of Markov processes, 46
linear families, 47
nonlinear families, 48
Ferromagnetism, 140, 188
Finite difference schemes, 29
Fluctuating force, 1
Langevin force, 2, 37
Fluctuation–dissipation theorem, 70
Fourier and moment expansions, 29
Free energy
K-model, 120
K_A-model, 126
bounded $B(A)$-model, 126
bounded $B(M_1)$-model, 152
compensated $B(A)$-model, 122
Desai–Zwanzig model, 148
dynamical Takatsuji model, 140
entropy case, 213
Fermi and Bose systems
energy phase space, 289
KSS model, 158
with sine coupling, 160
Maier–Saupe model, 197
mean field HKB model, 168
negative stochastic feedback model, 356
Shimizu–Yamada model, 136
Free energy principle, 75

GENERIC, 87
Gross–Pitaevskii equation, 3
Group behavior, 140, 206

H-Theorem
for stochastic processes, 43
free energy case, 77
linear nonequilibrium thermodynamics, 89
multistable variability model, 336
Hartree–Fock theory, 2
Heat bath, 1
Hitchhiker processes, 50
HKB potential, 167
Hysteresis, 200

Ito-Langevin equation, *see* Langevin equation

Kinetical interaction principle, 28, 281, 323
Kramers equation, 92
Fermi and Bose systems, 285
linear nonequilibrium thermodynamics, 95
Tsallis entropy, 288
with generalized thermostatistics, 284
Kramers–Moyal coefficients, 34

Lagrange multiplier, 77
Landau theory, 4
Langevin equation, 36
bunched particle beams, 345, 349
dynamical Takatsuji model, 147
Fermi and Bose systems, 282
energy phase space, 292
Gaussian entropy model, 278
Maier–Saupe model, 198
multistable variability model, 341
negative stochastic feedback model, 356
noise generator, 350
numerics, 39
Plastino–Plastino model, 257
self-consistent, 38
two-layered, 36
Vlasov–Fokker–Planck equation, 344
Langevin force, 2, 37
Linear nonequilibrium thermodynamics, 80
Liquid crystals, 195
Lyapunov functionals, 78, 112
$\delta^2 S$, 215

derivation of local Lyapunov
    functionals, 312
derivation of Lyapunov functionals,
    317
free energy case, 78
Plastino–Plastino model, 255
stability analysis, 112

Master equation, 28, 369
Maximum entropy principle, 76
Mean field theory, 2
Metal electron model, 292
Mobility coefficient, 83
Moving frame transformations, 306
Multistability, 25
    bounded $B(M_1)$-model, 152
    Desai–Zwanzig model, 148
    dynamical Takatsuji model, 140
    KSS model, 157
    Maier–Saupe model, 195
    mean field HKB model, 167
    multistable variability, 332
    perception-action systems, 209
    strongly nonlinear case, 43
    symmetries, 186
Muscular contraction, 202

Noise
    noise amplitude, 1
    noise generator, 349
    noise source, 1
Nonlinear Fokker–Planck equations
    classifications, 26
    entropy case, 213
    Fermi and Bose systems
        energy phase space, 290
    free energy case, 73
    Landau form, 341
    strongly nonlinear case, 31
    Vlasov type, 343
    with Boltzmann statistics, 109
Nonlinearity dimension, 25
Norm for perturbations, 132
Numerics
    Fokker–Planck equations, 28
    Langevin equations, 39

Order parameters, 4

Parabolic partial differential equations,
    365
Path integral solutions, 28
Planck's radiation formula, 294
Plasma physics, 341
Polytropic gas, 248
Population dynamics, 296
Porous medium equation, 246
Propagation of molecular chaos, 55

Reentrant phase transitions, 329
Relativity theory, 296
Rhythmic single limb movements, 210

Self-consistency equations, 24
    K-model, 121
    K_A-model, 124
    bounded $B(M_1)$ model, 152
    bounded $B(A)$-model, 125
    compensated $B(A)$-model, 123
    Desai–Zwanzig model, 149
    dynamical Takatsuji model, 143
    KSS model with sine coupling, 161
    Maier–Saupe model, 198
    mean field HKB model, 169
        symmetric case, 170, 173
    multistable variability model, 333
    reentrant phase transitions, 330
    Shimizu–Yamada model, 139
    stability analysis, 115
Self-organization, 3
Smoluchowski limit, 284, 323
Stability analysis, 110
    K-model, 120
    K_A-model, 124
    bounded $B(A)$-model, 125
    bounded $B(M_1)$-model, 153
    compensated $B(A)$-model, 122
    Desai–Zwanzig model, 149
    dynamical Takatsuji model, 144
    KSS model, 159, 163
    linear stability analysis
        free energy case, 113
        nonoscillatory instabilities, 299
    Lyapunov's direct method, 112
    Maier–Saupe model, 198
    mean field HKB model, 169
    multiplicative noise, 130
    multistable variability model, 333

self-consistency equations, 115
Shiino's decomposition of perturbations, 118
Shimizu–Yamada model, 139
Stationary solutions, 23
  K-model, 121
  K_A-model, 124
  bounded $B(A)$-model, 125
  bounded $B(M_1)$ model, 152
  compensated $B(A)$-model, 123
  Desai–Zwanzig model, 149
  dynamical Takatsuji model, 143
  entropy case, 214
  existence of stationary solutions, 153, 215
  Fermi and Bose systems, 281
    energy phase spaces, 290
  free energy case, 75
    with Boltzmann statistics, 109
  KSS model
    free energy case, 158
    with sine coupling, 161
  linear nonequilibrium thermodynamics, 88
  Maier–Saupe model, 198
  mean field HKB model, 169
    symmetric case, 170
  negative stochastic feedback model, 353
  Plastino–Plastino model, 251
  reentrant phase transitions, 330
  Sharma–Mittal model, 263
  Shimizu–Yamada model, 135
  strongly nonlinear case, 42
  uniqueness of stationary solutions, 215
Stochastic feedback, 5, 34
Stochastic processes, 19
  ensemble averaging, 19
  hierarchy of distributions, 19
  Markov diffusion processes, 2, 33
  Ornstein–Uhlenbeck processes, 62
  purely random processes, 61
  statistical ensemble, 19
  stochastic trajectory, 19
  Wiener processes, 62
Stratonovich
  Fokker-Planck equation, 21
  Langevin equation, 86
Superstatistics, 14
Synchronization, 191
  applause, 191
  brain activity, 193
  planar rotators and phase oscillators, 194
  tree crickets and fireflies, 192

Temperature fluctuations, 14
Thermodynamic flux, 81
Thermodynamic force, 81
Transient solutions, 22
  Barenblatt–Pattle solutions, 249
  Gaussian entropy model, 277
  Gaussians, 63
  Plastino–Plastino model, 253
  Sharma–Mittal model, 268
  Shimizu–Yamada model, 66
  strongly nonlinear case, 55
Transition probability densities
  Gaussian entropy model, 276
  Gaussians, 57
  linear nonequilibrium thermodynamics, 86
  Ornstein–Uhlenbeck processes, 62
  Shimizu–Yamada model, 66
  strongly nonlinear case, 55
  Wiener processes, 62
Traveling waves, 324

Springer Series in Synergetics

Synergetics An Introduction 3rd Edition
By H. Haken

Synergetics A Workshop
Editor: H. Haken

Synergetics Far from Equilibrium
Editors: A. Pacault, C. Vidal

Structural Stability in Physics
Editors: W. Güttinger, H. Eikemeier

Pattern Formation by Dynamic Systems and Pattern Recognition
Editor: H. Haken

Dynamics of Synergetic Systems
Editor: H. Haken

Problems of Biological Physics
By L. A. Blumenfeld

Stochastic Nonlinear Systems in Physics, Chemistry, and Biology
Editors: L. Arnold, R. Lefever

Numerical Methods in the Study of Critical Phenomena
Editors: J. Della Dora, J. Demongeot, B. Lacolle

The Kinetic Theory of Electromagnetic Processes By Yu. L. Klimontovich

Chaos and Order in Nature
Editor: H. Haken

Nonlinear Phenomena in Chemical Dynamics Editors: C. Vidal, A. Pacault

Handbook of Stochastic Methods for Physics, Chemistry, and the Natural Sciences 3rd Edition
By C. W. Gardiner

Concepts and Models of a Quantitative Sociology The Dynamics of Interacting Populations By W. Weidlich, G. Haag

Noise-Induced Transitions Theory and Applications in Physics, Chemistry, and Biology By W. Horsthemke, R. Lefever

Physics of Bioenergetic Processes
By L. A. Blumenfeld

Evolution of Order and Chaos in Physics, Chemistry, and Biology
Editor: H. Haken

The Fokker-Planck Equation
2nd Edition By H. Risken

Chemical Oscillations, Waves, and Turbulence By Y. Kuramoto

Advanced Synergetics
2nd Edition By H. Haken

Stochastic Phenomena and Chaotic Behaviour in Complex Systems
Editor: P. Schuster

Synergetics – From Microscopic to Macroscopic Order Editor: E. Frehland

Synergetics of the Brain
Editors: E. Başar, H. Flohr, H. Haken, A. J. Mandell

Chaos and Statistical Methods
Editor: Y. Kuramoto

Dynamics of Hierarchical Systems An Evolutionary Approach
By J. S. Nicolis

Self-Organization and Management of Social Systems Editors: H. Ulrich, G. J. B. Probst

Non-Equilibrium Dynamics in Chemical Systems
Editors: C. Vidal, A. Pacault

Self-Organization Autowaves and Structures Far from Equilibrium
Editor: V. I. Krinsky

Temporal Order Editors: L. Rensing, N. I. Jaeger

Dynamical Problems in Soliton Systems
Editor: S. Takeno

Complex Systems – Operational Approaches in Neurobiology, Physics, and Computers Editor: H. Haken

Dimensions and Entropies in Chaotic Systems Quantification of Complex Behavior 2nd Corr. Printing
Editor: G. Mayer-Kress

Selforganization by Nonlinear Irreversible Processes
Editors: W. Ebeling, H. Ulbricht

Instabilities and Chaos in Quantum Optics
Editors: F. T. Arecchi, R. G. Harrison

Nonequilibrium Phase Transitions in Semiconductors Self-Organization Induced by Generation and Recombination Processes By E. Schöll

Temporal Disorder in Human Oscillatory Systems
Editors: L. Rensing, U. an der Heiden, M. C. Mackey

The Physics of Structure Formation
Theory and Simulation
Editors: W. Guttinger, G. Dangelmayr

Computational Systems – Natural and Artificial Editor: H. Haken

From Chemical to Biological Organization Editors: M. Markus, S. C. Müller, G. Nicolis

Information and Self-Organization
A Macroscopic Approach to Complex Systems   2nd Edition   By H. Haken

Propagation in Systems Far from Equilibrium   Editors: J. E. Wesfreid, H. R. Brand, P. Manneville, G. Albinet, N. Boccara

Neural and Synergetic Computers
Editor: H. Haken

Cooperative Dynamics in Complex Physical Systems   Editor: H. Takayama

Optimal Structures in Heterogeneous Reaction Systems   Editor: P. J. Plath

Synergetics of Cognition
Editors: H. Haken, M. Stadler

Theories of Immune Networks
Editors: H. Atlan, I. R. Cohen

Relative Information Theories and Applications   By G. Jumarie

Dissipative Structures in Transport Processes and Combustion
Editor: D. Meinköhn

Neuronal Cooperativity
Editor: J. Krüger

Synergetic Computers and Cognition
A Top-Down Approach to Neural Nets
2nd edition   By H. Haken

Foundations of Synergetics I
Distributed Active Systems   2nd Edition
By A. S. Mikhailov

Foundations of Synergetics II
Complex Patterns   2nd Edition
By A. S. Mikhailov, A. Yu. Loskutov

Synergetic Economics   By W.-B. Zhang

Quantum Signatures of Chaos
2nd Edition   By F. Haake

Rhythms in Physiological Systems
Editors: H. Haken, H. P. Koepchen

Quantum Noise   3rd Edition
By C. W. Gardiner, P. Zoller

Nonlinear Nonequilibrium Thermodynamics I   Linear and Nonlinear Fluctuation-Dissipation Theorems
By R. Stratonovich

Self-organization and Clinical Psychology Empirical Approaches to Synergetics in Psychology
Editors: W. Tschacher, G. Schiepek, E. J. Brunner

Nonlinear Nonequilibrium Thermodynamics II   Advanced Theory
By R. Stratonovich

Limits of Predictability
Editor: Yu. A. Kravtsov

On Self-Organization
An Interdisciplinary Search for a Unifying Principle
Editors: R. K. Mishra, D. Maaß, E. Zwierlein

Interdisciplinary Approaches to Nonlinear Complex Systems
Editors: H. Haken, A. Mikhailov

Inside Versus Outside
Endo- and Exo-Concepts of Observation and Knowledge in Physics, Philosophy and Cognitive Science
Editors: H. Atmanspacher, G. J. Dalenoort

Ambiguity in Mind and Nature
Multistable Cognitive Phenomena
Editors: P. Kruse, M. Stadler

Modelling the Dynamics of Biological Systems
Editors: E. Mosekilde, O. G. Mouritsen

Self-Organization in Optical Systems and Applications in Information Technology   2nd Edition
Editors: M.A. Vorontsov, W. B. Miller

Principles of Brain Functioning
A Synergetic Approach to Brain Activity, Behavior and Cognition
By H. Haken

Synergetics of Measurement, Prediction and Control   By I. Grabec, W. Sachse

Predictability of Complex Dynamical Systems
By Yu. A. Kravtsov, J. B. Kadtke

Interfacial Wave Theory of Pattern Formation
Selection of Dentritic Growth and Viscous Fingerings in Hele–Shaw Flow   By Jian-Jun Xu

Asymptotic Approaches in Nonlinear Dynamics
New Trends and Applications
By J. Awrejcewicz, I. V. Andrianov, L. I. Manevitch

Brain Function and Oscillations
Volume I: Brain Oscillations.
Principles and Approaches
Volume II: Integrative Brain Function.
Neurophysiology and Cognitive Processes
By E. Başar

Asymptotic Methods for the Fokker–Planck Equation and the Exit Problem in Applications
By J. Grasman, O. A. van Herwaarden

Analysis of Neurophysiological Brain Functioning   Editor: Ch. Uhl

Phase Resetting in Medicine and Biology
Stochastic Modelling and Data Analysis
By P. A. Tass

Self-Organization and the City   By J. Portugali

Critical Phenomena in Natural Sciences
Chaos, Fractals, Selforganization and Disorder:
Concepts and Tools   2nd Edition   By D. Sornette

Spatial Hysteresis and Optical Patterns
By N. N. Rosanov

Nonlinear Dynamics of Chaotic and Stochastic Systems Tutorial and Modern Developments
By V. S. Anishchenko, V. V. Astakhov,
A. B. Neiman, T. E. Vadivasova,
L. Schimansky-Geier

Synergetic Phenomena in Active Lattices
Patterns, Waves, Solitons, Chaos
By V. I. Nekorkin, M. G. Velarde

Brain Dynamics
Synchronization and Activity Patterns
in Pulse-Coupled Neural Nets with Delays
and Noise   By H. Haken

From Cells to Societies
Models of Complex Coherent Action
By A. S. Mikhailov, V. Calenbuhr

Brownian Agents and Active Particles
Collective Dynamics in the Natural and Social
Sciences   By F. Schweitzer

Nonlinear Dynamics of the Lithosphere and Earthquake Prediction
By V. I. Keilis-Borok, A. A. Soloviev (Eds.)

Nonlinear Fokker–Planck Equations
Fundamentals and Applications
By T. D. Frank

Printing: Krips bv, Meppel
Binding: Litges & Dopf, Heppenheim